Innovative and Hybrid Technologies for Wastewater Treatment and Recycling

Innovative and Hybrid Technologies for Wastewater Treatment and Recycling investigates the biological and non-biological features of the treatment process for wastewater and emphasizes the benefits that these aspects bring for sustainable engineering. It discusses several approaches that are based on biological and non-biological processes and examines the fundamental principles, practical applications, current achievements, future aspects, and associated limits. Further, it provides a wide range of innovative research on the treatment of wastewater, as well as the applications in the treatment, remediation, and pollution prevention processes.

- Explains the principles and concepts of the most recent and innovative treatment processes for wastewater remediation.
- Examines emerging nanofiber technology for the purification of wastewater.
- Provides an overview of the most cutting-edge environmentally friendly technologies.

Innovative and Hybrid Technologies for Wastewater Treatment and Recycling

Edited by
Pervez Alam, Afzal Husain Khan, Juliana Heloisa Pinê Américo-Pinheiro, and Kafeel Ahmad

CRC Press
Taylor & Francis Group
Boca Raton London New York

CRC Press is an imprint of the
Taylor & Francis Group, an informa business

Designed cover image: Shutterstock

First edition published 2025
by CRC Press
2385 NW Executive Center Drive, Suite 320, Boca Raton FL 33431

and by CRC Press
4 Park Square, Milton Park, Abingdon, Oxon, OX14 4RN

CRC Press is an imprint of Taylor & Francis Group, LLC

ISBN: 978-1-032-59326-5 (hbk)
ISBN: 978-1-032-59328-9 (pbk)
ISBN: 978-1-003-45419-9 (ebk)

DOI: 10.1201/9781003454199

Typeset in Times
by Newgen Publishing UK

Contents

Preface

Worldwide, communities are confronted with the dual challenges of declining water quality and limited access. The rapid expansion of urban populations, industrial activities, and agricultural practices has led to the generation of vast quantities of wastewater. Addressing this issue has become a critical environmental challenge, necessitating the development of efficient and innovative technologies for wastewater treatment, recycling, and reuse. Eliminating harmful organic and inorganic impurities in water bodies and recycling wastewater as a secondary water supply are both critical steps in solving the problem of water shortage. There is substantial social and economic backing, in both developed and developing countries, for efforts to increase water quality. If they are to contribute significantly to the development of industry and the betterment of humanity, methods for filtering wastewater should be both cost-effective and environmentally friendly. Applying cutting-edge technologies would considerably improve present wastewater treatment processes. Over the past two decades, academics have given increased attention to the application of new technology in environmental contexts. The application of innovative and hybrid technologies in water and wastewater treatment has tremendous promise for augmenting present procedures.

This book, *Innovative and Hybrid Technologies for Wastewater Treatment and Recycling*, serves as a comprehensive guide to current advancements and emerging solutions in the field. It brings together contributions from experts across the globe, each presenting state-of-the-art research on various aspects of wastewater treatment. The chapters collectively address a wide spectrum of topics, from physicochemical and biological treatment technologies to the integration of nanotechnology and machine learning into modern wastewater management practices. The book also explores innovative methods tailored for specific industries, such as pharmaceuticals and chemical manufacturing, where wastewater management presents unique challenges.

Chapter 1 begins with a foundational overview of wastewater treatment and the key strategies for recycling and reuse. As the chapters progress, the reader is introduced to advanced physicochemical techniques, biological nutrient removal processes, and the role of membrane technologies. Cutting-edge research on the degradation of phenolic compounds and the removal of heavy metals through nanotechnology provides further insights into addressing complex wastewater pollutants. In addition to technological advancements, the book emphasizes the need for a sustainable approach to wastewater treatment. This is reflected in the discussions on the circular economy, where wastewater recycling and reuse are viewed as essential components of future environmental stewardship.

We believe that this book will be an invaluable resource for researchers, engineers, policymakers, and practitioners in the field of environmental science and wastewater management. By providing a detailed exploration of both established and emerging technologies, this volume aims to inspire new solutions for one of the most pressing environmental challenges of our time.

We appreciate all the authors' and contributors' tremendous efforts in presenting their most recent research and innovative technologies in the form of book chapters. The editors would also like to thank our friends, colleagues, collaborators, students, and family members for their assistance in bringing this book to its present state and format. Our sincere gratitude goes to all the expert reviewers who took the time to read the whole book and provide useful comments and criticism without conflict of interest. Their skilled and quick assistance has enabled the book to meet global standards. Editors' special gratitude to CRC Press editorial staff (Suriya Rajasekar, Stuart Murrayand others), who assisted us at every phase of the book's commencement, functioning, and execution, including the production.

Pervez Alam,
Afzal Husain Khan,
Juliana Heloisa Pinê Américo-Pinheiro,
Kafeel Ahmad

Contributors

Kafeel Ahmad
Department of Civil Engineering,
Jamia Millia Islamia (A Central University),
New Delhi, India

Kafeel Ahmad
Department of Natural and Applied Sciences,
TERI School of Advance Studies,
New Delhi, India

Shoeb Ahmad
School of Biosciences and Biotechnology,
Baba Ghulam Shah Badshah University,
Rajouri, Jammu and Kashmir, India

Sirajuddin Ahmed
Department of Environmental Science,
Jamia Millia Islamia (A Central University),
New Delhi, India

Pervez Alam
Department of Civil Engineering,
Baba Ghulam Shah Badshah University,
Rajouri, Jammu and Kashmir, India

Motasem Y.D. Alazaiza
Department of Civil and Environmental Engineering,
College of Engineering, A'Sharqiyah University Ibra,
Oman

Ahmed Albahnasawi
Department of Environmental Engineering,
Gebze Technical University
Kocaeli, Turkey

Sahar Alhout
Department of Pharmacy and Biotechnology
University of Palestine,
Palestine

Dilek Şenol Arslan
Department of Materials Science & Nanotechnology Engineering,
Abdullah Gül University,
Kayseri, Turkey

Mohammad Vikas Ashraf
School of Biosciences and Biotechnology
Baba Ghulam Shah Badshah University,
Rajouri, Jammu and Kashmir, India

Stany Balakumar
School of Biosciences and Technology,
Vellore Institute of Technology,
Tamil Nadu, India

Mark A.T. Blaskovich
Centre for Superbug Solutions, Institute for Molecular Bioscience,
The University of Queensland,
St Lucia Australia

Aline Frumi Camargo
Laboratory of Microbiology and Bioprocesses,
Federal University of Fronteira Sul,
Erechim, Brazil
Federal University of Santa Catarina,
Graduate Program in Biotechnology and Biosciences,
Florianópolis, Brazil

Ruchi Chauhan
Food Science Institute,
Kansas State University,
Manhattan, Kansas, USA

Zeynep Ciğeroğlu
Department of Chemical Engineering
Faculty of Engineering and Natural Sciences
Usak University,
Usak, Turkey

Tapas Das
Department of Chemical Engineering,
National Institute of Technology (NIT)
Rourkela, Odisha, India

Murat Eyvaz
Department of Environmental Engineering,
Gebze Technical University
Kocaeli, Turkey

Thozhan Gajendran
Department of Biotechnology,
Bharathidasan Institute of Technology,
Anna University, Tiruchirappalli,
Tamilnadu, India

Jordana Georgin
Department of Civil and Environmental
Universidad de la Costa,
Barranquilla, Atlántico, Colombia

Bidisha Ghosh
Department of Biotechnology, School of Life Sciences,
Swami Vivekananda University,
Barrackpore, West Bengal, India

Semanti Ghosh
Department of Microbiology, School of Life Sciences,
Swami Vivekananda University,
Barrackpore, West Bengal, India

Ercan Gurbulak
Department of Environmental Engineering,
Gebze Technical University,
Kocaeli, Turkey

Maryam El Hajam
Advanced Structures and Composites Center,
University of Maine,
Orono, USA

Mohammed Majeed Hameed
Department of Civil Engineering,
Al-Maarif University,
Ramadi City, Iraq

Khaoula Houssini
School of Environmental Science and Engineering Shanghai,
Jiao Tong University,
Shanghai, China

Mansuur Husein
School of the Environment and Safety Engineering,
Jiangsu University,
Zhejiang, China,
Department of Water and Environmental Engineering, Faculty of Engineering,
Tamale Technical University,
Tamale, Ghana

Shahid Ul Islam
Department of Civil Engineering,
Baba Ghulam Shah Badshah University,
Rajouri, Jammu and Kashmir,
India

Marut Jain
The University of Queensland
Indian Institute of Technology Delhi Academy of Research (UQIDAR) &
Department of Chemical Engineering,
Indian Institute of Technology Delhi,

New Delhi, India
Centre for Superbug Solutions, Institute for Molecular Bioscience,
The University of Queensland,
St Lucia, Australia

Joseline Kateita
National Water & Sewerage Corporation,
Kampala, Uganda

Emine Sena Kazan-Kaya
Gebze Technical University,
Chemical Engineering Department,
Kocaeli, Turkey

M.A. Hannan Khan
Department of Zoology
Baba Ghulam Shah Badshah University,
Rajouri, Jammu and Kashmir, India

Afzal Husain Khan
Civil and Architectural Engineering Department, College of Engineering and ComputerSciences, Jazan University, Saudi Arabia

Ramalingam Kowsalya
Department of Biochemistry,
Vivekanandha College of Arts and Sciences for Women, Tiruchengode
Namakkal,
Tamilnadu, India

Ashish Kumar
Bihar Engineering University,
Department of Science and Technology,
Government of Bihar, India

Ravi Kumar
Department of Chemical Engineering,
Indian Institute of Technology (ISM),
Dhanbad, Jharkhand, India

O.S. Kushwaha
Chemical Engineering Department,
Indian Institute of Technology,
Madras, Chennai, India

Abdellah Lacherai
Laboratory of Applied Chemistry and Environment, Faculty of Sciences,
Ibn Zohr University,
Agadir, Morocco

Vitoria Dassoler Longo
Laboratory of Microbiology and Bioprocesses,
Federal University of Fronteira Sul,
Erechim, Brazil

Murugan Rajamehala
Department of Biotechnology,
Vivekanandha College of Engineering for Women (Autonomous),
Tiruchengode, Namakkal,
Tamilnadu, India

Adil Masood
Department of Natural and Applied Sciences,
TERI School of Advance Studies,
New Delhi, India

Mohd. Aamir Mazhar
Department of Civil Engineering,
Jamia Millia Islamia (A Central University),
New Delhi, India

Noureddine El Messaoudi
Laboratory of Applied Chemistry and Environment, Faculty of Sciences,
Ibn Zohr University,
Agadir, Morocco

Deepti Mishra
The University of Queensland
Indian Institute of Technology Delhi Academy of Research (UQIDAR) &
Department of Chemical Engineering,
Indian Institute of Technology Delhi,
New Delhi India

Shatakshi Mishra
School of Biosciences and Technology,
Vellore Institute of Technology
Tamil Nadu, India

H. Mittal
University School of Chemical Technology,
Guru Gobind Singh Indraprastha University,
Delhi, India

Momina
Department of Civil Engineering
Jamia Millia Islamia (A Central University),
New Delhi, India

Innocent Tayari Mwizerwa
College of Environmental Science & Engineering,
Donghua University, North Renmin Road,
Shanghai, China

Kamal Kishore Pant
Department of Chemical Engineering,
Indian Institute of Technology, Delhi,
New Delhi, India

Nandhi Prabhu
Department of Biotechnology,
Vivekanandha College of Engineering for Women (Autonomous),
Tiruchengode, Namakkal,
Tamilnadu, India

Darakhshan Naaz
Jamia Millia Islamia,
New Delhi, India

Harshala S. Naik
The Institute of Science, 15 Madam Cama Road, Mumbai
Constituent and Lead Institute of Dr. Homi Bhabha State University,
Mumbai, India

Nurullah Özdoğan
Department of Environmental Engineering,
Bursa Uludag University,
Bursa, Turkey

Juliana Heloisa Pinê Américo-Pinheiro
Department of Forest Science, Soils and Environment,
School of Agronomic Sciences,
São Paulo State University (UNESP),
Botucatu, Brazil;
College of Engineering,
Department of Civil Engineering
São Paulo State University (UNESP),
Botucatu, Brazil

Susan Praise
Independent Research Scientist

Mayana Cardoso Raimundi
Laboratory of Microbiology and Bioprocesses,
Federal University of Fronteira Sul,
Erechim, Brazil

Yawar Raina
Department of Civil Engineering,
Baba Ghulam Shah Badshah University,
Rajouri, Jammu and Kashmir, India

Shabir Ahmad Rather
School of Biosciences and Biotechnology
Baba Ghulam Shah Badshah University,
Rajouri, Jammu and Kashmir, India

Rajesh W. Raut
The Institute of Science, 15 Madam Cama Road, Mumbai
Constituent and Lead Institute of Dr. Homi Bhabha State University,
Mumbai, India

Siti Fatin Mohd Razali
Department of Civil Engineering, Faculty of Engineering and Built Environment,
Universiti Kebangsaan,
Malaysia (UKM), Bangi, Selangor, Malaysia

Larissa Capeletti Romani
Laboratory of Microbiology and Bioprocesses,
Federal University of Fronteira Sul,
Erechim, Brazil

Parvindar M. Sah
The Institute of Science, 15 Madam Cama Road, Mumbai
Constituent and Lead Institute of Dr. Homi Bhabha State University,
Mumbai, India

Abhisek Sahoo
Department of Chemical Engineering,
Indian Institute of Technology Delhi,
New Delhi, India

Suranjana Sarkar
Department of Biotechnology, School of Life Sciences,
Swami Vivekananda University,
Barrackpore, West Bengal

Zeynep Mine Şenol
Cumhuriyet University, Faculty of Health Sciences,
Department of Nutrition and Diet,
Sivas, Turkey

Ram Sharan Singh
Department of Chemical Engineering & Technology,
Indian Institute of Technology (BHU), Varanasi,
Uttar Pradesh, India

Aman Srivastava
Department of Civil Engineering,
Indian Institute of Technology (IIT)
Kharagpur, West Bengal, India

Sangeetha Subramanian
School of Biosciences and Technology,
Vellore Institute of Technology, Tamil Nadu, India

Ganesh Swain
Department of Chemical and Biochemical Engineering,
Indian Institute of Technology (IIT), Patna,
Bihar, India

Mohd. Saquib Tanweer
Environmental Science Research Lab,
Department of Applied Sciences & Humanities,
Faculty of Engineering and Technology,
Jamia Millia Islamia,
New Delhi, India

Abhinay Thakur
Department of Chemistry,
School of Chemical Engineering and Physical Sciences,
Lovely Professional University,
Phagwara, Punjab, India

Himanshu Tiwari
Department of Chemical Engineering & Technology,
Indian Institute of Technology (BHU),
Varanasi, Uttar Pradesh, India

Helen Treichel
Laboratory of Microbiology and Bioprocesses,
Federal University of Fronteira Sul,
Erechim, Brazil & Federal University of Santa Catarina,
Graduate Program in Biotechnology and Biosciences,
Florianópolis, Brazil

Ebubekir Yuksel
Department of Environmental Engineering,
Gebze Technical University,
Kocaeli, Turkey

Zyta Maria Ziora
Centre for Superbug Solutions, Institute for Molecular Bioscience,
The University of Queensland,
St Lucia, Australia

1 Overview of Wastewater Treatment, Recycling, and Reuse

Pervez Alam, Shahid Ul Islam, Darakhshan Naaz, Yawar Raina, Mohd. Aamir Mazhar, and Afzal Husain Khan

1.1 INTRODUCTION

Wastewater treatment, recycling, and reuse are indispensable pillars of modern water management strategies that are crucial for ensuring environmental sustainability, public health, and the conservation of precious freshwater resources (Chen et al., 2012, Hameed et al., 2015). As the global population burgeons and industrial activities intensify, the generation of wastewater has escalated, presenting challenges and opportunities for communities worldwide (Tucker, 2016; In this introduction, we embark on a journey to explore the intricate nexus of wastewater treatment, recycling, and reuse, shedding light on the processes involved, the benefits accrued, and the imperative of adopting sustainable practices in water management. Wastewater, often perceived as a burden or a pollutant is, in fact, a valuable resource that, when appropriately managed, can be transformed into a source of replenishment and renewal (Kurian et al., 2005, Karthikeyan et al, 2016). At its core, wastewater embodies myriad contaminants, ranging from organic matter and nutrients to pathogens and toxic compounds, posing risks to aquatic ecosystems and public health if left untreated (Aluko et al., 2003;). However, through a systematic approach encompassing various treatment technologies and reuse schemes, wastewater can be purified, repurposed, and reintegrated into the water cycle, fostering resilience and resourcefulness in the face of mounting water challenges. The wastewater journey begins with its collection from residential, commercial, and industrial sources, traversing through intricate networks of sewers and pipelines to treatment facilities where its transformation unfolds. Here, a symphony of physical, chemical, and biological processes is the removal of impurities, sediments, and microorganisms, gradually rendering the wastewater fit for discharge or reuse (Pohland and Harper, 1985). From the initial screening and sedimentation in primary treatment to the biological degradation and filtration in secondary and tertiary treatment, each stage plays a pivotal role in purifying the wastewater and safeguarding environmental health (US EPA, 1984). Further, it extends into resource recovery, where wastewater metamorphoses into a reservoir of nutrients, energy, and water, waiting to be harnessed for myriad beneficial purposes. Through innovative

DOI: 10.1201/9781003454199-1

technologies and strategic planning, wastewater treatment plants evolve into resource recovery hubs, extracting biogas from organic matter, reclaiming nutrients for agricultural fertilizers, and producing recycled water for irrigation, industrial processes, and even potable use in select cases (Jha et al., 2003; Kansal, 2002). This paradigm shift towards resource recovery enhances the sustainability of wastewater management and aligns with the principles of circular economy, where waste is minimized, and resources are maximized. As we embark on this exploration of wastewater treatment, recycling, and reuse, we are confronted with a tapestry of challenges and opportunities that underscore the urgency of adopting sustainable water management practices (Nagarajan et al., 2012). From public perception and regulatory hurdles to infrastructural constraints and technological innovations, the path to a water-resilient future is fraught with complexities, yet is promising (Mukherjee et al., 2015). By delving into the nuances of wastewater management, we strive to unravel the intricacies of this vital, yet often overlooked, facet of modern civilization, advocating for a holistic approach that embraces innovation, collaboration, and stewardship in safeguarding our most precious resource – water (BIS, 2012). The evolution of wastewater treatment, recycling, and reuse technologies has been marked by significant milestones and advancements over the years. In 1984, the U.S. Environmental Protection Agency (EPA) introduced the Groundwater Protection Strategy, laying the groundwork for comprehensive wastewater management practices (US EPA, 1984). Subsequent studies, such as those conducted by Biswas et al. (2010), have provided insights into the environmental quality surrounding municipal solid waste dumpsites, prompting further research into sustainable waste management solutions. Chen et al. (2012) contributed to the field by critically reviewing the sustainability assessment of recycled water schemes, shedding light on the potential of water reuse strategies. Moreover, Hameed et al. (2015) explored isotopic characterization and mass balance techniques to reveal groundwater recharge patterns, offering valuable insights into groundwater management. Ennouri et al. (2016) delved into the effect of thermal pre-treatment on biogas production and microbial communities, advancing anaerobic digestion technologies for wastewater treatment. The evolution of wastewater treatment, recycling, and reuse technologies has witnessed significant milestones, but there remain gaps in the literature regarding comprehensive approaches to address emerging contaminants, optimize resource recovery, and integrate decentralized treatment solutions. This chapter aims to bridge these gaps by providing a comprehensive overview of recent advancements, challenges, and future directions in wastewater management. Through critical analysis and synthesis of existing research, this chapter seeks to elucidate the current state of knowledge, identify areas for further investigation, and propose strategies to enhance the efficiency, resilience, and sustainability of wastewater treatment, recycling, and reuse practices.

1.2 FUNDAMENTALS OF WASTEWATER TREATMENT

Wastewater treatment is a complex process that involves the removal of contaminants and pollutants from water before it is discharged into the environment or reused for various purposes (Biswas et al., 2010). Understanding the fundamentals of wastewater

TABLE 1.1
Overview of the Fundamental Aspects of Wastewater Treatment, Categorized for Clarity and Easy Reference

Aspect	Description
Wastewater Composition	- Generated from residential, commercial, and industrial activities - Contains organic matter, nutrients, pathogens, heavy metals, etc.
Primary Treatment	- Physical processes - Removes large solids and debris - Methods: include screening, grit chambers, sedimentation
Secondary Treatment	- Biological processes - Degrades organic matter, removes pathogens - Activated sludge, trickling filters, sequencing batch reactors - Microorganisms break down pollutants through aerobic/anaerobic processes
Tertiary Treatment	- Advanced processes - Further improves water quality - Filtration (sand, membrane), chemical treatment (coagulation, disinfection), advanced oxidation processes
Disinfection	- Kills or inactivates remaining pathogens - Common methods: chlorination, UV irradiation, ozonation
Sludge Treatment	- Byproduct of treatment - Contains solid residues, microbial biomass - Treatment methods: thickening, dewatering, digestion, drying
Monitoring and Control	- Continuous monitoring of crucial critical parameters (pH, DO, turbidity, nutrients) - Control systems and automation ensure efficient operation and regulatory compliance

Source: Compiled by the authors.

treatment is essential for effectively managing water resources and safeguarding environmental and public health, as shown in Table 1.1. In this section, we delve into the fundamental principles and processes that underpin wastewater treatment (see Figure 1.1).

1.2.1 Wastewater Composition

Wastewater is generated from various sources, including residential, commercial, and industrial activities. It contains diverse contaminants, including organic matter, nutrients (nitrogen and phosphorus), pathogens, heavy metals, and various chemical pollutants (Khalid et al. 2022).

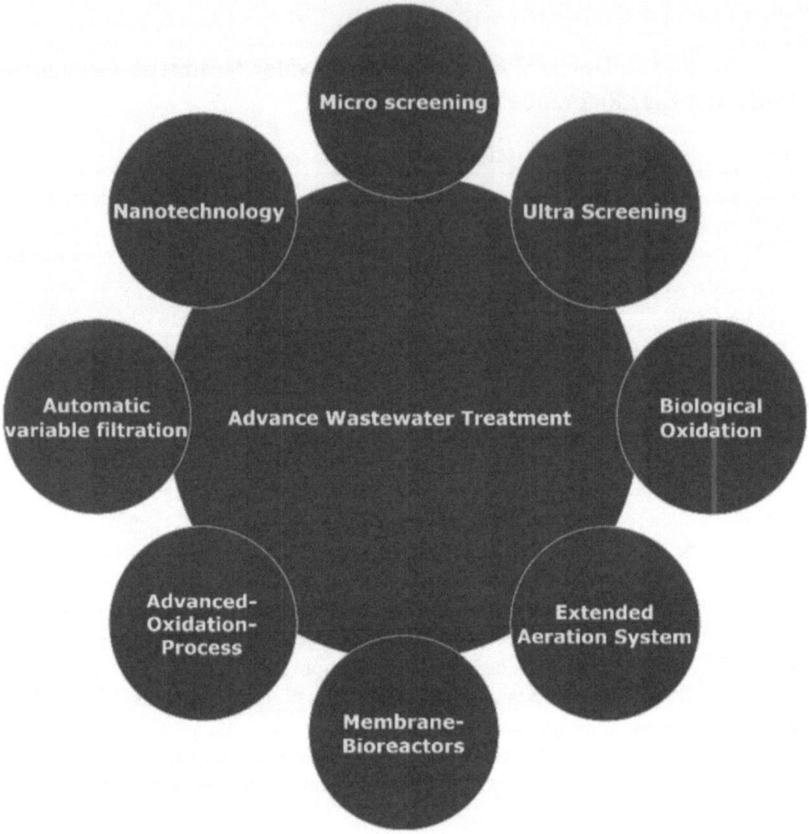

FIGURE 1.1 Overview of advanced treatment technologies of wastewater (compiled by the authors).

1.2.2 Primary Treatment

Primary treatment involves physical processes removing large solids and floating debris from wastewater. Standard primary treatment methods include screening to remove large objects, grit chambers to settle out sand and gravel, and sedimentation tanks to allow suspended solids to settle.

1.2.3 Secondary Treatment

Secondary treatment focuses on the biological degradation of organic matter and the removal of pathogens. Biological processes, such as activated sludge, trickling filters, and sequencing batch reactors, are commonly used for secondary treatment (Field and Sullivan, 2002). Microorganisms, including bacteria and protozoa, decompose organic pollutants into more straightforward, stable compounds through aerobic or anaerobic processes (Amir, 2018).

1.2.4 Tertiary Treatment

Tertiary treatment involves advanced processes to improve the quality of treated wastewater. Methods such as filtration (e.g., sand filtration, membrane filtration), chemical treatment (e.g., coagulation, disinfection), and advanced oxidation processes (e.g., ozonation, UV irradiation) are employed to remove remaining contaminants and pathogens. Disinfection is a critical step in wastewater treatment to kill or inactivate remaining pathogens before discharge or reuse. Standard disinfection methods include chlorination, ultraviolet (UV) irradiation, and ozonation.

1.3 SLUDGE TREATMENT AND DISPOSAL

Sludge, a byproduct of wastewater treatment, contains solid residues and microbial biomass removed during treatment (Chatterjee, 2010). Sludge is typically treated through thickening, dewatering, digestion, and drying before disposal or beneficial reuse (e.g., as fertilizer or soil conditioner).

1.4 ADVANCED TREATMENT TECHNOLOGIES

Advanced treatment technologies represent the forefront of wastewater treatment innovation, offering enhanced efficiency, reliability, and environmental sustainability. Membrane bioreactors (MBRs) integrate biological treatment with membrane filtration, effectively removing suspended solids, pathogens, and organic matter from wastewater to produce high-quality effluent suitable for reuse or discharge into sensitive environments (Kitonga, 2018) (see Table 1.2). Advanced oxidation processes (AOPs), such as ozonation and UV/H2O2, utilize chemical oxidation to degrade recalcitrant pollutants and emerging contaminants, generating highly reactive hydroxyl radicals that oxidize pollutants into simpler, less harmful compounds. Nutrient removal technologies target the removal of nitrogen and phosphorus to prevent eutrophication in receiving water bodies, employing biological nutrient removal (BNR), chemical precipitation, and enhanced biological phosphorus removal (EBPR) processes. Membrane filtration, including microfiltration, ultrafiltration, and reverse osmosis, employs semi-permeable membranes to physically separate contaminants, ensuring high-quality treated effluent for water reuse and potable water production. These advanced treatment technologies are vital in meeting stringent water quality standards, mitigating environmental pollution, and promoting sustainable water management practices (see Figure 1.1).

1.5 RESOURCE RECOVERY FROM WASTEWATER

Resource recovery from wastewater involves extracting and utilizing valuable materials and energy from sewage and other wastewater streams, thereby turning what was once considered waste into a useful, valuable, beneficial, and helpful resource (Gude, 2021). This process typically includes technologies such as anaerobic digestion, which produces biogas for energy generation, and nutrient recovery methods like struvite precipitation for reclaiming phosphorus and nitrogen. Additionally, advanced

TABLE 1.2
Description of Advanced Treatment Technologies for Wastewater

Technology	Description
Membrane Bioreactors (MBRs)	- Integrates biological treatment with membrane filtration - Removes suspended solids, pathogens, and organic matter from wastewater - Produces high-quality effluent suitable for reuse or discharge into sensitive environments - Compact design, high treatment efficiency, and minimal footprint - Requires energy for membrane operation and periodic membrane maintenance - Commonly used in decentralized wastewater treatment systems and water reuse applications
Advanced Oxidation Processes (AOPs)	- Utilizes chemical oxidation to degrade organic and inorganic contaminants in wastewater - Generates highly reactive hydroxyl radicals to oxidize pollutants into simpler, less harmful compounds - Common AOPs include ozonation, UV/H2O2, and Fenton's reaction - Effective in removing recalcitrant pollutants, pharmaceuticals, and emerging contaminants - Requires careful control of operating conditions and may generate harmful byproducts under certain circumstances - Often employed as a polishing step in tertiary treatment for water reuse applications
Nutrient Removal Technologies	- Targets the removal of nitrogen and phosphorus from wastewater to prevent eutrophication in receiving water bodies - Methods include biological nutrient removal (BNR), chemical precipitation, and enhanced biological phosphorus removal (EBPR) - BNR processes utilize specific microbial communities to convert nitrogen and phosphorus into nitrogen gas and phosphate minerals - Chemical precipitation involves the addition of chemicals (e.g., alum, ferric chloride) to precipitate phosphorus and nitrogen compounds - EBPR utilizes polyphosphate-accumulating organisms to store and release phosphorus under specific conditions - Integration of multiple nutrient removal processes enhances treatment efficiency and compliance with water quality regulations
Membrane Filtration	- Employs semi-permeable membranes to separate contaminants from wastewater physically - Common membrane types include microfiltration (MF), ultrafiltration (UF), nanofiltration (NF), and reverse osmosis (RO) - Removes suspended solids, colloids, microorganisms, and dissolved pollutants with high efficiency and reliability - Used in both tertiary treatment for water reuse and direct potable water production - Requires energy for membrane operation, periodic cleaning, and membrane replacement as part of maintenance

Source: Compiled by the authors.

TABLE 1.3
Overview and Specific Methods and Applications for Resource Recovery from Wastewater

Resource	Recovery Method	Examples/Uses
Water	Filtration, Reverse Osmosis, Distillation	Reuse for irrigation, industrial processes, or potable water supply
Nutrients (Nitrogen, Phosphorus)	Biological or Chemical Treatment	Fertilizer production, agricultural use
Energy (Biogas, Thermal Energy)	Anaerobic Digestion, Thermal Treatment	Power generation, heating
Metals (e.g., Silver, Gold)	Chemical Precipitation, Ion Exchange	Precious metal recovery
Organic Matter	Anaerobic Digestion, Composting	Biofuel production, soil amendment
Biosolids	Dewatering, Composting	Soil amendment, fertilizer

Source: Compiled by the authors.

treatment processes can remove contaminants and produce high-quality water suitable for reuse in industrial or agricultural settings. Environmental pollution can be mitigated through resource recovery from wastewater, and significant economic and ecological benefits can be realized, contributing to a more sustainable and circular approach to water management (see Table 1.3).

1.6 WASTEWATER RECYCLING FOR VARIOUS APPLICATIONS

Wastewater recycling offers a multifaceted solution to address water scarcity and environmental sustainability across diverse applications. In agriculture, treated wastewater serves as a valuable resource for irrigation, reducing reliance on freshwater sources while providing essential nutrients to crops (see Figure 1.2). Industrial processes benefit from recycled wastewater, cutting down on freshwater consumption and minimizing environmental impact through efficient water management (Wang et al., 2013). Urban landscapes thrive on recycled wastewater for irrigation, fostering green spaces without limiting water supplies. Groundwater recharge initiatives replenish aquifers, safeguarding against depletion and saltwater intrusion in coastal regions (see Table 1.4). From innovative toilet flushing systems to recreational uses like golf course irrigation, treated wastewater finds myriad applications, conserving water resources and promoting eco-friendly practices Aquifer storage and recovery methods ensure water availability during fluctuating demand. At the same time, advanced treatment technologies enable the safe production of potable water from wastewater, ensuring a sustainable water supply for communities worldwide (Sangu and Sharma 1987).

TABLE 1.4
Application of Wastewater in Various Sectors

Application	Description
Agricultural Irrigation	Reusing treated wastewater for irrigation purposes in agriculture reduces the demand for freshwater resources.
Industrial Processes	Utilizing recycled wastewater in industrial processes such as cooling, cleaning, or manufacturing reduces reliance on freshwater sources.
Urban Landscape Irrigation	Reusing treated wastewater for watering parks, gardens, and other urban green spaces, conserving freshwater for different uses
Groundwater Recharge	Injecting treated wastewater into underground aquifers replenishes groundwater levels and prevents saltwater intrusion in coastal areas.
Toilet Flushing Systems	Using treated wastewater for toilet flushing in commercial or residential buildings reduces potable water consumption.
Recreational Purposes	Reusing treated wastewater for activities like golf course irrigation, artificial lakes, or decorative fountains in public spaces.
Aquifer Storage and Recovery (ASR)	Storing treated wastewater in underground aquifers during times of surplus for later recovery during periods of high demand.
Potable Water Supply	Treating wastewater to a high standard suitable for human consumption, providing a sustainable source of drinking water.

Source: Compiled by the authors.

1.7 WASTEWATER REUSE FOR VARIOUS OPERATIONS

Wastewater reuse is a pivotal strategy in addressing water scarcity and promoting sustainable resource management across diverse operations. In agriculture, treated wastewater serves as a valuable alternative for irrigation, offering essential nutrients to crops while mitigating the demand for freshwater resources. Industries benefit significantly by integrating recycled wastewater into their processes, reducing reliance on potable water and minimizing environmental impact. Urban areas harness the potential of treated wastewater for landscape irrigation, fostering green spaces without compromising water conservation goals. Additionally, wastewater reuse plays a crucial role in replenishing groundwater levels, preventing depletion, and maintaining the ecological balance of aquifers. From industrial cooling to toilet flushing systems, the versatility of recycled wastewater extends to various operations, promoting a circular and responsible approach to water usage. As technological advancements improve treatment methods, the potential for safe and efficient wastewater reuse expands, contributing to a more sustainable and resilient water future (see Figure 1.3).

1.8 ENVIRONMENTAL IMPACTS AND HEALTH CONSIDERATIONS

Wastewater poses significant environmental impacts and health considerations due to its complex composition and potential contaminants. Discharging untreated or

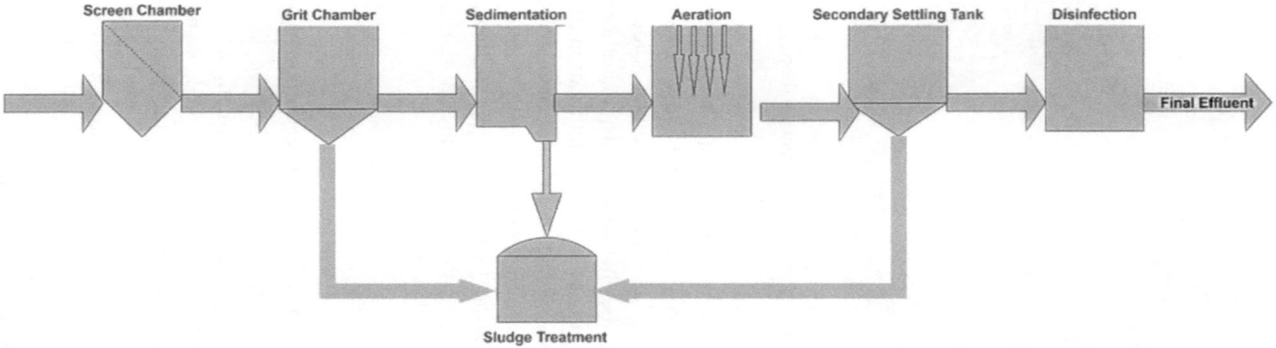

FIGURE 1.2 Schematic diagram of wastewater treatment plant (compiled by the authors).

FIGURE 1.3 Reuse of wastewater across diverse sectors (compiled by the authors).

inadequately treated wastewater into natural water bodies can lead to eutrophication (García et al., 2014 APHA-AWWA-WPCF, 1994 and Vasanthi et al., 2008), disrupting aquatic ecosystems and causing harmful algal blooms. Additionally, wastewater may contain pathogens, heavy metals, pharmaceuticals, and other pollutants, posing risks to human health through direct exposure or contamination of drinking water sources (Moo-young et al., 2004). Inadequate wastewater management can also contribute to the spread of waterborne diseases, compromising public health in communities without access to safe sanitation facilities. Proper treatment and management of wastewater are essential to mitigate these environmental and health risks, safeguarding both ecosystems and human well-being.

1.9 CONCLUSION

In conclusion, this chapter, "Overview of Wastewater Treatment, Recycling, and Reuse," underscores the critical importance of effective wastewater management in addressing global water challenges. If advanced treatment technologies and innovative recycling and reuse strategies are implemented, significant progress can be made toward conserving water resources, reducing pollution, and promoting sustainable development. However, achieving these goals requires concerted efforts from policymakers, industry stakeholders, and the public to prioritize investment in wastewater infrastructure, research, and education. By embracing a holistic approach that integrates engineering solutions with environmental stewardship principles, we can create a future where wastewater is no longer viewed as a liability but as a valuable resource for meeting the growing water demands of society. Continuing collaboration and innovation will be essential to realize the full potential of wastewater treatment, recycling, and reuse in building a more resilient and water-secure world.

Effective wastewater management is paramount for global water sustainability. Advanced treatment technologies and recycling strategies offer promising avenues for conserving water resources and reducing pollution. However, concerted efforts from policymakers, industries, and the public are essential for success. Ongoing research, investment in infrastructure, and public awareness campaigns are crucial for realizing the full potential of wastewater treatment, recycling, and reuse. Integrating innovative solutions and responsible practices can create a more resilient and water-secure world for future generations.

REFERENCES

Amir, M.S.I.I., and Khan, M.S.A. (2018) 'An innovative technique of tidal river sediment management to solve the waterlogging problem in southwestern Bangladesh', *Elsevier eBooks*, pp. 165–199. https://doi.org/10.1016/b978-0-12-810473-6.00011-x

Aluko, O.O., Sridhar, M.K. and Oluwande, P.A. (2003) 'Characterization of leachates from a municipal solid waste landfill site in Ibadan, Nigeria', *Journal of Environmental Health Research*, Vol. 2, No. 1, pp. 32–37.

APHA-AWWA-WPCF (1994) *Standard Methods for Examining Water and Wastewater*, 15th ed., American Public Health Association, Washington, DC, USA.

Biswas, B., Kumar, A., Babu, S., Bhattacharyya, S. and Chakrabarti, J. (2010) 'Studies on environmental quality in and around municipal solid waste dumpsite', *Journal of Resources, Conservation and Recycling*, Vol. 55, No. 2, pp. 129–134.

Bureau of Indian Standards (BIS) (2012) *IS 10500:2012, Indian Standard Specification for Drinking Water*, 4th ed., India.

Chatterjee, R. (2010) 'Municipal solid waste management in Kohima city, India', *Journal of Environmental Health Science and Engineering*, Vol. 7, No. 2, pp. 173–180.

Chen, Z., Ngo, H.H. and Guo, W. (2012) 'A critical review on sustainability assessment of recycled water schemes', *Science of the Total Environment*, Vol. 426, No. Supplement C, pp. 13–31.

Ennouri, H., Miladi, B., Diaz, S.Z., Güelfo, L.A.F., Solera, R., Hamdi, M. and Bouallagui, H. (2016) 'Effect of thermal pre-treatment on the biogas production and microbial communities balance during anaerobic digestion of urban and industrial waste activated sludge', *Bioresource Technology*, Vol. 214, No. 1, pp. 184–191.

Field, R., & Sullivan, D. (2002). *Wet-weather flow in the urban watershed: technology and management*. CRC press.

García, L.A.F., Lloréns, M.D.C.E., Lorenzo, E.V., Venta, M.O.B., Torres, M.L., Rodriguez, Y.R., Torres, I.F., Aerrechea, A.P. and Hung, Y.T. (2014) 'Assessment of leachates from a Cuban municipal solid waste landfill: characterisation and treatment', *International Journal of Environment and Waste Management*, Vol. 13, No. 3, pp. 291–316.

Gude, V. G. (Ed.). (2021). *Resource Recovery from Wastewater: Toward Sustainability*. CRC Press.

Hameed, A.S., Resmi, T.R., Suraj, S., Warrier, C.U., Sudheesh, M. and Deshpande, R.D. (2015) 'Isotopic characterization and mass balance reveals groundwater recharge pattern in Chaliyar river basin, Kerala, India', *Journal of Hydrology: Regional Studies*, Vol. 4, No. 1, pp. 48–58.

Jha, M.K., Sondhi, O.A.K. and Pansare, M. (2003) 'Solid waste management – a case study', *Indian Journal of Environmental Protection*, Vol. 23, No. 10, pp. 1153–1160.

Kansal, A. (2002) 'Solid waste management strategies for India', *Indian Journal of Environmental Protection*, Vol. 22, No. 4, pp. 444–448.

Karthikeyan, O. P., Muthu, S. S., & Heimann, K. (2016). *Recycling of solid waste for biofuels and bio-chemicals*. Springer.

Khalid, M. A., Mumtaz, N., Izhar, T., Ahmad, S. A., & Ali, N. (2022). *Handbook of environmental materials management*. Springer.

Kitonga, L. M., Moturi, W. N., Mwonga, S. M., & Taabu, I. (2018) 'Assessment of physico-chemical characteristics of groundwater among different farm sizes in Ainabkoi sub-county, Uasin Gishu County, Kenya' , *African Journal of Environmental Science and Technology*, Vol.12, Nos. 11, pp. 408-416. http://41.89.96.81:8080/xmlui/handle/123456789/2626

Kurian, J., Esakku, S., Nagendran, R. and Visvanathan, C. (2005) 'A decision making tool for dumpsite rehabilitation in developing countries', *Proceedings Sardinia, Tenth International Waste Management and Landfill Symposium*, S. Margherita di Pula, Cagliari, Italy, pp. 3–7.

Lema, J.M., Mendez, R. and Blazquez, R. (1988) 'Characteristics of landfill leachates and alternatives for their treatment: a review', *Water, Air, and Soil Pollution*, Vol. 40, Nos. 3–4, pp. 223–250.

Moo-Young, H., Johnson, B., Johnson, A., Carson, D., Lew, C., Liu, S. and Hancock, K. (2004) 'Characterization of infiltration rates from landfills: supporting groundwater modeling efforts', *Environmental Monitoring and Assessment*, Vol. 96, Nos. 1–3, pp. 283–311.

Mukherjee, S., Mukhopadhyay, S., Hashim, M.A. and Gupta, B.S. (2015) 'Contemporary environmental issues of landfill leachate: assessment and remedies', *Critical Reviews in Environmental Science and Technology*, Vol. 45, No. 5, pp. 472–590.

Nagarajan, R., Thirumalaisamy, S. and Lakshumanan, E. (2012) 'Impact of leachate on groundwater pollution due to non-engineered municipal solid waste landfill sites of erode city, Tamil Nadu, India', *Iranian Journal of Environmental Health Science and Engineering*, Vol. 9, No. 1, pp. 1–12.

Pohland, F.G. and Harper, S.R. (1985) 'Leachate generation and control at landfill disposal sites', *Water Pollution Research Journal of Canada*, Vol. 20, No. 3, pp. 10–24.

Sangu, R.P.S. and Sharma, S.K. (1987) 'An assessment of water quality of river Ganga at Garhmukteshwar', *Indian Journal of Ecology*, Vol. 14, No. 20, pp. 278–287.

Tucker, G.S. (2016) *Food Preservation and Biodeterioration*, 2nd ed., John Wiley & Sons, UK.

United States Environmental Protection Agency (US EPA) (1984) *Office of Drinking Water, A Groundwater Protection Strategy for the Environmental Protection Agency*, 3rd ed., Vol. 11, USA.

Vasanthi, P., Kaliappan, S. and Srinivasaraghavan, R. (2008) 'Impact of poor solid waste management on ground water', *Environmental Monitoring and Assessment*, Vol. 143, Nos. 1–3, pp. 227–238.

Wang, J., Liu, G., Liu, H., & Lam, P. K. (2013) 'Multivariate statistical evaluation of dissolved trace elements and a water quality assessment in the middle reaches of Huaihe River, Anhui, China" *The Science of the Total Environment*, Vol. 583, pp. 421–431, https://doi.org/10.1016/j.scitotenv.2017.01.088

2 Treatment of Urban Wastewater with a View towards Recycling and Reuse in India

Marut Jain, Ruchi Chauhan, Deepti Mishra, Abhisek Sahoo, Kamal Kishore Pant, Zyta Maria Ziora, and Mark A.T. Blaskovich

ABBREVIATIONS

AL	Aerated Lagoon
AMRUT	Atal Mission for Rejuvenation and Urban Transformation
ASP	Advanced Sludge Process
BCM	Billion Cubic Metre
BOD	Biochemical Oxygen Demand
CETP	Combined Effluent Treatment Plants
COD	Chemical Oxygen Demand
CPCB	Central Pollution Control Board
CPHEEO	Central Public Health and Environmental Engineering Organization
CWC	Central Water Commission
DEWATS	Decentralized Wastewater Treatment System
EA	Extended Aeration
ETP	Effluent Treatment Plants
FAB	Fluidized Aerobic Bed Reactor
FSSM	Faecal Sludge and Septage Management
IHSDP	Integrated Housing and Slum Development Program
JNNURM	Jawaharlal Nehru National Urban Renewal Mission
MBBR	Moving Bed Biofilm Reactor
MBR	Membrane Bioreactor
MLD	Million Liters per Day
MOEF	Ministry of Environment, Forest, and Climate Change
MoHUA	Ministry of Housing and Urban Affairs
MoWR	Ministry of Water Resources,
NCT	National Capital Region
NRCD	National River Conservation Directorate

DOI: 10.1201/9781003454199-2

OP	Oxidation Pond
PUB	Public Utilities Board
RD-GR	River Development and Ganga Rejuvenation
RO	Reverse Osmosis
SBR	Sequential Batch Reactor
SPCBs	State Pollution Control Boards
STPs	Sewage Wastewater Treatment Plants
TDS	Total Dissolved Solids
TF	Trickling Filter
TSS	Total Suspended Solids
UASB	Up-flow Anaerobic Sludge Blanket
UF	Ultrafiltration
UIDSSMT	Urban Infrastructure Development Scheme for Small and Medium Towns
UMCs	Urban Municipal Councils
UV	Ultraviolet
UWW	Urban Wastewater
WWT	Wastewater treatment

2.1 INTRODUCTION

India, a country with profound contrasts and enduring traditions, is at the cusp of a transformative phase. With increasing population growth, India faces a range of challenges, with one of the most alarming problems being the lack of treatment of urban wastewater (UWW). With the rapid pace of urbanization comes an unprecedented surge in wastewater generation, making it a critical issue that demands innovative, sustainable solutions. In response to this environmental imperative, the following chapter explores a visionary landscape of wastewater treatment and recycling in urban India, where traditional wisdom meets contemporary technology to address the difficulties of the twenty-first century.

Urban centers across India are the breeding ground of economic dynamism and the epicenter of resource consumption. This duality presents a complex problem where burgeoning urban populations exert tremendous pressure on the environment, water resources, and public health. Wastewater, once perceived as a nuisance, has emerged as a valuable resource, and the paradigm is shifting from 'waste management' to 'resource recovery.' This chapter explores the multifaceted aspects of this transformation, focusing on treating UWW with a specific lens on recycling and reuse. The treatment of UWW is an essential step towards achieving sustainable development. Based on the descriptive bibliometric analysis, this chapter evaluates the statistical viability of urban wastewater treatment, recycling, and water utilization in India. Data are provided on a wide range of potentially harmful contaminants and limitations in urban wastewater treatment, as well as projections for the most promising future solutions. Furthermore, reutilization and reuse of UWW in agricultural applications, landscape irrigation, industrial uses, and urban non-irrigation are also well discussed to provide a comprehensive scenario of the current situation.

The chapter explores the global best practices in urban wastewater management. Acknowledging that solutions to India's urban wastewater crisis can draw inspiration from the collective wisdom of the international community, a myriad of treatment technologies, regulatory frameworks, and successful case studies can provide invaluable insights into what works best under diverse conditions. India must amalgamate these lessons with its indigenous knowledge, fostering a unique approach tailored to its needs and resources. The focus of this chapter is the exploration of innovative and sustainable treatment technologies. Various methods for treating UWW are examined, assessing their environmental impact, cost-effectiveness, and applicability in the Indian context. Importantly, we explore conventional methods, cutting-edge technologies, and nature-based solutions that may hold immense promise. Our focus extends to the vital domain of decentralized wastewater treatment, which empowers local communities to manage their wastewater effectively and is an essential strategy in remote regions or areas without significant infrastructure. Reutilization, this narrative's fulcrum, emerges as the vanguard of water sustainability. The myriad opportunities and challenges associated with wastewater recycling are explored, scrutinizing the socio-economic, environmental, and health benefits of implementing such practices.

The main goal of this chapter is to provide a comprehensive understanding of how India manages urban wastewater. In this chapter, wastewater generation and statistics present the amount of sewage produced in urban areas as well as insights into the current technologies and practices used for wastewater treatment. The chapter focuses on policies and regulations governing wastewater management. The chapter also discusses the challenges in wastewater treatment, ensuring water quality meets safety standards.

The chapter provides recommendations for sustainable and effective urban wastewater management practices in India, guiding stakeholders toward better solutions.

2.2 URBAN WASTEWATER IN INDIA: GENERATION AND STATISTICS

The rapid increase in India's population has put significant pressure on the country's water resources. The population is expected to exceed 1.5 billion by 2050 [1]. The enormous rise in population is accompanied by a shift to urbanization, with the urban population building from 11 percent of the total population in 1901 to 35 percent in 2021 [1]. At this growth rate, 50 percent of the country's population (877 million) will be settled in urban cities by 2050 [2]. The combination of the expanding population with fast urbanization, industrialization, and agricultural growth has exacerbated water supply–demand imbalances. As per estimates, the total annual water demand in India for domestic, industrial, and farming sectors will reach 1447 billion cubic meters (BCM) by 2050 [2]. This is significantly more than the current estimate of total annual usable water in the country (1123 BCM), out of which 433 BCM is total replenishable groundwater (but only 395 BCM is utilizable), and 690 BCM is estimated utilizable surface water [3]. Therefore, the expansion of urban areas has

TABLE 2.1
Techniques for Wastewater Treatment in India

Wastewater Treatment Technology	Type of Wastewater	Reference
Sequential batch reactor	Municipal & industrial wastewater	[49]
Activated sludge process	Municipal & industrial wastewater	[50]
Waste stabilization ponds	Aerobic system	[51]
Oxidation pond system	Aerobic system	[52]
Moving bed biofilm reactor	Domestic & industrial wastewater, removal of organic matter & nutrients	[53]
Up-flow Anaerobic Sludge	Anaerobic system	[54]
Extended aeration system	Aerobic system	[55]
Fluidized aerobic bed reactor	Bioremediation	[38]
Trickling filters	Pharmaceuticals and personal care products removal	[56]

led to a rise in the generation of UWW, including contaminated water from domestic, commercial, and industrial sources.

About 380 BCM of wastewater is generated annually across the world. Asia generates the most significant volumes of wastewater, estimated to be about 42 percent (159 BCM) of the total wastewater globally. This value is expected to exceed 44 percent by the year 2030. Notably, high-income countries collect 82 percent of the wastewater they generate and treat 74 percent of it, which can be attributed to better regulation and infrastructure. On the other hand, low-income countries can only collect 9 percent and treat 4 percent of their total wastewater generated, highlighting the challenges faced in resource-constrained countries [2]. India stands as the highest wastewater generator among South Asian countries, with approximately 80 percent of the water supplied for domestic use being generated as wastewater [4]. Despite the large volume of urban sewage generated in the country – 72,368 million liters per day (MLD) – only 28 percent of this total generated urban sewage (20,236 MLD) is treated [1]. The current installed sewage treatment capacity is 31,841 MLD with 26,869 MLD operational capacity [1]. Therefore, vast amounts of generated wastewater remain untreated, finding its way into soil and surface water bodies and then leaching into the groundwater, causing long-term health, environmental, and climate-related hazards. Table 2.1 comprises different techniques for wastewater treatment in India.

2.3 URBAN WASTEWATER GENERATION

Urban wastewater is mainly generated through domestic/municipal, industrial, and rain/storm water runoffs. Municipal or sewage wastewater primarily includes wastewater generated through daily household activities, such as cooking, bathing, laundry, toilet flushing, lawn watering, and commercial and institutional facilities. The quantity and quality of domestic wastewater vary significantly based on the water availability, community habits, and water sanitation systems in India. Mega and metro cities in India generally have higher per capita water consumption levels, with sewage systems for wastewater collection and treatment. In contrast, rural areas have lower per capita water consumption and lack adequate sewage treatment systems. Over the past four decades, there has been a significant increase in domestic wastewater generation based on available census population data. In 1978–1979, class-I cities (with a population over 1 million) generated 7,067 MLD, which increased to 12,145 MLD (1988–1989), 16,662 MLD (1999–2000), and 72,638 MLD (2020–2021) [1]. States like Maharashtra, Uttar Pradesh, and Tamil Nadu produced the most significant quantity of sewage wastewater, 9,107 MLD, 8,263 MLD, and 6,421 MLD, respectively, during 2020–2021 [1]. Sewage wastewater predominantly contains dissolved solids, suspended solids, organic matter, and inorganic compounds. Proteins and carbohydrates mainly constitute organic matter, while inorganic components include phosphorus, nitrogen, and salts like carbonates, bicarbonates, chlorides, and sulfates [5]. Phosphorus mainly originates from soaps, detergents, and oil/fat residues, while nitrogen comes from proteins present in sewage water. Careful monitoring of the salts formed by these elements is crucial, especially when considering water reuse practices.

Apart from sewage wastewater, 13,468 MLD of industrial wastewater is generated, and only 60 percent is treated (CPCB 2005) [6]. Industrial wastewater comes from various manufacturing units established in the country, such as chemical, pesticide, fertilizer, construction, metal, textile, petrochemical, engineering, mining, steel, and food processing industries [7].

India's chemical and petrochemical sector has a current market size estimated at $178 Bn and is expected to grow significantly to $300 Bn by 2025. The chemical industry sector includes bulk, specialty, agrochemicals, petrochemicals, polymers, and fertilizers. There was a 3.27 percent rise in the production of significant chemicals in 2022–2023 compared to 2021–2022, with substantial growth in subsectors, including pesticides and insecticides (10.51%), alkali chemicals (6.93%), and inorganic chemicals (1.56%) [8]. The rapid growth of these industries is accompanied by the release of their waste and effluents, further exacerbating the challenges already faced in managing wastewater and protecting water resources in India.

Discharging untreated or partially treated industrial waste and effluents can be in the solid, liquid, or gaseous form. Industrial effluents contain both organic and inorganic pollutants such as nitrogen, phosphorus, nitrates, nitrites, inorganic phosphorus, arsenic, iron, chromium, cadmium, fluoride, heavy grease, and oil [7]. Effluents from food processing and agricultural industries contain primarily nitrogen and phosphorus, which can contribute to eutrophication of water bodies, excessive growth of microorganisms and aquatic plants, and severe negative impacts on the ecosystem. The wastewater generated by leather industries often contains many salts, chromium, sulfides, and lime [8]. The effluents from wastewater treatment plants, resulting from the treatment of municipal sewage and industrial wastes, often contain elevated concentrations of undesirable nitrogen (N), phosphorus (P), and heavy metals. When this treated wastewater is reused for irrigation, it not only introduces these harmful contaminants into the soil but also poses significant toxicity risks to crops and surrounding water resources. This contamination can accumulate over time, leading to detrimental health effects such as cancer and other serious diseases in humans, as highlighted by Hussain et al. (2023) and Shivarajappa et al. (2023). Thus, it is imperative to implement proper treatment and disposal measures for industrial wastewater to address the potential adverse effects on ecosystems [9].

2.4 CURRENT SCENARIO OF URBAN WASTEWATER TREATMENT

According to the Central Pollution Control Board (CPCB), there are 1,631 sewage treatment plants (STPs) in India, with a total capacity of 36,668 MLD [1, 10]. Out of these, only 1,093 STPs are in operating condition, while 102 are non-operational, 274 are under construction, and 162 are proposed for construction. With sewage generation of 72,368 MLD in urban centers, the installed sewage treatment capacity serves only 43.9 percent (31,841 MLD) of total sewage generated. Out of the installed capacity, operational capacity was only 84 percent (23,869 MLD). The actual utilized capacity was only 75 percent (20,235 MLD) of 72,368 MLD. In total, ten states in the country, including NCT Delhi, Haryana, Punjab, Uttar Pradesh, Rajasthan, Gujarat, Madhya Pradesh, Maharashtra, Tamil Nadu, and Karnataka, contribute to about 86 percent of the total installed treatment capacity. Additionally, five states, including NCT Delhi, Uttar Pradesh, Gujarat, Maharashtra, and Karnataka, contribute to 60.5 percent (19,250 MLD) of the total installed treatment capacity in the country. The actual quantity of wastewater treated as a percentage of total sewage generated was highest (125%) for Chandigarh. The states of Haryana, NCT Delhi, and Punjab treated about 70 percent of the total waste generated in 2020–2021. On the other hand, Andaman and Nicobar Islands, Arunachal Pradesh, Assam, Bihar, Chhattisgarh, Lakshadweep, Manipur, Meghalaya, Mizoram, Nagaland did not treat wastewater at all. The north-eastern regions like Arunachal Pradesh, Andaman and Nicobar Islands, Lakshadweep, Manipur, Meghalaya, and Nagaland have not installed STPs [1].

The current urban wastewater treatment in India involves on-site and off-site treatment systems. Approximately 60 percent of the population of India relies on on-site sewage management systems that handle sewage at the source of generation. On-site treatment involves the storage of domestic sewage near toilets in septic tanks/pits with periodic removal and treatment of fecal sludge/septage.

In the off-site system, sewage is transported through a network of sewers to STPs and then treated. Treated wastewater from the off-site system can either be reused for various purposes such as irrigation, potable water, industrial uses, and fish farming or can be disposed into water bodies with solids going to drying beds. The off-site treatment system comprises three significant steps: primary, secondary, and tertiary (advanced) treatment.

Primary treatment involves removing suspended solids from the wastewater, including screens, grit chambers, and skimming tanks to remove larger particles and debris from sewage. Further, wastewater is subjected to primary sedimentation in sedimentation tanks or primary settling tanks/clarifiers to remove inorganic suspended materials and organic and residual inorganic solids. Chemical coagulation and flocculation processes generate chemical flocs and floating debris released in this step. This treatment generally results in the reduction of BOD values by 25–50 percent and a reduction in TSS by 50–70 percent.

The secondary method treats the wastewater output of primary treatment by removing dissolved and colloidal biodegradable organic matter. It reduces BOD, TSS, and TDS values by up to 85 percent [1]. Biological unit processes are further classified into two categories: aerobic attached growth systems (trickling filters/biofilters, rotating biological contactors) and aerobic suspended growth systems (activated sludge process, aerated lagoons, oxidation ditches/ponds). Aerobic attached growth processes are much more straightforward, requiring less energy consumption and less equipment maintenance than suspended growth systems [11].

Generally, pretreatment methods are implemented to get rid of suspended solids before the application of tertiary treatment. The tertiary method treats wastewater by removing residual TSS and nutrients using membrane and chemical precipitation-based technologies. Membrane-based technology includes Membrane Bioreactor (MBR), microfiltration, and ultra filtration. Chemical precipitation removes heavy metals, phosphorus, and salts from the sewage. Disinfection can be applied after tertiary treatments using ultra-violet light, chlorination, and other agents. Treated wastewater is then further used as potable water and for industrial applications.

The analysis of treated wastewater collected from STPs is characterized by the physical, chemical, and biological parameters to assess water quality and its impacts on human terrestrial and aquatic life. Physical and chemical tests (acidity, alkalinity, sulfate, chloride, trace elements (e.g., heavy metals) are performed to measure the effectiveness of treated wastewater along with microbial evaluation of some STPs.

2.5 CURRENT TECHNOLOGIES AND PRACTICES IN URBAN WASTEWATER

The state of urban wastewater treatment in India encompasses a variety of conventional, natural, and advanced technologies to manage the ever-increasing wastewater generated in urban centers. WWT in metropolitan cities and densely populated urban areas heavily relies on traditional technologies such as Sequential Batch Reactor (SBR) and Activated Sludge Process (ASP) due to their affordable cost and less space requirements [2]. Notably, 490 STPs are based on SBR, while 321 STPs utilize ASP technology, both commonly adopted for removing organic matter and nutrients from municipal wastewater in mega and metro cities. Some natural wastewater treatment systems like the Waste Stabilization Pond (WSP) with 67 STPs and Oxidation Pond (OP) with 61 STPs are implemented for smaller towns/cities that have land availability for treatment plants [2].

Advanced methods include Moving Bed Biofilm Reactor (MBBR) with 201 STPs, Up-flow Anaerobic Sludge Blanket (UASB) with 76 STPs, Extended Aeration (EA) with 30 STPs, Fluidized Aerobic Bed Reactor (FAB) with 21 STPs [1]. Additionally, 364 STPs include several technologies like Trickling Filter (TF), Aerated Lagoon (AL), electrocoagulation, Bio tower, and so forth.

Bioremediation technology has also emerged as a practical approach to treat UWW in India. For example, blue-green algae-mediated bioremediation improved water quality more efficiently than conventional stabilization pond systems in wetlands. This is because bacteria degrades organic matter and leaches inorganic nutrients from wastewater, which are then further utilized by algae [12]. However, large-scale treatment facilities face challenges with energy consumption, which significantly impacts the cost of the process.

2.6 POLICIES AND REGULATIONS FOR WASTEWATER MANAGEMENT

Despite these advancements, wastewater treatment infrastructure in India faces significant challenges due to poorly developed underground sewage networks, the presence of industrial wastes and effluents, and inadequate maintenance and operation of STPs. Moreover, non-compliance with discharge standards further leads to water quality concerns and environmental impacts. The escalation of the growth of industries has resulted in approximately 11 percent of these enterprises consistently breaching pollution-control standards in India from 2011 to 2018 [13].

No single policy entirely covers wastewater management at the central level in India. Instead, various laws and policies are distributed across different levels of governance. The Schedule Seven of the Indian Constitution designates "water" as a state subject. States have jurisdiction over water resources within their territories, while Parliament oversees laws related to interstate waters. The Prevention and Control of Pollution Act (1974) established Central and State Pollution Control Boards responsible for preventing and managing water pollution by implementing measures on water disposal. CPCB assesses the quantity and quality of sewage generation and its treatment and an inventory of STPs in association with the State Pollution Control Boards, Pollution Control Committees, and local bodies. The Water Cess Act (1977), the Environment Protection Act (1986), and the Waste Management Rules (1989) also play vital roles in regulating and managing water pollution in India. The 74th Constitutional Amendment Act of 1993 further decentralized the control and management of water supply and sanitation services from the state governments to Urban Municipal Councils (UMCs).

The Central and State Pollution Control Boards set minimum acceptable discharge standards for water quality parameters, including Biochemical Oxygen Demand (BOD), Chemical Oxygen Demand (COD), Total Suspended Solids, total nitrogen, and total phosphorus [2,14]. Additionally, urban areas must recycle or reuse at least 20 percent of sewage wastewater, per the Ministry of Housing and Urban Affairs (MoHUA) Service Level Benchmarks. To maintain the availability of safe water in urban regions, industries, and STPs must comply with wastewater discharge standards. This necessitates proper planning and strict monitoring by the Central Pollution Control Board, state Pollution Control Board, and other organizations to ensure appropriate wastewater collection, treatment, disposal, and reuse.

The Department of Water Resources, under the Ministry of Water Resources, River development and Ganga Rejuvenation (MoWR, RD-GR), is primarily responsible for formulating policies and programs for water development, conservation, and management in India. It encompasses water laws and legislations, water planning, inter-state and trans-boundary water issues, water quality assessment, and initiatives for rejuvenating the river Ganga and its tributaries. Additionally, it is entrusted with subjects related to technical oversight of irrigation and flood control projects, flood-proofing, and dam safety. The Central Ground Water Board (CGWB) under MoWR, RD-GR, is responsible for the country's scientific management and development of groundwater resources. It also develops, monitors, and implements national policies, programs, and practices and disseminates information and technologies for the protection, conservation, and equitable distribution of groundwater resources. The Central Water Commission (CWC), also under MoWR, RD-GR, is responsible for initiating, coordinating, and implementing schemes in consultation with relevant State Governments to control, conserve, and utilize water resources nationwide.

Pollution of rivers is one of the numerous challenges of UWW management in India. The National River Conservation Directorate (NRCD), under the Ministry of Environment and Forests, Government of India, oversees river conservation plans in the country. Its primary objective is to establish sewage management and treatment facilities, including individual Effluent Treatment Plants (ETP) or common and Combined Effluent Treatment Plants (CETP), to mitigate pollution from domestic and industrial sources [15]. Several initiatives for water quality monitoring, such as the Global Environment Monitoring System, Monitoring of Indian National Aquatic Resources System, Yamuna Action Plan, and National River Conservation Plan are in place to preserve the water quality of Indian rivers and water resources. NRCD has established effluent standard parameters focused on removing organic matter and suspended solids, reducing pathogenic bacteria, and eliminating nutrients.

The Government of India has been making efforts to implement several urban development schemes and programs to prevent and control water pollution and address water scarcity, reuse, and recycling issues. These include the Jawaharlal Nehru National Urban Renewal Mission (JNNURM, 2005), Urban Infrastructure Development Scheme for Small and Medium Towns (UIDSSMT, 2005), Integrated Housing and Slum Development Program (IHSDP, 1993), Urban Solid Waste Management and Rajiv Gandhi Urban Development Mission, Clean India Mission (2014), and Clean Ganga Mission (2011) [16]. A combination of laws and policies address wastewater management by imposing legal restrictions on wastewater discharges into wells, lakes, sewage, groundwater, and land, as shown in Table 2.2.

2.7 CHALLENGES ASSOCIATED WITH TREATING WASTEWATER

Availability of fresh water is one of the critical issues facing the globe, and a significant portion of drinking water requirements of the world is fulfilled from surface sources such as lakes, canals, rivers, and dams [17]. Unfortunately, these water resources also act as a medium for releasing multiple industrial/domestic wastes. Treatment of wastewater is an essential solution to address the growing depletion of water resources triggered by climate change and the utilization of water by human activities. Nonetheless, broad consumption of treated wastewater in various fields necessitates enhancement in wastewater treatment and management processes to increase the purity of the treated water. Treatment/reprocessing of contaminated water is a chief task in attaining water protection and environmental fortification. The Swachh Bharat Abhiyan (raised as the Clean India Mission, initiated by the Government of India for Clean and Health India) would only be efficacious alone when water is accessible in the lavatories which have been built as a division of this

TABLE 2.2
List of Policies and Legal Regulations by the Government of India

S. No.	Acts and Regulations	Year	Policy Goals
1	Water Act/Prevention and Control of pollution Act	1974	It aims to prevent and control water pollution by establishing Central and State Pollution Control Boards (CPCB and SPCBs).
2	The Water Cess Act	1977	Imposed cess (tax) on industrial water consumption with a provision to incentivize wastewater treatment.
3	The National Water Quality Monitoring Program of India	1978	Advises central and state governments on effective measures to prevent and control water pollution while setting stream and well water quality standards.
4	The Environment (Protection) Act	1986	Grants the central administration authority to establish sewage and effluent management regulations. Includes The Environment (Protection) Rules of 1989 for industrial emission and effluent disposal.
5	The Coastal Regulation Zone Notification	1991	It aims to reduce waste disposal in coastal areas.
6	National Water Policy	2002	Effluents must undergo proper treatment to acceptable limits before being discharged into natural resources.
7	The National Environment Policy	2006	Emphasizes the reuse of urban wastewater prior to discharge into water sources. Sets guidelines for metropolitan areas to implement
8	National Urban Sanitation Policy	2008	It focuses on policies directing state governments to develop urban sanitation strategies with reuse and recycling and safe waste disposal.
9	National Water Mission	2011	Promotes sewage wastewater reuse to fulfill urban water demand.
10	The Prohibition of Employment as Manual Scavengers and their Rehabilitation Act;	2013	It aims to restrict manual cleaning of septic tanks and sewage, eradicate unhygienic toilets, and provide alternative employment opportunities for recognized manual scavengers.
11	The National Guidelines on Zero Liquid Discharge developed by CPC	2013	Emphasizes zero effluent discharge from industries and prevents groundwater contamination.
12	The Model Bill for Regulation of Groundwater Development:	2016	Established ground water development acts in many states, including regulations on chemical fertilizers, pesticides, burial injection, and industrial waste to ensure groundwater quality in urban and rural regions.

TABLE 2.2 (Continued)
List of Policies and Legal Regulations by the Government of India

S. No.	Acts and Regulations	Year	Policy Goals
13	Power's Tariff Policy	2016	Thermal power stations within 50 km of STPs must use treated municipal wastewater.
14	National Fecal Sludge and Septage Management Policy	2017	Emphasizes safe fecal sludge and sewerage disposal and robust discharge regulations directing towards decentralized systems for wastewater management.

Source: NITI Aayog (2022).

scheme.[18]. This effort is critical to reducing environmental pollution and upholding human health from the associated risks. Research on wastewater treatment is generally categorized into three categories: primary, secondary, and tertiary treatment, as discussed earlier [19,20]. However, many challenges are linked with the conduction of wastewater treatment, including a hierarchy of processes.

There are a lot of factors affecting wastewater treatment, including quality and reuse conditions, regulatory considerations encompassing the lawful, recognized institutional and technological agendas, forms of treatment schemes cost (economic), and the institutional system for successively operating the plants. Finally, for future water treatment capability, available funding, and the marketplace for treated contaminated water are essential factors to consider for new technology.

2.7.1 Quality

Water quality comprises physical, chemical, and biological characteristics [21]. Wastewater effluents are the chief contributors to a range of water contamination crises. Evidently, if polluted water is discharged directly into freshwater streams without adequate treatment, an imbalance in the marine ecosystems would be created, posing a critical danger to all the living organisms along the food sequence. The quality of contaminated water is usually defined by the amount of contaminants, which generally encompasses numerous microorganisms, heavy metal ions, radionuclides, nutrients, and pharmacological and personal care products [22]. Consuming wastewater may also impact wildlife and their habitat. Therefore, assessments of water quality are essential for monitoring levels of pollution. The Central and State Pollution Control Boards (the CPCB and SPCBs) operate water quality examinations in India, which examine and implement the relevant guidelines [23,24]. The Act intends to reduce and prevent water pollution and maintain and reinstate water integrity by monitoring and enforcing the guidelines. The Act identifies the constituents of the board and the terms/situations of provision of their representatives. In addition, the Water (Prevention and Control of Pollution) Cess Act of 1977 (corrected 2003) provides the rates/gathering of a cess on water utilized by individuals, supporting the activities of the CPCB and SPCBs [24]. This enactment provides waste guidelines and consequences for non-acquiescence with appropriate waste-discharge. CPCB guides the administration on any concern relating to the restriction and regulation of water pollution, directs pollution control conduct, and specifies technical assistance and management. To prevent water contamination and measure effective prevention and control, CPCBs and SPCBs are also involved in collecting, compiling, and publishing technical and statistical data [25]. They also prepare guides and codes related to the remedy and removal of sewerage and trade waste and distribute knowledge of wastewater management [26]. The following objectives of the CPCB and the SPCBs are designed to improve water quality.

- Sensible preparation of pollution management approaches and their arrangement
- Measuring the features and magnitude of pollution checks necessary in numerous water forms
- Assessing the efficiency of contamination regulation already in action
- Investigating drifts in water quality over time
- Identifying the ecological future of numerous contaminants
- Measuring the appropriateness of water for several usages.

2.7.2 Institutional

The CPCB water quality control plan includes overseeing water quality objectives, monitoring water quality, identifying the extent and nature of pollution, assessing pollution sources, gathering data on water quality, selecting technologies for effluent control, financing waste management, maintaining sewage treatment plants, and regulating industrial effluents. This encompasses resource recovery and recycling, adopting clean technologies, and establishing effluent release standards and costs for residual contamination [14]. Moreover, to guard river water quality, the NRCD was founded by the Indian Administration below the Ministry of Environment, Forest, and Climate Change (MOEF) to deliver sufficient aid to the state government to enhance the sewage handling capability of municipalities that are discharging sewage into rivers [14,27].

While considerable efforts have been made to establish institutional and legal frameworks nationwide for restricting water contamination, there is a notable absence of specific policies at both the central and state levels to regulate the management and prevention of waste discharge into water bodies, the treatment of contaminated water, or the ecological restoration of polluted water bodies. Nevertheless, a particular provision in the Water (Prevention and Control of Pollution) Act of 1974 has raised much apprehension over the successful rehabilitation of polluted hydraulic systems. The analysis emphasized that the reduced focus on the repercussions of not complying, together with the State's lack of strict enforcement of the Act, has resulted in a scenario where the expense of not complying is far lower than the expense of complying. Furthermore, the State Pollution Control Board's liberal supervision techniques guarantee that the fines for infractions are significantly less onerous than complying with the rules, therefore compromising the Act's efficacy in discouraging and managing water pollution.

Moreover, there are also immense troubles in setting up pollution control bodies at the institutional level due to several issues. These include a lack of, or a limited, workforce and resources with the SPCBs to check on water quality (with a sensible regularity and topographical spread of gathering, testing, and understanding) in all water systems. Furthermore, a lack of facilities or capabilities in labs to measure the exact organic pollutants and parameters leads to delays in, or absence of, testing of samples. These factors lead to an inaccurate depiction of what degree of water pollution exists in the country. Consequently, the information appears to portray a depiction far more favorable than what exists. Secondly, if the SPCBs do not possess the authority and power to penalize the contaminators/polluters, they cannot initiate legal proceedings against them. Unfortunately, they only keep restricted administrative capacities to take up legal proceedings against wrongdoers of contamination management norms or individuals who disagree with fulfilling the effluent dumping standards, often accompanied by delays in official action.

2.7.3 Regulatory

Currently, no policy obligation related to wastewater management is implemented at the central level in India. The lack of any policy and the absence of any specific regulatory framework has led to the maladministration of water reserves [28,29]. Contaminated sewage and industrial waste are the chief sources of the pollution of groundwater and running surface water in India. The Water Act (Prevention and Control of Pollution), adopted in 1974, has been the primary governmental regulation explicitly focusing on water contamination and maintenance in the country. It deals with the release of wastewater as a source of pollution [28].

Further, this Act aimed to launch SPCBs and CPCBs accountable for controlling and monitoring water contamination. It also sets punishments/fines for disturbing the water stream by releasing harmful materials into water courses, wells, and drains. In practice, the attempts of SPCBs are broader as they also examine drain water and industry wastes and wastewater management plants, along with evaluations and establishment of guidelines.

To control and manage water contamination, the Water Act is accompanied by the Water (Prevention and Control of Pollution) Act (1977), which was passed by Congress in 1992 [25]. The Act additionally encourages the construction of wastewater treatment plants by permitting an allowance of 25 percent on the allocated cess [30].

According to the 2019 report [23], many sewage treatment plants found under the Ganga Action Proposal and Yamuna Action Proposal are not adequately operational. Out of the 33,000 million liters per day (MLD) of effluents formed, only 7,000 MLD is treated [23]. The report emphasizes that the General Policy on Faecal Sludge and Septage Management (FSSM) was implemented under the Atal Mission for Rejuvenation and Urban Transformation (AMRUT) in the year 2017 for the reason that only 64 percent of India's 846 metropolitan sewage treatment plants were effective. Consequently, they can only handle 37 percent of the daily humanoid waste produced in municipal India [30,31]. Furthermore, according to the Indian administration figures, 62.5 percent of generated wastewater in the metropolitan part of India continued to be untouched or only partially treated [19,32].

As a result of solid gaps in policies, Schedule 7 of the Indian statute pinpoints water as a state concern. Nevertheless, it jeopardized the obligations stated in the Union List [19,32]. Schedule 7 allows the Government to establish control and acquire inter-state waters when there is a significant public concern. The state retains independence to outline rules concerning water usage inside the state on matters such as water supply, irrigation, drainage and embankments, water storage, and so on.

The existing legal systems have created power disparities between the central Government (Centre) and the individual states. Unfortunately, this has led to uncertainty in jurisdictional matters. Additionally, the lack of clearly defined roles and responsibilities for all involved parties, and overlapping and confusing areas of authority based on water types, has resulted in ineffective regulation. Specifically, when it comes to wastewater organization, one state's inactivity alters the welfare of one or the other states and triggers disagreements.

Wastewater must be perceived as an environmental contamination problem and a water division issue that all significant state and regional authorities must negotiate. India's National Water Policy, 2012 recommended that the mutually combined view for planning, improving, and administrating water reserves is rarely accomplished on the ground [33]. The policy identifies the incorporation and implementation of metropolitan water distribution and sewage treatment proposals should occur concurrently. It furthermore emphasizes the need to incentivize decentralized sewerage treatment plants, recover and reclaim remedied water via organized tariff arrangements, and subsidize the treatment of industrial wastes.

2.7.4 Economic

Economic inducements, specifically wastewater charges/tariffs, are fundamental to succeeding in developing minimum-cost wastewater administration and management. Efficient project design is achieved when cost-effective wastewater tariffs are implemented, eliminating the need for further analysis [34]. Economic analysis is more effectively concentrated on the complete organization and management of the wastewater management framework rather than on an individual project. The management of wastewater treatment is a matter of cumulative returns to scale, and in this regard, economic efficacy needs significant, centralized wastewater collection and treatment on a large scale. In other words, wastewater treatment is an actual monopoly. Therefore, the involvement of society is rational to the extent of controlling the organization that has control of the operation of the treatment plant. In urban areas, effluents or wastewater, both treated and untreated, are generally directly poured into open water bodies such as rivers, streams, or lakes. Surface water is characteristically an accessible and available admittance resource: a natural resource that society may use without any restrictions, similar to the public good, but which cannot be used without at least partly detracting from other people's use [26]. However, all open-access natural resources should be considered collective property controlled by formal or informal government oversight for social and economic reasons. Economic analysis is most important in making decisions on focuses/targets and is remarkably beneficial in expanding and creating preferred wastewater strategies. The economic method should be applied to select tactics to attain efficient water quality objectives at minimum cost. This analysis should acknowledge at least two components of community participation in wastewater treatment: that is, through controlling or managing treatment plants and managing the water bodies that receive the plants' discharge. Each aspect influences the planning and assessment of wastewater treatment missions and projects.

Economic analysis can select the most essential factors in deciding objectives and setting targets. For example, the political process has declared that goals should be determined by considering standard benefit-cost analysis. This means that an administration would need to gather comprehensive information regarding the financial value of the benefits that would extend from improving the existing level of water quality to a better standard [29,35]. As for deciding on the level of water quality to be targeted, governance plays a crucial role in determining the most cost-effective means that meet the target value. While most economic experts emphasize the standard benefit-cost analysis, other approaches to social choice are available, such as those that rely on economic theory. The author also suggests that objectives that do not directly enhance net profits can still provide social significance not captured by economic analysis [20].

To summarize the above content, pollution control at a minimum cost can be achieved for the sewage system and wastewater treatment plant, and it will be economically justifiable if (i) the utility assesses a tariff set at the utility's average cost; (ii) the treatment plant that conducts the treatment of wastewater and discharge should run at the average cost, and (iii) all kind of waste has to be considered in the sewerage system that would otherwise discharge into the same stream as the wastewater treatment plant.

2.7.5 Social

In wastewater treatment management and facilities, technical and financial aspects have always been considered the top priority, without considering other social considerations. Hence, no proper methodology has been developed to assess social benefits or harms. A widespread trait among rational social selection approaches is that all are constructed based on people's preferences. A convincing approach to social choice measures people's predilections and then combines these preferences by some means to obtain shared preferences. A common choice is completed through an elected governmental procedure and will be guided directly by discrete predilections. Still, community choice attained through an administrative procedure must reflect the public's preferences. Public participation is, therefore, the key to ensuring social value in social choice [18]. This holds for any method of social choice, including standard benefit-cost analysis.

Rigorous environmental laws and examination for compliance counteract excessive pollution of freshwater sources. The rapid population increase and improved living standards for more significant portions of the human populace enhance the need for better water quality using the currently available technology. This leads to the hypothesis of sustainable development as a chief controlling principle.

Social and economic requirements are often divergent for those different groups of people living above and below the poverty line, and we may need to pursue other ideas /solutions for the diverse populations [21,36]. People above the deprivation line generally consume healthful water streams and enjoy the clean atmosphere around them. For people of this segment, there is an emphasis on expanding water supply and managing wastewater schemes while looking for methods to decrease source consumption. Indeed, such methodologies are being researched. On the other hand, for people living below the poverty mark, there is an urgent requirement to create better-quality tactics/methods to supply water and to manage wastewater that can be employed in the reserve-constrained atmosphere in which these communities are surviving.

To serve lower-income communities, two essential points may be helpful to follow. The first point concerns constructing progressive approaches to supply the water and managing the wastewater they can apply for themselves. It is often seen that poor people are required to finance a substantial segment of their resources and find time to acquire water and manage the generated waste. As a result, this situation puts a great deal of burden on them. If they knew of options that could assist them in creating/generating safe water and efficiently managing their generated waste, they would see significant benefits and have enhanced productivity. Better technologies can be erected to deliver better health for this section of humanity. In this way, better health would advance to intense efficiency and the capability for beneficial enhancement, with long-term economic benefits.

This guides towards the subsequent notion of the stage of provision. Suppose we become capable of achieving this very basic level of protection in terms of public health. In that case, the only obstacle left to deal with is continuously enhancing the service level to meet the demand as the standard of living for these communities increases. In other words, after achieving the basic level of public health security for these individuals, the next vital problem to address is to allow for progressively accessible assistance as earnings levels (and equivalently, capability to pay) improve.

However, it is well established that the decentralized wastewater treatment system (DEWATS) affords an economically feasible and proficient solution for wastewater treatment in a developing country like India [37,38]. The techniques used for wastewater treatment are composed of various stages, such as treatment, discharge, or reuse of wastewater near the sources where the waste is generated. Natural or advanced treatment systems like DEWATS are usually employed in developing countries like India. In contrast, in developed countries, commonly used treatment systems comprise filtration using various membranes and batch reactors. A rigorous assessment of site circumstances should be completed to decide on what specialized treatment system to use. DEWATS affords treatment of both domestic and industrial sources, with treatment efficiency ranges for the organic wastewater streams from 1 to 1000 m^3/day. However, this does not consider additional processes needed for essential chemical pretreatment of wastewater from industries [23,28,30].

The DEWATS structure in India's countryside and semi-urban areas often fails to fulfill effluent standards over time. The primary reasons are the lack of communal involvement and the acceptance of technology-restricted systems for repairs and maintenance. However, given that national-level guidelines and databases that provide overall strategies and motivations are already in place in India, better organization of DEWATS is essential. It is likewise cardinal to create a central consolidated authority model that centralizes on a marketplace supremacy system, management amongst administration organizations, and committed contribution from all significant participants.

Construction of new wastewater treatment facilities, transformations of sewer systems, expansion of current wastewater treatment approaches, and attention to economic feasibility are all necessary for the circular economy of water. The idea of the 'Swachh Bharat' or 'Clean India' movement would be further productive in focusing more attention on water reclamation from integrated toilets and source reclamation of the generated wastewater.

2.7.6 Technological

Technologies like membrane filtration play a significant role in water treatment/reclamation. A "Modern Classic" using these technologies is NeWater, which was established by the Singapore Public Utilities Board (PUB). The PUB has improved the traditional secondary effluent treatment through technologies such as ultrafiltration (UF), which was subsequently followed by the reverse osmosis (RO) process and ultraviolet (UV) disinfection process [17,21,39,40]. They have reported that treated water that undergoes these processes is safer and more suitable than water from a regular water source like a river or well. Therefore, the treatment process is highly recommendable for industrial-level applications. PUB has moved forward to establish a complete range of NeWater Plants to serve at the industry level and fulfill the populace's water supply. "NeWater" plants are now being established all over the globe and are claimed to be generating high-quality reclaimed water to reuse for various applications. The next chapter of technology development involves replacing traditional biological-derived treatment and ultrafiltration with a membrane bioreactor (MBR) [34,40,41]. To illustrate how quickly this technology is being accepted, the Gippsland (Australia) Water Company is using an MBR intended to integrate widespread biological nutrient elimination tailored by RO to generate the water that can be employed for a pulp and paper mill [20,42]. Treatment alternatives should also be investigated: resource separation is being examined as an alternative methodology for controlling domestic wastewater. For example, toilets that can separately collect waste matter, such as urine, exist, with studies on combining them into complete wastewater management systems in development [18,22,35].

However, in India, technologies are identified based on previous experience, without any realistic framework to help choose between different wastewater treatment technologies [18,41]. Identifying a suitable technology is critical to achieving sustainability, but this is not easy to measure. Numerous quantitative/qualitative principles and attributes/indicators, incorporating technical, economic, technological, and societal importance, have been established by investigators to measure the sustainability of technologies.

Asolekar et al. delineated the three essential tests of "affordability, acceptability, and manageability" to meet the criteria of the treatment process, such as "Appropriate Technology" for wastewater treatment [43]. The following are the critical tests:

- The *Affordability Criteria* discourses the people's economic capacity and the land and energy constraint readiness with the population for which the treatment plant will be structured.
- The *Acceptance Criteria* justify the considerations such as treatment efficacy, implementation of the expertise, and operating concerns (such as odor difficulties in the regional area).
- The *Manageability Criteria* explain factors such as the plant's procedure and preservation and the plant operator's necessary expertise rank.

2.8 REUSE OF WATER: AN INTEGRAL APPROACH FOR SURVIVAL

An essential requirement for water reuse is the use of decentralized systems. The common reutilization of water projects involves a system for irrigation activities with drip emitters, habitat development, indoor reutilization, and groundwater recharge. The treatment systems for decentralized wastewater utilize local soil to spread effluents. The phenomenon of wastewater uptake in plants or waste nutrients is not affected by systems used in soil infiltration. Therefore, the arrival of untreated wastewater in groundwater and then wastewater absorbed in infiltration systems is assumed to be an unplanned reuse of water. The planned system for reuse of decentralized water system utilization for landscape irrigation is likely to occur in various locations because it leads to a decrease in the water demand, it is a safe process, and the reusability value is approximately equal to the generation of reclaimed water. For water reuse in landscape irrigation, wastewater systems are being designed which can be utilized for recycled water. The nutrient rate should be enough for plant uptake and used for drip irrigation under effluent distribution methods.

Wastewater from homes, commercial facilities, and buildings can be kept in a septic tank or separation units for solid. It can then be used for drip irrigation purposes, such as watering trees because nutrients in this wastewater are beneficial. Septic tank effluent is used after passing it through effluent filters to remove solid particles whose particles are coarse and size greater than 2–3 nm [44].

Wastewater derived from uses such as washing and bathing does not contain concentrated waste and is known as greywater. Untreated greywater can be used for subsurface irrigation, but it should be treated before any application. Treatment technologies include disinfection, biological treatment, and removal of suspended particles. If untreated greywater is stored over time, anaerobic processes, like those in the septic tank, will eventually occur. Untreated greywater can be used for the irrigation of trees by using dispersal systems [45]. The concentrations of pathogens, trace constituents, organic nutrients, and organic matter are lower than in wastewater. The sodium concentration depends on the presence and nature of detergent and home cleaning agents [44]. The treatment processes that use plant and ecological sources are known to be natural sources for treatment systems.

2.8.1 Reutilization for Agricultural Applications

The evolution of using wastewater in agricultural irrigation arises from ancient times when untreated sewage wastewater was directly added to farmlands and crops. This practice is still followed in some areas of the world despite the potential for critical health and environmental hazards. India uses 84 percent of its water for agricultural purposes and approximately 4 percent for drinking. The remaining 12 percent is allocated for industrial and other purposes. The potential for reusing treated wastewater keeps increasing as urban centers expand due to rapid urbanization. As the largest freshwater user, agriculture is the most appropriate sector for impactful wastewater reuse [46]. About 66 percent of treated wastewater is reused in agriculture globally, representing approximately 4 percent of total agricultural water consumption. Benefits of using treated wastewater in agriculture include enhancing water availability for irrigation, improving soil fertility due to nutrient content in wastewater, supporting sustainable agricultural practices, and mitigating water scarcity challenges in arid regions. In summary, promoting the productive reuse of treated wastewater in agriculture can contribute significantly to India's water security and sustainable development.

2.8.2 Reutilization for Landscape Irrigation

Notably, a large-scale groundwater recharge project in Kolar district, Karnataka, using treated wastewater from Bengaluru, has improved groundwater quality and agricultural production, serving as a model for other states [46]. Urban India generates 72,368 million litters per day (MLD) of sewage; however, only 28 percent of this sewage is treated and reused. The remaining untreated wastewater flows into groundwater, rivers, and lakes, causing pollution and harm to wildlife. A study estimated that 8,603 million cubic meters (MCM) of treated wastewater could have been reused for irrigation. This could have irrigated approximately 1.38 million hectares of land and produced over 28 million metric tonnes of horticultural crops. The potential revenue generated from these crops is estimated at Rs. 966 billion based on retail prices [46]. In summary, promoting the reuse of treated wastewater for landscape irrigation can alleviate pressure on water bodies, reduce pollution, and enhance water security in the face of climate-induced challenges.

2.9 REUTILIZATION FOR INDUSTRIAL USES

Most treated water is used by various industrial users, including refineries and power plants related to manufacturing units, where water is utilized for chilling. Water quality depends on dissolved oxygen, total dissolved solids, and chlorides. These are important because they create potential scaling and corrosion issues in the heat exchanger and piping systems. Organic matter contributes to the growth of organic microorganisms in equipment such as cooling towers, evaporation towers, and heat exchangers. Water quality is essential for industrial processes, and additional treatment is being used. [44].

The market value of the total available treated wastewater in 2021 would have been approximately INR 630 million if there were mechanisms to sell treated wastewater to different sectors for reuse. The market value is projected to increase substantially to over INR 830 million in 2025 and INR 1.9 billion in 2050 at the current market rate [37,47]. The available treated wastewater in India for the irrigation sector could have irrigated an area nine times the size of New Delhi in 2021. The revenue generated from the agricultural yield produced from this land area would have been approximately INR 966 billion [47]1. Over 6,000 metric tonnes (MT) of nutrients could have been recovered from the available treated wastewater in 2021, leading to more than INR 50 million savings due to reduced synthetic fertilizer use. Reusing treated wastewater in irrigation could have reduced greenhouse gas (GHG) emissions by 1.3 million tonnes in 2021 [47]. In summary, promoting the reuse of treated wastewater in industrial applications can lead to economic benefits, resource conservation, and environmental sustainability.

2.8.3 Reutilization for Urban Non-irrigation

Non-irrigation urban activities comprise many applications, such as cooling water conditioning, ornamental water features, fire protection, toilet/urinal flushing, and maintenance. In urban areas, washing cars and commercial laundries with reclaimed water is common. It has limited use in apartments, offices, and buildings, which have more economic advantages due to the installation of distribution systems for potable and treated water. The dual distribution supply and system for potable and reclaimed water leads to high infrastructure costs and limitations related to cross-connection between the two different water supplies. [48]. A report by the Council on Energy, Environment, and Water (CEEW) highlights that 80 percent of wastewater generated by urban India has the potential to be treated and reused for non-potable purposes [47]. Non-irrigation uses, such as industrial processes, sanitation, and other urban needs, can benefit from treated wastewater [46]. Reusing treated wastewater in non-irrigation contexts can relieve pressure on natural water bodies, lower pollution levels, and enhance water security, especially in the face of climate-induced weather events that affect water availability [46]. In summary, promoting the reuse of treated wastewater in urban non-irrigation settings can contribute to sustainable water management and address water scarcity challenges.

2.9 FUTURE RECOMMENDATIONS

It is sufficient to say that many new methodologies are being investigated around the globe. Further investment in research and development would significantly expand these advances. To be straightforward, many outsides of business are extending resolutions to the problem statements mentioned above. However, the question always arises whether to include them or enlist them. Enlisting all of them will be fruitful as it provides a more significant number of partners that we require to address the issue regarding wastewater generation, and it would lead to the first step towards the foundation of Vision 2050 of more justifiable wastewater administration systems. However, technology always comes into the picture to grant a strong foundation for such approaches, and this idea must consist of all kinds of urban water, containing water supply, management of wastewater, and stormwater collection. Furthermore, it incorporates not only the water quality and quantity but also addresses the various technological problems and restrictions, along with official challenges and hurdles.

To achieve this vision, investing in research and development (R&D) is essential, constantly developing new tools and approaches needed to evaluate the technology. The support of governmental agencies and organizations to provide the necessary funding and facilities is critical to accomplish this concept. From an overall economic prospect, R&D investments may pay more valuable disbursements than direct investments to finance wastewater organizations. Federal assets should include a more significant portion and assistance for R&D in developing new methods related to water wastewater infrastructure. Such investments substantially impact the attainment of enormous environmental performance from metropolitan water management organizations and contribute to economic growth by expanding ecological industries.

The development and implementation of technology require joint action by all professional organizations. Addressing the metropolitan water requirements of those existing above and below the poverty line is essential. In this way, it can be imagined that a sequence of building blocks to achieve such approaches can be identified, together with instruments to evaluate them.

2.10 CONCLUSION

In conclusion, this chapter comprehensively presents the challenges and opportunities within India's urban wastewater management landscape. Addressing critical aspects such as generation statistics, current treatment technologies, and regulatory frameworks lays a solid foundation for understanding the system's intricacies. Examining challenges spanning quality, institutional, regulatory, economic, social, and technological realms provides a nuanced perspective. Emphasizing the importance of water reuse, the chapter underscores its integral role in sustaining diverse sectors, from agriculture to urban applications. The future recommendations strategically guide stakeholders toward sustainable solutions. This chapter serves as a valuable reference for academics, policymakers, and practitioners, offering insights that can inform effective strategies to tackle the complexities of urban wastewater management in India and contribute to a more sustainable and resilient water future.

REFERENCES

[1] CPCB. National Inventory of Sewage Treatment Plants (2021), Central Pollution Control Board, Delhi. 2021.
[2] Niti Aayog. Urban Wastewater Scenario in India. 2022;1–75.
[3] Ministry of Jal Shakti. Central Water Commission Annual Report 2021–2022. 2022;1–366.

[4] Supply, W., Drains, S.W., Roads, U. and Lighting, S., Report on Indian Urban Infrastructure and Services.. Indian Council for Research on International Economic Relations. 2011;284.
[5] Lakshmi KS, Reddy MA. Wastewater – An overview. *Int J Contemp Res Rev.* 2017;8:20254–20262.
[6] Central Pollution Control Board. Performance Status of Common Effluent Treatment Plants in India. 2005;68.
[7] Kumar V, Kumar P, Singh J, et al. Current status of water pollution by integrated industrial hubs (IIHs) in India. In *Environmental Degradation: Causes and Remediation Strategies*. Edited by Vinod Kumar Jogendra Singh and Pankaj Kumar. Agriculture and Environmental Science Academy, 2020;104–119.
[8] Chemical Industry in India – Investment, FDI Policy, & Market. www.investindia.gov.in/
[9] Hussain A, Priyadarshi M, Qureshi F, et al. Risk assessment on human health with effect of heavy metals. In *Applied Water Science Vol. 1 Fundamentals and Applications*. Editor by Inamuddin, Mohd Imran Ahamed, Rajender Boddula, and Tauseef Ahmad Rangreez. Scrivener Publishing, 2023;1:401–419.
[10] Shivarajappa, Surinaidu L, Gupta PK, et al. Impact of urban wastewater reuse for irrigation on hydro-agro-ecological systems and human health risks: A case study from Musi river basin, South India. *HydroResearch*. 2023;6:122–129.
[11] Patel AK, Joun J, Sim SJ. A sustainable mixotrophic microalgae cultivation from dairy wastes for carbon credit, bioremediation and lucrative biofuels. *Bioresour Technol.* 2020;313:123681.
[12] Mukherjee S, Paul S, Das R, et al. A comparative analysis of treating urban wastewater using bioremediation method in Kolkata, India. *J Indian Chem Soc.* 2019;96:523–526.
[13] Pandey K. Grossly polluting industries more than doubled in 8 years: SOE in Figures. *DownToEarth.* 2019. www.downtoearth.org.in/pollution/grossly-polluting-industries-more-than-doubled-in-8-years-soe-in-figures-64962
[14] CPCB. Surface Water Quality Criteria. 2019.
[15] Singh Asiwal R, Kumar Sar S, Singh S, et al. Wastewater treatment by effluent treatment plants. *Int J Civ Eng.* 2016;3:19–24.
[16] Vij S, Moors E, Kujawa-Roeleveld K, et al. From pea soup to water factories: wastewater paradigms in India and the Netherlands. *Environ Sci Policy* [Internet]. 2021 [cited 2023 October 23];115:16–25. Available from: https://research.wur.nl/en/publications/from-pea-soup-to-water-factories-wastewater-paradigms-in-india-an
[17] Villarín MC, Merel S. Paradigm shifts and current challenges in wastewater management. *J Hazard Mater.* 2020;390:122139.
[18] Starkl M, Anthony J, Aymerich E, et al. Interpreting best available technologies more flexibly: A policy perspective for municipal wastewater management in India and other developing countries. *Environ Impact Assess Rev.* 2018;71:132–141.
[19] Di Giacomo G, Romano P. Evolution and prospects in managing sewage sludge resulting from municipal wastewater purification. *Energies.* 2022;15:5633.
[20] Soares A. Wastewater treatment in 2050: Challenges ahead and future vision in a European context. *Environ Sci Ecotechnol.* 2020;2:100030.
[21] Zang J, Kumar M, Werner D. Real-world sustainability analysis of an innovative decentralized water system with rainwater harvesting and wastewater reclamation. *J Environ Manage.* 2021;280:111639.
[22] Agarwal S, Darbar S, Saha S. Challenges in management of domestic wastewater for sustainable development. (Chapter 25) In *Urban Water Crisis and Management*

(1st ed.). Editors by Sughosh Madhav, Abhishek Kumar Bhardwaj, Eugenia Valsami-Jones, and Arun Lal Srivastav. Elsevier Inc.; 2022.
[23] Starkl M, Amerasinghe P, Essl L, et al. Potential of natural treatment technologies for wastewater management in India. *J Water Sanit Hyg Dev*. 2013;3:500–511.
[24] Kumar MD, Tortajada C. *Springer Briefs in Water Science and Technology. Assessing Wastewater Management in India.* Springer; 2020.
[25] Murty MN, Kumar S. India Infrastructure Report 2011 – Water Pollution in India. Oxford University Press, 2011.
[26] Vymazal J, Zhao Y, Mander Ü. Recent research challenges in constructed wetlands for wastewater treatment: A review. *Ecol Eng*. 2021;169:106318.
[27] Mekala GD, Davidson B, Samad M, et al. *Wastewater Reuse and Recycling Systems: A Perspective into India and Australia.* International Water Management Institute (IWMI); 2008.
[28] Goyal K, Kumar A. A comprehensive view of existing policy directives and future interventions for water reuse in India. *Water Policy*. 2022;24:1195–1207.
[29] Qadir M, Wichelns D, Raschid-Sally L, et al. The challenges of wastewater irrigation in developing countries. *Agric Water Manage*. 2010;97:561–568.
[30] Schellenberg T, Subramanian V, Ganeshan G, et al. Wastewater discharge standards in the evolving context of urban sustainability – The case of India. *Front Environ Sci.* 2020;8:30.
[31] Sharma R, Tanwar S, Rizvi S. Growth of urbanization in Himachal Pradesh: A statistical analysis. *Int J Eng Technol.* 2018;7:39–46.
[32] Pandey AK, Reji Kumar R, Kalidasan B, et al. Utilization of solar energy for wastewater treatment: Challenges and progressive research trends. *J Environ Manage.* 2021;297:113300.
[33] Flaxman EW. National water policies. *J Prof Issues Eng*. 1988;114:408–412.
[34] Gu Y, Li Y, Li X, et al. The feasibility and challenges of energy self-sufficient wastewater treatment plants. *Appl Energy*. 2017;204:1463–1475.
[35] Reymond P, Chandragiri R, Ulrich L. Governance arrangements for the scaling up of small-scale wastewater treatment and reuse systems – Lessons from India. *Front Environ Sci*. 2020;8:72.
[36] Kalbar PP. Hybrid treatment systems: a paradigm shift to achieve sustainable wastewater treatment and recycling in India. *Clean Technol Environ Policy*. 2021;23:1365–1373.
[37] Kuttuva P, Lele S, Mendez GV. Decentralized wastewater systems in Bengaluru, India: Success or failure? *Water Econ Policy*. 2018;4:1–22.
[38] Singh NK, Kazmi AA. Performance and cost analysis of decentralized wastewater treatment plants in northern India: Case study. *J Water Resour Plan Manage.* 2018;144:1–9.
[39] Al-Hazmi HE, Mohammadi A, Hejna A, et al. Wastewater reuse in agriculture: Prospects and challenges. *Environ Res*. 2023;236. https://doi.org/10.1016/j.envres.2023.116711
[40] Bui X-T, Chiemchaisri C, Fujioka T, et al. *Water and Wastewater Treatment Technologies*. Springer; 2019.
[41] Mizyed NR. Challenges to treated wastewater reuse in arid and semi-arid areas. *Environ Sci Policy*. 2013;25:186–195.
[42] Massoud MA, Tarhini A, Nasr JA. Decentralized approaches to wastewater treatment and management: Applicability in developing countries. *J Environ Manage.* 2009;90:652–659.
[43] Kalbar PP, Karmakar S, Asolekar SR. Technology assessment for wastewater treatment using multiple-attribute decision-making. *Technol Soc*. 2012;34:295–302.

[44] Metcalf & Eddy, Inc., Asano T, Burton F, et al. *Water Reuse: Issues, Technologies, and Applications* [Internet]. McGrawHill; 2007 [cited 2021 August 11];1570. Available from: www.accessengineeringlibrary.com/content/book/9780071459273

[45] Ludwig A. *Create an Oasis with Greywater* [Internet]. Oasis Design; 1997 [cited 2021 August 11]. Available from: https://oasisdesign.net/greywater/createanoasis/

[46] Verma K, Manisha M, Santrupt RM, et al. Assessing groundwater recharge rates, water quality changes, and agricultural impacts of large-scale water recycling. *Sci Total Environ.* 2023;877:162869.

[47] CEEW. Reuse of Treated Wastewater in India [Internet]. 2023. Available from: www.ceew.in/publications/reuse-treated-wastewater-india

[48] Eslamian S, Harooni Y, Ostad-Ali-Askari K, et al. Natural purification of industrial waste water used as the water resources by using hyper accumulator plants. *Int J Emerg Eng Res Technol.* 2018;6:7–11.

[49] Mane SS, Munavalli GR. Sequential batch reactor–Application to wastewater–A review. *Proceedings of International Conference SWRDM*. Shivaji University, 2012;121–128.

[50] Sustarsic M. Wastewater treatment: Understanding the activated sludge process. *Chem Eng Prog*. 2009;26–29.

[51] Liu XY, Hong Y. Microalgae-based wastewater treatment and recovery with biomass and value-added products: A brief review. *Curr Pollut Rep*. 2021;7:227–245.

[52] Rawat I, Ranjith Kumar R, Mutanda T, et al. Dual role of microalgae: Phycoremediation of domestic wastewater and biomass production for sustainable biofuels production. *Appl Energy*. 2011;88:3411–3424.

[53] Liu J, Yu J, Li D, et al. Reduction of bromate in a biological activated carbon filter under high bulk dissolved oxygen conditions and characterization of bromate-reducing isolates. *Biochem Eng J*. 2012;65:44–50.

[54] Sherafatmand M, Ng HY. Using sediment microbial fuel cells (SMFCs) for bioremediation of polycyclic aromatic hydrocarbons (PAHs). *Bioresour Technol.* 2015;195:122–130.

[55] Luo Y, Le-Clech P, Henderson RK. Simultaneous microalgae cultivation and wastewater treatment in submerged membrane photobioreactors: A review. *Algal Res.* 2017;24:425–437.

[56] Kasprzyk-Hordern B, Dinsdale RM, Guwy AJ. The removal of pharmaceuticals, personal care products, endocrine disruptors and illicit drugs during wastewater treatment and its impact on the quality of receiving waters. *Water Res*. 2009;43:363–380.

3 Advanced Physicochemical Techniques for Wastewater Treatment

Noureddine El Messaoudi, Jordana Georgin, Zeynep Ciğeroğlu, Zeynep Mine Şenol, Emine Sena Kazan-Kaya, Dilek Şenol Arslan, Maryam El Hajam, and Abdellah Lacherai

3.1 INTRODUCTION

Wastewater refers to any water that has been contaminated by various human activities and is no longer suitable for its original purpose. It is generated from a wide range of sources, including households, industries, commercial establishments, and agricultural practices (Koul, Yadav, et al., 2022). Wastewater contains a mixture of pollutants, contaminants, and organic matter that can have adverse effects on human health and the environment if left untreated (Deblonde et al., 2011; El Messaoudi et al., 2016; Gul et al., 2022; Khalid et al., 2018).

Human activities such as bathing, washing, toilet flushing, and laundry contribute to domestic wastewater. Industrial processes generate wastewater that can contain a diverse array of chemicals, heavy metals (El Mouden et al., 2022; El Mouden, El Messaoudi, El Guerraf, Bouich, Mehmeti, Lacherai, Jada, & Sher, 2023), and toxins (Ciğeroğlu et al., 2023; El Khomri, El Messaoudi, Dbik, Bentahar, Lacherai, Faska, et al., 2022; El Messaoudi, El Khomri, Chegini, Dbik, et al., 2022). Additionally, runoff from agricultural fields, urban areas, and construction sites can carry sediments, fertilizers, pesticides, and other pollutants into the water bodies (Das & Poater, 2021; Rad et al., 2022).

Wastewater is typically classified into two main types: sanitary wastewater and industrial wastewater. Sanitary wastewater, also known as sewage, originates from domestic sources like residences, offices, and public facilities. It contains organic matter, nutrients, pathogens, and other contaminants associated with human waste and household activities. Industrial wastewater, on the other hand, is produced by industrial

DOI: 10.1201/9781003454199-3

and commercial operations and can contain a wide range of specific pollutants based on the industry or process involved (Sathya et al., 2022; Lacherai, 2021).

If left untreated or discharged improperly, wastewater can have severe consequences. It can contaminate water sources, leading to the spread of waterborne diseases and the degradation of aquatic ecosystems. Nutrient-rich wastewater, when released into water bodies, can cause excessive growth of algae, depleting oxygen levels and leading to eutrophication. Toxic chemicals and heavy metals in wastewater can accumulate in the environment, posing risks to human health and disrupting ecosystems (Abdel-Raouf et al., 2012).

To mitigate these risks and protect public health and the environment, wastewater treatment is essential. Wastewater treatment is the process of removing contaminants, pollutants, and harmful substances from wastewater to make it safe for disposal or reuse (Figure 3.1). The treatment involves physical, chemical, and biological processes that aim to reduce or eliminate pollutants, organic matter, and pathogens present in the wastewater.

Wastewater treatment plays a critical role in maintaining water quality, preventing pollution, and conserving water resources. It helps safeguard public health by minimizing the transmission of waterborne diseases and ensuring that water sources are safe for various uses, including drinking water supply. Additionally, treated wastewater can be recycled for irrigation, industrial processes, and replenishing natural water bodies, promoting sustainable water management practices.

As population growth, urbanization, and industrialization continue to increase the volume and complexity of wastewater generated, effective wastewater management becomes paramount. Governments, communities, and industries are increasingly recognizing the importance of investing in robust wastewater treatment infrastructure,

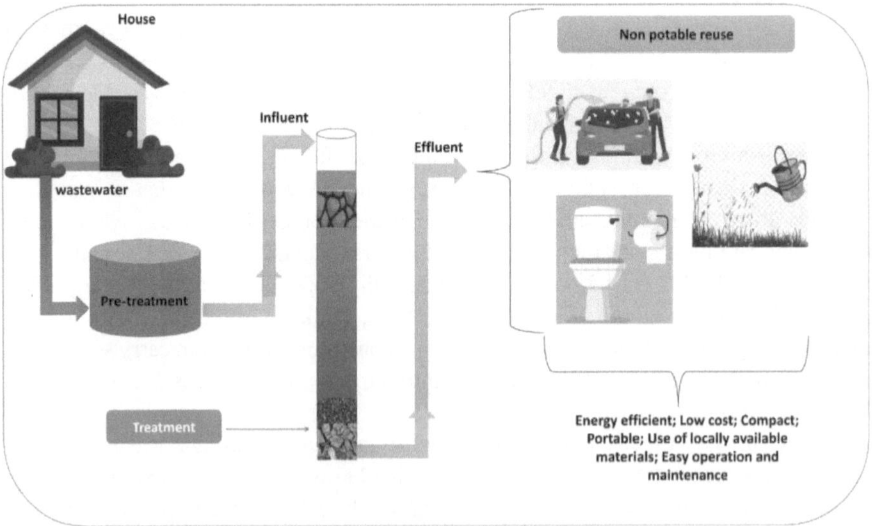

FIGURE 3.1 Wastewater treatment systems and their reuse possibilities. (Compiled by the authors.)

implementing stringent regulations, and adopting sustainable practices to minimize the environmental impact of wastewater discharge (Z. Chen et al., 2017).

In recent years, the global increase in population, urbanization, and industrialization has led to a significant rise in wastewater generation. The discharge of untreated or inadequately treated wastewater poses a severe threat to the environment and public health (Wear et al., 2021). Consequently, there is a growing need for innovative and efficient wastewater treatment techniques to mitigate the negative impacts of wastewater pollution.

The chapter begins by providing an overview of the conventional physicochemical processes (CPCPs) commonly employed in wastewater treatment. It discusses the principles and mechanisms behind these techniques, including coagulation, flocculation, adsorption, filtration, and chemical precipitation. A thorough understanding of these processes is essential for designing and optimizing conventional treatment systems (Ammar et al., 2019; El Messaoudi et al., 2021; Tóth et al., 2022).

This chapter explores the field of APCTs for wastewater treatment, which involves the application of various physical and chemical processes to remove contaminants from wastewater. These techniques have gained considerable attention due to their effectiveness in treating complex and recalcitrant pollutants that are challenging to remove using conventional APCTs (Basile et al., 2023).

The chapter concludes by highlighting the future prospects and research directions in the field of APCTs for wastewater treatment. It emphasizes the need for continued innovation and development to enhance treatment efficiency, reduce operational costs, and promote environmental sustainability.

Next, the chapter delves into specific APCTs and their applications in wastewater treatment. It explores emerging technologies such as membrane techniques, electrochemical methods, sonochemical treatment, microwave-assisted processes, hybrid processes, and green chemistry approaches. The chapter also presents case studies and discusses the advantages, limitations, and future prospects of APCTs.

Overall, this chapter provides a comprehensive overview of APCTs for wastewater treatment. It serves as a valuable resource for researchers, engineers, and professionals working in the field of environmental engineering and water management, offering insights into cutting-edge technologies and their potential for addressing the challenges posed by wastewater pollution.

3.2 CONVENTIONAL PHYSICOCHEMICAL PROCESSES (CPCPS)

3.2.1 Coagulation and Flocculation

In water treatment, flocculation and coagulation are used to remove suspended solids through a process that destabilizes the balance of suspended fine particles in water solutions. Suspended particles differ in origin, load, particle size, shape, and density. These characteristics allow coagulation and flocculation to be applied effectively (Koul, Bhat, et al., 2022). Suspended solids in water have a negative charge and repel each other when they approach each other because they have the same type of surface charge. Therefore, unless proper coagulation and flocculation are used, suspended solids remain in suspension and do not settle together (Pillai & Thombre, 2023).

Coagulation and flocculation are the basic processes in the majority of water and complex wastewater treatment systems. The difference is that coagulation is controlled by electrical attraction forces, whereas in flocculation, a physical bridge is built with the polymer added to the medium, bringing the particles together (El-taweel et al., 2023).

In summary, coagulation is a chemical process, while flocculation is a physical process. During coagulation, coagulants such as inorganic aluminum or iron salts neutralize suspended particles, while in flocculation, organic polymers are used to bridge and stabilize the flocs.

3.2.1.1 Coagulation

Coagulation is a chemical water treatment technique used to remove solid particles from water by altering the electrostatic charges of suspended particles in the water (J. Q. Jiang, 2015). This technique involves the destabilization (aggregation) of the suspension (Tarleton & Wakeman, 2007).

Colloids in a liquid medium always contain an electric charge, usually a negative charge (Liang et al., 2007). The concentration of opposite ions on the particle surface is extremely high. As the particles move away from the surface, the concentration of opposite ions decreases. Therefore, an electrical potential is formed between the particle surface and the liquid solution. This potential on the particle surface is called the zeta potential. This potential is related to the stability of the colloidal suspension (Kovalchuk et al., 2009). The ability of colloids to maintain their structure for a long time indicates that they are stable. This equilibrium state, which is the precipitation of colloids, is an undesirable condition. So, we need to neutralize this charge to destabilize particles. This can be achieved by adding an oppositely charged ion to the colloidal suspension. Since most colloids in water are negatively charged, the addition of positively charged coagulant ions reduces the total charge (Kabsch-Korbutowicz, 2005). The experimental steps of coagulation for wastewater treatment are shown in Figure 3.2.

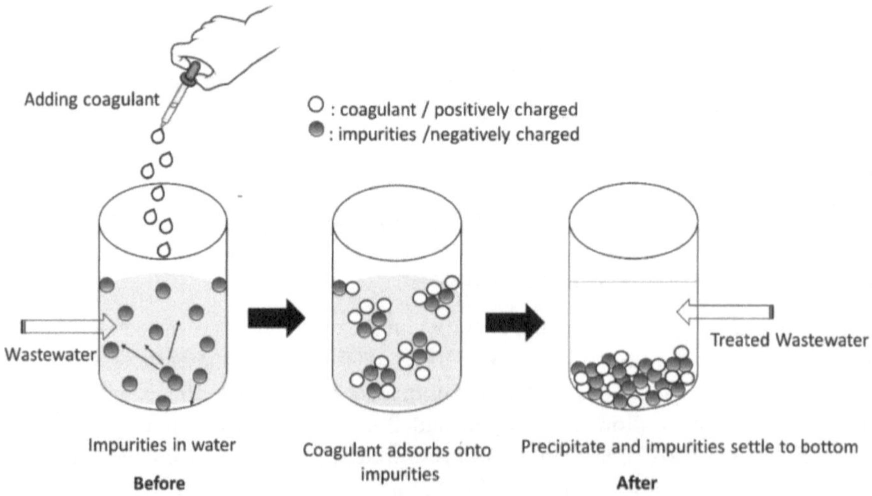

FIGURE 3.2 Schematic diagram of coagulation process for wastewater treatment. (Compiled by the authors.)

As can be seen in Figure 3.2, coagulation is a process used to remove impurities from water by manipulating the electrostatic charges of suspended particles. This process is performed by adding coagulants, positively charged molecules, to the water to destabilize the charges on suspended particles, colloids, or oily materials (Li et al., 2023). Coagulation is the precipitation of substances that cause turbidity and color in water as coarse particles with the help of coagulants (Shabanizadeh & Taghavijeloudar, 2023). There are numerous coagulants, such as higher valence cationic salts (Al^{3+}, Fe^{3+}, etc.), that are routinely employed in the treatment of various types of wastewater (Licskó, 1997; Papić et al., 2000).

Coagulants have four different mechanisms that can act individually or simultaneously:

1. Compression of the electrical double layer
2. Electrostatic adsorption
3. Adsorption and bridging
4. Precipitation

3.2.1.2 Flocculation

Flocculation is a water treatment process where fine particulates aggregate together like a floc, which can be effectively removed from wastewater by sedimentation (López-Maldonado et al., 2014). With this technique, a physical bridge is formed between the particles by the addition of high-molecular-weight polymers to the medium, and clustering of the particles is achieved. The pH value and temperature of the suspension, ionic forces, zeta potential of the particle surface, solubility of minerals in water, polymer molecular weight and structure, and charge of the active group in the suspension determine the efficiency of the flocculation process (López-Maldonado et al., 2014). The pH value and temperature of the suspension, ionic forces, zeta potential of the particle surface, solubility of minerals in water, polymer molecular weight and structure, and charge of the active group in the suspension determine the efficiency of the flocculation process (Zhao et al., 2021).

In the flocculation process, the particles are aggregated according to two common mechanisms (Lu et al., 2006):

1. Surface charge neutralization: patch model: reducing the zeta potential of the particles by using flocculants and binding the particles together by Van der Waals attraction forces.
2. Bridging: high-molecular-weight flocculants adsorb on particle surfaces. Due to their large sizes, the flocculants bridge particles and later generate large flocs. The process of surface charge neutralization and bridging proceeds as shown in Figure 3.3.

Figure 3.3 shows a high-molecular-weight polymer is added to the wastewater, and impurities are brought together to form a physical bridge. Polymers bind the agglomerated flocs together so that they do not easily disperse, even when the water is gently agitated. After flocculation is complete, large solid particles settle to the bottom and can be removed from the wastewater stream.

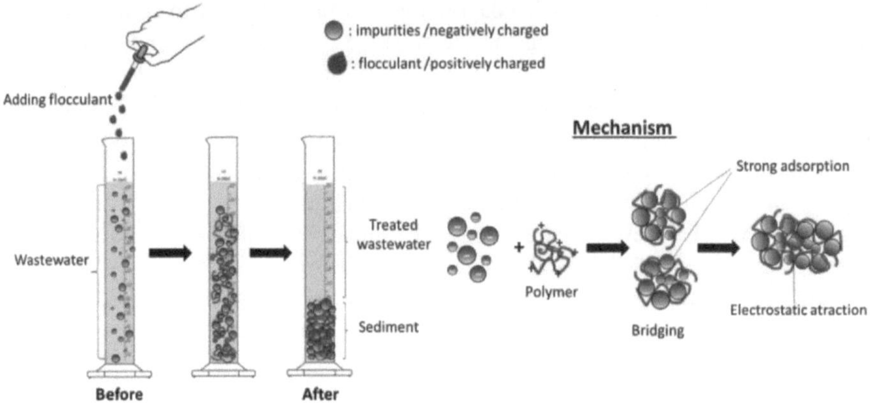

FIGURE 3.3 Schematic diagram of flocculation process for wastewater treatment. (Compiled by the authors.)

TABLE 3.1
Differences between Coagulation and Flocculation

Variables	Coagulation	Flocculation
Types of process	Chemical	Physical
Treated solid	Fine particles	Large particles
Mechanism	Coagulation is the destabilization of colloids by addition of chemicals that neutralize the charges on suspended particles.	Flocculation is the agglomeration of unstable particles into flocs, which are large particles, with the addition of polymer.
Chemicals	Coagulant: Inorganic chemicals Al^{+3}, Fe^{+3}, $Al_2(SO_4)_3$, $Fe_2(SO_4)_3$, etc.	Floculant: Polymers such as polyacrylamide, sodium polyacrylate, polyoxyethylene, polyvinylamine, polyvinyl sulfonate, etc.
Speed	Rapid	Slow
Time in process	Seconds	Minutes
Equipment	Mechanical or hydraulic rapid-mixing devices	Gentle mixing using blade mixers, turbines or other devices

Source: Compiled by the authors.

The mechanisms of coagulation and flocculation are clearly different. Coagulation is based on charge neutralization, while flocculation is based on floc formation. Table 3.1 below shows the main differences between coagulation and flocculation.

3.2.2 Adsorption

Adsorption is defined as the condensation and accumulation of ions or molecules in one phase on the surface of another phase (El Khomri, El Messaoudi, Dbik, Bentahar, Lacherai, Chegini, et al., 2022; Vermeulen, 1958). Adsorption occurs between two phases, such as liquid-liquid, liquid-gas, liquid-solid, or gas-solid. The surfaces separating these two phases are called "interfaces". The type of adsorption used in water and wastewater treatment processes is liquid-solid adsorption, and the accumulation of dissolved substances at the interface depends on the relative attraction forces between the adsorbate and the solvent. In the adsorption process, the substance that adheres to the surface is called "adsorbate", and the solid on which adsorption takes place is called "adsorbent". The process of returning the adsorbed material to the environment, that is, reducing the concentration on the surface, is called "desorption". The desorption process is defined as the reverse of the adsorption process (Messaoudi et al., 2021).

The adsorption method for removing pollutants and impurities in wastewater is a preferred treatment method due to its low cost and environmental friendliness (Amdeha, 2023). The selection of the appropriate adsorbent is vital to the adsorption method. A good adsorbent should be non-toxic, environmentally safe, inexpensive, easily obtainable and recoverable, contain functional groups, be insoluble in water, and have a large surface area. Activated carbon, zeolite, clay, ion exchange resins, and activated silica are the most well-known adsorbent materials (Aryee et al., 2023; El Messaoudi, El Khomri, El Mouden, et al., 2022).

In wastewater treatment, the process of adsorbing pollutants from the aqueous solution by the adsorbent takes place in four stages (El Messaoudi, Ciğeroğlu, Şenol, Bouich, et al., 2023). (i) migration of pollutants to the adsorbent surface (bulk solution transport), (ii) diffusion of pollutants to the surface through the boundary layer (film diffusion), (iii) adsorption of pollutants to the adsorbent surface (surface adsorption), and (iv) diffusion of pollutants into the micropores of the adsorbent (intraparticle diffusion) occur (Figure 3.4).

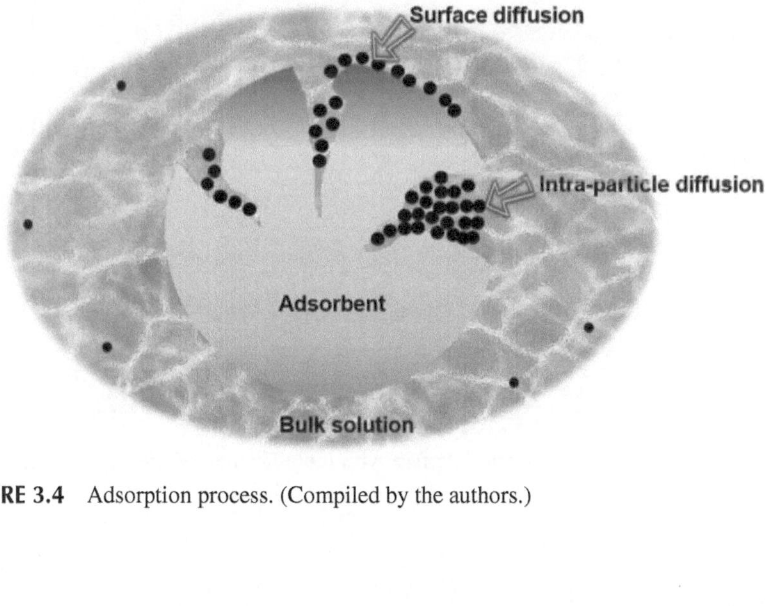

FIGURE 3.4 Adsorption process. (Compiled by the authors.)

3.2.3 Filtration

When analyzing a water treatment process, energy efficiency must be taken into account. In this sense, physical processes are usually energy efficient; this includes granular and multimedia media filters. These systems usually occupy small spaces, which is an important factor, especially in large urban centers where territorial occupation is a major problem. The use of materials available in the analyzed region is the great advantage of applying filters with granular media (S. Singh et al., 2021). In the literature, it is possible to observe several studies using different filter media in the treatment of wastewater, such as gravel, sand, and volcanic rock. For the filtration process to be efficient, optimization of operational parameters is essential. In this sense, most of the studies analyze the organic load rate (OLR), the depth, the hydraulic load rate (HLR), and the media size. Therefore, this topic evaluated the resources of filter media and their potential in wastewater treatment. Each of the parameters that affect the process is described separately, namely: medium size, operation mode, and depth. Together with the parameters, the characteristics of the wastewater and the performance of the filters in media and multimedia will be analyzed.

3.2.3.1 Influence of the Parameters on the Filtration Media

The efficiency of the filter column depends on several factors, as mentioned earlier, including the physical and chemical characteristics of the effluent, loading rate, medium depth, operating mode, and loading rate, as summarized in Figure 3.5. Laboratory-scale studies used materials with non-corrosive characteristics, such as polyvinyl chloride (PVC), acrylic, glass, plastic, concrete, and unplasticized PVC (Al-Hamaiedeh & Bino, 2010; Antonopoulou et al., 2013; Babaei et al., 2019;

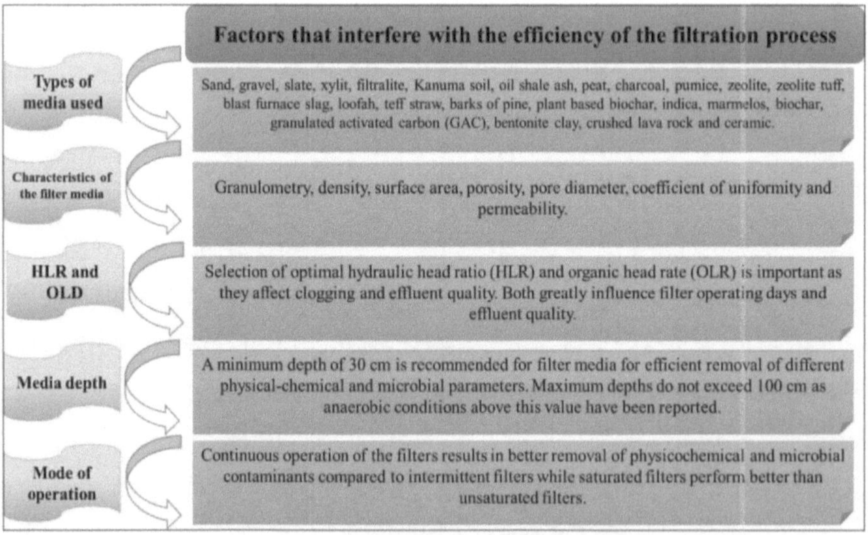

FIGURE 3.5 Main parameters that interfere with the efficiency of media and multimedia filters. (Compiled by the authors.)

Sabogal-Paz et al., 2020). Dead pockets on the edges are usually present in filters with rectangular shapes, so laboratory-scale studies are usually built in cylindrical shapes and pilot-scale studies with concrete material and rectangular shapes (Patil & Munavalli, 2016).

In the literature, it is possible to find a wide variety of materials that were used in filter media, namely: slate, sand, kanuma soil, xylite, pumice stone, filtrate, zeolite, shale ash, pine bark, indica, blast furnace, biocharcoal, teff straw, quinces, zeolite tuff, loofah, and charcoal (Al-Zou'by et al., 2017; Assayed et al., 2014; Bahrami et al., 2020; Bernardes et al., 2017; Dalahmeh et al., 2014; Dubey et al., 2010; Emslie et al., 2022; Itayama et al., 2006; Karabelnik et al., 2012; Katukiza et al., 2014; Spychala et al., 2019; Ushijima et al., 2013; Yaseen & Scholz, 2019; Zipf et al., 2016). In addition to these materials, it is possible to find filters that use granular activated carbon (GAC), ceramics, biochar, crushed lava rock, and bentonite clay. Carbonaceous materials such as activated carbons have a high renewal potential, high mechanical strength and chemical resistance, and high porosity. Sandy materials are also frequently used in filtering layers; however, due to the great diversity in granulometry, studies have analyzed volcanic tuff, river, construction, quartz and silica, pool, and dune sand. The granulometry and chemical composition are what differentiate one sandy material from another and may interfere with the removal efficiency (Alsulaili et al., 2017; Assayed et al., 2015; Bolton & Randall, 2019; Chaillou et al., 2011; Friedler et al., 2006). The choice of filter media significantly impacts filter efficiency, and understanding the material properties of these media is crucial for optimizing the water treatment process.

Material properties such as gravity, permeability, surface area, porosity, and media size can have a significant effect on filter efficiency. Filter media with an increased average particle size tend to be less efficient in removing nutrients, microorganisms, organic material, and solids. Turbidity was also affected, with a 10 percent reduction when increasing the media size from 0.8 mm to 1.18 mm (S. Singh et al., 2021). This result corroborates the study by Xu et al. (Xu et al., 2020), who, by increasing the average from 1 mm to 8 mm, observed a reduction of almost 30 percent in the removal of turbidity in wastewater. Using media containing sand and gravel to treat graywater, Abdel-Shafy et al. (Abdel-Shafy et al., 2014) also observed a reduction of almost 30 percent in the removal of turbidity. When using ceramic filters with three different particle sizes, the authors observed that the smallest size (0.25mm) was the most efficient in reducing TSS, TOC, COD, and turbidity in residential water in Oman (Ochoa et al., 2015). Confirming that pollutants have greater adhesion to filtering media formed by smaller particles. It should be noted that the effluent pH value is not altered by the media size, however, the presence of free hydrogen ions in the solution alters the pH (Xu et al., 2020). Another factor is the ion exchange potential of the materials, normally, situations with high exchange capacity raise the pH of the treated effluent (Aghakhani et al., 2013).

The organic charge ratio (OLR) and the hydraulic charge ratio (HLR) are fundamental to the efficiency of the filter. The composition as well as the quality of the effluent depend on the OLR and HLR values. Studies in the literature have evaluated filter performance with OLR (7.83–496.00 gBOD/m^2/day) and HLR (3.2 cm/day-42.1 m/h) (Abdel-Shafy et al., 2014; Al-Zou'by et al., 2017; Dalahmeh et al., 2014;

Katukiza et al., 2014). These studies indicate that the quality of the filtered effluent as well as the clogging during the process depends on optimal rates of OLR and HLR. With an HLR of 8.27 m/h, the authors observed that the filter only operated for about 16 min (Friedler et al., 2006). With an HLR lower than 7 cm/day, the process operated for about 200 days (Ghaitidak & Yadav, 2016). In general, it is observed that the increase in these two parameters considerably reduced the efficiency of the filters in removing microbes, organic matter, and nutrients.

Among the parameters that interfere with the efficiency of the filter, the depth exerts a significant interference on the removal. When analyzing the surveys, it is observed that most use a depth of 30 cm, which is recommended by Ellis and Wood (Ellis, 1985). Studies show that this value results in good removal of microorganisms and various nutrients, improving the physical and chemical characteristics of the effluent (Aizenchtadt et al., 2009; Sharaf & Liu, 2021; Thompson et al., 2020; Zipf et al., 2016). However, it is possible to observe studies that used depths of less than 30 cm (Ward et al., 2015). Where it is observed that a good part of the contaminants is removed between 10 cm and 30 cm of the filter media. It is also possible to find studies that used depths greater than 30 cm, but not greater than 100 cm, since above this value, unfavorable anaerobic conditions are created for the process (Katukiza et al., 2014; Shaikh & Ahammed, 2021a).

The system operating mode influences the quality of the treated effluent, so the filters can operate intermittently or continuously, in vertical or horizontal flow, in saturated or unsaturated conditions, and in descending or ascending flow mode. In continuous filtration systems, over time, the contact does not vary, whereas in intermittent systems, with the decrease in height above the filtering medium, the contact time increases with the course of the operation. Continuous systems have demonstrated an improvement in the quality of the treated effluent. Thus, saturated filters are more efficient than unsaturated ones. Turbidity is removed more effectively in a continuous system than in an intermittent system (Sabogal-Paz et al., 2020). On the other hand, COD and BOD were relatively lower in the effluent treated with a continuous filter when compared to the intermittent one (Shaikh & Ahammed, 2021b). The authors state that continuous feeding provides a longer retention time, which favors the removal of nutrients. Regarding the removal of PO_4-P and NH_4-N, continuous filtration showed higher removal values when compared to intermittent filtration (Shaikh & Ahammed, 2021a). In the study by Verma et al. (Verma et al., 2016), the clearance of E. coli was also higher in the continuous system. The intermittent system, however, showed a greater effect favorable to the removal with the increase in OLR and HLR than the continuous mode (S. Singh et al., 2021).

3.2.4 Chemical Precipitation

Conventional precipitation techniques involve the addition of chemical reagents to wastewater in order to generate insoluble precipitates that can be easily separated, thereby eliminating pollutants. Sulphides, hydroxides, and carbonates are just a few examples of chemicals that can react with organic and inorganic contaminants in wastewater to create insoluble precipitates, which is a simple cleanup procedure (Q. Chen et al., 2018).

3.2.4.1 Hydroxide Precipitation

The removal of impurities and pollutants from wastewater is frequently accomplished using the hydroxide precipitation method. It entails adding hydroxide substances, such as sodium hydroxide, calcium hydroxide (lime), or magnesium hydroxide, to the wastewater (Meng et al., 2022). The pollutants in the water react with these hydroxide compounds, forming insoluble compounds that precipitate. Through settling or filtration procedures, these precipitates can then be easily separated from the water (Pohl, 2020). In the hydroxide precipitation process, hydroxide ions (OH^-) are added to a solution, causing metal cations to form insoluble metal hydroxide precipitates. The following Equation 3.1 is the general chemical equation for hydroxide precipitation:

$$M^{+n}+ + nOH^- \rightarrow M(OH)^n\downarrow \quad (3.1)$$

Peng et al. (Peng et al., 2017) investigated the precipitation procedure for copper and ammonia removal from wastewater. They evaluated the impacts of pH on the precipitation process using sodium hydroxide as a precipitating agent (Peng et al., 2017). As it is known, metal hydroxide solubility depends on pH. Some metal hydroxides may be insoluble in neutral or alkaline environments, leading to precipitation. Their results showed that the optimal pH was found to be 6 for both copper and ammonia. For the successful separation and precipitation of suitable metal ions, proper pH management is of the utmost importance throughout the hydroxide precipitation method. Ultimately, the hydroxide precipitation method is a widely utilized technology in the wastewater treatment industry for the elimination of pollutants and contaminants. The hydroxide compound used, as well as the optimization of process variables such as pH and reaction time, are all affected by the particular pollutants existing in the effluent and the intended treatment objectives.

3.2.4.2 Sulphide Precipitation

To remove heavy metals from wastewater solutions, sulfide precipitation is a common technique. According to Zainuddin et al. (Ain Zainuddin et al., 2019), it entails using sulfides to precipitate heavy metals from the plating wastewater. In their study, samples are used to assess the efficacy of sulphide and hydroxide precipitation in removing nickel, zinc, and copper from industrial wastewater. Compared to hydroxide precipitation, sulfuric acid precipitation was more effective at removing heavy metals. Due to the poor solubility of metal sulphides, which enables efficient metal removal, sulfide precipitation has several advantages (Peng et al., 2017). Sulphide precipitation, however, is not without its difficulties. The toxicity of the gases created during precipitation, such as sulfides and hydrogen sulfide (H_2S), is a problem (Peng et al., 2017). To protect the environment and anyone working in the process, these harmful byproducts must be handled carefully. Additionally, the technology may become more expensive and less useful due to the separation process and the slow precipitation of metal ions (Peng et al., 2017). Sulphide precipitation has advantages over other precipitation processes, such as hydroxide precipitation, that make it stand out. In addition to having greater sludge dewatering properties, it can operate well over a wider pH range (Lewis, 2010). The special needs of the wastewater, however, as well as the desired results, must be taken into account.

The fundamental chemical formula for hydroxide precipitation is Equation 3.2, which reads as follows:

$$M^{+n}+ + S^{n-} \rightarrow (MS)^{n} \downarrow \tag{3.2}$$

3.2.4.3 Carbonate Precipitation

Carbonate precipitation is used in the treatment of wastewater for the removal of heavy metals such as copper and cobalt. According to Quiton et al.'s work from 2022, this process involves adding carbonate ions to the wastewater, which react with heavy metal ions such as copper and cobalt to form insoluble metal carbonate precipitates. Their findings showed that the degree of cobalt and copper removal is determined by significant pH dependence. Also, the precipitant-to-metal ratio had distinct effects on the removal of cobalt and copper, depending on whether the streams were polluted simultaneously or separately. Moreover, when the precipitation methods were compared, carbonate precipitation generated more sludge than hydroxide precipitation. Briefly, it can be said that the carbonate precipitation method is widely used in the industrial industry because it is straightforward to use, effective, and relatively inexpensive to operate (Raha & Ahmaruzzaman, 2022).

Equation 3.3 gives the basic chemical formula for carbonate precipitation as follows:

$$M^{+n}+ + CO_3^{n-} \rightarrow MCO_3 \downarrow \tag{3.3}$$

3.2.4.4 Phosphate Precipitation

A typical technique for removing phosphate from wastewater is phosphate precipitation, especially the creation of magnesium ammonium phosphate (struvite) ($MgNH_4PO_4{\cdot}6H_2O$), Kstruvite ($KMgPO_4{\cdot}6H_2O$), hydroxyapatite (HAP,$Ca_5(PO_4)_3OH$), vivianite ($Fe_3(PO_4)_2{\cdot}8H_2O$) (Borojovich et al., 2010; Y. Wu et al., 2019). Struvite can form when the pH is more than 7.5 and equal amounts of magnesium (Mg^{2+}), ammonium (NH^{4+}), and phosphate (PO_4^{3-}) are present (De-Bashan & Bashan, 2004). To encourage the development of struvite crystals, this process is frequently triggered by altering the pH of the wastewater (Borojovich et al., 2010). To fully comprehend and perfect the circumstances for effective phosphate recovery through struvite precipitation, the actual precipitation process can be altered by a number of variables, such as temperature and pH.

Struvite ($MgNH_4PO_46H_2O$) is a precipitate that forms when Mg^{2+}, NH^{4+}, PO_4^{3-}, and $6H_2O$ combine (F. Wang et al., 2019). This reaction was shown in Equation 3.4:

$$Mg^{2+} + NH^{4+} + PO_4^{3-} + 6H_2O \rightarrow MgNH_4PO_4{\cdot}6H_2O \text{ (struvite)} \tag{3.4}$$

This equation describes how the reaction of magnesium ions (Mg^{2+}), ammonium ions (NH_4^{+}), and phosphate ions (PO_4^{3-}) in the presence of water results in the precipitation of struvite.

3.3 Advanced Physicochemical Techniques (APCTs)

3.3.1 Sonochemical Treatment (SCT)

Wastewater contamination poses significant environmental and public health challenges, necessitating the development of advanced treatment methods. SCT, leveraging the principles of sonochemistry, has emerged as a promising and versatile technique for wastewater remediation. This innovative approach utilizes ultrasonic waves to induce chemical and physical transformations in the liquid medium, offering efficient and sustainable solutions to address diverse water quality issues (Bao et al., 2023). At the heart of SCT is the phenomenon of cavitation. Ultrasonic waves cause the rapid formation and collapse of microbubbles in the liquid, generating intense local conditions characterized by high temperatures and pressures. This cavitation process leads to the production of reactive species, including hydroxyl radicals (•OH), which play a key role in the degradation and transformation of pollutants (Dehghani et al., 2023).

SCT is particularly effective in breaking down organic pollutants in wastewater. The generated hydroxyl radicals initiate oxidation reactions, leading to the degradation of complex organic compounds into simpler, less harmful substances. The mechanical effects of cavitation, such as shock waves and microjets, can disrupt the cellular structures of microorganisms (Yasuda, 2021). SCT serves as a potent disinfection method for inactivating bacteria, viruses, and other pathogens in wastewater (Ohrdes et al., 2018). SCT facilitates the removal of heavy metals from wastewater through various mechanisms. The cavitation process induces the formation of metal oxides or hydroxides, aiding in their subsequent separation (Bao et al., 2023). Sonochemistry is employed to synthesize nanostructures with unique properties. In wastewater treatment, the creation of nanomaterials, such as nanoparticles, enhances adsorption capacities and catalytic activities for pollutant removal (El Mouden, El Messaoudi, El Guerraf, Bouich, Mehmeti, Lacherai, Jada, & Pinê Américo-Pinheiro, 2023; Irshad et al., 2022). Volatile organic compounds (VOCs) in wastewater can be decomposed through sonochemical treatment. The cavitation process effectively breaks down these compounds into less volatile or inert by-products (Z. Wu et al., 2020). SCT aids in the formation of flocs, promoting the coagulation of suspended particles. This enhances flocculation and facilitates subsequent sedimentation processes in wastewater treatment (Agunbiade et al., 2016). SCT shows promise in addressing emerging contaminants, including pharmaceuticals and personal care products. The technique contributes to the degradation and removal of these persistent compounds (Liu et al., 2023).

SCT operates at ambient temperatures, making it a non-thermal process. This is advantageous for applications involving temperature-sensitive materials and avoids the need for excessive energy consumption (Bhargava et al., 2021). Sonochemical reactions occur at exceptionally high rates, allowing for the rapid and efficient treatment of wastewater. This is beneficial for applications where swift pollutant removal is crucial (Savun-Hekimoğlu, 2020). Sonochemistry enables the selective degradation of specific pollutants, providing a level of control over the treatment process. This selectivity can be tailored to the characteristics of the wastewater

(Bao et al., 2023). In certain applications, sonochemical treatment can reduce overall chemical consumption, contributing to a more sustainable and environmentally friendly approach to wastewater remediation. The cavitation process in sonochemical treatment enhances mass transfer by creating microflows and turbulence. This ensures better contact between contaminants and reactive species, improving overall treatment efficiency.

SCT can be energy-intensive, and optimizing the energy efficiency of the process is a consideration for practical implementation. Scaling up from laboratory studies to full-scale applications may present engineering challenges and require careful design considerations (Manickam et al., 2023). Designing and maintaining sonochemical systems can be complex, necessitating skilled engineering and a thorough understanding of ultrasonic technology. Sonochemical treatment may generate by-products, and the identification and management of these by-products are essential to ensuring environmental safety. Achieving precise control over sonochemical processes requires advanced monitoring and control systems, adding to the overall complexity (Adewuyi, 2001).

SCT continues to be an area of active research and innovation in the field of wastewater remediation. Ongoing efforts focus on addressing challenges, optimizing energy efficiency, and expanding the scope of applications. As technology advances, sonochemical treatment is expected to play a pivotal role in achieving sustainable and effective solutions for the increasingly complex challenges associated with wastewater pollution.

3.3.2 Microwave-assisted Processes (MAPs)

Wastewater treatment is a critical aspect of environmental stewardship, and the integration of advanced technologies has become essential for efficient and sustainable remediation. MAPs have emerged as innovative and promising techniques, harnessing the power of microwave radiation to enhance various physicochemical and biological processes involved in wastewater treatment.

MAPs leverage microwave radiation, typically operating at frequencies of 915 MHz or 2.45 GHz, to induce selective and rapid heating within a substance. This selective energy absorption primarily targets water molecules, leading to volumetric and uniform heating (Hueso et al., 2023). The unique heating profile of microwaves has paved the way for their application in accelerating various wastewater treatment processes.

Microwave-assisted oxidation involves the synergistic combination of microwave energy and oxidizing agents to expedite the breakdown of organic pollutants. This process enhances the efficiency of oxidative reactions, contributing to the removal of complex organic compounds (Tian et al., 2022). Adsorption processes, such as the use of activated carbon or other adsorbents (El Messaoudi, Mouden, et al., 2022), benefit from microwave-assisted heating. This approach improves adsorption capacities for contaminants like heavy metals, dyes, and organic pollutants (Amdeha, 2023; Arslan et al., 2023). Microwave energy can be coupled with photocatalytic materials, such as titanium dioxide (TiO_2), to accelerate photocatalytic reactions (Cho et al., 2018). This results in improved removal of organic pollutants and inactivation of microorganisms in wastewater. Chemical precipitation processes for heavy metal and phosphorus removal are enhanced by microwave heating. The controlled temperature increase facilitates the precipitation and subsequent separation of metal hydroxides or phosphates (Liao et al., 2005; S. Wang et al., 2023). Microwave energy aids in the desorption of pollutants from adsorbent surfaces, expediting the regeneration of adsorbents and enabling their reuse in wastewater treatment (Oladejo et al., 2020; Şenol, Messaoudi, et al., 2023). Biological treatment processes, such as anaerobic digestion or sludge treatment, can benefit from controlled microwave heating. This enhances the digestion of organic matter and contributes to the disinfection of biosolids (Lizama et al., 2017). The application of microwave energy in Fenton and photo-Fenton reactions enhances the production of hydroxyl radicals, promoting the advanced oxidation of refractory pollutants.

MAPs offer rapid and volumetric heating, resulting in shorter treatment times compared to conventional methods. The selective nature of microwave heating allows for targeted and controlled heating, leading to enhanced process efficiency. Microwave heating induces internal heating within the material, promoting improved mass transfer in processes like adsorption and desorption (El Khomri, El Messaoudi, Dbik, Bentahar, Fernine, et al., 2022). Selective and volumetric heating often leads to reduced overall energy consumption compared to traditional heating methods (Goyal et al., 2022). MAPs offer precise temperature control, allowing for adjustments tailored to specific treatment requirements. The rapid and energy-efficient nature of microwave-assisted processes aligns with sustainability goals, providing an environmentally friendly approach to wastewater treatment (Vialkova et al., 2021).

Designing and implementing microwave systems for large-scale wastewater treatment can involve higher initial costs. Achieving uniform heating throughout complex matrices in wastewater can be a challenge and requires careful system design (Vialkova et al., 2021). The compatibility of materials with microwave radiation must be considered to avoid issues such as overheating or damage to equipment. Scaling up microwave-assisted processes from laboratory-scale to industrial-scale operations requires careful consideration of engineering and process optimization.

MAPs continue to be a dynamic area of research and innovation in wastewater treatment. Ongoing developments aim to address challenges, optimize processes, and expand applications. As technology advances, MAPs are expected to play a crucial role in the development of sustainable, energy-efficient, and effective solutions for wastewater remediation.

3.3.3 Membrane Techniques

Membrane technology processes imply the separation of chemical contaminants through a membrane interphase by the difference in the transport rate, depending on the concentration of each single species, the driving force, and their mobility. Basically, a membrane is a barrier that helps to separate two phases by controlling and blocking the movement of chemical species through it in a selective way (Takht Ravanchi et al., 2009). The molecular weight of the solute, the structure and morphology of the membrane, and the chemical affinity are the key parameters that affect the separation process of chemical components (Amit & Rupali, 2004; Jhaveri &

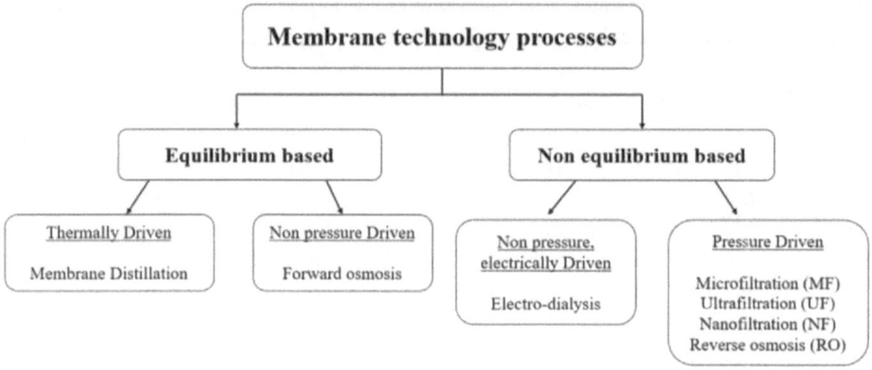

FIGURE 3.6 Diagram representing different categories of membrane processes used in wastewater treatment industries. (Compiled by the authors.)

Murthy, 2016). Characteristically, membrane processes are usually classified based on the driving force, as shown in Figure 3.6.

3.3.3.1 Pressure Driven Membrane Processes

Pressure-driven membrane processes are known to be the most widely used techniques in wastewater treatment. These processes aim to reconcentrate the dilute solution, depending on hydraulic pressure, to achieve the phases' separation. Basically, there are four principal types of these processes: microfiltration (MF), ultrafiltration (UF), nanofiltration (NF), and reverse osmosis (RO). The pore size of their membrane is considered the main difference between these types of membrane processes (Martha Noro Chollo, 2014).

Claudia Muro et al. (Muro et al., 2012) reported that among these four processes, reverse osmosis (RO) is the most efficient one for dissolved solids and small particle removal, including bacteria and monovalent ions like sodium and chloride ions, up to 99.5 percent. RO is only permeable to the molecules of water, and during this process; a strong hydrostatic pressure should be generated to exceed the intrinsic feed's pressure, which is against the com mon osmosis process.

In order to avoid or minimize the reverse osmosis (RO) membrane and to boost the constant flux's maintenance, usually microfiltration (MF), ultrafiltration (UF), and nanofiltration (NF) are set out as pretreatment steps prior to RO. This is also considered a multi-membrane treatment for the treatment of wastewater (Bartels et al., 2005).

Pressure-driven membrane technology has obviously made water reuse from wastewater a great alternative. However, the energy requirements due to the applied pressure are the biggest remaining challenge for these processes.

3.3.3.2 Temperature Driven Membrane Process

Membrane distillation technology is a promising separation process for seawater desalting and wastewater treatment as a potentially low-carbon process. In this method, hydrophobic membranes are highly efficient at removing almost all macromolecules, colloids, volatile and nonvolatile substances, and salts compared to hydrophilic membranes (Bera et al., 2022). The application of this separation process leads to critical recalcitrant degradation; therefore, less production of sludge results in a lowered footprint from this method, despite providing better effluent quality. In addition to that, the commercialization of this technology is limited because of its various options: hollow fiber versus flat sheet, direct contact membrane distillation versus vacuum membrane distillation. It is probably that one of these hybrid combinations needs to dominate before commercial takeoff, this will direct membrane distillation towards niche applications, for example, produced water, and it could be a good technology for zero liquid discharge through the membrane distillation-crystallization process (G. Chen et al., 2014).

3.3.3.3 Forward Osmosis Process

The forward osmosis (FO) process is a natural occurrence that follows the natural osmosis process, where the solvent molecules (water) move from the solution of lower concentration to the solution of higher concentration through a semipermeable or permeable membrane (Ong et al., 2017). In this case, the highly concentrated draw solution is the one that provides a concentration gradient to draw solvent molecules from the feed solution and then produces the needed osmotic pressure difference to guide water molecules from the feed solution to the draw solution until an equilibrium of chemical potential is attained (Suwaileh et al., 2018).

FO is found to be a promising and efficient method of separation with low-rate production of brine. As this method promises to solve water problems, it has been given serious consideration and has been well studied; however, the high cost of the draw solution's regeneration during this process limits its use (Blandin et al., 2016). FO is known for its several advantages; for example, it does not need external pressure; in other words, the energy consumption during the FO process is lower compared to a pressure-driven process. Water treatment is easier due to the applied osmotic pressure during this process. Moreover, the possibility of multiple reuses of the draw solution is excellent for reducing the process cost.

FO also some weaknesses, such as the low permeate flux because of the concentration polarization and the high energy requirement during the regeneration of draw solutes (Shaffer et al., 2015; Wang & Chung, 2015; Y. N. Wang et al., 2018).

3.3.3.4 Electro-dialysis (ED) Process

Electro-dialysis was among the first large-scale wastewater treatment processes using membranes and is still universally used. It is a membrane process that uses electricity and ion-permeable membranes in combination for the selective separation of ionic components from water. This technology uses electric potential to move the dissolved ion from a dilute solution to a concentrated solution across an ideal permeable membrane (Bolong et al., 2009; Campione et al., 2018). Basically, two types of ion exchange membranes are used during electro-dialysis: the first one is permeable to cations and rejects anions, and the opposite is true for the second one. There are also two streams of solutions: the dilute solution and the concentrated solution.

Ion exchange membranes (IEMs) are prepared from polymeric materials using fixed ionic charge groups in their polymeric matrix. There are two classes of IEMs: anion-exchange membranes (AEMs) and cation-exchange membranes (CEMs). AEMs are characterized by positively charged groups, while CEMs have negatively charged groups in their polymer matrix. In this process, dissolved ions move from the dilute solution to the concentrated one, and that happens after the electric current is passed through the system. In other words, the cations are retained by the positively charged anion-exchange membrane (AEM), while the anions are retained by the cation-exchange membrane (CEM) (anions go to the anode and cations go to the cathode) (Galama et al., 2014; Strathmann, 2004).

3.3.4 Hybrid Processes (HPs)

Wastewater treatment often requires a multifaceted approach to effectively address diverse pollutants and contaminants. HPs, combining two or more distinct treatment methods, have emerged as innovative solutions that synergistically enhance treatment efficiency, flexibility, and overall performance. Hybrid processes integrate different wastewater treatment technologies, each contributing specific strengths to create a synergistic effect. The combination of complementary processes aims to maximize pollutant removal, improve overall treatment performance, and address the limitations of individual methods.

Membrane Bioreactor (MBR) combines biological treatment (the activated sludge process) with membrane filtration (Rahman et al., 2023). The biological reactor breaks down organic matter, and membranes (microfiltration or ultrafiltration) separate solids from treated water. This results in high-quality effluent suitable for various applications (Al-Asheh et al., 2021). This hybrid involves combining adsorption processes, such as activated carbon adsorption, with membrane filtration. Adsorption removes specific pollutants (Şenol, Elma, et al., 2023), while membranes provide additional removal of suspended solids and microbes, resulting in improved water quality. Integrating AOPs like ozonation or UV/H_2O_2 with biological treatment enhances the breakdown of recalcitrant pollutants. AOPs generate highly reactive species that augment the biological degradation of organic compounds (Kumari & Kumar, 2023).This hybrid process combines chemical coagulation (using coagulants like ferric chloride or alum) with membrane filtration. Coagulation aids in the aggregation of particles, and membranes provide fine filtration, resulting in improved turbidity and pathogen removal (Koul, Bhat, et al., 2022). Ozone is introduced to the wastewater to oxidize and break down pollutants, and the treated effluent undergoes subsequent biological treatment for the removal of residual organics (Lim et al., 2022). This hybrid enhances overall organic and pathogen removal. Combining the activated sludge process with constructed wetlands leverages the biological treatment capabilities of both systems. Activated sludge provides nutrient removal, and constructed wetlands offer additional polishing and natural filtration (Sakurai et al., 2021).

HPs leverage the strengths of different treatment methods, creating a synergistic effect that enhances overall treatment efficiency. The combination of diverse technologies provides versatility in addressing various pollutants and allows for the flexibility to adapt to different wastewater compositions. Hybrid processes often result in treated water of higher quality, meeting stringent regulatory standards, and making it suitable for various reuse applications (Yang et al., 2020). Hybrid systems can potentially reduce the physical footprint of treatment plants by combining processes that might otherwise require separate infrastructure. Integration of biological processes with other treatment methods enhances the removal of nutrients such as nitrogen and phosphorus, addressing eutrophication concerns.

Designing and operating hybrid treatment systems can be complex, requiring careful consideration of compatibility and optimization of each component (Kim et al., 2011). The implementation of HPs may involve higher capital and operational costs compared to individual treatment methods. Cost-effectiveness should be carefully evaluated (Farghali et al., 2023). The maintenance of hybrid systems may be more demanding due to the integration of multiple components. Regular monitoring and upkeep are essential for sustained performance. Scaling up HPs from laboratory or pilot-scale to full-scale operations requires thorough engineering considerations to ensure seamless integration and performance (Naidu et al., 2020).

HPs are at the forefront of wastewater treatment innovation, and their role is expected to expand in the quest for more efficient, sustainable, and versatile solutions. Ongoing research and technological advancements will likely focus on optimizing hybrid configurations, addressing challenges, and developing tailored approaches for specific wastewater types and treatment goals. As the demand for high-quality treated water continues to grow, the integration of HPs will play a pivotal role in shaping the future of wastewater treatment.

3.3.5 Electrochemical Treatments

There are approaches to creating more environmentally friendly solutions for water treatment applications using electrochemical wastewater treatment techniques. Electricity is the primary reactant used in many electrochemical water treatment techniques. Emerging techniques that exhibit strong preconditions for their usage in large-scale applications can be used in addition to established electrochemical water treatment techniques (Sillanpää & Shestakova, 2017). These techniques have the tendency to boost additional pollution removal rates, do away with problems, expand the use of current electrochemical water treatment techniques, and raise cost effectiveness (Yakamercan et al., 2023).

The electrode, which contains a fluid or solid electrolyte and a metal or semiconducting substance, is used in the electrochemical reaction. Chemicals, in particular radicals produced at the electrode surface, are in charge of decomposing the contaminants during electrochemical operations. The connection between an electric charge and a chemical reaction is explained by the electrolysis process. In 1889, the electrochemical method was used for the first time to treat wastewater. Due to their simplicity of use, high efficiency, minimal maintenance requirements, and lack of chemical addition, electrochemical applications are now often utilized in wastewater treatment for resource recovery. As is known today, leachate, industrial wastewater, and municipal wastewater are cleaned up using electrochemical techniques.

Additionally, these procedures have the benefits of simple operation, adaptable equipment, and quick response times. The electrochemical process depends heavily on electricity; thus, it uses a great amount of electricity. Therefore, the method's primary flaw is its high use of power (Chaplin, 2019; Sun et al., 2022).

Prior to choosing an electrochemical treatment method, it is important to identify the characteristics of the pollutants to be treated, perform a cost-benefit analysis, and choose a treatment method that will have the fewest secondary pollutants and toxic side effects. Environmental systems successfully remediate wastewater using electrochemical methods. Electrocoagulation (EC), electro-Fenton (EFN), electrooxidation (EO), electrodialysis (ED), and electroflotation (EF) are important examples of commonly utilized electrochemical processes. Each approach has benefits and drawbacks, so it is important to choose one that will effectively breakdown sewage while taking the amount of water pollution and treatment costs into account.

3.3.5.1 Electrocoagulation Technique

The removal of pollutants from wastewater is accomplished by the EC technique, which uses sacrificial electrodes and an electric current supply. It is a very fast wastewater treatment procedure (Hasanah et al., 2023). The electrochemical method of EC, which removes small particles from wastewater without the use of chemicals, uses an electric current to cleanse sewage. Additionally, EC's inexpensive initial installation and maintenance costs, the minimal quantity of sludge produced after treatment, a fast-settling time, and strong pollutant removal effectiveness made it progressive and widely used (Demirci & Özbeyaz, 2019).

The EC procedure offers various benefits over conventional forms of treatment. Without using chemicals, electrocoagulation can treat a variety of industrial wastes. Previous research has shown that the electrocoagulation technique is successful in eliminating total suspended solids (TSS), heavy metals, phosphate, chemical oxygen demand (COD), nitrogen components, and color from wastewater. Simple coagulation processes have the drawback of using too many chemicals, which might increase the chemical pollution in the treated, decolored effluent (Ebba et al., 2022; Yaqub et al., 2023). Due to its absence of secondary pollutants, high treatment efficacy, high active ingredient concentration, inexpensive equipment, and ease of automated control implementation, EC is steadily evolving into an environmentally benign water treatment process (Abdul Rahman et al., 2023; Emamjomeh & Sivakumar, 2009; Núñez et al., 2019).

Electrodes of the stainless steel (SS), aluminum (Al), and iron (Fe) varieties are employed in the EC process for wastewater treatment. Very good ionic and colloidal matter removal efficiency. The price of an electrode is modest. It is likely that the operation will run continuously. The muck is created when running. The electrode must be replaced since it has disintegrated. can only distinguish contaminants (Arabameri et al., 2023; Elazzouzi et al., 2018; J.-Y. Wang et al., 2023).

3.3.5.2 Electrooxidation Technique

The electrochemical advanced oxidation process (AOP) is known as electrooxidation and generally includes the electrolytic reactions occurring on the electrode surface, such as the elimination of organics, the creation of anodic metal and metal hydroxide cations in water, as well as the adsorption of soluble or colloidal pollutants on the surface of metal hydroxides (El Messaoudi et al., 2024). Here, the active substance hydroxyl radical with strong oxidizing ability is generated to make the pollutant in wastewater low-toxic or non-toxic at a small molecular level (Yin et al., 2023). For the treatment of home, commercial, and agrozootechnical wastewater, electrooxidation is a green technique. As a result of the electrooxidation process's reduction reaction, organic molecules undergo oxidation at the anode's interface with an aqueous solution. How well electrooxidation performs as a pretreatment for wastewater is contingent upon the anodes used to mediate the oxidation of the existing organic molecules (Feijoo et al., 2023; Gad et al., 2022; Garcia-Segura et al., 2018; Zhang et al., 2023).

Due to their advantages of flexible operation, high efficiency, high modularity, and consistent performance, electrochemical advanced oxidation technologies are receiving more and more attention in wastewater treatment. Organic matter can either be eliminated directly by an anodic reaction or by powerful oxidizing species, such a hydroxyl radical, that can be produced by the anodic reaction in order to efficiently and totally breakdown contaminants in line with the needs of environmental protection. Additionally, there is little to no secondary pollution and relatively basic equipment required for this method. The anode material, however, is what makes this method effective in degrading organic pollutants. Conventional anodes have good stability, significant oxidation capacities, and high electrochemical activity; nonetheless, they are costly and incline toward "passivation" from surface organic debris, making them challenging to employ industrially on a wide scale. (Tang et al., 2023) In engineering applications, economical and stable Ti-based anodes with outstanding electrochemical activity have taken the lead as electrooxidation anode materials. EAOPs can fully mineralize organic contaminants without the use of chemicals due to the oxidation of the hydroxyl free radical formed through the anodic oxidation of water and a direct electron transfer process (DET) on the "non-active anode". The electrodes based on PbO_2 or SnO_2, the boron-doped diamond (BDD) anode, and titanium suboxides in the magnéli phase have all been shown to be effective "non-active" anodes for the removal of organic contaminants. As a result, the material used for anodes plays a crucial role when it comes to how AOPs are applied in the environment (Gomez-Ruiz et al., 2017; R. Wang et al., 2023).

3.3.5.3 Electrodialysis Technique

Due to its high removal efficiency, simple operation, and low process cost, electrodialysis (ED) is a well-respected method for the treatment of persistent pollutants. It has been used in industries such as wastewater treatment, brine and saltwater desalination, environmental protection, and intermediate pre-desalination (T. Chen et al., 2022). The use of this substance for the treatment of wastewaters, particularly those containing heavy metals, is underway due to its good desalination qualities and helpful resource reclamation nature. In the electroplating and finishing sector, chromium is a common material utilized; however, it is extremely carcinogenic and non-biodegradable. Chromium species have been recovered from wastewater using electrodialysis (C. Jiang et al., 2018).

Ion exchange membranes and electricity work in tandem during ED. It is categorized as an electrochemical and membrane filtering technique. The switch membrane is capable of handling many types of wastewaters and does so without the need for chemical detergents, depending on whether it is anionic or cationic. In ED cells, both anode and cathode electrodes exist, yet they are separated by parallel membranes that act as buffers to block or transistorize certain ions. These membranes are separated into concentrated and dilute partitions within the electrodialysis cell by a spacer gasket based on their electric charge, which contains the electrolyte solution pumped via these partitions (AlJaberi et al., 2023; Sedighi et al., 2023; Xu et al., 2019).

The electric current passing through the cell at the anode initiates an oxidation process, which culminates in the creation of protons in the anion partition, where anions are recovered using cation membranes (Jia et al., 2023; C. Jiang et al., 2018; Salehmin et al., 2023; Voutetaki et al., 2023; S. Wang et al., 2023). The reduction process generates hydroxide ions at the cathode subsequently pass through the membranes that separate cations from one another. By the transferring cationic ions to the cathode, and anionic ions to the anode via cation exchange membranes and anion exchange membranes, respectively, electrical equilibrium is attained, resulting in the generation of concentrated and diluted effluents (Zhou et al., 2023).

Electrodialysis has been utilized to process wastewater includes a range of inorganic and organic compounds in a number of studies. To reach the required capacity, this technology has worked effectively both on its own and in hybrid systems with other technologies.

3.3.5.4 Electroflotation (EF) Technique

A useful method for treating wastewater with varied impurities is EF. The type of pollutant will determine how well the pollutant layer can be removed using skimming. Because of the small bubbles created, these pollutants, such as organic contaminants, can subsequently be eliminated by lifting their light organic pollutants to the solution's surface through buoyant force. Solution electrolysis takes place during EF in the chloride medium when current flows through the electrochemical cell, resulting in small bubbles of gases like O_2 and H_2 generated from the cathode and anode (Cuprys et al., 2020; da Silva et al., 2023; Naha et al., 2023; J. Wang et al., 2023).

With the same metal electrode, H_2 flotation equipment generates smaller H_2 bubbles in a neutral solution than O_2 bubbles, in contrast to conventional air flotation equipment that produces larger bubbles. The vast amount of H_2 produced is more effective in removing light pollution particles while they are in the floating phase because of the increased surface area of the small bubbles. The size and quantity of bubbles are, however, greatly influenced by a variety of factors, including the electrolyte, the pH of the solution, the kind and arrangement of the electrodes, and the current density (Aghaei et al., 2023; Xie et al., 2022). Reduction and oxidation reactions produce hydrogen and oxygen, respectively. As a result of the metal concentration of the used electrodes and/or electrolyte, different gases are also generated in the electrochemical cell.

3.3.6 Green Chemistry (GC) Approaches

Green chemistry, also known as sustainable or environmentally friendly chemistry, is a field that emphasizes the design and implementation of processes that minimize the use and generation of hazardous substances. In the context of wastewater treatment, green chemistry approaches aim to develop methods that are not only effective in removing pollutants but also reduce the environmental impact of the treatment process.

Green adsorbents, derived from renewable resources such as agricultural by-products or natural polymers, are used to remove pollutants from wastewater through adsorption (El Messaoudi, El Mouden, et al., 2023). Utilizing green adsorbents reduces the reliance on synthetic adsorbents and minimizes the environmental footprint of the treatment process. Constructed wetlands and phytoremediation involve using plants to naturally treat wastewater (Hassan et al., 2021). Plants take up nutrients and pollutants, providing a green and sustainable approach to water purification. It requires minimal energy input, promotes biodiversity, and offers an aesthetically pleasing and ecologically friendly solution (Mihalakakou et al., 2023). Implementing environmentally friendly oxidation processes, such as ozonation or hydrogen peroxide treatment, to degrade organic pollutants in wastewater. It minimizes the use of harsh chemicals, reduces the formation of harmful by-products, and enhances the overall sustainability of the treatment (Kumari & Kumar, 2023). Using bio-based polymers, derived from natural sources, as coagulants for the removal of suspended solids and colloidal particles in water treatment. It decreases dependence on synthetic coagulants, reduces sludge production, and enhances the biodegradability of residual chemicals (Kolya & Kang, 2023). Utilizing green solvents, such as supercritical carbon dioxide or natural extractants, in extraction processes for the removal of organic pollutants from wastewater. It reduces the use of toxic organic solvents, minimizes environmental impact, and enhances the overall sustainability of extraction methods (Mishra et al., 2023). Harnessing the power of microorganisms for biological treatment processes, such as activated sludge, anaerobic digestion, or bioaugmentation. reduces the need for harsh chemicals, promotes natural processes, and minimizes the environmental impact associated with chemical additives (Rahimi et al., 2020). Developing environmentally friendly nanomaterials for various applications, such as nanoadsorbents or green synthesis of nanoparticles, to enhance pollutant removal in wastewater (El Messaoudi, Ciğeroğlu, Şenol, El Hajam, et al., 2023). It focuses on sustainable and non-toxic nanomaterials, minimizing potential environmental and health risks associated with conventional nanotechnology (Mir et al., 2023). Adopting a preventive approach by controlling pollutants at their source reduces or eliminates the need for end-of-pipe treatment. It addresses pollution at its root, minimizing the use of treatment chemicals, energy, and resources (Lee & Bi, 2019).

Achieving the right balance between effectiveness and environmental impact can be challenging in green chemistry approaches. Some green technologies may have higher initial costs, necessitating a careful economic evaluation. Ensuring that green chemistry approaches comply with regulatory standards and guidelines (Ekins & Zenghelis, 2021).

GC approaches in wastewater treatment are continuously evolving as researchers and practitioners seek more sustainable and environmentally friendly solutions. The future will likely see increased integration of green technologies into mainstream wastewater treatment practices, driven by a growing awareness of environmental issues and a commitment to creating more sustainable and resilient water treatment systems.

3.4 ADVANTAGES OF APCTS

One of the primary advantages of APCTs is their superior removal efficiency compared to conventional methods. These techniques are designed to target specific pollutants, leading to more effective and thorough treatment. Advanced techniques are capable of treating complex and challenging pollutants, including recalcitrant organic compounds, emerging contaminants, and persistent pollutants that are difficult to address with traditional methods (Kumari & Kumar, 2023). These techniques are versatile and can be applied to a wide range of contaminants, making them suitable for diverse industrial and municipal wastewater streams. Their adaptability contributes to a more comprehensive approach to water treatment. APCTs often allow for the selective removal of specific pollutants. This selectivity is advantageous for tailoring treatment processes to the characteristics of a particular wastewater stream, resulting in optimized results (Barakat, 2011). Some advanced techniques result in reduced generation of sludge or residuals compared to conventional processes. This minimizes the environmental impact associated with the disposal and management of waste generated during the treatment process (Gherghel et al., 2019). Advanced techniques are effective at treating low concentrations of contaminants, providing reliable performance even when dealing with diluted wastewater streams. This is crucial for meeting stringent water quality standards. Certain advanced techniques support the recovery of valuable resources from wastewater, such as nutrients, metals, or energy. This contributes to the concept of a circular economy by turning waste into valuable assets (B. J. Singh et al., 2023). Some APCTs require fewer or more selective chemicals, leading to a reduction in overall chemical usage. This is advantageous for both cost-effectiveness and environmental sustainability. Many advanced techniques offer rapid treatment processes, allowing for quicker response times to environmental challenges. This is particularly important for industries requiring timely and efficient wastewater treatment (Shewa & Dagnew, 2020). Some advanced techniques, such as advanced oxidation processes (AOPs) and disinfection methods, effectively inactivate pathogens, ensuring the production of treated water that meets health and safety standards (Kokkinos et al., 2021). APCTs can be seamlessly integrated with conventional treatment processes, providing a hybrid approach that capitalizes on the strengths of both methods. This integration enhances overall treatment performance. The high removal efficiency of advanced techniques contributes to improved water quality, making treated water more suitable for reuse purposes (Belete et al., 2023). This is essential in regions facing water scarcity or where water reuse is a priority. In some cases, APCTs exhibit environmental sustainability by minimizing the use of energy, reducing chemical consumption, and lowering the ecological footprint associated with wastewater treatment (Tóth et al., 2022). While APCTs offer numerous advantages, their application should be carefully considered based on the specific characteristics of the wastewater and treatment goals. Additionally, ongoing research and development are essential to further optimize these techniques and address emerging challenges in wastewater treatment.

3.5 LIMITATIONS OF APCTS

Despite their numerous advantages, APCTs for wastewater treatment also come with certain limitations. Understanding these limitations is crucial for implementing these techniques effectively and addressing challenges that may arise. Many advanced techniques require sophisticated equipment, specialized materials, and precise operating conditions, leading to higher capital and operational costs compared to conventional methods. Some advanced physicochemical processes, such as AOPs, may be energy-intensive, especially when high-energy inputs are required to generate reactive species for pollutant degradation. Implementing and maintaining advanced techniques can be complex, requiring skilled personnel and intricate process control. This complexity may pose challenges for smaller wastewater treatment facilities or regions with limited resources. Certain advanced techniques may necessitate the use of specific chemicals or catalysts, which can be expensive or introduce additional challenges related to chemical storage, handling, and disposal. Some APCTs may face challenges when scaled up for large-scale industrial applications. Issues related to uniformity, reactor design, and overall process efficiency may arise during the transition from laboratory-scale studies to full-scale implementation. The breakdown of pollutants in advanced processes may result in the generation of by-products, some of which may be toxic or require additional treatment. Proper management of these by-products is essential to preventing secondary environmental impacts. While selectivity is an advantage, it can also be a limitation. Advanced techniques may not effectively target every pollutant in a complex wastewater stream, leading to the need for additional treatment steps or complementary processes. Skilled personnel are required for the operation and maintenance of advanced physicochemical systems.

3.6 FUTURE PROSPECTS OF APCTS

Advances in material science are anticipated to lead to the development of novel materials with enhanced properties, such as improved catalysts for AOPs and more resilient membranes for filtration technologies. These innovations can contribute to increased efficiency and durability in wastewater treatment (Asif & Zhang, 2021). The integration of nanotechnology holds great potential for improving the performance of APCTs. Nanomaterials, such as nanoparticles and nanocomposites, can be tailored to exhibit specific properties, enhancing adsorption, catalysis, and membrane filtration capabilities (El Messaoudi, El Khomri, Chegini, Bouich, et al., 2022; Peralta et al., 2020). The future of wastewater treatment involves the incorporation of smart technologies and automation. APCTs will benefit from real-time monitoring, data analytics, and automated control systems, optimizing process parameters for improved efficiency and resource utilization. Research and development efforts are expected to focus on reducing the energy intensity of APCTs. The future is likely to witness the integration of APCTs with other treatment methods, forming hybrid approaches. Combining physicochemical processes with biological treatment or nature-based solutions can provide comprehensive and synergistic treatment outcomes. Advances in resource recovery from wastewater are expected to align with circular economy principles. Technologies that enable the extraction of valuable resources, such as nutrients and metals, from treated wastewater will play a significant role in sustainable and circular wastewater management. Artificial intelligence (AI) and machine learning applications will become integral to optimizing and predicting the performance of APCTs. AI algorithms can analyze complex data sets, identify patterns, and offer insights for better process control and decision-making. Future advancements will likely focus on tailoring advanced techniques to target specific emerging contaminants, pharmaceuticals, and microplastics. Customized approaches will ensure that treatment processes are effective against the evolving landscape of water pollutants. Advances in miniaturized and decentralized treatment technologies will offer solutions for localized and remote wastewater treatment needs. These systems can be modular, adaptable, and easier to implement in regions lacking centralized infrastructure. The future of APCTs will place a greater emphasis on green and sustainable approaches. This includes the development of processes that minimize chemical usage, reduce environmental impact, and align with broader sustainability goals. Collaborative efforts among researchers, industries, and regulatory bodies will be crucial for driving innovation in advanced physicochemical techniques. Research initiatives focused on interdisciplinary approaches and holistic solutions will shape the future of wastewater treatment. APCTs will play a pivotal role in supporting water reuse initiatives. As these advancements continue to unfold, the potential for more efficient, cost-effective, and environmentally friendly wastewater treatment solutions will become increasingly apparent.

3.7 CONCLUSION

This chapter represents a significant leap forward in the quest for efficient, sustainable, and comprehensive solutions to address the complex challenges posed by contaminated water sources. The amalgamation of various APCTs has ushered in a new era of wastewater treatment, characterized by enhanced contaminant removal, improved process control, and an eco-friendlier approach. These advanced techniques excel at providing unparalleled efficiency and precision in the removal of diverse contaminants from wastewater. The versatility of these techniques allows for the effective treatment of wastewater with varying compositions. Whether addressing organic pollutants, heavy metals, or emerging contaminants, the adaptability of APCTs makes them a versatile tool in the water treatment toolkit. A notable aspect is the conscious effort to minimize the environmental impact of wastewater treatment. Green chemistry principles, the use of sustainable materials, and the reduction of chemical consumption underscore a commitment to environmentally friendly practices. The integration of APCTs goes beyond mere pollutant removal; it embraces the concept of resource recovery. Hybrid approaches that synergize different physicochemical techniques showcase a holistic and integrated strategy. Acknowledging the challenges, such as energy consumption and scale-up complexities, is vital. However, these challenges present opportunities for further research, innovation, and optimization. Addressing these hurdles will be crucial for the widespread adoption of APCTs. While recognizing the potential higher costs associated with advanced techniques, it is essential to emphasize that investments in these technologies contribute to their long-term economic viability. The benefits of improved water quality, reduced health risks, and resource recovery position these techniques as strategic investments in environmental and public health. The landscape of wastewater treatment is dynamic, with continuous innovation pushing the boundaries of what is achievable. The future promises even more sophisticated techniques, improved materials, and a deeper understanding of the interconnected processes involved in wastewater treatment. In essence, this chapter stands at the forefront of the global effort to ensure access to clean water resources while minimizing the environmental footprint of treatment processes. This paradigm shift towards advanced technologies reflects a commitment to a sustainable and resilient water future for peoples worldwide.

REFERENCES

Abdel-Raouf, N., Al-Homaidan, A. A., & Ibraheem, I. B. M. (2012). Microalgae and wastewater treatment. *Saudi Journal of Biological Sciences*, *19*(3), 257–275. https://doi.org/10.1016/J.SJBS.2012.04.005

Abdel-Shafy, H. I., El-Khateeb, M. A., & Shehata, M. (2014). Greywater treatment using different designs of sand filters. *Desalination and Water Treatment*, *52*(28–30), 5237–5242. https://doi.org/10.1080/19443994.2013.813007

Abdul Rahman, N., Albania Linus, A., Jose Jol, C., Abdul Jalal, N. S., Kwong Ming, C., Wan Borhan, W. W. S., Baharuddin, N., Samsul, S. N. A., & Abdul Mutalip, N. (2023). Kinetic modelling of peat water treatment with continuous electrocoagulation using aluminium electrodes. *Journal of Environmental Chemical Engineering*, *11*(3), 109559. https://doi.org/https://doi.org/10.1016/j.jece.2023.109559

Adewuyi, Y. G. (2001). Sonochemistry: Environmental science and engineering applications. *Industrial and Engineering Chemistry Research*, *40*(22), 4681–4715. https://doi.org/10.1021/IE010096L

Aghaei, A., Suresh, K., Firouzjaei, M. D., Elliott, M., Rahimpour, A., & Sadrzadeh, M. (2023). 14 – Hybrid/integrated treatment technologies for oily wastewater treatment. In A. Basile, A. Cassano, M. R. Rahimpour, & M. A. Makarem (eds.), *Advanced Technologies in Wastewater Treatment. Oily Wastewaters* (pp. 377–419). Elsevier. https://doi.org/https://doi.org/10.1016/B978-0-323-99916-8.00002-X

Aghakhani, A., Mousavi, S. F., & Mostafazadeh-Fard, B. (2013). Desalination of saline water with single and combined adsorbents. *Desalination and Water Treatment*, *51*(7–9), 1928–1935. https://doi.org/10.1080/19443994.2012.714731

Agunbiade, M. O., Pohl, C. H., & Ashafa, A. O. T. (2016). A review of the application of biofloccualnts in wastewater treatment. *Polish Journal of Environmental Studies*, *25*(4), 1381–1389. https://doi.org/10.15244/pjoes/61063

Ain Zainuddin, N., Azwan Raja Mamat, T., Imam Maarof, H., Wahidah Puasa, S., & Rohana Mohd Yatim, S. (2019). Removal of nickel, zinc and copper from plating process industrial raw effluent via hydroxide precipitation versus sulphide precipitation. *IOP Conference Series: Materials Science and Engineering*, *551*(1), 012122. https://doi.org/10.1088/1757-899X/551/1/012122

Aizenchtadt, E., Ingman, D., & Friedler, E. (2009). Analysis of the long-term performance of an on-site greywater treatment plant using novel statistical approaches. *Urban Water Journal*, *6*(5), 341–354. https://doi.org/10.1080/15730620902795210

Al-Asheh, S., Bagheri, M., & Aidan, A. (2021). Membrane bioreactor for wastewater treatment: A review. *Case Studies in Chemical and Environmental Engineering*, *4*, 100109. https://doi.org/10.1016/J.CSCEE.2021.100109

Al-Hamaiedeh, H., & Bino, M. (2010). Effect of treated grey water reuse in irrigation on soil and plants. *Desalination*, *256*(1–3), 115–119. https://doi.org/10.1016/j.desal.2010.02.004

AlJaberi, F. Y., Ahmed, S. A., Makki, H. F., Naje, A. S., Zwain, H. M., Salman, A. D., Juzsakova, T., Viktor, S., Van, B., Le, P.-C., La, D. D., Chang, S. W., Um, M.-J., Ngo, H. H., & Nguyen, D. D. (2023). Recent advances and applicable flexibility potential of electrochemical processes for wastewater treatment. *Science of The Total Environment*, *867*, 161361. https://doi.org/https://doi.org/10.1016/j.scitotenv.2022.161361

Alsulaili, A. D., Hamoda, M. F., Al-Jarallah, R., & Alrukaibi, D. (2017). Treatment and potential reuse of greywater from schools: A pilot study. *Water Science and Technology*, *75*(9), 2119–2129. https://doi.org/10.2166/wst.2017.088

Al-Zou'by, J. Y., Al-Zboon, K. K., & Al-Tabbal, J. A. (2017). Low-cost treatment of grey water and reuse for irrigation of home garden plants. *Environmental Engineering and Management Journal*, *16*(2), 351–359. https://doi.org/10.30638/eemj.2017.035

Amdeha, E. (2023). Biochar-based nanocomposites for industrial wastewater treatment via adsorption and photocatalytic degradation and the parameters affecting these processes. *Biomass Conversion and Biorefinery*. https://doi.org/10.1007/s13399-023-04512-2

Amit, S., & Rupali, G. (2004). Developments in waste-water treatment technology. *Desalination*, *167*, 55–63.

Ammar, S. H., Ismail, N. N., Ali, A. D., & Abbas, W. M. (2019). Electrocoagulation technique for refinery wastewater treatment in an internal loop split-plate airlift reactor. *Journal of Environmental Chemical Engineering*, *7*(6), 103489. https://doi.org/10.1016/J.JECE.2019.103489

Antonopoulou, G., Kirkou, A., & Stasinakis, A. S. (2013). Quantitative and qualitative greywater characterization in Greek households and investigation of their treatment using physicochemical methods. *Science of the Total Environment*, *454–455*, 426–432. https://doi.org/10.1016/j.scitotenv.2013.03.045

Arabameri, A., Alavi Moghaddam, M. R., Azadmehr, A. R., & Karamati-Niaragh, E. (2023). Determination of optimal operating conditions for AC-powered electrocoagulation process coupling green additive Tartaric Acid to remove Ni^{2+}: Pyomo and RSM approach. *Journal of Environmental Management*, *330*, 117152. https://doi.org/https://doi.org/10.1016/j.jenvman.2022.117152

Arslan, D. Ş., Ertap, H., Şenol, Z. M., El Messaoudi, N., & Mehmeti, V. (2023). Preparation of polyacrylamide titanium dioxide hybrid nanocomposite by direct polymerization and its applicability in removing crystal violet from aqueous solution. *Journal of Polymers and the Environment*, *32*, 573–587. https://doi.org/10.1007/S10924-023-03004-8

Aryee, A. A., Liu, Y., Han, R., & Qu, L. (2023). Bimetallic adsorbents for wastewater treatment: a review. *Environmental Chemistry Letters*, *21*(3), 1811–1835. https://doi.org/10.1007/s10311-023-01566-6

Asif, M. B., & Zhang, Z. (2021). Ceramic membrane technology for water and wastewater treatment: A critical review of performance, full-scale applications, membrane fouling and prospects. *Chemical Engineering Journal*, *418*, 129481. https://doi.org/10.1016/J.CEJ.2021.129481

Assayed, A., Chenoweth, J., & Pedley, S. (2014). Drawer compacted sand filter: A new and innovative method for on-site grey water treatment. *Environmental Technology (United Kingdom)*, *35*(19), 2435–2446. https://doi.org/10.1080/09593330.2014.909886

Assayed, A., Chenoweth, J., & Pedley, S. (2015). Assessing the efficiency of an innovative method for onsite greywater treatment: Drawer compacted sand filter – A case study in Jordan. *Ecological Engineering*, *81*, 525–533. https://doi.org/10.1016/j.ecoleng.2015.04.042

Babaei, F., Ehrampoush, M. H., Eslami, H., Ghaneian, M. T., Fallahzadeh, H., Talebi, P., Fard, R. F., & Ebrahimi, A. A. (2019). Removal of linear alkylbenzene sulfonate and turbidity from greywater by a hybrid multi-layer slow sand filter microfiltration ultrafiltration system. *Journal of Cleaner Production*, *211*, 922–931. https://doi.org/10.1016/j.jclepro.2018.11.255

Bahrami, M., Amiri, M. J., & Badkubi, M. (2020). Application of horizontal series filtration in greywater treatment: A semi-industrial study. *Australian Journal of Water Resources*, *24*(2), 236–247. https://doi.org/10.1080/13241583.2020.1824610

Bao, J., Guo, S., Fan, D., Cheng, J., Zhang, Y., & Pang, X. (2023). Sonoactivated nanomaterials: A potent armament for wastewater treatment. *Ultrasonics Sonochemistry*, *99*, 106569. https://doi.org/10.1016/J.ULTSONCH.2023.106569

Barakat, M. A. (2011). New trends in removing heavy metals from industrial wastewater. *Arabian Journal of Chemistry*, *4*(4), 361–377. https://doi.org/https://doi.org/10.1016/j.arabjc.2010.07.019

Bartels, C. R., Wilf, M., Andes, K., & Iong, J. (2005). Design considerations for wastewater treatment by reverse osmosis. *Water Science and Technology*, *51*(6–7), 473–482. https://doi.org/10.2166/wst.2005.0670

Basile, A., Cassano, A., Rahimpour, M. R., & Makarem, M. A. (2023). *Advanced Technologies in Wastewater Treatment: Oily Wastewaters* (pp. 1–499). Elsevier. https://doi.org/10.1016/C2021-0-00545-9

Belete, B., Desye, B., Ambelu, A., & Yenew, C. (2023). Micropollutant removal efficiency of advanced wastewater treatment plants: A systematic review. *Environmental Health Insights*, *17*, 1–11. https://doi.org/10.1177/11786302231195158/ASSET/IMAGES/LARGE/10.1177_11786302231195158-FIG4.JPEG

Bera, S. P., Godhaniya, M., & Kothari, C. (2022). Emerging and advanced membrane technology for wastewater treatment: A review. *Journal of Basic Microbiology*, *62*(3–4), 245–259. https://doi.org/10.1002/jobm.202100259

Bernardes, F. S., de Oliveira Diniz, R. C., Araújo, J. C., & Paulo, P. L. (2017). Studies of filter media for zero-discharge systems collecting light greywater. *Environmental Technology (United Kingdom)*, *38*(17), 2173–2184. https://doi.org/10.1080/09593330.2016.1249414

Bhargava, N., Mor, R. S., Kumar, K., & Sharanagat, V. S. (2021). Advances in application of ultrasound in food processing: A review. *Ultrasonics Sonochemistry*, *70*, 105293. https://doi.org/10.1016/J.ULTSONCH.2020.105293

Blandin, G., Verliefde, A. R. D., Comas, J., Rodriguez-Roda, I., & Le-Clech, P. (2016). Efficiently combining water reuse and desalination through forward osmosis-reverse osmosis (FO-RO) hybrids: A critical review. *Membranes*, *6*(3), 1–24. https://doi.org/10.3390/membranes6030037

Bolong, N., Ismail, A. F., Salim, M. R., & Matsuura, T. (2009). A review of the effects of emerging contaminants in wastewater and options for their removal. *Desalination*, *239*(1–3), 229–246. https://doi.org/10.1016/j.desal.2008.03.020

Bolton, C. R., & Randall, D. G. (2019). Development of an integrated wetland microbial fuel cell and sand filtration system for greywater treatment. *Journal of Environmental Chemical Engineering*, *7*(4), 103249. https://doi.org/10.1016/j.jece.2019.103249

Borojovich, E. J. C., Münster, M., Rafailov, G., & Porat, Z. (2010). Precipitation of ammonium from concentrated industrial wastes as struvite: A search for the optimal reagents. *Water Environment Research*, *82*(7), 586–591. https://doi.org/10.2175/106143009X12529484815755

Campione, A., Gurreri, L., Ciofalo, M., Micale, G., Tamburini, A., & Cipollina, A. (2018). Electrodialysis for water desalination: A critical assessment of recent developments on process fundamentals, models and applications. *Desalination*, *434*(October 2017), 121–160. https://doi.org/10.1016/j.desal.2017.12.044

Chaillou, K., Gérente, C., Andrès, Y., & Wolbert, D. (2011). Bathroom greywater characterization and potential treatments for reuse. *Water, Air, and Soil Pollution*, *215*(1–4), 31–42. https://doi.org/10.1007/s11270-010-0454-5

Chaplin, B. P. (2019). The prospect of electrochemical technologies advancing worldwide water treatment. *Accounts of Chemical Research*, *52*(3), 596–604. https://doi.org/10.1021/acs.accounts.8b00611

Chen, G., Lu, Y., Krantz, W. B., Wang, R., & Fane, A. G. (2014). Optimization of operating conditions for a continuous membrane distillation crystallization process with zero salty water discharge. *Journal of Membrane Science*, *450*, 1–11. https://doi.org/10.1016/j.memsci.2013.08.034

Chen, Q., Yao, Y., Li, X., Lu, J., Zhou, J., & Huang, Z. (2018). Comparison of heavy metal removals from aqueous solutions by chemical precipitation and characteristics of precipitates. *Journal of Water Process Engineering*, *26*, 289–300. https://doi.org/10.1016/J.JWPE.2018.11.003

Chen, T., Bi, J., Ji, Z., Yuan, J., & Zhao, Y. (2022). Application of bipolar membrane electrodialysis for simultaneous recovery of high-value acid/alkali from saline wastewater: An in-depth review. *Water Research*, *226*, 119274. https://doi.org/https://doi.org/10.1016/j.watres.2022.119274

Chen, Z., Wu, Q., Wu, G., & Hu, H. Y. (2017). Centralized water reuse system with multiple applications in urban areas: Lessons from China's experience. *Resources, Conservation and Recycling*, *117*, 125–136. https://doi.org/10.1016/J.RESCONREC.2016.11.008

Cho, E. C., Chang-Jian, C. W., Zheng, J. H., Huang, J. H., Lee, K. C., Ho, B. C., & Hsiao, Y. S. (2018). Microwave-assisted synthesis of TiO_2/WS_2 heterojunctions with enhanced photocatalytic activity. *Journal of the Taiwan Institute of Chemical Engineers*, *91*, 489–498. https://doi.org/10.1016/J.JTICE.2018.05.025

Ciğeroğlu, Z., Kazan-Kaya, E. S., El Messaoudi, N., Fernine, Y., Américo-Pinheiro, J. H. P., & Jada, A. (2023). Remediation of tetracycline from aqueous solution through adsorption on g-C_3N_4-ZnO-$BaTiO_3$ nanocomposite: Optimization, modeling, and theoretical calculation. *Journal of Molecular Liquids*, *369*, 120866. https://doi.org/10.1016/J.MOLLIQ.2022.120866

Cuprys, A., Thomson, P., Ouarda, Y., Suresh, G., Rouissi, T., Kaur Brar, S., Drogui, P., & Surampalli, R. Y. (2020). Ciprofloxacin removal via sequential electro-oxidation and enzymatic oxidation. *Journal of Hazardous Materials*, *389*, 121890. https://doi.org/https://doi.org/10.1016/j.jhazmat.2019.121890

da Silva, A. F. V., da Silva, J., Vicente, R., Ambrosi, A., Zin, G., Di Luccio, M., & de Oliveira, J. V. (2023). Recent advances in surface modification using polydopamine for the development of photocatalytic membranes for oily wastewater treatment. *Journal of Water Process Engineering*, *53*, 103743. https://doi.org/https://doi.org/10.1016/j.jwpe.2023.103743

Dalahmeh, S. S., Pell, M., Hylander, L. D., Lalander, C., Vinnerås, B., & Jönsson, H. (2014). Effects of changing hydraulic and organic loading rates on pollutant reduction in bark, charcoal and sand filters treating greywater. *Journal of Environmental Management*, *132*, 338–345. https://doi.org/10.1016/j.jenvman.2013.11.005

Das, T. K., & Poater, A. (2021). Review on the use of heavy metal deposits from water treatment waste towards catalytic chemical syntheses. *International Journal of Molecular Sciences*, *22*(24), 13383. https://doi.org/10.3390/IJMS222413383

De-Bashan, L. E., & Bashan, Y. (2004). Recent advances in removing phosphorus from wastewater and its future use as fertilizer (1997–2003). *Water Research*, *38*(19), 4222–4246. https://doi.org/10.1016/J.WATRES.2004.07.014

Deblonde, T., Cossu-Leguille, C., & Hartemann, P. (2011). Emerging pollutants in wastewater: A review of the literature. *International Journal of Hygiene and Environmental Health*, *214*(6), 442–448. https://doi.org/10.1016/J.IJHEH.2011.08.002

Dehghani, M. H., Karri, R. R., Koduru, J. R., Manickam, S., Tyagi, I., Mubarak, N. M., & Suhas. (2023). Recent trends in the applications of sonochemical reactors as an advanced oxidation process for the remediation of microbial hazards associated with water and wastewater: A critical review. *Ultrasonics Sonochemistry*, *94*, 106302. https://doi.org/10.1016/J.ULTSONCH.2023.106302

Demirci, Y., & Özbeyaz, A. (2019). Wastewater treatment in electrocoagulation systems: Investigation of the impact of temperature using a fuzzy logic control algorithm. *Environmental Science and Pollution Research*, *26*, 30893–30906. https://doi.org/10.1007/s11356-019-06279-4

Dubey, S. P., Dwivedi, A. D., Sillanpää, M., & Gopal, K. (2010). *Artemisia vulgaris*-derived mesoporous honeycomb-shaped activated carbon for ibuprofen adsorption. *Chemical Engineering Journal*, *165*(2), 537–544. https://doi.org/10.1016/j.cej.2010.09.068

Ebba, M., Asaithambi, P., & Alemayehu, E. (2022). Development of electrocoagulation process for wastewater treatment: Optimization by response surface methodology. *Heliyon*, *8*(5), e09383. https://doi.org/https://doi.org/10.1016/j.heliyon.2022.e09383

Ekins, P., & Zenghelis, D. (2021). The costs and benefits of environmental sustainability. *Sustainability Science*, *16*(3), 949–965. https://doi.org/10.1007/S11625-021-00910-5/FIGURES/5

El Khomri, M., El Messaoudi, N., Dbik, A., Bentahar, S., Fernine, Y., Lacherai, A., & Jada, A. (2022). Optimization based on response surface methodology of anionic dye desorption from two agricultural solid wastes. *Chemistry Africa*, *7*(4), 1083–1095. https://doi.org/10.1007/S42250-022-00395-4

El Khomri, M., El Messaoudi, N., Dbik, A., Bentahar, S., Lacherai, A., Chegini, Z. G., & Bouich, A. (2022). Removal of Congo red from aqueous solution in single and binary mixture systems using Argan nutshell wood. *Pigment and Resin Technology*, *51*(5), 477–488. https://doi.org/10.1108/PRT-04-2021-0045

El Khomri, M., El Messaoudi, N., Dbik, A., Bentahar, S., Lacherai, A., Faska, N., & Jada, A. (2022). Regeneration of argan nutshell and almond shell using HNO_3 for their reusability to remove cationic dye from aqueous solution. *Chemical Engineering Communications*, *209*(10), 1304–1315. https://doi.org/10.1080/00986445.2021.1963960

El Messaoudi, N., Ciğeroğlu, Z., Şenol, Z. M., Bouich, A., Kazan-Kaya, E. S., Noureen, L., & Américo-Pinheiro, J. H. P. (2023). Green synthesis of nanoparticles for remediation organic pollutants in wastewater by adsorption. *Advances in Chemical Pollution, Environmental Management and Protection.10*, 304–345. https://doi.org/10.1016/BS.APMP.2023.06.016

El Messaoudi, N., Ciğeroğlu, Z., Şenol, Z. M., El Hajam, M., & Noureen, L. (2023). A comparative review of the adsorption and photocatalytic degradation of tetracycline in aquatic environment by g-C_3N_4-based materials. *Journal of Water Process Engineering*, *55*, 104150. https://doi.org/10.1016/J.JWPE.2023.104150

El Messaoudi, N., Ciğeroğlu, Z., Şenol, Z. M., Kazan-Kaya, E. S., Fernine, Y., Gubernat, S., & Lopicic, Z. (2024). Green synthesis of $CuFe_2O_4$ nanoparticles from bioresource extracts and their applications in different areas: A review. *Biomass Conversion and Biorefinery, 2024*, 1–22. https://doi.org/10.1007/S13399-023-05264-9

El Messaoudi, N., El Khomri, M., Chegini, Z. G., Bouich, A., Dbik, A., Bentahar, S., Labjar, N., Iqbal, M., Jada, A., & Lacherai, A. (2022). Dye removal from aqueous solution using nanocomposite synthesized from oxalic acid-modified agricultural solid waste and $ZnFe_2O_4$ nanoparticles. *Nanotechnology for Environmental Engineering, 7*(1), 797–811. https://doi.org/10.1007/S41204-021-00173-6

El Messaoudi, N., El Khomri, M., Chegini, Z. G., Dbik, A., Bentahar, S., Iqbal, M., Jada, A., & Lacherai, A. (2022). Desorption of crystal violet from alkali-treated agricultural material waste: An experimental study, kinetic, equilibrium and thermodynamic modeling. *Pigment & Resin Technology, 51*(3), 309–319. https://doi.org/10.1108/prt-02-2021-0019

El Messaoudi, N., El Khomri, M., Chlif, N., Chegini, Z. G., Dbik, A., Bentahar, S., & Lacherai, A. (2021). Desorption of Congo red from dye-loaded *Phoenix dactylifera* date stones and *Ziziphus lotus* jujube shells. *Groundwater for Sustainable Development, 12*, 100552. https://doi.org/10.1016/j.gsd.2021.100552

El Messaoudi, N., El Khomri, M., Dbik, A., Bentahar, S., & Lacherai, A. (2016). Selective and competitive removal of dyes from binary and ternary systems in aqueous solutions by pretreated jujube shell (*Zizyphus lotus*). *Journal of Dispersion Science and Technology, 38*(8), 1168–1174. https://doi.org/10.1080/01932691.2016.1228070

El Messaoudi, N., El Khomri, M., El Mouden, A., Bouich, A., Jada, A., Lacherai, A., Iqbal, H. M. N., Mulla, S. I., Kumar, V., & Américo-Pinheiro, J. H. P. (2022). Regeneration and reusability of non-conventional low-cost adsorbents to remove dyes from wastewaters in multiple consecutive adsorption–desorption cycles: A review. *Biomass Conversion and Biorefinery 2022*, 1–18. https://doi.org/10.1007/S13399-022-03604-9

El Messaoudi, N., El Mouden, A., El Khomri, M., Bouich, A., Fernine, Y., Ciğeroğlu, Z., Américo-Pinheiro, J. H. P., Labjar, N., Jada, A., Sillanpää, M., & Lacherai, A. (2022). Experimental study and theoretical statistical modeling of acid blue 25 remediation using activated carbon from *Citrus sinensis* leaf. *Fluid Phase Equilibria, 563*, 113585. https://doi.org/10.1016/J.FLUID.2022.113585

El Messaoudi, N., El Mouden, A., Fernine, Y., El Khomri, M., Bouich, A., Faska, N., Ciğeroğlu, Z., Américo-Pinheiro, J. H. P., Jada, A., & Lacherai, A. (2023). Green synthesis of Ag_2O nanoparticles using *Punica granatum* leaf extract for sulfamethoxazole antibiotic adsorption: Characterization, experimental study, modeling, and DFT calculation. *Environmental Science and Pollution Research, 30*(34), 81352–81369. https://doi.org/10.1007/S11356-022-21554-7

El Mouden, A., El Guerraf, A., El Messaoudi, N., Haounati, R., Ait El Fakir, A., & Lacherai, A. (2022). Date stone functionalized with 3-aminopropyltriethoxysilane as a potential biosorbent for heavy metal ions removal from aqueous solution. *Chemistry Africa, 5*(3), 745–759. https://doi.org/10.1007/S42250-022-00350-3

El Mouden, A., El Messaoudi, N., El Guerraf, A., Bouich, A., Mehmeti, V., Lacherai, A., Jada, A., & Pinê Américo-Pinheiro, J. H. (2023). Removal of cadmium and lead ions from aqueous solutions by novel dolomite-quartz@Fe_3O_4 nanocomposite fabricated as nanoadsorbent. *Environmental Research, 225*, 115606. https://doi.org/10.1016/J.ENVRES.2023.115606

El Mouden, A., El Messaoudi, N., El Guerraf, A., Bouich, A., Mehmeti, V., Lacherai, A., Jada, A., & Sher, F. (2023). Multifunctional cobalt oxide nanocomposites for efficient removal of heavy metals from aqueous solutions. *Chemosphere, 317*, 137922. https://doi.org/10.1016/J.CHEMOSPHERE.2023.137922

Elazzouzi, M., El Kasmi, A., Haboubi, K., & Elyoubi, M. S. (2018). A novel electrocoagulation process using insulated edges of Al electrodes for enhancement of urban wastewater treatment: Techno-economic study. *Process Safety and Environmental Protection, 116*, 506–515. https://doi.org/https://doi.org/10.1016/j.psep.2018.03.006

Ellis, K. V. (1985). Slow sand filtration. *Critical Reviews in Environmental Control, 15*(4), 315–354. https://doi.org/10.1080/10643388509381736

El-taweel, R. M., Mohamed, N., Alrefaey, K. A., Husien, S., Abdel-Aziz, A. B., Salim, A. I., Mostafa, N. G., Said, L. A., Fahim, I. S., & Radwan, A. G. (2023). A review of coagulation explaining its definition, mechanism, coagulant types, and optimization models; RSM, and ANN. *Current Research in Green and Sustainable Chemistry, 6*, 100358. https://doi.org/10.1016/J.CRGSC.2023.100358

Emamjomeh, M. M., & Sivakumar, M. (2009). Review of pollutants removed by electrocoagulation and electrocoagulation/flotation processes. *Journal of Environmental Management, 90*(5), 1663–1679. https://doi.org/https://doi.org/10.1016/j.jenvman.2008.12.011

Emslie, D., Siddiqua, S., Crawford, B., & Teece, W. (2022). Biofilm formation and effectiveness of biosand filtration systems with typical and innovative filter media. *Geotechnical and Geological Engineering, 40*(3), 1251–1266. https://doi.org/10.1007/s10706-021-01960-3

Farghali, M., Osman, A. I., Chen, Z., Abdelhaleem, A., Ihara, I., Mohamed, I. M. A., Yap, P. S., & Rooney, D. W. (2023). Social, environmental, and economic consequences of integrating renewable energies in the electricity sector: A review. *Environmental Chemistry Letters, 21*(3), 1381–1418. https://doi.org/10.1007/S10311-023-01587-1

Feijoo, S., Estévez, S., Kamali, M., Dewil, R., & Moreira, M. T. (2023). Scale-up modelling and life cycle assessment of electrochemical oxidation in wastewater treatment. *Chemical Engineering Journal, 455*, 140627. https://doi.org/https://doi.org/10.1016/j.cej.2022.140627

Friedler, E., Kovalio, R., & Ben-Zvi, A. (2006). Comparative study of the microbial quality of greywater treated by three on-site treatment systems. *Environmental Technology, 27*(6), 653–663. https://doi.org/10.1080/09593332708618674

Gad, E. S., Abbas, M. A., Abdelkreem, M., & Ahmed, A. H. (2022). Electrochemical treatment of industrial effluent and its impact on stainless steel corrosion. *International Journal of Electrochemical Science, 17*(12), 221211. https://doi.org/https://doi.org/10.20964/2022.12.14

Galama, A. H., Saakes, M., Bruning, H., Rijnaarts, H. H. M., & Post, J. W. (2014). Seawater predesalination with electrodialysis. *Desalination, 342*, 61–69. https://doi.org/10.1016/j.desal.2013.07.012

Garcia-Segura, S., Ocon, J. D., & Chong, M. N. (2018). Electrochemical oxidation remediation of real wastewater effluents – A review. *Process Safety and Environmental Protection, 113*, 48–67. https://doi.org/https://doi.org/10.1016/j.psep.2017.09.014

Ghaitidak, D. M., & Yadav, K. D. (2016). Greywater treatment for reuse: Comparison of reuse options using analytic hierarchy process. *Journal of Water Reuse and Desalination, 6*(1), 108–124. https://doi.org/10.2166/wrd.2015.177

Gherghel, A., Teodosiu, C., & De Gisi, S. (2019). A review on wastewater sludge valorisation and its challenges in the context of circular economy. *Journal of Cleaner Production, 228*, 244–263. https://doi.org/10.1016/J.JCLEPRO.2019.04.240

Gomez-Ruiz, B., Gómez-Lavín, S., Diban, N., Boiteux, V., Colin, A., Dauchy, X., & Urtiaga, A. (2017). Boron doped diamond electrooxidation of 6:2 fluorotelomers and perfluorocarboxylic acids. Application to industrial wastewaters treatment. *Journal of Electroanalytical Chemistry, 798*, 51–57. https://doi.org/https://doi.org/10.1016/j.jelechem.2017.05.033

Goyal, H., Chen, T. Y., Chen, W., & Vlachos, D. G. (2022). A review of microwave-assisted process intensified multiphase reactors. *Chemical Engineering Journal*, *430*, 133183. https://doi.org/10.1016/J.CEJ.2021.133183

Gul, U. D., Senol, Z. M., & Tastan, B. E. (2022). Treatment of the Allura Red food colorant contaminated water by a novel cyanobacterium *Desertifilum tharense*. *Water Science and Technology*, *85*(1), 279–290. https://doi.org/10.2166/WST.2021.615

Hasanah, M., Susilawati, S., & Ramadhan, A. (2023). Performance optimization of CuO–ZnO ceramic electrode on the electrocoagulation of wastewater. *Materials Science for Energy Technologies*, *6*, 7–14. https://doi.org/https://doi.org/10.1016/j.mset.2022.11.001

Hassan, I., Chowdhury, S. R., Prihartato, P. K., & Razzak, S. A. (2021). Wastewater treatment using constructed wetland: Current trends and future potential. *Processes*, *9*(11), 1917. https://doi.org/10.3390/PR9111917

Hueso, J. L., Mallada, R., & Santamaria, J. (2023). Gas–solid contactors and catalytic reactors with direct microwave heating: Current status and perspectives. *Catalysis Today*, *423*, 113927. https://doi.org/10.1016/J.CATTOD.2022.10.009

Irshad, Z., Bibi, I., Ghafoor, A., Majid, F., Kamal, S., Ezzine, S., Elqahtani, Z. M., Alwadai, N., El Messaoudi, N., & Iqbal, M. (2022). Ni doped $SrFe_{12}O_{19}$ nanoparticles synthesized via micro-emulsion route and photocatalytic activity evaluation for the degradation of crystal violet under visible light irradiation. *Results in Physics*, *42*, 106006. https://doi.org/10.1016/J.RINP.2022.106006

Itayama, T., Kiji, M., Suetsugu, A., Tanaka, N., Saito, T., Iwami, N., Mizuochi, M., & Inamori, Y. (2006). On site experiments of the slanted soil treatment systems for domestic gray water. *Water Science and Technology*, *53*(9), 193–201. https://doi.org/10.2166/wst.2006.290

Jhaveri, J. H., & Murthy, Z. V. P. (2016). A comprehensive review on anti-fouling nanocomposite membranes for pressure driven membrane separation processes. *Desalination*, *379*, 137–154. https://doi.org/10.1016/j.desal.2015.11.009

Jia, Y., Wu, X., Xu, S., Zhang, Y., & Wang, S. (2023). Experimental investigation on dyeing wastewater treatment and by-product hydrogen with a reverse electrodialysis flocculator. *International Journal of Hydrogen Energy*, *48*(50), 19022–19032. https://doi.org/https://doi.org/10.1016/j.ijhydene.2023.02.032

Jiang, C., Chen, H., Zhang, Y., Feng, H., Shehzad, M. A., Wang, Y., & Xu, T. (2018). Complexation electrodialysis as a general method to simultaneously treat wastewaters with metal and organic matter. *Chemical Engineering Journal*, *348*, 952–959. https://doi.org/https://doi.org/10.1016/j.cej.2018.05.022

Jiang, J. Q. (2015). The role of coagulation in water treatment. *Current Opinion in Chemical Engineering*, *8*, 36–44. https://doi.org/10.1016/j.coche.2015.01.008

Kabsch-Korbutowicz, M. (2005). Effect of Al coagulant type on natural organic matter removal efficiency in coagulation/ultrafiltration process. *Desalination*, *185*(1–3), 327–333. https://doi.org/10.1016/j.desal.2005.02.083

Karabelnik, K., Kõiv, M., Kasak, K., Jenssen, P. D., & Mander, Ü. (2012). High-strength greywater treatment in compact hybrid filter systems with alternative substrates. *Ecological Engineering*, *49*, 84–92. https://doi.org/10.1016/j.ecoleng.2012.08.035

Katukiza, A. Y., Ronteltap, M., Niwagaba, C. B., Kansiime, F., & Lens, P. N. L. (2014). Grey water treatment in urban slums by a filtration system: Optimisation of the filtration medium. *Journal of Environmental Management*, *146*, 131–141. https://doi.org/10.1016/j.jenvman.2014.07.033

Khalid, S., Shahid, M., Natasha, Bibi, I., Sarwar, T., Shah, A. H., & Niazi, N. K. (2018). A review of environmental contamination and health risk assessment of wastewater use for crop irrigation with a focus on low and high-income countries. *International Journal of Environmental Research and Public Health*, *15*(5), 895. https://doi.org/10.3390/IJERPH15050895

Kim, K., Von Spakovsky, M. R., Wang, M., & Nelson, D. J. (2011). A hybrid multi-level optimization approach for the dynamic synthesis/design and operation/control under uncertainty of a fuel cell system. *Energy*, *36*(6), 3933–3943. https://doi.org/10.1016/J.ENERGY.2010.08.024

Kokkinos, P., Venieri, D., & Mantzavinos, D. (2021). Advanced oxidation processes for water and wastewater viral disinfection. A systematic review. *Food and Environmental Virology*, *13*(3), 283–302. https://doi.org/10.1007/S12560-021-09481-1/FIGURES/3

Kolya, H., & Kang, C. W. (2023). Bio-based polymeric flocculants and adsorbents for wastewater treatment. *Sustainability*, *15*(12), 9844. https://doi.org/10.3390/SU15129844

Koul, B., Bhat, N., Abubakar, M., Mishra, M., Arukha, A. P., & Yadav, D. (2022). Application of natural coagulants in water treatment: A sustainable alternative to chemicals. *Water*, *14*(22), 3751. https://doi.org/10.3390/W14223751

Koul, B., Yadav, D., Singh, S., Kumar, M., & Song, M. (2022). Insights into the Domestic Wastewater Treatment (DWWT) Regimes: A review. *Water*, *14*(21), 3542. https://doi.org/10.3390/W14213542

Kovalchuk, N., Starov, V., Langston, P., & Hilal, N. (2009). Formation of stable clusters in colloidal suspensions. *Advances in Colloid and Interface Science*, *147–148*(C), 144–154. https://doi.org/10.1016/j.cis.2008.11.001

Kumari, P., & Kumar, A. (2023). Advanced oxidation process: A remediation technique for organic and non-biodegradable pollutant. *Results in Surfaces and Interfaces*, *11*, 100122. https://doi.org/10.1016/J.RSURFI.2023.100122

Lee, S., & Bi, X. (2019). Can adoption of pollution prevention techniques reduce pollution substitution? *PLOS ONE*, *14*(11), e0224868. https://doi.org/10.1371/JOURNAL.PONE.0224868

Lewis, A. E. (2010). Review of metal sulphide precipitation. *Hydrometallurgy*, *104*(2), 222–234. https://doi.org/10.1016/J.HYDROMET.2010.06.010

Li, J., How, Z. T., Benally, C., Sun, Y., Zeng, H., & Gamal El-Din, M. (2023). Removal of colloidal impurities by thermal softening-coagulation-flocculation-sedimentation in steam assisted gravity drainage (SAGD) produced water: Performance, interaction effects and mechanism study. *Separation and Purification Technology*, *313*, 123484. https://doi.org/10.1016/j.seppur.2023.123484

Liang, Y., Hilal, N., Langston, P., & Starov, V. (2007). Interaction forces between colloidal particles in liquid: Theory and experiment. *Advances in Colloid and Interface Science*, *134–135*, 151–166. https://doi.org/10.1016/j.cis.2007.04.003

Liao, P. H., Wong, W. T., & Lo, K. V. (2005). Release of phosphorus from sewage sludge using microwave technology. *Journal of Environmental Engineering and Science*, *4*(1), 77–81. https://doi.org/10.1139/S04-056

Licskó, I. (1997). Realistic coagulation mechanisms in the use of aluminium and iron(III) salts. *Water Science and Technology*, *36*(4), 103–110. https://doi.org/10.1016/S0273-1223(97)00426-5

Lim, S., Shi, J. L., von Gunten, U., & McCurry, D. L. (2022). Ozonation of organic compounds in water and wastewater: A critical review. *Water Research*, *213*, 118053. https://doi.org/10.1016/J.WATRES.2022.118053

Liu, T., Aniagor, C. O., Ejimofor, M. I., Menkiti, M. C., Tang, K. H. D., Chin, B. L. F., Chan, Y. H., Yiin, C. L., Cheah, K. W., Ho Chai, Y., Lock, S. S. M., Yap, K. L., Wee, M. X. J., & Yap, P. S. (2023). Technologies for removing pharmaceuticals and personal care products (PPCPs) from aqueous solutions: Recent advances, performances, challenges and recommendations for improvements. *Journal of Molecular Liquids*, *374*, 121144. https://doi.org/10.1016/J.MOLLIQ.2022.121144

Lizama, A. C., Figueiras, C. C., Herrera, R. R., Pedreguera, A. Z., & Ruiz Espinoza, J. E. (2017). Effects of ultrasonic pretreatment on the solubilization and kinetic study of biogas production from anaerobic digestion of waste activated sludge. *International Biodeterioration & Biodegradation*, *123*, 1–9. https://doi.org/10.1016/J.IBIOD.2017.05.020

López-Maldonado, E. A., Oropeza-Guzman, M. T., Jurado-Baizaval, J. L., & Ochoa-Terán, A. (2014). Coagulation–flocculation mechanisms in wastewater treatment plants through zeta potential measurements. *Journal of Hazardous Materials*, *279*, 1–10. https://doi.org/10.1016/j.jhazmat.2014.06.025

Lu, S., Pugh, R. J., & Forssberg, E. (2006). Interfacial separation of particles. *China Particuology*, *4*(5), 258. https://doi.org/10.1016/s1672-2515(07)60272-3

Manickam, S., Camilla Boffito, D., Flores, E. M. M., Leveque, J. M., Pflieger, R., Pollet, B. G., & Ashokkumar, M. (2023). Ultrasonics and sonochemistry: Editors' perspective. *Ultrasonics Sonochemistry*, *99*, 106540. https://doi.org/10.1016/J.ULTSONCH.2023.106540

Chollom, M. N. (2014). Treatment and reuse of reactive dye effluent from textile industry using membrane technology. July, 1–144. https://doi.org/10.51415/10321/1388

Meng, S., Wen, S., Han, G., Wang, X., & Feng, Q. (2022). Wastewater treatment in mineral processing of non-ferrous metal resources: A review. *Water*, *14*(5), 726. https://doi.org/10.3390/W14050726

Messaoudi, N. El, Khomri, M. El, Chegini, Z. G., Chlif, N., Dbik, A., Bentahar, S., Iqbal, M., Jada, A., & Lacherai, A. (2021). Desorption study and reusability of raw and H_2SO_4 modified jujube shells (*Zizyphus lotus*) for the methylene blue adsorption. *International Journal of Environmental Analytical Chemistry*, *103*(16), 3762–3778. https://doi.org/10.1080/03067319.2021.1912338

Mihalakakou, G., Souliotis, M., Papadaki, M., Menounou, P., Dimopoulos, P., Kolokotsa, D., Paravantis, J. A., Tsangrassoulis, A., Panaras, G., Giannakopoulos, E., & Papaefthimiou, S. (2023). Green roofs as a nature-based solution for improving urban sustainability: Progress and perspectives. *Renewable and Sustainable Energy Reviews*, *180*, 113306. https://doi.org/10.1016/J.RSER.2023.113306

Mir, T. ul G., Katoch, V., Angurana, R., Wani, A. K., Shukla, S., El Messaoudi, N., Sher, F., Mulla, S. I., & Américo-Pinheiro, J. H. P. (2023). Environmental and toxicological concerns associated with nanomaterials used in the industries. *Nanomaterials for Bioreactors and Bioprocessing Applications*. Elsevier 141–193. https://doi.org/10.1016/B978-0-323-91782-7.00010-2

Mishra, K., Siwal, S. S., Nayaka, S. C., Guan, Z., & Thakur, V. K. (2023). Waste-to-chemicals: Green solutions for bioeconomy markets. *Science of The Total Environment*, *887*, 164006. https://doi.org/10.1016/J.SCITOTENV.2023.164006

Muro, C., Riera, F., & Carmen Diaz, M. del. (2012). Membrane separation process in wastewater treatment of food industry. *Food Industrial Processes – Methods and Equipment*, *14*, 254–279. https://doi.org/10.5772/31116

Naha, A., Antony, S., Nath, S., Sharma, D., Mishra, A., Biju, D. T., Madhavan, A., Binod, P., Varjani, S., & Sindhu, R. (2023). A hypothetical model of multi-layered cost-effective wastewater treatment plant integrating microbial fuel cell and nanofiltration technology: A comprehensive review on wastewater treatment and sustainable remediation. *Environmental Pollution*, *323*, 121274. https://doi.org/https://doi.org/10.1016/j.envpol.2023.121274

Naidu, G., Tijing, L., Johir, M. A. H., Shon, H., & Vigneswaran, S. (2020). Hybrid membrane distillation: Resource, nutrient and energy recovery. *Journal of Membrane Science*, *599*, 117832. https://doi.org/10.1016/J.MEMSCI.2020.117832

Núñez, J., Yeber, M., Cisternas, N., Thibaut, R., Medina, P., & Carrasco, C. (2019). Application of electrocoagulation for the efficient pollutants removal to reuse the treated wastewater in the dyeing process of the textile industry. *Journal of Hazardous Materials*, *371*, 705–711. https://doi.org/https://doi.org/10.1016/j.jhazmat.2019.03.030

Ochoa, S. I. C., Ushijima, K., Hijikata, N., & Funamizu, N. (2015). Treatment of domestic greywater by geotextile filter and intermittent sand filtration bioreactor. *Journal of Water Reuse and Desalination*, *5*(1), 39–49. https://doi.org/10.2166/wrd.2014.042

Ohrdes, H., Ille, I., Twiefel, J., Wallaschek, J., Nogueira, R., & Rosenwinkel, K. H. (2018). A control system for ultrasound devices utilized for inactivating *E. coli* in wastewater. *Ultrasonics Sonochemistry*, *40*, 158–162. https://doi.org/10.1016/J.ULTSONCH.2017.04.017

Oladejo, J., Shi, K., Chen, Y., Luo, X., Gang, Y., & Wu, T. (2020). Closing the active carbon cycle: Regeneration of spent activated carbon from a wastewater treatment facility for resource optimization. *Chemical Engineering and Processing – Process Intensification*, *150*, 107878. https://doi.org/10.1016/J.CEP.2020.107878

Ong, C. S., Al-Anzi, B., Lau, W. J., Goh, P. S., Lai, G. S., Ismail, A. F., & Ong, Y. S. (2017). Anti-fouling double-skinned forward osmosis membrane with zwitterionic brush for oily wastewater treatment. *Scientific Reports*, *7*(1), 1–11. https://doi.org/10.1038/s41598-017-07369-4

Papić, S., Koprivanac, N., & Božić, A. L. (2000). Removal of reactive dyes from wastewater using Fe(III) coagulant. *Coloration Technology*, *116*(11), 352–358. https://doi.org/10.1111/j.1478-4408.2000.tb00013.x

Patil, Y. M., & Munavalli, G. R. (2016). Performance evaluation of an integrated on-site greywater treatment system in a tropical region. *Ecological Engineering*, *95*, 492–500. https://doi.org/10.1016/j.ecoleng.2016.06.078

Peng, C., Chai, L. Y., Tang, C. J., Min, X. B., Ali, M., Song, Y. X., & Qi, W. M. (2017). Feasibility and enhancement of copper and ammonia removal from wastewater using struvite formation: A comparative research. *Journal of Chemical Technology & Biotechnology*, *92*(2), 325–333. https://doi.org/10.1002/JCTB.5009

Peralta, M. E., Ocampo, S., Funes, I. G., Medina, F. O., Parolo, M. E., & Carlos, L. (2020). Nanomaterials with tailored magnetic properties as adsorbents of organic pollutants from wastewaters. *Inorganics*, *8*(4), 24. https://doi.org/10.3390/INORGANICS8040024

Pillai, S. B., & Thombre, N. V. (2023). Coagulation, flocculation, and precipitation in water and used water purification. *Handbook of Water and Used Water Purification*. Editor by Josef Lahnsteiner. Springer, 1–25. https://doi.org/10.1007/978-3-319-66382-1_63-1

Pohl, A. (2020). Removal of heavy metal ions from water and wastewaters by sulfur-containing precipitation agents. *Water, Air, and Soil Pollution*, *231*(10), 1–17. https://doi.org/10.1007/S11270-020-04863-W/TABLES/2

Rad, S. M., Ray, A. K., & Barghi, S. (2022). Water pollution and agriculture pesticide. *Clean Technologies*, *4*(4), 1088–1102. https://doi.org/10.3390/CLEANTECHNOL4040066

Raha, S., & Ahmaruzzaman, M. (2022). ZnO nanostructured materials and their potential applications: Progress, challenges and perspectives. *Nanoscale Advances*, *4*(8), 1868–1925. https://doi.org/10.1039/D1NA00880C

Rahimi, S., Modin, O., & Mijakovic, I. (2020). Technologies for biological removal and recovery of nitrogen from wastewater. *Biotechnology Advances*, *43*, 107570. https://doi.org/10.1016/J.BIOTECHADV.2020.107570

Rahman, T. U., Roy, H., Islam, M. R., Tahmid, M., Fariha, A., Mazumder, A., Tasnim, N., Pervez, M. N., Cai, Y., Naddeo, V., & Islam, M. S. (2023). The advancement in membrane bioreactor (MBR) technology toward sustainable industrial wastewater management. *Membranes*, *13*(2), 181. https://doi.org/10.3390/MEMBRANES13020181/S1

Sabogal-Paz, L. P., Campos, L. C., Bogush, A., & Canales, M. (2020). Household slow sand filters in intermittent and continuous flows to treat water containing low mineral ion concentrations and Bisphenol A. *Science of the Total Environment*, *702*, 135078. https://doi.org/10.1016/j.scitotenv.2019.135078

Sakurai, K. S. I., Pompei, C. M. E., Tomita, I. N., Santos-Neto, Á. J., & Silva, G. H. R. (2021). Hybrid constructed wetlands as post-treatment of blackwater: An assessment of the removal of antibiotics. *Journal of Environmental Management*, *278*, 111552. https://doi.org/10.1016/J.JENVMAN.2020.111552

Salehmin, M. N. I., Hil Me, M. F., Daud, W. R. W., Mohd Yasin, N. H., Abu Bakar, M. H., Sulong, A. B., & Lim, S. S. (2023). Construction of microbial electrodialysis cells equipped with internal proton migration pathways: Enhancement of wastewater treatment, desalination, and hydrogen production. *Science of The Total Environment*, *855*, 158527. https://doi.org/https://doi.org/10.1016/j.scitotenv.2022.158527

Sathya, K., Nagarajan, K., Carlin Geor Malar, G., Rajalakshmi, S., & Raja Lakshmi, P. (2022). A comprehensive review on comparison among effluent treatment methods and modern methods of treatment of industrial wastewater effluent from different sources. *Applied Water Science*, *12*(4), 1–27. https://doi.org/10.1007/S13201-022-01594-7

Savun-Hekimoğlu, B. (2020). A review on sonochemistry and its environmental applications. *Acoustics*, *2*(4), 766–775. https://doi.org/10.3390/ACOUSTICS2040042

Sedighi, M., Behvand Usefi, M. M., Ismail, A. F., & Ghasemi, M. (2023). Environmental sustainability and ions removal through electrodialysis desalination: Operating conditions and process parameters. *Desalination*, *549*, 116319. https://doi.org/https://doi.org/10.1016/j.desal.2022.116319

Şenol, Z. M., Elma, E., El Messaoudi, N., & Mehmeti, V. (2023). Performance of cross-linked chitosan-zeolite composite adsorbent for removal of Pb^{2+} ions from aqueous solutions: Experimental and Monte Carlo simulations studies. *Journal of Molecular Liquids*, *391*, 123310. https://doi.org/10.1016/J.MOLLIQ.2023.123310

Şenol, Z. M., Messaoudi, N. El, Fernine, Y., & Keskin, Z. S. (2023). Bioremoval of rhodamine B dye from aqueous solution by using agricultural solid waste (almond shell): Experimental and DFT modeling studies. *Biomass Conversion and Biorefinery*, *14*, 17927–17940. https://doi.org/10.1007/S13399-023-03781-1

Shabanizadeh, H., & Taghavijeloudar, M. (2023). A sustainable approach for industrial wastewater treatment using pomegranate seeds in flocculation–coagulation process: Optimization of COD and turbidity removal by response surface methodology (RSM). *Journal of Water Process Engineering*, *53*(October 2022), 103651. https://doi.org/10.1016/j.jwpe.2023.103651

Shaffer, D. L., Werber, J. R., Jaramillo, H., Lin, S., & Elimelech, M. (2015). Forward osmosis: Where are we now? *Desalination*, *356*, 271–284. https://doi.org/10.1016/j.desal.2014.10.031

Shaikh, I., & Ahammed, M. M. (2021a). Coagulation followed by continuous sand filtration for treatment of graywater. *Journal of Hazardous, Toxic, and Radioactive Waste*, *25*(4), 1–8. https://doi.org/10.1061/(asce)hz.2153-5515.0000640

Shaikh, I. N., & Ahammed, M. M. (2021b). Effect of operating mode on the performance of sand filters treating greywater. *Environmental Science and Pollution Research*, *28*(28), 38209–38223. https://doi.org/10.1007/s11356-021-13413-8

Sharaf, A., & Liu, Y. (2021). Mechanisms and kinetics of greywater treatment using biologically active granular activated carbon. *Chemosphere*, *263*, 128113. https://doi.org/10.1016/j.chemosphere.2020.128113

Shewa, W. A., & Dagnew, M. (2020). Revisiting chemically enhanced primary treatment of wastewater: A review. *Sustainability*, *12*(15), 5928. https://doi.org/10.3390/SU12155928

Sillanpää, M., & Shestakova, M. (2017). Chapter 3 – Emerging and combined electrochemical methods. In M. Sillanpää & M. Shestakova (eds.), *Electrochemical Water Treatment Methods*. Butterworth-Heinemann, pp. 131–225. https://doi.org/https://doi.org/10.1016/B978-0-12-811462-9.00003-7

Singh, B. J., Chakraborty, A., & Sehgal, R. (2023). A systematic review of industrial wastewater management: Evaluating challenges and enablers. *Journal of Environmental Management*, *348*, 119230. https://doi.org/10.1016/J.JENVMAN.2023.119230

Singh, S., Ahammed, M. M., & Shaikh, I. N. (2021). Combined coagulation and intermittent sand filtration for on-site treatment of greywater. *IOP Conference Series: Materials Science and Engineering*, *1114*(1), 012031. https://doi.org/10.1088/1757-899x/1114/1/012031

Spychala, M., Nieć, J., Zawadzki, P., Matz, R., & Nguyen, T. H. (2019). Removal of volatile solids from greywater using sand filters. *Applied Sciences (Switzerland)*, *9*(4), 770. https://doi.org/10.3390/app9040770

Strathmann, H. (2004). Assessment of electrodialysis water desalination process costs. *Proceedings of the International Conference on Desalination Costing, Lemassol, Cyprus, December 6–8, 2004, January 2004*, pp. 32–54.

Sun, L., Mo, Y., & Zhang, L. (2022). A mini review on bio-electrochemical systems for the treatment of azo dye wastewater: State-of-the-art and future prospects. *Chemosphere*, *294*, 133801. https://doi.org/https://doi.org/10.1016/j.chemosphere.2022.133801

Suwaileh, W., Johnson, D., Sarp, S., & Hilal, N. (2018). Advances in forward osmosis membranes: Altering the sub-layer structure via recent fabrication and chemical modification approaches. *Desalination*, *436*, 176–201.

Takht Ravanchi, M., Kaghazchi, T., & Kargari, A. (2009). Application of membrane separation processes in petrochemical industry: A review. *Desalination*, *235*(1–3), 199–244. https://doi.org/10.1016/j.desal.2007.10.042

Tang, C., Cui, D., Li, Z., Yu, L., Xue, J., & Yin, X. (2023). Electrooxidation degradation of hydroxychloroquine in wastewater using a long-acting Ti-based PbO_2 anode with an arc-sprayed (Ti,Zr)N interlayer. *Chemosphere*, *335*, 139074. https://doi.org/https://doi.org/10.1016/j.chemosphere.2023.139074

Tarleton, E. S., & Wakeman, R. J. (2007). Pretreatment of suspensions. In *Solid/Liquid Separation*. Edited by E.S. Tarleton and R.J. Wakeman. Elsevier, 126–151. https://doi.org/10.1016/b978-185617421-3/50003-1

Thompson, K. A., Valencia, E. W., Scott Summers, R., & Cook, S. M. (2020). Sorption, coagulation, and biodegradation for graywater treatment. *Water Science and Technology*, *81*(10), 2152–2162. https://doi.org/10.2166/wst.2020.273

Tian, L., Lv, G., Liu, M., Lei, X., Rao, W., & Liao, L. (2022). Reviews: Microwave-induced oxidation technology and its applications. *Progress in Natural Science: Materials International*, *32*(6), 665–673. https://doi.org/10.1016/J.PNSC.2022.10.005

Tóth, A. J., Fózer, D., Mizsey, P., Varbanov, P. S., & Klemeš, J. J. (2022). Physicochemical methods for process wastewater treatment: Powerful tools for circular economy in the chemical industry. *Reviews in Chemical Engineering*, *39*(7), 1123–1151. https://doi.org/10.1515/revce-2021-0094

Ushijima, K., Ito, K., Ito, R., & Funamizu, N. (2013). Greywater treatment by slanted soil system. *Ecological Engineering*, *50*, 62–68. https://doi.org/10.1016/j.ecoleng.2012.04.028

Verma, A., Kumar, S., & Kumar, S. (2016). Biosorption of lead ions from the aqueous solution by *Sargassum filipendula*: Equilibrium and kinetic studies. *Journal of Environmental Chemical Engineering*, *4*(4), 4587–4599. https://doi.org/10.1016/J.JECE.2016.10.026

Vermeulen, T. (1958) Separation by adsorption methods. *Advances in Chemical Engineering*, *2*, 147–208. https://doi.org/10.1016/s0065-2377(08)60228-8

Vialkova, E., Obukhova, M., & Belova, L. (2021). Microwave irradiation in technologies of wastewater and wastewater sludge treatment: A review. *Water*, *13*(13), 1784. https://doi.org/10.3390/W13131784

Voutetaki, A., Plakas, K. V, Papadopoulos, A. I., Bollas, D., Parcharidis, S., & Seferlis, P. (2023). Pilot-scale separation of lead and sulfate ions from aqueous solutions using electrodialysis: Application and parameter optimization for the battery industry. *Journal of Cleaner Production*, *410*, 137200. https://doi.org/https://doi.org/10.1016/j.jclepro.2023.137200

Wang, F., Fu, R., Lv, H., Zhu, G., Lu, B., Zhou, Z., Wu, X., & Chen, H. (2019). Phosphate recovery from swine wastewater by a struvite precipitation electrolyzer. *Scientific Reports*, *9*(1), 1–10. https://doi.org/10.1038/s41598-019-45085-3

Wang, J., Chen, B., Lin, H., & Li, R. (2023). Efficient oily wastewater treatment by a novel electroflotation-membrane separation system consisting a Ni-Cu-P membrane prepared by electroless nickel plating. *Science of The Total Environment*, *881*, 163135. https://doi.org/https://doi.org/10.1016/j.scitotenv.2023.163135

Wang, J.-Y., Kadier, A., Peralta-Hernández, J. M., Liang, C.-G., & Ma, P.-C. (2023). An electrocoagulation (EC) process with a novel edge insulated electrode for separation of oil-in-water emulsions. *Case Studies in Chemical and Environmental Engineering*, *7*, 100354. https://doi.org/https://doi.org/10.1016/j.cscee.2023.100354

Wang, P., & Chung, T. S. (2015). Recent advances in membrane distillation processes: Membrane development, configuration design and application exploring. *Journal of Membrane Science*, *474*, 39–56. https://doi.org/10.1016/j.memsci.2014.09.016

Wang, R., Wan, H., Wang, B., Zhang, K., Shi, H., Zhang, W., Tang, J., Wang, H., Gao, S., & Huang, Q. (2023). Mechanistic study of electrooxidation of coexisting chloramphenicol and natural organic matter: Performance, DFT calculation and removal route. *Separation and Purification Technology*, *306*, 122584. https://doi.org/https://doi.org/10.1016/j.seppur.2022.122584

Wang, S., Wu, X., Xu, S., Leng, Q., Jin, D., Wang, P., Dong, F., & Wu, D. (2023). Energetic evaluation of phenol wastewater treatment by reverse electrodialysis reactor using different anodes. *Journal of Environmental Management*, *329*, 117089. https://doi.org/https://doi.org/10.1016/j.jenvman.2022.117089

Wang, Y. N., Goh, K., Li, X., Setiawan, L., & Wang, R. (2018). Membranes and processes for forward osmosis-based desalination: Recent advances and future prospects. *Desalination*, *434*(October), 81–99. https://doi.org/10.1016/j.desal.2017.10.028

Ward, L., Page, M., Jurevis, J., Nelson, A., Rivera, M., Hernandez, M., Chappell, M., & Dusenbury, J. (2015). Assessment of biologically active GAC and complementary technologies for gray water treatment. *Journal of Water Reuse and Desalination*, *5*(3), 239–249. https://doi.org/10.2166/wrd.2015.088

Wear, S. L., Acuña, V., McDonald, R., & Font, C. (2021). Sewage pollution, declining ecosystem health, and cross-sector collaboration. *Biological Conservation*, *255*, 109010. https://doi.org/10.1016/J.BIOCON.2021.109010

Wu, Y., Luo, J., Zhang, Q., Aleem, M., Fang, F., Xue, Z., & Cao, J. (2019). Potentials and challenges of phosphorus recovery as vivianite from wastewater: A review. *Chemosphere*, *226*, 246–258. https://doi.org/10.1016/J.CHEMOSPHERE.2019.03.138

Wu, Z., Abramova, A., Nikonov, R., & Cravotto, G. (2020). Sonozonation (sonication/ozonation) for the degradation of organic contaminants – A review. *Ultrasonics Sonochemistry*, *68*, 105195. https://doi.org/10.1016/J.ULTSONCH.2020.105195

Xie, A., Ladner, D. A., & Popat, S. C. (2022). Electrocoagulation-electroflotation for primary treatment of animal rendering wastewater to enable recovery of fats. *Chemical Engineering Journal*, *431*, 133910. https://doi.org/https://doi.org/10.1016/j.cej.2021.133910

Xu, L., Chen, L., Zhang, M., Zhu, A., Zheng, X., Zhao, M., Kong, H., & Zheng, X. (2020). Treating graywater using quartz sand filters: The effect of particle size, substrate combinations, and reflux ratio. *Desalination and Water Treatment*, *197*, 131–138. https://doi.org/10.5004/dwt.2020.25976

Xu, L., Xu, X., & Wu, D. (2019). Initial dissolved oxygen-adjusted electrochemical generation of sulfate green rust for cadmium removal using a closed-atmosphere Fe-electrocoagulation system. *Chemical Engineering Journal*, *359*, 1411–1418. https://doi.org/https://doi.org/10.1016/j.cej.2018.11.032

Yakamercan, E., Bhatt, P., Aygun, A., Adesope, A. W., & Simsek, H. (2023). Comprehensive understanding of electrochemical treatment systems combined with biological processes for wastewater remediation. *Environmental Pollution*, *330*, 121680. https://doi.org/https://doi.org/10.1016/j.envpol.2023.121680

Yang, J., Monnot, M., Ercolei, L., & Moulin, P. (2020). Membrane-based processes used in municipal wastewater treatment for water reuse: State-of-the-art and performance analysis. *Membranes*, *10*(6), 131. https://doi.org/10.3390/MEMBRANES10060131

Yaqub, A., Raza, H., Ajab, H., Shah, S. H., Shad, A., & Bhatti, Z. A. (2023). Decolorization of reactive blue-2 dye in aqueous solution by electrocoagulation process using aluminum and steel electrodes. *Journal of Hazardous Materials Advances*, *9*, 100248. https://doi.org/https://doi.org/10.1016/j.hazadv.2023.100248

Yaseen, D. A., & Scholz, M. (2019). Textile dye wastewater characteristics and constituents of synthetic effluents: A critical review. *International Journal of Environmental Science and Technology*, *16*(2), 1193–1226. https://doi.org/10.1007/s13762-018-2130-z

Yasuda, K. (2021). Sonochemical green technology using active bubbles: Degradation of organic substances in water. *Current Opinion in Green and Sustainable Chemistry*, *27*, 100411. https://doi.org/10.1016/J.COGSC.2020.100411

Yin, C., Ma, J., Qiu, J., Liu, R., & Ba, L. (2023). Mass-producible low-cost flexible electronic fabrics for azo dye wastewater treatment by electrocoagulation. *Chinese Journal of Chemical Engineering*, *59*, 222–230. https://doi.org/https://doi.org/10.1016/j.cjche.2023.02.011

Zhang, X., Ao, K., & Daoud, W. A. (2023). Nano-sphere RuO_2 embedded in MOF-derived carbon arrays as a dual-matrix anode for cost-effective electrochemical wastewater treatment. *Science of The Total Environment*, *865*, 161126. https://doi.org/https://doi.org/10.1016/j.scitotenv.2022.161126

Zhao, C., Zhou, J., Yan, Y., Yang, L., Xing, G., Li, H., Wu, P., Wang, M., & Zheng, H. (2021). Application of coagulation/flocculation in oily wastewater treatment: A review. *Science of the Total Environment*, *765*, 142795. https://doi.org/10.1016/j.scitotenv.2020.142795

Zhou, Z., Zhang, M., Xia, Q., Zhao, X., Ming, Q., & Zeng, L. (2023). Effects of nanofiltration on desalination of flue gas desulfurization wastewater by electrodialysis: Treatment effect, fouling property and techno-economic analysis. *Separation and Purification Technology*, *316*, 123768. https://doi.org/https://doi.org/10.1016/j.seppur.2023.123768

Zipf, M. S., Pinheiro, I. G., & Conegero, M. G. (2016). Simplified greywater treatment systems: Slow filters of sand and slate waste followed by granular activated carbon. *Journal of Environmental Management*, *176*, 119–127. https://doi.org/10.1016/j.jenvman.2016.03.035

4 An Overview of Innovative Oxidation Technologies for Wastewater Treatment

Ram Sharan Singh and Himanshu Tiwari

4.1 INTRODUCTION

Water is a vital resource for all forms of life, playing a crucial role in sustaining ecosystems, supporting agriculture, and ensuring human well-being. The increase in wastewater discharge containing anthropogenic pollutants deteriorates the water quality. Both households and industries generate a gigantic amount of wastewater with different chemical compositions, which also reduces the availability of clean water. The wastewater predicament is significantly intensified by the textile and related chemical industries, further contributing to the issue of water pollution [1]. The wastewater emanating from these industries contains a wide range of organic compounds, including complex organic dyes, bleaching agents, solvents, heavy metals, and other hazardous substances, imposing significant challenges to effective treatment [2]. These dyes are often resistant to conventional wastewater treatment methods, resulting in their release into aquatic biota. Additionally, the chemicals used in textile processing, such as bleaching agents and finishing agents, can also contribute to the pollution of water resources [3]. The discharge of untreated or inadequately treated wastewater can introduce harmful substances into rivers, lakes, and other water bodies. These substances not only contaminate water resources but also pose risks to aquatic life, wildlife, and even to human health [4].

Various wastewater treatment techniques are employed for wastewater pollution remediation and to protect our freshwater resources [5]. These techniques encompass a range of physical, chemical, and biological processes that aim to remove pollutants and contaminants from wastewater before its release into the environment [6]. Among these techniques, advanced oxidation techniques (AOTs) have gained increasing importance due to their effectiveness in treating recalcitrant pollutants and emerging contaminants (ECs), [7]. The importance of implementing innovative AOTs lies in their ability to effectively break down and mineralize persistent pollutants into non-toxic by-products. AOTs typically involve the generation of highly reactive hydroxyl radicals ($HO^{\bullet}$), which have exceptional oxidative capabilities and can react with various organic compounds, leading to their degradation [8].

DOI: 10.1201/9781003454199-4

The implementation of innovative AOTs in wastewater treatment offers several key advantages. Initially, these methods prove highly adept at removing an extensive array of pollutants, such as pharmaceuticals, pesticides, dyes, ECs, and municipal wastewater. These substances often present challenges for removal through traditional treatment approaches [9]. This ensures the wastewater quality parameters and minimizes the potential risks associated with these pollutants. Secondly, AOTs not only enhance treatment efficiency significantly but also ensure a more comprehensive removal of pollutants in comparison to conventional methods. The strong oxidizing properties of these techniques result in the complete degradation of organic compounds, converting them into simpler, less harmful substances, such as carbon dioxide and water [10]. This complete degradation ensures a sustainable removal of contaminants, reducing the risk of residual pollutants entering the environment.

The implementation of innovative AOTs exhibits effective and sustainable wastewater treatment. This chapter critically highlights the Ozonation, Fenton's reagent, Non-Thermal Plasma (NTP), and Photocatalysis as revolutionizing AOTs by effectively degrading and eliminating complex organic contaminants. The attractiveness of these methods lies in their innovative approaches and ability to tackle diverse contaminants.

4.2 IMPORTANCE OF WASTEWATER TREATMENT

Wastewater treatment plays a crucial role in preserving the environment, protecting human health, ensuring the lucrative reuse of water resources, and promoting sustainable development. It is a critical process that involves removing pollutants including organic and inorganic substances, and pathogens from wastewater before discarding directly into the environment or reusing them for various purposes. By treating wastewater to a suitable standard, it can be reused for various non-potable purposes such as irrigation, industrial processes, and environmental restoration. This reduces the demand for freshwater resources and alleviates the strain on water supplies, particularly in water-scarce regions. Moreover, wastewater treatment is necessary nowadays for mitigating the adverse environmental impact of industries and human activities. As industrialization, urbanization, and globalization surge, freshwater scarcity is becoming a growing concern worldwide [11]. Therefore, the proper treatment and reuse of wastewater become necessary to suppress the excessive pressure on groundwater consumption. Moreover, wastewater treatment plays a significant role in mitigating climate change. The decomposition of organic matter in untreated wastewater releases greenhouse gases, such as methane, contributing to global warming. Implementing treatment processes that capture and utilize these gases, such as anaerobic digestion, can minimize the environmental impact and harness renewable energy sources. Thus, emphasizing the significance of wastewater treatment and investing in its advancement is essential for the sustenance of our future generation and aquatic life.

Wastewater treatment plant (WWTP) sludge refers to the residual material generated during the treatment of wastewater in a dedicated facility. This sludge is a byproduct of the wastewater treatment process, which involves the removal of pollutants and contaminants from both industrial and domestic wastewater streams.

In situations where industrial and domestic wastewater is mixed, the resulting wastewater treatment process aims to address a diverse range of contaminants originating from various sources. The sludge produced in such a scenario is a composite of the residuals from both industrial and domestic wastewater treatment. The composition of the sludge may vary depending on the characteristics of the incoming wastewater, the treatment technologies employed, and the specific pollutants present. Industrial wastewater often contains unique and sometimes more concentrated contaminants compared to domestic wastewater. Therefore, the combined sludge from mixed wastewater treatment reflects this amalgamation of pollutants from diverse sources. It is important to note that the treatment of mixed wastewater and the subsequent management of the generated sludge require a comprehensive approach. The treatment process must be capable of effectively addressing the combined pollutants from industrial and domestic sources to meet regulatory standards and environmental guidelines. Additionally, the management of the resulting sludge involves considerations for proper disposal or beneficial reuse, taking into account its potential environmental impact.

Disposal norms for wastewater vary significantly from country to country due to differences in regulatory frameworks, environmental conditions, and local infrastructure. It is important to note that each nation has established its own set of guidelines and standards for the disposal of wastewater into sewer lines or other treatment systems. These norms are shaped by factors such as the country's environmental policies, water quality objectives, and public health considerations. In general, countries typically have regulatory bodies responsible for overseeing water quality and managing wastewater disposal. These organizations set specific guidelines regarding the permissible levels of various contaminants in wastewater and prescribe the appropriate treatment processes. The disposal norms may address parameters such as chemical oxygen demand (COD), biological oxygen demand (BOD), suspended solids, nutrient content, and potentially harmful substances.

4.3 ADVANCED OXIDATION PROCESSES (AOPS)

Advanced Oxidation Processes (AOPs) are a group of innovative techniques used in water and wastewater treatment to remove persistent organic pollutants and contaminants that are difficult to degrade using conventional methods. AOPs involve the generation of highly reactive hydroxyl radicals ($HO^{\bullet}$) or other powerful oxidants to break down and oxidize organic compounds into simpler, less harmful substances [12]. Hydroxyl radicals have a high oxidation potential (2.80 V) and can react with a wide range of contaminants, including pharmaceuticals, pesticides, dyes, and ECs. These reactions typically involve the transfer of an electron from the contaminant to the hydroxyl radical, resulting in the degradation of the pollutant into smaller, more biodegradable molecules. Various oxidizing agents are generated during a typical oxidation process having enough oxidation potential, as illustrated in Figure 4.1.

The efficiency and effectiveness of AOPs depend on several factors, such as the type of oxidants used, reaction conditions, and the presence of catalysts or energy sources. Various AOPs have been developed and employed in wastewater treatment,

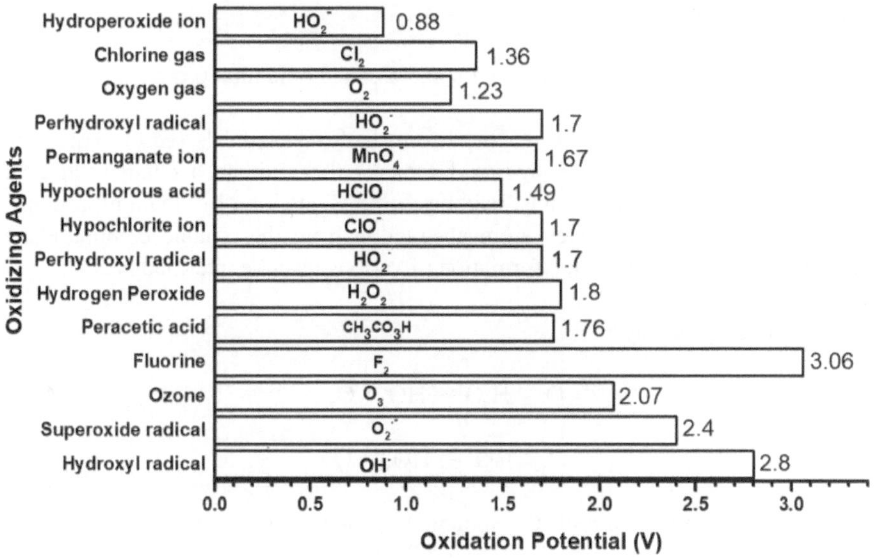

FIGURE 4.1 An illustration of various oxidizing agents generated during the advanced oxidation process along with respective oxidation potential.

including methods such as Ozonation, Fenton's reagent (a combination of Hydrogen Peroxide and Ferrous Iron) [13], Photocatalysis, and NTP [7,14–16]. Each technique has its unique mechanisms for generating hydroxyl radicals or other powerful oxidants, providing flexibility and adaptability to different wastewater compositions and treatment requirements. These processes are known for their versatility and efficiency in treating complex wastewater streams and have gained significant attention in recent years due to their potential to address emerging environmental challenges. AOPs offer enhanced treatment efficiency, flexibility in handling different wastewater compositions, and the potential for on-site treatment applications. These advanced techniques play a vital role in addressing the growing challenges of water pollution and ensuring the availability of clean and safe water resources.

4.4 OZONATION

Ozonation is an advanced oxidation technique that involves the use of ozone (O_3) to degrade and eliminate organic compounds in wastewater. It is a powerful and efficient method for the removal of various pollutants, including persistent organic compounds, taste and odor-causing substances, and micropollutants [17]. O_3 is a powerful oxidant (Oxidation Potential 2.07 V) that readily reacts with organic compounds, initiating a series of chemical reactions [18]. One of the primary reactions is direct O_3 oxidation, where O_3 molecules directly react with organic compounds, leading to the formation of intermediate products. This reaction involves the transfer of O_2 atoms from O_3 to the organic compound, resulting in the cleavage of carbon–carbon and carbon–hydrogen bonds.

Direct Ozone Attack:

$$O_3 + \text{Organic Compound} \rightarrow \text{Ozone-assisted oxidation products} \quad (4.1)$$

In the presence of water, O_3 reacts to form hydroxyl radicals ($HO^{\bullet}$), which are highly reactive species. These hydroxyl radicals then initiate a chain of radical reactions with organic compounds, leading to the formation of oxidation products. The reactions involve hydrogen atom abstraction from the organic compound by hydroxyl radicals, leading to the formation of various reactive intermediates.

Indirect Radical Pathways:

$$O_3 + H_2O \rightarrow HO^{\bullet} + O_2 \quad (4.2)$$

$$O_3 + HO^{\bullet} \rightarrow HO_2^{\bullet} + O_2 \quad (4.3)$$

$$HO_2^{\bullet} + \text{Organic Compound} \rightarrow \text{Oxidation Products} \quad (4.4)$$

$$HO^{\bullet} + \text{Organic Compound} \rightarrow \text{Oxidation Products} \quad (4.5)$$

The reaction of O_3 with organic compounds results in the formation of reactive intermediates, such as carbonyl compounds, aldehydes, and peroxides. These intermediates undergo further reactions and decomposition, ultimately leading to the mineralization of organic pollutants into carbon dioxide (CO_2), water (H_2O), and other simple, less harmful by-products.

The ozonation process, while effective in degrading organic molecules in wastewater, does have certain limitations that can hinder its complete removal of all pollutants. These limitations stem from factors such as ozone's stability and solubility in water, incomplete oxidation of organic compounds, mass transfer limitations, and low reaction kinetics [19]. One of the challenges with ozonation is the relatively low solubility of ozone in water, which affects its effectiveness in reaching all areas of wastewater. This can result in incomplete contact between ozone and organic molecules, limiting the overall degradation efficiency. Additionally, ozone's stability can be compromised over time, reducing its oxidative potential and leading to reduced degradation rates.

Furthermore, the reaction kinetics of ozone with organic compounds can vary, depending on factors such as the nature of the organic molecules and the reaction conditions. Some compounds may exhibit slower reaction rates, requiring longer contact times with ozone for complete degradation. This can pose challenges in large-scale wastewater treatment systems, where efficient reaction kinetics are crucial. Mass transfer limitations can also impact the efficiency of ozonation. In wastewater treatment processes, it is important to ensure proper mixing and distribution of ozone throughout the water matrix. Insufficient mass transfer can result in localized areas with limited ozone exposure, leading to incomplete degradation of organic pollutants. Enhancements in mass transfer mechanisms and the use of additives or catalysts can also improve the overall effectiveness of the ozonation process [17].

Homogeneous and heterogeneous catalytic ozonation processes are two variations of ozonation techniques that involve the use of catalysts to enhance the efficiency and effectiveness of the ozonation process [20]. In homogeneous catalytic ozonation, a soluble catalyst is added to the ozonation system, where it readily dissolves in the water phase. The catalyst participates directly in the ozone reactions, facilitating the degradation of organic compounds [21]. Common catalysts used in homogeneous catalytic ozonation include metal ions, such as iron (Fe^{2+}/Fe^{3+}), manganese (Mn^{2+} /Mn^{3+}), cobalt (Co^{2+}), silver (Ag^{+}), and cerium (Ce^{4+}) [22], as well as complexing agents like EDTA (Ethylenediaminetetraacetic Acid). The catalyst acts as a mediator, accelerating ozone decomposition and promoting the formation of more reactive hydroxyl radicals. This results in enhanced degradation rates and improved removal of organic pollutants. In heterogeneous catalytic ozonation, the catalyst is in a solid phase and remains separate from the water phase. Typically, the catalyst is supported on a solid material, such as activated carbon, zeolites, or metal oxides like TiO_2 (Titanium Dioxide) or Al_2O_3 (Alumina). The wastewater passes through or over the solid catalyst, and the catalyst's surface facilitates the decomposition of ozone and the generation of reactive species [23]. The adsorption and reaction mechanisms on the catalyst surface contribute to the degradation of organic compounds. Heterogeneous catalytic ozonation offers advantages such as easy catalyst separation, reuse of the catalyst, and improved stability compared to homogeneous catalytic systems [24].

Both homogeneous and heterogeneous catalytic ozonation processes offer potential advantages over conventional ozonation. The presence of a catalyst enhances the degradation kinetics, increases the formation of reactive species, and promotes the complete mineralization of organic compounds. The choice between homogeneous and heterogeneous catalytic ozonation depends on factors such as the nature of the wastewater, the desired treatment objectives, catalyst stability, ease of separation, and economic considerations.

4.5 FENTON'S REAGENT

Fenton's reagent is an innovative and powerful chemical treatment method used for the degradation of organic pollutants in wastewater. It involves the combination of Hydrogen Peroxide (H_2O_2) and ferrous iron (Fe^{2+}) under acidic conditions to generate highly reactive hydroxyl radicals ($HO^\bullet$) [25]. This process begins with the oxidation of Fe (II) by hydrogen peroxide, leading to the formation of Fe (III) and a hydroxyl radical. Simultaneously, hydrogen peroxide undergoes reduction, converting another Fe (III) ion back to Fe (II) and producing a hydroperoxyl radical ($HOO^\bullet$) and a proton [26]. As a result, Fenton's reagent facilitates the simultaneous presence of two distinct oxygen-radical species derived from hydrogen peroxide. This unique characteristic is crucial for the efficient degradation of complex organic compounds present in wastewater. These hydroxyl radicals exhibit strong oxidative properties, allowing for the efficient breakdown of complex organic compounds. The overall reaction mechanism can be summarized as follows:

$$Fe^{2+} + H_2O_2 \rightarrow Fe^{3+} + HO^\bullet + OH^- \quad (4.6)$$

$$Fe^{3+} + H_2O_2 \rightarrow Fe^{2+} + HOO^{\bullet} + H^+ \quad (4.7)$$

$$2\,H_2O_2 \rightarrow HO^{\bullet} + HOO^{\bullet} + H_2O \quad (4.8)$$

The process can be optimized by adjusting the dosage of H_2O_2 and Fe^{2+} to achieve the desired oxidation efficiency. Moreover, the reaction conditions, such as pH and temperature, can be controlled to enhance the degradation performance.

One of the key advantages of Fenton's reagent is its ability to efficiently degrade recalcitrant and persistent organic compounds that are typically found in textile wastewater [27]. It offers an innovative approach for treating wastewater streams that contain a complex mixture of dyes, chemicals, and organic pollutants. Fenton's reagent provides an effective solution to tackle these persistent pollutants and enhance the overall treatment efficiency [26]. Implementing Fenton's reagent for textile wastewater treatment also aligns with the principles of sustainability. Utilizing a chemical reaction that occurs under mild conditions, reduces the need for energy-intensive processes. Moreover, the use of inexpensive and readily available reagents, such as hydrogen peroxide and ferrous iron, makes Fenton's reagent a cost-effective and viable option for large-scale textile wastewater treatment facilities. Furthermore, the by-products of the Fenton reaction, such as ferric iron (Fe^{3+}) and water, are non-toxic and environmentally benign, minimizing the impact on the ecosystem.

4.6 NON-THERMAL PLASMA (NTP)

Non-Thermal Plasma (NTP), also known as cold plasma/non-equilibrium plasma, is a unique state of matter (partially or fully ionized gas state) that exists at relatively low temperatures. Unlike thermal plasmas, which are characterized by high temperatures and ionization of gas, NTP operates at near-room temperatures and maintains a substantial population of neutral gas species. It consists of a collection of highly energetic electrons, ions, excited atoms, and molecules that exhibit unique chemical and physical properties.

In NTP, the electric discharge causes collisions between electrons and gas molecules, leading to the formation of a plasma cloud, as shown in Figure 4.2a. This plasma cloud consists of various energetic species, including electrons, ions, and radicals, but the overall gas temperature remains relatively low [28]. This can be achieved through various methods, such as Dielectric Barrier Discharge (DBD) [29], Corona Discharge (CD) [30], or Gliding Arc Discharge (GAD) [31]. The electrons in the plasma possess high energy and can collide with gas molecules, causing dissociation, excitation, and ionization [32]. These processes generate a range of reactive species, such as ozone, hydrogen peroxide, hydroxyl radicals ($HO^{\bullet}$), and atomic oxygen (O), which have strong oxidizing properties [33]. These techniques create a highly reactive environment with a combination of electric fields, charge carriers, and gas molecules.

The reactive species produced in NTP play a crucial role in various applications, including air and water purification, surface treatment, and pollutant degradation. In wastewater treatment, NTP is employed as an advanced oxidation process (AOP) to effectively degrade and remove various organic pollutants. When the plasma is brought into contact with wastewater, the reactive species generated in the plasma chamber react with organic compounds, breaking them down into smaller, less harmful molecules, as illustrated in Figure 4.2b. It provides an efficient means of breaking down complex organic compounds that are recalcitrant to conventional treatment methods, without the need for chemical additives [29]. The highly reactive species in NTP can directly attack and degrade organic pollutants, transforming them into simpler, less harmful by-products. Furthermore, NTP has the potential for the inactivation of pathogens and the destruction of toxic compounds, making it valuable for water disinfection applications. It can be used in the treatment of industrial wastewater, municipal wastewater, and even drinking water to ensure the removal of harmful contaminants [34].

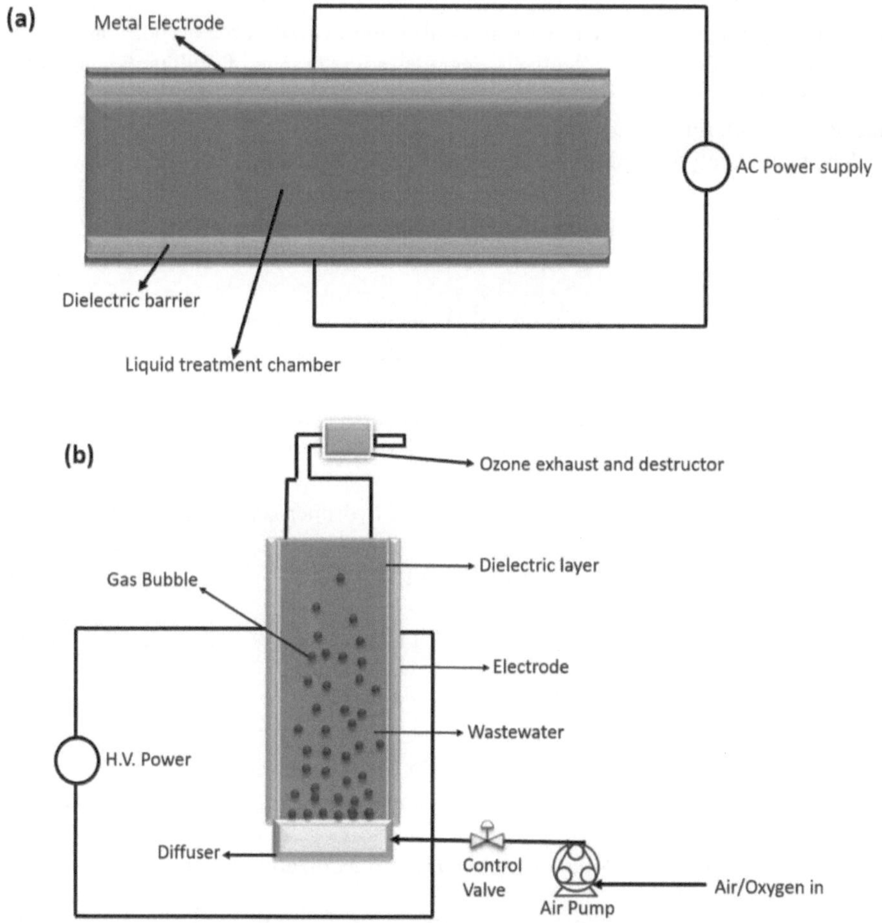

FIGURE 4.2 (a) Schematic representation of the Non-Thermal Plasma (NTP) process illustration; (b) NTP reactor specifically designed for efficient wastewater treatment.

NTP technology continues to be an area of active research and development, with ongoing efforts to optimize plasma reactors, improve energy efficiency, and explore its application in different wastewater treatment scenarios, as highlighted in Table 4.1. With its unique capabilities, NTP presents a promising and innovative solution for efficient and sustainable wastewater treatment.

4.7 PHOTOCATALYSIS

Photocatalysis is an innovative, sustainable advanced oxidation technique that utilizes light energy and a catalyst to degrade and remove organic pollutants in wastewater. Specifically, photocatalysis involves the use of a photocatalyst, typically a semiconductor material such as Titanium (IV) oxide (TiO_2), which typically absorbs UV light energy and creates electron-hole pairs. These photo-induced electron-hole pairs undergo redox reactions with water and oxygen present in the surrounding environment, generating highly reactive species such as hydroxyl radicals ($HO^{\bullet}$), superoxide radicals ($O_2^{\cdot-}$), and holes (h^+) that can oxidize and degrade organic pollutants [15]. The photocatalytic process effectively degrades a wide range of pollutants, including dyes, pesticides, pharmaceuticals, and organic contaminants, as well as some heavy metals and bacteria.

The photocatalytic oxidation mechanism can be summarized as given below [8]:

$$\text{Photoexcitation: } TiO_2 + h\nu \text{ (UV irradiation)} \rightarrow TiO_2 \text{ } (e^- \text{ (CB)} + h^+ \text{ (VB)}) \tag{4.9}$$

$$\text{Ionization of water: } H_2O \text{ (ads)} + h^+ \text{ (VB)} \rightarrow OH^{\bullet} \text{ (ads)} + H^+ \text{ (ads)} \tag{4.10}$$

$$\text{Ionosorption of Oxygen and superoxide radical generation: } O_2 + e^- \text{ (CB)} \rightarrow O_2^{\cdot\bullet} \text{ (ads)} \tag{4.11}$$

Superoxide protonation and formation of hydroperoxyl radical ($HO_2^{\bullet}$):

$$O_2^{\cdot\bullet} \text{ (ads)} + H^+ \leftrightarrow HOO^{\bullet} \text{ (ads)} \tag{4.12}$$

$$2HOO^{\bullet} \text{ (ads)} \rightarrow H_2O_2 \text{ (ads)} + O_2 \tag{4.13}$$

$$H_2O_2 \text{ (ads)} \rightarrow 2OH^{\bullet} \text{ (ads)} \tag{4.14}$$

$$\text{Organic Pollutants} + OH^{\bullet} \rightarrow CO_2 + H_2O \text{ (Formation of intermediate compounds)} \tag{4.15}$$

$$\text{Organic Pollutants} + h^+ \text{ (VB)} \rightarrow \text{Oxidized Product} \tag{4.16}$$

$$\text{Organic Pollutants} + e^- \text{ (CB)} \rightarrow \text{Reduced Product} \tag{4.17}$$

TABLE 4.1
Comprehensive Analysis of Non-Thermal Plasma (NTP) for the Abatement of Various Pollutants in Wastewater

S.N.	Organic Compound	Reactor	Operating Conditions	Working gas	Initial Concentration	Pollutant removal/ Mineralization	References
1.	Levofloxacin	Dielectric Barrier Discharge (DBD)	Voltage (V) = 10 kV, 18 min	Air	20 mg/L	63%, 11% per 18 min	[35]
2.	Thiamphenicol	Corona Discharge (CD)	V = 19 kV, 50Hz, 60 min	4 L/min Air	30 mg/L	27.5%, 27.5% per 60 min	[36]
3.	Chlorpyrifos	DBD	V = 80 kV, 2 min	Atmospheric Air	5 µg/mL	62.7%	[37]
4.	Diazinon	CD	150 W Power, 300 s	Air	100 mg/L	99.5%	[38]
5.	Lindane	DBD	220 V, 50 Hz, 14.8 min	Oxygen	1 mg/L	92.12%	[39]
6.	Crystal Violet	CD	2.5 W Power, 12 min	Ar	40.8 mg/L	100%	[40]
7.	Orange II	DBD	13.6 W, 5 min, 1kHz	Air	35 mg/L	99.9%	[41]
8.	Congo Red	CD	1.2 kV, 20 min, 16 Hz	Ar	200 mg/L	95%	[42]
9.	Acid Blue 25	Gliding arc discharge	15 kV, 60 min	Air	20 mg/L	78%	[43]

Source: Compiled by the authors.

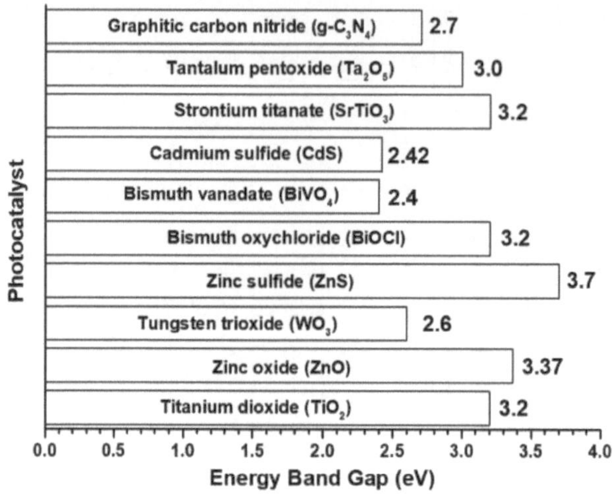

FIGURE 4.3 A typical illustration of various semiconductor-based photocatalysts along with their energy band gaps.

Some commonly used Photocatalysts along with their energy band gap are illustrated in Figure 4.3.

Furthermore, photocatalysis operates under mild reaction conditions, typically at room temperature and ambient pressure. This not only reduces energy consumption but also minimizes the formation of unwanted by-products or secondary pollutants. Additionally, photocatalytic systems can be easily integrated into existing wastewater treatment processes, providing a cost-effective solution for improving overall treatment efficiency. Another advantage of photocatalysis is its ability to degrade pollutants directly in aqueous solutions, eliminating the need for additional chemical reagents or the production of excessive sludge. This makes it a cleaner and more environmentally friendly approach to wastewater treatment [44].

Moreover, photocatalytic systems can be coupled with other treatment technologies, such as adsorption or membrane processes, to enhance the overall treatment performance [45]. By combining different techniques, synergistic effects can be achieved, resulting in improved pollutant removal efficiency and water quality. Therefore, photocatalysis is a promising technology for wastewater treatment due to its effectiveness in degrading organic pollutants, mild operating conditions, the potential for mineralization, and utilization of renewable energy sources [46]. Its environmentally friendly nature, versatility, and potential for integration with existing treatment processes make it an attractive option for sustainable and effective water purification.

4.8 COMPARATIVE ANALYSIS OF INNOVATIVE OXIDATION TECHNOLOGIES

The comparative analysis of innovative oxidation technologies provides insights into their operational principles, advantages, and limitations for wastewater treatment.

Fenton's reagent offers effective pollutant degradation but needs additional chemical additions and sludge management. NTP provides efficient degradation and disinfection but poses serious challenges in scalability and ozone by-product control. Ozonation exhibits strong oxidizing power but has limitations in solubility and stability. Photocatalysis has broad applicability and environmental friendliness, but its efficiency depends on the choice of photocatalyst and light source. Understanding these factors will improve the selection of the superior oxidation technology as per the particular wastewater treatment (see Table 4.2).

4.9 FUTURE PROSPECTS AND CHALLENGES OF INNOVATIVE OXIDATION TECHNOLOGIES

Innovative oxidation technologies, such as Ozonation, Fenton's reagent, NTP, and Photocatalysis, have been adopted nowadays for advanced wastewater treatment. These technologies impose an enhanced pollutant degradation potential compared to conventional methods. However, their widespread implementation faces both prospects and challenges that need to be considered for their successful integration into practical applications.

4.9.1 Ozonation

Ozonation has shown significant potential in the removal of recalcitrant organic pollutants. Its prospects lie in its ability to effectively degrade a wide range of organic compounds and ECs, including pharmaceuticals and personal care products. However, the integration of ozonation with other treatment processes, such as biological treatment or activated carbon adsorption, can lead to synergistic effects and improved overall treatment efficiency. Keynote challenges are the massive energy consumption during the process, the formation of toxic intermediate compounds, the optimization of reactor design, and efficient ozone utilization.

4.9.2 Fenton's Reagent

Fenton's reagent, which utilizes a combination of H_2O_2 and Fe^{2+} ions, exhibits an excellent oxidation potential for the degradation of organic pollutants. Its prospects lie in the development of novel catalysts or modifying the existing Fenton systems to enhance efficiency and reduce the consumption of chemicals. Additionally, the integration of Fenton's reagent with other treatment methods, such as ultrasound or photocatalysis, leads to an improved mineralization of the recalcitrant pollutant. Severe challenges include the optimization of Fenton's reagent dosage, control of pH and reaction conditions, and the management of the generated sludge containing iron precipitates.

4.9.3 NTP

Nowadays, NTP has become a desirable innovative oxidation technology for wastewater treatment. Its prospects lie in its ability to effectively mineralize persistent

TABLE 4.2
Comparative Analysis of Innovative Oxidation Technologies for Wastewater Treatment

S.N.	Oxidation Technology	Principle of Operation	Advantages	Limitations
1.	Fenton's Reagent	Involves the reaction between H_2O_2 and Fe^{2+} to produce $HO^{\bullet}$ that oxidize pollutants.	Effective for the degradation of various organic pollutants. Can operate under mild conditions.	Requires the addition of chemicals (H_2O_2 and Fe^{2+}). Generates sludge that needs proper disposal. pH dependency for optimal performance.
2.	Non-Thermal Plasma	Creates a highly reactive mixture of electrons, ions, excited species, and free radicals by applying high-voltage electrical discharges to a gas or gas mixture.	Efficient degradation of complex organic compounds. Can remove pathogens and microorganisms. Low energy consumption compared to thermal processes.	Limited scalability for large-scale wastewater treatment. Equipment complexity and maintenance requirements. Generation of ozone as a by-product, which needs to be controlled
3.	Ozonation	Involves the application of O_3 to oxidize and degrade pollutants.	Strong oxidizing power for the degradation of recalcitrant organic compounds. Decomposes into oxygen without leaving harmful residues. Can disinfect wastewater.	Limited solubility and stability in water. Requires appropriate ozone generation and dissolution systems. Limited efficiency for the degradation of certain pollutants.
4.	Photocatalysis	Utilizes a photocatalyst, activated by light energy, to generate electron-hole pairs that initiate redox reactions with adsorbed pollutants.	Broad applicability for the degradation of various organic and inorganic pollutants. Environmentally friendly and operates under mild conditions. Self-regeneration capability. Can be combined with other treatment methods.	Dependence on UV or visible light sources. The efficiency is influenced by the choice of photocatalyst and reaction conditions. Potential catalyst deactivation over time.

Source: Compiled by the authors.

organic pollutants and ECs, as well as for water disinfection. The development of scalable and energy-efficient NTP reactor design will enhance its wider implementation. Major process challenges comprise the optimization of operating conditions, efficient generation of reactive species, and the potential for electrode fouling and reactor scaling-up.

4.9.4 Photocatalysis

Photocatalysis has mammoth potential for the degradation of organic pollutants in wastewater using a light energy source and a suitable photocatalyst. Its prospects lie in the development of more efficient and stable photocatalysts, such as novel nanostructured materials and co-catalyst systems. Furthermore, the incorporation of visible light for photocatalysis leads an improved process applicability and becomes more energy efficient. Numerous challenges that govern photocatalysis include limited photocatalytic activity under natural sunlight, catalyst separation, recovery management, and photocatalyst fouling or deactivation control.

The prospects of these innovative oxidation technologies are crucial for advanced wastewater treatment. The continuous efforts focusing on improving efficiency, minimizing energy consumption, and addressing recent challenges will thrive on their successful implementation. Integration of these technologies with conventional treatment processes, that is, the development of hybrid approaches are more sustainable and efficacious solutions for water pollution control, as also contributes to the preservation of natural water resources and ensures environmental sustainability.

4.10 CONCLUSIONS

Innovative oxidation technologies have emerged as promising solutions for advanced wastewater treatment, offering enhanced capabilities in the removal of recalcitrant organic pollutants. Ozonation, Fenton's reagent, NTP, and photocatalysis have shown great potential in the degradation of various pollutants, including emerging contaminants, synthetic dyes, pharmaceuticals, and personal care products. These technologies have gained significant attention in recent years due to their effectiveness and compatibility with existing treatment processes. Ozonation has shown great efficacy in degrading a wide range of organic compounds, exhibiting numerous applicability in diverse wastewater treatment. Its combination with other treatment processes and optimization of reactor design enhanced the process efficiency. Fenton's reagent, with its powerful oxidative capabilities, offers effective degradation of pollutants, however, the development of novel catalysts and system modification will enhance the process efficiency and reduce chemical consumption.

NTP technology, with its ability to generate reactive species for pollutant degradation, become a sustainable and energy-efficient technique. Meanwhile, advancements in NTP reactor design and optimization of operating conditions will proliferate its practical applicability toward efficacious wastewater treatment. Photocatalysis on the other hand utilizes light energy and photocatalyst, alleviating the pollutant degradation potential. The development of more efficient and stable photocatalysts, along with the utilization of visible light, can proliferate its applicability for recalcitrant pollutant mineralization. The integration of these innovative oxidation technologies with conventional treatment processes and the development of hybrid approaches alleviate the integrated process efficiency and become a sustainable and cost-effective wastewater treatment solution. Continuous research and development efforts focusing on improving efficiency, minimal energy requirements, and addressing current challenges are essential for the successful implementation of these technologies.

REFERENCES

[1] Tiwari H, Sonwani RK, Singh RS. Bioremediation of dyes: A brief review of bioreactor performance. *Environ Technol Rev.* 2023;12:83–128. https://doi.org/10.1080/21622515.2023.2184276

[2] Tiwari H, Sonwani RK, Singh RS. Biodegradation and detoxification study of triphenylmethane dye (Brilliant Green) in a recirculating packed-bed bioreactor by bacterial consortium. *Environ Technol.* 2022;45: 1–42. https://doi/abs/10.1080/09593330.2022.2131469

[3] Chaturvedi A, Rai BN, Singh RS, et al. Comparative toxicity assessment using plant and luminescent bacterial assays after anaerobic treatments of dyeing wastewater in a recirculating fixed bed bioreactor. *J Environ Chem Eng.* 2021;9:1–9.

[4] Masood AS, Ali MS, Manzar MS, et al. Current situation of pharmaceutical wastewater around the globe. In *The Treatment of Pharmaceutical Wastewater. Innovative Technologies and the Adaptation of Treatment Systems.* Edited by Afzal Husain Khan, Nadeem A. Khan, Mu. Naushad, and Hamidi Abdul Aziz. 2023;19–52. Elsevier.

[5] Husain Khan A, Abdul Aziz H, Palaniandy P, et al. Pharmaceutical residues in the ecosystem: Antibiotic resistance, health impacts, and removal techniques. *Chemosphere.* 2023;339:139647.

[6] Tiwari H, Singh RS. Biotechnological approaches for microbial treatment of textile wastewater and resource recovery: Opportunities, challenges, and future perspectives. *Microb Technol Wastewater Recycl Manag.* 2022;269–279.

[7] Bilińska L, Gmurek M. Novel trends in AOPs for textile wastewater treatment. Enhanced dye by-products removal by catalytic and synergistic actions. *Water Resour Ind.* 2021;26:100160.

[8] Tiwari H, Sonwani RK, Singh RS. A comprehensive evaluation of the integrated photocatalytic-fixed bed bioreactor system for the treatment of Acid Blue 113 dye. *Bioresour Technol.* 2022;364:128037.

[9] Pirsaheb M, Moradi N. A systematic review of the sonophotocatalytic process for the decolorization of dyes in aqueous solution: Synergistic mechanisms, degradation pathways, and process optimization. *J Water Process Eng.* 2021;44:102314.

[10] Donkadokula NY, Kola AK, Naz I, et al. A review on advanced physico-chemical and biological textile dye wastewater treatment techniques. *Rev Environ Sci Biotechnol.* 2020;19:543–560.

[11] Raza N, Rizwan M, Mujtaba G. Bioremediation of real textile wastewater with a microalgal-bacterial consortium: An eco-friendly strategy. *Biomass Convers Biorefinery.* 2022;14:1–13.

[12] Feijoo S, Yu X, Kamali M, et al. Generation of oxidative radicals by advanced oxidation processes (AOPs) in wastewater treatment: A mechanistic, environmental and economic review. *Rev Environ Sci Bio/Technol.* 2023;22:205–248. https://10.1007/s11157-023-09645-4

[13] Khan NA, Khan AH, Tiwari P, et al. New insights into the integrated application of Fenton-based oxidation processes for the treatment of pharmaceutical wastewater. *J Water Process Eng*. 2021;44:102440. https://doi.org/10.1016/j.jwpe.2021.102440
[14] Liu L, Chen Z, Zhang J, et al. Treatment of industrial dye wastewater and pharmaceutical residue wastewater by advanced oxidation processes and its combination with nanocatalysts: A review. *J Water Process Eng*. 2021;42:102122.
[15] Junploy P, Janta R, Wongchai P, et al. Photodegradation of organic dyes and antibacterial activity of *Escherichia coli* and *Staphylococcus aureus* by ZnO nanoparticles under UVA radiation. *Mater Technol*. 2021;37:789–797.
[16] Hafeez A, Javed F, Fazal T, et al. Intensification of ozone generation and degradation of azo dye in non-thermal hybrid corona-DBD plasma micro-reactor. *Chem Eng Process – Process Intensif*. 2021;159:108205. https://doi.org/10.1016/j.cep.2020.108205
[17] Li X, Fu L, Chen F, et al. Application of heterogeneous catalytic ozonation in wastewater treatment: An overview. *Catalysts*. 2023;13:342.
[18] Epelle EI, Macfarlane A, Cusack M, et al. Ozone application in different industries: A review of recent developments. *Chem Eng J*. 2023;454:140188.
[19] Christian D, Gaekwad A, Dani H, et al. Recent techniques of textile industrial wastewater treatment: A review. *Mater Today Proc*. 2022;77:277–285.
[20] Manna M, Sen S. Advanced oxidation process: A sustainable technology for treating refractory organic compounds present in industrial wastewater. *Environ Sci Pollut Res Int*. 2022;30:25477–25505. https://10.1007/s11356-022-19435-0
[21] Yang Z, Yang H, Liu Y, et al. Heterogeneous catalytic ozonation for water treatment: Preparation and application of catalyst. *Ozone: Sci Eng*. 2023;45:147–173. https://doi.org/10.1080/01919512.2022.2050183
[22] Jin X, Wu C, Fu L, et al. Development, dilemma and potential strategies for the application of nanocatalysts in wastewater catalytic ozonation: A review. *J Environ Sci*. 2023;124:330–349.
[23] Chen W, He H, Liang J, et al. A comprehensive review on metal based active sites and their interaction with O_3 during heterogeneous catalytic ozonation process: Types, regulation and authentication. *J Hazard Mater*. 2023;443:130302.
[24] Mishra R, Pandey S, Fosso-Kankeu E. Role of heterogeneous catalysts for advanced oxidation process in wastewater treatment. In *Photoreactors in Advanced Oxidation Processes: The Future of Wasterwater Treatment*. Edited by Elvis Fosso-Kankeu, Sadanand Pandey, and Suprakas Sinha Ray. 2023;37–69. Wiley. https://doi/10.1002/9781394167289.ch2
[25] Bello MM, Abdul Raman AA, Asghar A. A review on approaches for addressing the limitations of Fenton oxidation for recalcitrant wastewater treatment. *Process Saf Environ Prot*. 2019;126:119–140.
[26] Kothai A, Sathishkumar C, Muthupriya R, et al. Experimental investigation of textile dyeing wastewater treatment using aluminium in electro coagulation process and Fenton's reagent in advanced oxidation process. *Mater Today Proc*. 2021;45:1411–1416.
[27] Ramos MDN, Santana CS, Velloso CCV, et al. A review on the treatment of textile industry effluents through Fenton processes. *Process Saf Environ Prot*. 2021;155:366–386.
[28] Coutinho NM, Silveira MR, Rocha RS, et al. Cold plasma processing of milk and dairy products. *Trends Food Sci Technol*. 2018;74:56–68.
[29] Kyere-Yeboah K, Bique IK, Qiao X-C. Advances of non-thermal plasma discharge technology in degrading recalcitrant wastewater pollutants. A comprehensive review. *Chemosphere*. 2023;320:138061.

[30] Naz MY, Shukrullah S, Rehman SU, et al. Optical characterization of non-thermal plasma jet energy carriers for effective catalytic processing of industrial wastewaters. *Sci Rep*. 2021;11:1–13. https://doi.org/10.1038/s41598-021-82019-4
[31] Gharagozalian M, Dorranian D, Ghoranneviss M. Water treatment by the AC gliding arc air plasma. *J Theor Appl Phys*. 2017;11:171–180.
[32] Misra NN, Pankaj SK, Segat A, et al. Cold plasma interactions with enzymes in foods and model systems. *Trends Food Sci Technol*. 2016;55:39–47.
[33] Li C, Rao Y, Zhang B, et al. Extraordinary catalysis induced by titanium foil cathode plasma for degradation of water pollutant. *Chemosphere*. 2019;214:341–348.
[34] Sanito RC, You SJ, Wang YF. Degradation of contaminants in plasma technology: An overview. *J Hazard Mater*. 2022;424:127390.
[35] Gong S, Sun Y, Zheng K, et al. Degradation of levofloxacin in aqueous solution by non-thermal plasma combined with Ag_3PO_4/activated carbon fibers: Mechanism and degradation pathways. *Sep Purif Technol*. 2020;250:117264. https:// /pii/S138358662031738X
[36] Guo H, Li Z, Xiang L, et al. Efficient removal of antibiotic thiamphenicol by pulsed discharge plasma coupled with complex catalysis using graphene-WO_3-Fe_3O_4 nanocomposites. *J Hazard Mater*. 2021;403:123673.
[37] Cong L, Huang M, Zhang J, et al. Effect of dielectric barrier discharge plasma on the degradation of malathion and chlorpyrifos on lettuce. *J Sci Food Agric*. 2021;101:424–432.
[38] Khan MSI, Lee SH, Kim YJ. A mechanistic and kinetic study of diazinone degradation under the influence of microplasma discharge water. *J Water Process Eng*. 2020;36:101310.
[39] Ansari M, Sharifian M, Farzadkia M. Removal of lindane in water by non-thermal plasma: Parametric optimization, kinetic study, energy yield evaluation, and toxicity assessment. *Sep Purif Technol*. 2022;299:121549.
[40] Taghvaei H, Kondeti VSSK, Bruggeman PJ. Decomposition of crystal violet by an atmospheric pressure RF plasma jet: The role of radicals, ozone, near-interfacial reactions and convective transport. *Plasma Chem Plasma Process*. 2019;39:729–749. https://doi.org/10.1007/s11090-019-09965-w
[41] Zhou R, Zhang T, Zhou R, et al. Underwater microplasma bubbles for efficient and simultaneous degradation of mixed dye pollutants. *Sci Total Environ*. 2021;750:142295.
[42] Attri P, Yusupov M, Park JH, et al. Mechanism and comparison of needle-type non-thermal direct and indirect atmospheric pressure plasma jets on the degradation of dyes. *Sci Rep*. 2016;6:1–14.
[43] Ghodbane H, Hamdaoui O, Vandamme J, et al. Degradation of AB25 dye in liquid medium by atmospheric pressure non-thermal plasma and plasma combination with photocatalyst TiO_2. *Open Chem*. 2015;13:325–331.
[44] Al-Mamun MR, Kader S, Islam MS, et al. Photocatalytic activity improvement and application of UV-TiO_2 photocatalysis in textile wastewater treatment: A review. *J Environ Chem Eng*. 2019;7:103248.
[45] Samsami S, Mohamadi M, Sarrafzadeh MH, et al. Recent advances in the treatment of dye-containing wastewater from textile industries: Overview and perspectives. *Process Saf Environ Prot*. 2020;143:138–163.
[46] Dihom HR, Al-Shaibani MM, Radin Mohamed RMS, et al. Photocatalytic degradation of disperse azo dyes in textile wastewater using green zinc oxide nanoparticles synthesized in plant extract: A critical review. *J Water Process Eng*. 2022;47:102705.

5 Innovative Treatment Technologies for Municipal Wastewater

Mohammad Vikas Ashraf, Shabir Ahmad Rather, M.A. Hannan Khan, and Shoeb Ahmad

5.1 INTRODUCTION

Innovative treatment technologies for municipal wastewater represent a pivotal shift in addressing the growing challenges of urban wastewater management. With the escalating demands of burgeoning populations and environmental concerns, traditional treatment methods are often inadequate in mitigating pollution and preserving water quality. However, emerging technologies offer promising solutions by integrating advanced processes, such as biological, physical, and chemical treatments, alongside cutting-edge techniques like membrane filtration, advanced oxidation, and resource recovery systems. These innovative approaches not only aim to enhance the efficiency of wastewater treatment but also strive to minimize environmental impacts, conserve resources, and promote sustainability. In this dynamic landscape of wastewater management, embracing these novel technologies heralds a transformative era in ensuring cleaner waterways and healthier communities.

5.1.1 Overview of the Importance of Municipal Wastewater Treatment

Municipal wastewater treatment plays a vital role in protecting public health, preserving the environment, and in maintaining sustainable management of water resources [1]. This overview provides a detailed account of the significance of municipal wastewater treatment, drawing upon scientific literature to underscore its importance.

5.1.1.1 Public Health Protection

A diverse range of contaminants are present in municipal wastewater, including nutrients, toxic chemicals, organic matter as well as pathogens, which pose a significant threat to public health, thus necessitating its treatment [2]. The pathogens include a wide range of bacteria, viruses, fungi as well as parasites, which have the ability to cause severe waterborne diseases, such as cholera, typhoid, gastroenteritis, leading to illness and even death [3]. Effective wastewater treatment processes, including

DOI: 10.1201/9781003454199-5

physical, chemical, and biological treatments, are essential for removing or inactivating these pathogens, thereby preventing the spread of waterborne diseases [3].

5.1.1.2 Environmental Conservation

Untreated or inadequately treated municipal sewage has detrimental effects on aquatic ecosystems and the environment [4]. Macronutrients, including nitrogen and phosphorus, present in wastewater, can lead to phenomena known as eutrophication of water bodies, causing algal blooms, increased BOD (Biological Oxygen demand), and fish kills [5]. Furthermore, organic matter and toxic pollutants can degrade water quality, impairing the health of aquatic organisms and disrupting ecological balance MISHRA 2023. By treating wastewater before discharge, municipalities mitigate these adverse impacts, preserving the health and integrity of aquatic habitats [6]

5.1.1.3 Protection of Water Resources

With growing population and urbanization, the demand for freshwater resources is increasing, placing strain on available water sources [7]. This makes it essential to treat the generated municipal wastewater, which allows judicious use of treated effluent for non-potable uses such as industrial processes, construction, agriculture and environmental restoration thus facilitates efficient management of available water resources [1]. This practice reduces the reliance on freshwater sources and helps in reducing concerns of water shortage, contributing to the sustainable use of water resources [8].

5.1.1.4 Compliance with Regulations

Order to protect public health and environment, regulatory agencies impose stringent standards and regulations governing the treatment and discharge of wastewater [9]. Municipalities are required to comply with these regulations to avoid legal repercussions and fines [10]. Effective wastewater treatment ensures that municipalities meet regulatory standards, demonstrating their commitment to environmental stewardship and regulatory compliance [6].

5.1.1.5 Socio-economic Benefits

Investments in wastewater infrastructure and treatment facilities yield socio-economic benefits for communities. Wastewater treatment supports economic activities such as agriculture, aquaculture, and tourism by maintaining water quality and environmental health [11]. Additionally, the wastewater sector creates employment opportunities, stimulates economic growth, and enhances the overall quality of life in urban areas [12].

Municipal wastewater treatment is essential to maintain water quality in order to protect public health and to preserve environment by complying with regulations thus promoting overall socio-economic development. By implementing effective wastewater treatment strategies and technologies, municipalities can achieve sustainable water management goals, thereby positively impacting both society and the environment. [1]

5.1.2 Brief History of Wastewater Treatment Methods

Throughout history, civilizations have grappled with the challenge of managing wastewater to protect public health and preserve the environment. Ancient civilizations, including Mesopotamia and Egypt, devised basic drainage systems to dispose of wastewater, while the Romans pioneered sophisticated sewage networks, such as the Cloaca Maxima in Rome, to transport wastewater away from populated areas [13]. However, during the Middle Ages, European cities faced sanitation crises, leading to outbreaks of waterborne diseases due to inadequate wastewater management [14]. The Industrial Revolution brought rapid urbanization and pollution, prompting the development of more advanced methods for treatment of waste water that includes screening and sedimentation [15]. During the latter part of the nineteenth century and the beginning of the twentieth century, the introduction of biological treatment processes, including the activated sludge process and trickling filters, revolutionized wastewater treatment by harnessing microbial activity to degrade organic pollutants [16]. The mid-twentieth century saw the adoption of secondary treatment processes and the implementation of regulatory standards for wastewater discharge, with the enactment of the Clean Water Act in the United States in 1972 driving investments in wastewater infrastructure [17]. Recent decades have witnessed advancements in tertiary treatment processes and the emergence of innovative technologies, such as advanced oxidation processes and membrane bioreactors, to address emerging contaminants and produce high-quality effluent [18]. The history of wastewater treatment methods underscores humanity's ongoing efforts to balance the need for sanitation with environmental sustainability and public health protection [1].

The need for innovative technologies in wastewater treatment is multifaceted and pressing [19]. With the global population rapidly increasing and urbanization intensifying, traditional treatment methods are facing unprecedented challenges in handling larger volumes of wastewater [20]. Moreover, emerging contaminants such as microplastics and pharmaceuticals present in wastewater pose significant environmental and health risks, necessitating advanced treatment technologies capable of removing these contaminants effectively [21]. Additionally, there is a growing recognition of the potential value in recovering resources from wastewater, including nutrients and energy, to enhance the sustainability of treatment processes and reduce environmental impact [12]. Furthermore, the energy-intensive nature of conventional treatment methods contributes to the emissions of gases that contribute to the greenhouse effect and the phenomenon of global climate change, underscoring the need for energy-efficient technologies and renewable energy integration in wastewater treatment facilities [22]. Aging infrastructure further compounds the challenges, highlighting the importance of innovative solutions such as smart sensors and predictive maintenance tools to optimize performance and minimize operational costs [23]. Finally, stringent regulatory standards for wastewater discharge necessitate continuous improvement and innovation in treatment technologies to ensure compliance and protect water quality [9]. In conclusion, addressing these complex and interconnected challenges requires ongoing research, development, and implementation of innovative technologies to achieve sustainable and resilient wastewater management practices.

5.2 CHALLENGES IN CONVENTIONAL WASTEWATER TREATMENT

5.2.1 Limitations of Conventional Methods of Wastewater Treatment

Conventional methods of wastewater treatment, while effective to a certain extent, have several limitations that hinder their ability to adequately address the complex composition of wastewater and emerging environmental challenges [24]. One significant limitation is the incomplete removal of emerging contaminants, which includes personal care items, microplastics, and pharmaceuticals, which are hard to separate and remove by conventional treatment [25]. Studies have shown that conventional water treatment methods, which includes sedimentation and activated sludge, fail to separate and eliminate these contaminants, leading to their release in the environment thus causing adverse impact on aquatic ecosystems and human health [26]. Furthermore, conventional treatment plants are typically designed to treat specific pollutants, such as nutrients and organic matter, and fail to remove other harmful pollutants – that includes heavy metals and pathogens – which can persist in treated effluent and pose risks to downstream ecosystems [27]. Additionally, conventional treatment processes are energy-intensive and require significant operational and maintenance costs, particularly for aeration and sludge handling, thus adding significantly to the overall carbon footprint [28]. Moreover, conventional treatment methods often produce large volumes of sludge as a byproduct, which must be disposed of or treated further, posing challenges in terms of environmental sustainability and resource management [29]. Finally, the aging infrastructure of many conventional treatment plants presents operational and reliability issues, requiring costly upgrades or replacements to maintain effectiveness and compliance with regulatory standards [30]. Therefore, while traditional methods for treating wastewater have been instrumental in protecting public health and the environment, their limitations underscore the need for continued research and innovation to develop more sustainable, efficient, and cost-effective methods that are capable of addressing the challenges of emerging pollutants along with mitigating the emerging environmental challenges.

5.2.2 Emerging Pollutants and Their Impact on Conventional Treatment Processes

Emerging pollutants pose significant challenges to conventional wastewater treatment processes due to their mobility, making their removal difficult, thus posing a threat to human health and the environment [31]. Pharmaceuticals and personal care products (PPCPs), such as antibiotics, hormones, and pain relievers, are continuously discharged into wastewater from various sources, including hospitals, households, and pharmaceutical manufacturing facilities [32]. These compounds are not often degraded by conventionally used treatment methods, such as activated sludge and sedimentation technologies, leading to their presence in treated effluent and receiving water bodies [33]. Studies have shown that exposure to trace levels of pharmaceuticals and PPCPs can disrupt endocrine systems, induce antibiotic resistance, and harm aquatic organisms, highlighting the importance of developing advanced treatment technologies capable of removing these emerging contaminants [34].

Industrial chemicals and flame retardants are another class of emerging pollutants that have become a cause of concern because of being widely used in manufacturing processes and potential environmental persistence [35]. Conventional treatment processes may not effectively remove these chemicals, leading to their accumulation in sludge and potential release into the environment through land application or incineration [36]. Pre- and Polyfluoalkyl substances (PFAS), in particular, have been associated with ill health effects such as immune system disorders, developmental delays and cancer, thus emphasizing the need for better treatment methods to eliminate these pollutants from wastewater [37].

Microplastics, with article diameter less than 5 millimetres, are known to be present ubiquitously in the environment. As a result, it poses severe negative effects on the ecosystem and to human health, thus have become a cause of major environmental concern worldwide [38]. Conventional treatment methods generally fail to efficiently remove microplastics from wastewater, allowing these particles to enter surface waters and marine environments [39]. Microplastics can adsorb and transport other contaminants, such as persistent organic pollutants and heavy metals, exacerbating their impacts on aquatic organisms and ecosystems [40].

Emerging pollutants present significant challenges to conventional wastewater treatment processes, requiring the development and application of advanced technologies of water treatment, which are capable of effectively removing these contaminants to safeguard public health and the environment [26].

5.2.3 Increasing Population and Urbanization Leading to Higher Volumes of Wastewater

The global population growth and rapid urbanization have resulted in significant challenges related to wastewater management, with higher volumes of wastewater being generated in urban areas [41]. As the population increases, particularly in urban centres, the demand for water and sanitation services escalates, resulting in a rise in the quantity of wastewater discharged into treatment systems [1]. Numerous studies have highlighted the correlation between urbanization and wastewater generation, demonstrating that urban areas tend to produce larger quantities of wastewater compared to rural areas [42]. This trend can be attributed to various factors associated with urban living, including higher levels of water consumption, increased industrial activities, and larger concentrations of commercial establishments and institutions [43].

Furthermore, urbanization is often accompanied by changes in lifestyle and consumption patterns, such as dietary preferences and personal hygiene practices, which can influence the composition and characteristics of wastewater [44]. For instance, the adoption of water-intensive diets and the use of water-intensive appliances, such as dishwashers and washing machines that add to water consumption, leading to overall increase in wastewater volume [45]. Furthermore, due to increase in urbanization, impermeable surfaces like pavements, roads, and buildings are added. This decreases natural infiltration and increases surface runoff, which increases the amount of wastewater produced during storm events [46]. The challenges posed by rising population leading to increased urbanization underscore the need for sustainable wastewater management strategies and infrastructure investments to accommodate the growing demand for water and sanitation services [47]. Studies have emphasized the importance of integrating decentralized wastewater treatment systems, water reuse practices, and smart water management technologies into urban planning and development initiatives to ensure the efficient use and treatment of wastewater [48]. Moreover, public awareness campaigns and education programs aimed at promoting water conservation and responsible wastewater disposal behaviours can help mitigate the adverse impacts of increasing population and urbanization on wastewater management [6].

The escalating volumes of wastewater associated with population growth and urbanization present significant challenges to wastewater management worldwide. Addressing these challenges requires collaborative efforts from policymakers, urban planners, researchers, and the public to develop innovative solutions and sustainable practices for effective and efficient wastewater treatment and judicious usage of available water resources.

5.3 EMERGING TECHNOLOGIES FOR THE TREATMENT OF MUNICIPAL WASTEWATER

Emerging technologies for the treatment of municipal wastewater are continuously evolving to address the increasing challenges posed by population growth, urbanization, and emerging contaminants. These technologies aim to improve treatment efficiency, reduce energy consumption, and to increase the efficiency of pollutant removal from wastewater [49]. Several innovative approaches have emerged in recent years, offering promising solutions for sustainable wastewater management. Here are some notable emerging technologies:

5.3.1 Membrane Bioreactors (MBRs)

Membrane Bioreactors (MBRs) (Figure 5.1) represent an innovative technology that integrates biological treatment processes with membrane filtration to achieve efficient wastewater treatment [50]. Compared to traditional treatment techniques, MBRs offer number of benefits, such as better effluent quality, lesser environmental impact, and increased treatment efficiency [51]. The core components of MBR systems consist of bioreactors, where microorganisms degrade organic pollutants, and membrane modules, which separate treated effluent from suspended solids and pathogens [52]. The biological treatment in MBRs relies on microbial communities, including bacteria, fungi, and protozoa, which metabolize organic matter and nutrients present in wastewater [53]. These microorganisms form biofilms on the surface of submerged membranes or suspended biomass within the bioreactor, facilitating the breakdown of organic contaminants through biological oxidation processes [54]. The continuous mixing and aeration in MBRs provide ideal conditions for microbial growth and activity which encourages the breakdown of complex organic molecules and the removal of contaminants [55]. The membrane filtration component of MBRs serves as a physical barrier to retain suspended solids, colloids, and microorganisms, allowing only treated effluent to pass through [56]. Membranes in MBR systems are typically made of polymeric materials, such as polyethylene, polypropylene, or polyvinylidene

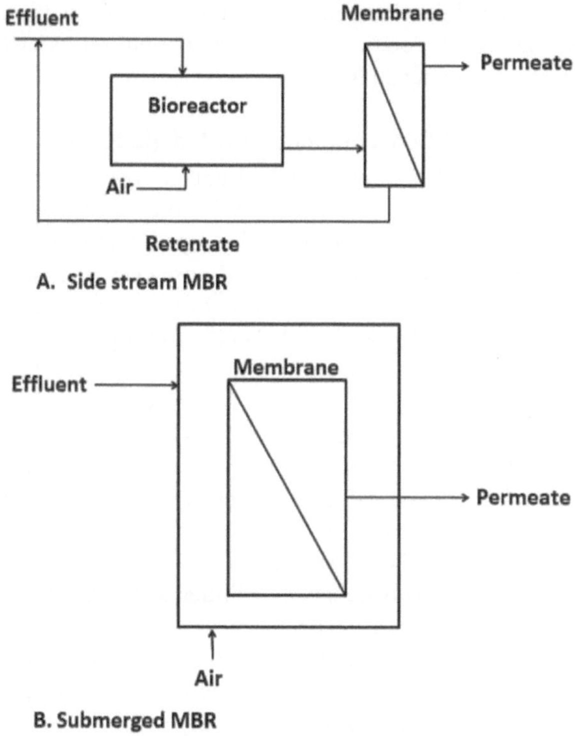

FIGURE 5.1 Types of membrane bioreactors. (Adapted from Ref. [56].)

fluoride, with pore sizes ranging from microfiltration to nanofiltration [52]. The membrane pore size determines the level of particle removal and defines the quality of the treated effluent.

Several factors influence the performance of MBRs, including membrane fouling, hydraulic conditions, and biomass characteristics. Membrane fouling, caused by the accumulation of organic matter, inorganic particles and microbial growth on the surface of the membrane, is a primary operational challenge in MBR systems [57]. Strategies to mitigate membrane fouling include membrane cleaning, backwashing, and the use of advanced membrane materials with antifouling properties [58].

Research on MBRs has focused on improving system design, optimizing operational parameters, and enhancing membrane performance to overcome operational challenges and maximize treatment efficiency. Recent advancements in MBR technology include the development of submerged membrane modules, membrane bioreactor configurations, and hybrid treatment processes [59]. Additionally, research efforts have explored the potential of MBRs for the management of wastewater, including community sewage, industrial effluents, and decentralized wastewater streams [30]. Membrane bioreactors (MBRs) are a promising technology that can be used to treat municipal wastewater because of its excellent effluent quality, small footprint, and high treatment efficiency [52].

Continued research and development efforts are necessary to enhance MBR efficacy, alleviate operational difficulties, and broaden the utilisation of MBRs for sustainable wastewater management. [50].

5.3.1.1 Advantages over Conventional Activated Sludge Systems

Membrane bioreactors (MBRs) offer several advantages over conventional activated sludge systems, making them an increasingly attractive option for wastewater treatment [59]. One significant advantage is the superior effluent quality achieved by MBRs. Studies have demonstrated that MBRs produce effluent with lower turbidity, suspended solids, and pathogens compared to conventional activated sludge systems [52]. The membrane barrier in MBRs effectively retains particulate matter and microorganisms, thereby producing high-quality treated discharge appropriate for either use or release into sensitive receiving waters [60].

Another advantage of MBRs is their smaller footprint relative to conventional treatment systems. By a concerted effort of biological treatment processes with membrane filtration technology, MBRs eliminate the need for secondary settling tanks, clarifiers, and tertiary filtration units required in conventional activated sludge systems [59]. This compact design allows MBRs to be installed in densely populated urban areas or retrofit into existing treatment facilities with limited space [61]. MBRs also offer enhanced process control and operational flexibility compared to traditional treatment methods. The continuous membrane filtration in MBRs maintains stable biomass concentrations and hydraulic conditions, minimizing fluctuations in treatment performance and effluent quality [50]. Additionally, MBRs can adapt to varying influent characteristics and loadings, making them suitable for treating wastewater with fluctuating flow rates, organic loads, and pollutant concentrations [59].

Furthermore, MBRs provide opportunities for advanced treatment and resource recovery, which are not feasible with conventional activated sludge systems. The membrane barrier in MBRs effectively removes trace contaminants, including microplastics, personal care products, and pharmaceuticals, which may pass through conventional treatment processes [62]. Additionally, MBRs make it possible to recover important resources from wastewater, like water and nutrients, by using membrane filtration and concentration techniques [50].

Compared to traditional activated sludge systems, Membrane Bioreactors (MBRs) have a number of benefits, including as smaller footprints, improved process control, better effluent quality, and potential for resource recovery and advanced treatment. These advantages make MBRs an attractive and sustainable option for wastewater treatment in various applications, from municipal sewage treatment to industrial wastewater recycling [52].

5.3.1.2 Applications and Case Studies

Membrane bioreactors (MBRs) have found extensive use in public and industrial sewage management owing to their high-quality effluent, small space requirements, and versatile operation capabilities [57]. Here are some key applications along with case studies showcasing their effectiveness:

5.3.1.2.1 *Municipal Wastewater Treatment*

MBRs are extensively used for municipal wastewater treatment in urban areas, where space constraints and stringent discharge standards are common [63]. One notable case study is the Kallang MBR Plant in Singapore, which is one of the largest MBR plants in the world. The Kallang MBR Plant treats approximately 80 million liters of municipal wastewater per day, producing high-quality effluent suitable for non-potable reuse [59]. The plant's compact footprint and robust performance have made it a model for sustainable urban water management [64].

5.3.1.2.2 *Industrial Wastewater Treatment*

MBRs are also applied in various industries, including food and beverage, pharmaceuticals, and petrochemicals, for treating complex industrial effluents. An exemplary case is the application of MBRs for textile wastewater treatment [65]. In another study, a pilot-scale MBR was employed to treat wastewater from a textile dyeing plant. The MBR effectively removed colour, organic pollutants, and total suspended solids from the textile wastewater, producing treated effluent compliant with regulatory standards. The study demonstrated the feasibility of using MBRs for sustainable treatment of textile industry effluents [65].

5.3.1.2.3 *Decentralized Wastewater Treatment*

MBRs are increasingly used for decentralized wastewater treatment in rural and remote areas, where centralized sewerage infrastructure is unavailable. One notable example is the Eco-Valley MBR Project in Malaysia, which utilizes decentralized MBR systems to treat wastewater from residential and commercial developments in a remote valley region. The decentralized MBRs provide reliable and cost-effective wastewater treatment solutions, minimizing environmental impacts and preserving water resources in the valley [48].

5.3.1.2.4 *Water Reuse and Recycling*

MBRs play a vital role in water reuse and recycling initiatives by producing high-quality reclaimed water for various non-potable applications, such as industrial processes, environmental restoration and irrigation [66]. A case study from Australia demonstrates the successful implementation of MBRs for water recycling in a residential community. The Western Corridor Recycled Water Scheme in Queensland utilizes MBR technology to treat secondary effluent from wastewater treatment plants, producing reclaimed water for landscape irrigation, toilet flushing, and industrial use [67]. The scheme has significantly reduced the demand for potable water and alleviated pressure on freshwater resources in the region.

Membrane Bioreactors (MBRs) offer versatile solutions for wastewater treatment in municipal, industrial, decentralized, and water reuse applications. Case studies from around the world demonstrate the effectiveness of MBRs in producing high-quality effluent, reducing environmental impact, and promoting sustainable water management practices [52].

5.3.2 Advanced Oxidation Processes (AOPs)

Advanced Oxidation Processes (AOPs), as depicted in Figure 5.2, encompass a diverse group of chemical treatment techniques that utilize highly reactive oxidants to degrade and remove organic and inorganic pollutants from wastewater [68]. AOPs offer a robust and versatile approach to wastewater treatment, capable of efficiently oxidizing a vast array of recalcitrant pollutants, such as pharmaceuticals, dyes, industrial chemicals, and pesticides [69]. AOPs typically involve the generation of powerful oxidizing agents, such as hydroxyl radicals (•OH), ozone (O_3), hydrogen peroxide (H_2O_2), and ultraviolet (UV) radiation, which react with organic pollutants present in wastewater to initiate degradation processes [70]. These oxidants produce extremely reactive substances like hydroxyl radicals, singlet oxygen, and superoxide radicals, which attack and break down organic molecules through oxidation-reduction reactions [71].

5.3.2.1 Mechanisms of AOPs

5.3.2.1.1 Hydroxyl Radical (•OH) Generation

Hydroxyl radicals (•OH) are the primary oxidizing species responsible for pollutant degradation in AOPs. The generation of hydroxyl radicals can occur through various mechanisms, including:

- Photolysis of hydrogen peroxide (H_2O_2) or ozone (O_3) under the influence of UV radiation and subsequent formation of hydroxyl radicals [72].
- Fenton and photo-Fenton reactions, where ferrous ions (Fe^{2+}) catalyse the breakdown of hydrogen peroxide (H_2O_2) to produce hydroxyl radicals (•OH) [73].

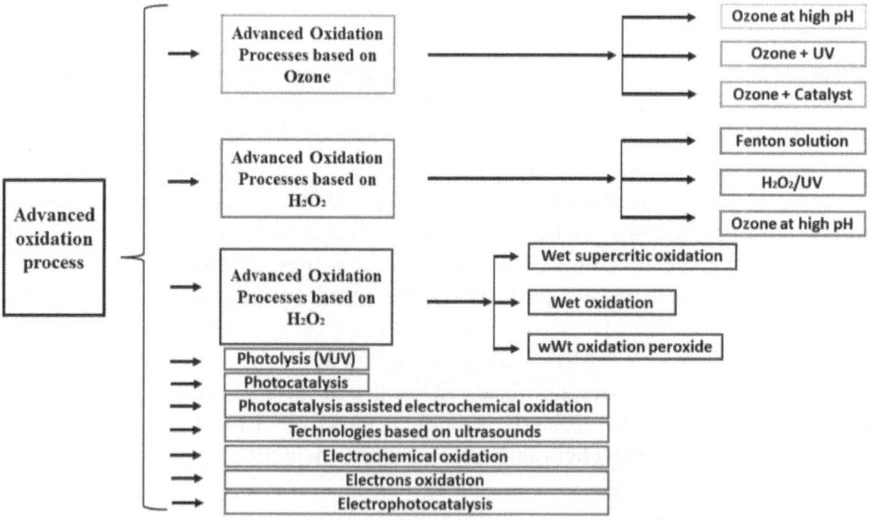

FIGURE 5.2 Classification of advanced oxidation processes. (Adapted from Ref. [42].)

- Ozonation, where ozone (O_3) reacts with water molecules to generate hydroxyl radicals (•OH) through the decomposition of ozone (O_3) into hydroperoxyl radicals ($HO_2^•$) and superoxide radicals ($•O_2^{-}$) [74].

5.3.2.1.2 Reaction with Organic Pollutants

Hydroxyl radicals (•OH) generated in AOPs react rapidly with organic pollutants present in wastewater, initiating oxidation reactions that degrade complex organic molecules into smaller, less toxic byproducts [70]. The hydroxyl radical attacks organic compounds at multiple sites, leading to the formation of reactive intermediates, such as carbonyl compounds, carboxylic acids, and hydroxylated derivatives [75]. These reactive intermediates undergo further degradation through successive oxidation reactions until complete mineralization to carbon dioxide (CO_2) and water (H_2O) is achieved [76].

5.3.2.1.3 Mineralization of Pollutants

AOPs promote the mineralization of organic pollutants by converting them into inorganic compounds, such as mineral salts, carbon dioxide and water. The oxidation of organic contaminants by hydroxyl radicals (•OH) results in the cleavage of carbon-carbon and carbon-heteroatom bonds, leading to the formation of simpler, more stable organic compounds [78]. Ultimately, the mineralization of pollutants to inorganic end-products ensures the removal of contaminants and reduces environmental impact [69].

Advanced Oxidation Processes (AOPs) present a powerful and effective method of treating wastewater by utilizing highly reactive oxidants to degrade and remove organic and inorganic pollutants. The mechanisms entail the production of hydroxyl radicals (•OH) along with various other reactive species, which initiate oxidation reactions leading to the mineralization of contaminants [70].

5.3.2.2 Examples of AOPs Such as Ozonation, UV Irradiation, and Photocatalysis

Advanced Oxidation Processes (AOPs) encompass a variety of treatment techniques that utilize highly reactive oxidants to degrade inorganic and organic pollutants in wastewater. Three prominent examples of AOPs are ozonation, UV irradiation, and photocatalysis, each offering unique mechanisms for pollutant removal [70].

5.3.2.2.1 Ozonation

Ozonation comprises the use of ozone (O_3), which is a strong oxidizing agent, to degrade organic pollutants present in wastewater. On-site ozone generators produce ozone, which is subsequently injected into the wastewater stream to react with organic pollutants via free radical and direct oxidation processes[79]. Ozonation results in the creation of hydroxyl radicals (•OH) and additional reactive oxygen species, which initiate oxidation reactions that degrade organic molecules into simpler, less toxic byproducts [77]. Another study investigated the usage of ozonation for the elimination of pharmaceutical compounds from wastewater. The researchers demonstrated that ozonation effectively degraded a variety of pharmaceuticals, including antibiotics, analgesics, and hormones, achieving significant removal efficiencies and reducing the environmental impact of pharmaceutical contamination in water bodies [32].

5.3.2.2.2 UV Irradiation

UV irradiation involves the exposure of wastewater to the ultraviolet spectrum, to generate hydroxyl radicals (•OH) through photolysis of water molecules [80]. UV radiation may be employed independently or in conjunction with hydrogen peroxide (UV/H_2O_2) to augment the production of hydroxyl radicals and improve treatment efficiency [81]. UV irradiation disrupts the chemical structure of organic contaminants, leading to the generation of reactive intermediates and ultimately mineralizing contaminants to carbon dioxide (CO_2) and water (H_2O) [76]. Another study investigated the application of UV/H_2O_2 for the degradation of phenol in wastewater. The researchers demonstrated that UV irradiation combined with hydrogen peroxide effectively degraded phenol and its intermediates, achieving complete mineralization and reducing the toxicity of the treated effluent. The study demonstrated how effectively UV-based AOPs remove persistent organic contaminants from wastewater [82].

5.3.2.2.3 Photocatalysis

Photocatalysis, as shown in Figure 5.3, involves the use of semiconductor materials, such as titanium dioxide (TiO_2), as catalysts to generate hydroxyl radicals (•OH) under UV or visible light irradiation [83]. When illuminated, semiconductor photocatalysts absorb photons and generate electron-hole pairs, which migrate to the surface and react with water and oxygen molecules to produce hydroxyl radicals (•OH) and other reactive species [76]. Photocatalysis facilitates the breakdown of organic pollutants through oxidative processes, leading to the mineralization of contaminants to harmless end-products [76]. Research conducted by Zhang and colleagues in 2020 examined

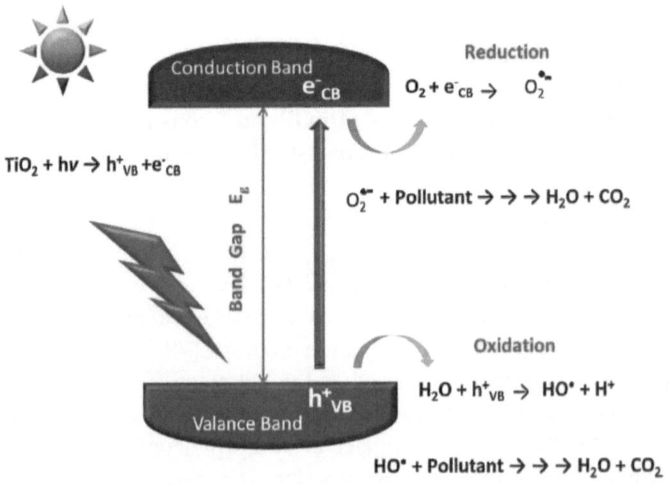

FIGURE 5.3 Indicative mechanism of photocatalysis on TiO_2. (Adapted from Ref. [84].)

the use of TiO_2 photocatalysis for removing emerging contaminants from effluents. The researchers demonstrated that TiO_2 photocatalysis effectively degraded a variety of pharmaceuticals and personal care products, achieving high removal efficiencies and reducing the environmental influence of these pollutants in aquatic ecosystems. This study highlighted the usage of photocatalysis for sustainable treatment technology of emerging pollutants [5].

In conclusion, advanced oxidation processes (AOPs) such as ozonation, UV irradiation, and photocatalysis provide practical and efficient ways to treat wastewater. These AOPs utilize highly reactive oxidants to degrade organic pollutants and promote the removal of contaminants, contributing to sustainable water management practices.

5.3.2.3 Removal of Emerging Contaminants Using AOPs

Emerging contaminants, including industrial chemicals, pesticides, pharmaceuticals and personal care products pose significant challenges to water quality and environmental health due to their persistence, bioaccumulation potential, and ill effects on aquatic ecosystems and human health [85]. Advanced Oxidation Processes (AOPs) have appeared as active measures for the degradation of emerging contaminants from wastewater, offering efficient degradation and mineralization of these pollutants [86]. AOPs utilize extremely reactive oxidants, such as hydroxyl radicals (•OH), ozone (O_3), and UV irradiation, to degrade and transform emerging contaminants into harmless end-products through oxidation-reduction reactions [70]. Hydroxyl radicals (•OH), generated through various AOPs, attack the chemical bonds of organic pollutants, which result in the decomposition of intricate molecules into less complex and less harmful byproducts [87]. The oxidation of emerging contaminants by AOPs results in the mineralization of pollutants to carbon dioxide (CO_2) inorganic ions and water (H_2O) thus ensuring their complete removal from water matrices [77].

5.3.2.4 Applications of AOPs for the Removal of the Following:

5.3.2.4.1 Pharmaceuticals

AOPs, such as photocatalysis and ozonation, have been successfully used for the elimination of pharmaceuticals from wastewater. Studies have demonstrated the efficacy of AOPs in degrading various classes of pharmaceuticals, including antibiotics, analgesics, hormones, and antidepressants, achieving high removal efficiencies and reducing the environmental impact of pharmaceutical contamination [88]. Another study investigated the breakdown of carbamazepine, a common pharmaceutical contaminant, using ozonation and UV irradiation. The researchers demonstrated that both AOPs effectively degraded carbamazepine, with ozonation achieving higher removal efficiencies compared to UV irradiation alone [89]. Another study examined the photocatalytic degradation of triclosan, a widely used antimicrobial agent, using TiO_2 photocatalysis. The researchers found that TiO_2 photocatalysis effectively degraded triclosan, achieving complete mineralization and reducing its toxicity. The study demonstrated the applicability of photocatalysis for the exclusion of personal care products from wastewater [90].

5.3.2.4.2 Personal Care Products (PCPs)

AOPs offer effective solutions for the elimination of PCPs, such as fragrances, UV filters, and antimicrobial drugs from wastewater. UV irradiation combined with hydrogen peroxide (UV/H_2O_2) and photocatalysis using semiconductor catalysts has been employed to degrade and mineralize personal care products, achieving a significant reduction in their concentrations and mitigating their ecological risks [91].

5.3.2.4.3 Pesticides and Industrial Chemicals

AOPs perform a vital role in the removal of pesticides and industrial chemicals from wastewater, offering efficient degradation and detoxification of these contaminants. Ozonation and UV/H_2O_2 have been used for the treatment of pesticide-contaminated wastewater, leading to the degradation of agricultural effluents and the elimination of their toxic effects on aquatic organisms [86].

The AOPs (Advanced Oxidation Processes) offer successful and sustainable solutions for the degradation of emerging contaminants from sewage, including pharmaceuticals, personal care products, pesticides, and industrial chemicals. AOPs utilize highly reactive oxidants to degrade and mineralize pollutants, ensuring the protection of water resources and environmental health.

5.3.3 Constructed Wetlands

Constructed wetlands (CWs) are human-made constructions intended to replicate the purpose of naturally occurring wetlands [92]. They utilize the complex interactions among plants, microorganisms, and substances to eliminate pollutants and enhance the quality of water. Constructed wetlands (CWs) have been acknowledged as economical and environmentally friendly substitutes for traditional wastewater treatment approaches. They are recognized for their capacity to deliver effective treatment while also providing habitat and promoting biodiversity [92].

Constructed wetlands typically consist of shallow, marshy areas filled with water-tolerant plants, such as cattails, reeds, and rushes, along with a substrate layer (e.g., gravel, sand) for supporting plant growth and microbial activity [93]. The wastewater flows through the wetland, where physical, chemical, and biological processes occur to remove pollutants. These constructed wetlands, as shown in Figure 5.4, can further be classified based on flow patterns. These includes horizontal flow (HF), vertical flow (VF), and hybrid systems combining both flow types [94].

5.3.3.1 Treatment Mechanisms

5.3.3.1.1 Physical Filtration

When wastewater flows through CWs, the substrate layer serves as a physical filter, capturing particle debris and suspended solids. This procedure improves water clarity and aids in the elimination of contaminants [95].

5.3.3.1.2 Biological Processes

Microorganisms present in the substrate and rhizosphere of wetland plants play a crucial role in the biodegradation and transformation of organic contaminants.

FIGURE 5.4 Constructed wetlands: A. Surface flow wetlands [*Eichhornia crassipes* (commonly known as water hyacinth)] B. Constructed wetland to treat runoff water [*Lemna* spp. (duckweed) and *Taxodium distichum* (baldcypress)] C. Constructed wetland established with *Nuphar lutea* (spatterdock), planned for tertiary treatment of 800,000 PE D. Surface flow CW [Macrophytes (mostly *Myriophyllum spicatum*, water milfoil)] E. Floating wetland established with *Cyperus alternifolius* F. surface flow CW [Mangroves (*Kandelia candel*)] for treating mariculture effluent. (Adapted from Ref. [95].)

These microorganisms break down organic matter through processes such as aerobic and anaerobic decomposition, nitrification, denitrification, and phosphorus uptake [96].

5.3.3.1.3 Plant Uptake

Wetland plants absorb nutrients and contaminants from the water column through their roots, thereby contributing to nutrient removal and pollutant uptake. This process, known as phytoremediation, enhances the overall treatment performance of CWs [97].

Constructed wetlands have been successfully applied for treating a wide array of effluents, such as municipal effluent, agricultural runoff, commercial effluents, and stormwater runoff [92]. Several studies have demonstrated the efficiency of CWs in eliminating contaminants such organic matter, suspended particles, heavy metals, nutrients (phosphorus and nitrogen), and pathogens [98]. CWs have also been shown to provide additional ecosystem services, including habitat creation, biodiversity enhancement, and aesthetic improvement [99].

Further progress on constructed wetlands continues to focus on optimizing design parameters, improving treatment efficiency, and expanding application areas [92]. Advances in modelling, monitoring, and operation and maintenance practices are also contributing to the advancement of CW technology, making it a viable and sustainable option for decentralized wastewater treatment and ecological restoration efforts [98].

Constructed wetlands offer a natural solution for wastewater management, utilizing natural processes to remove contaminants and improve water quality. With their ability to provide effective treatment, habitat creation, and ecosystem services, CWs represent a promising approach for sustainable water management and environmental conservation.

5.3.4 Anaerobic Digestion

Anaerobic digestion (AD) is a biological process that degrades organic matter in the absence of oxygen. It breaks down complex organic compounds primarily into methane (CH_4) and carbon dioxide (CO_2), along with trace amounts of other gases [100]. AD occurs naturally in anaerobic environments such as wetlands, marshes, and the digestive systems of animals. In engineered systems, AD is utilized for organic waste treatment and biogas production, offering environmental and energy benefits. There are four primary phases of anaerobic digestion: hydrolysis, acidogenesis, acetogenesis and methanogenesis [101]. In the hydrolysis stage, complex organic compounds such as carbohydrates, lipids and proteins are broken down by various enzymes into simple molecules, including sugars, fatty acids and amino acids. These simple soluble compounds are then fermented by acidogenic bacteria in the acidogenesis stage, producing volatile fatty acids (VFAs), carbon dioxide (CO_2) and hydrogen (H_2) [102]. This process of anaerobic waste water treatment is illustrated in Figure 5.5.

In the acetogenesis stage, acetogenic bacteria utilizes VFAs, H_2, and CO_2 to produce acetic acid, carbon dioxide and hydrogen. Finally, in the methanogenesis stage, methanogenic archaea convert acetic acid, carbon dioxide and hydrogen into methane (CH_4) and carbon dioxide (CO_2). The overall process results in the production of biogas, which is a mixture of methane (CH_4) and carbon dioxide (CO_2), that can be utilized as a renewable energy [104].

Biogas produced during anaerobic digestion primarily contains methane (CH_4) (approximately 50–70%) and carbon dioxide (CO_2) (30–50%). Some trace amount of other gases such as hydrogen sulfide (H_2S) and ammonia (NH_3) are also produced [105]. Biogas is a renewable energy source that can be used to fuel vehicles, produce heat, and generate electricity. Biogas can be used in internal combustion engines or gas turbines to generate electricity and heat, thereby offering an inexhaustible

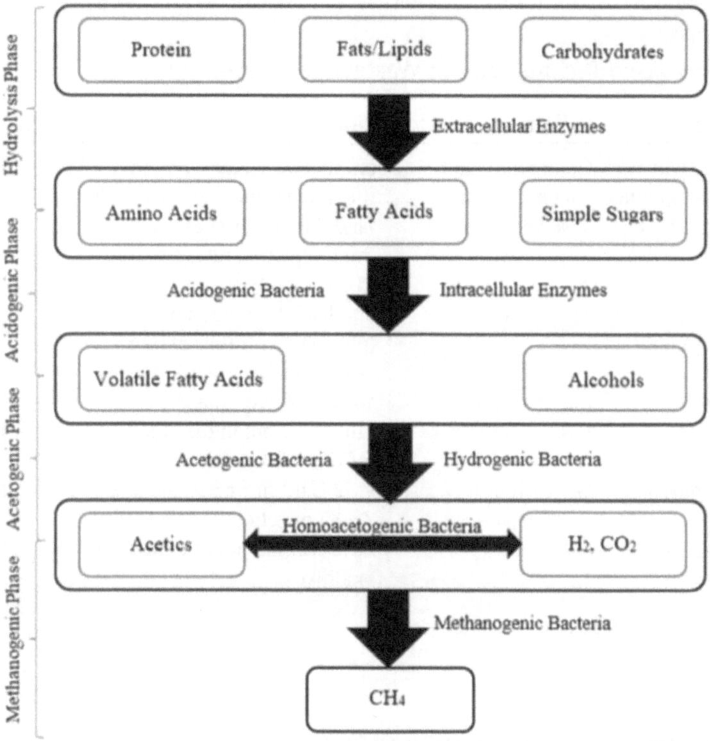

FIGURE 5.5 Methanogenic fermentation pathways for anaerobic wastewater treatment. (Adapted from Ref. [103].)

energy source which, in turn, diminishes emissions and dependence on conventional fuels [106].

Anaerobic digestion can be combined with other treatment processes to improve treatment efficiency and resource recovery. Co-digestion involves treating multiple organic waste streams together to optimize biogas production and nutrient recycling [107]. Co-digestion of sewage sludge with organic wastes such as food waste, agricultural residues, and industrial effluents can enhance biogas yield, improve digestate quality, and reduce overall treatment costs [108].

Additionally, by combining anaerobic digestion with nutrient recovery techniques like ammonia stripping and struvite precipitation, essential nutrients can be extracted from digestate and reused as fertilizers [109]. These integrated approaches support resource conservation and the circular economy by promoting resource recovery, waste valorization, and sustainable wastewater treatment techniques [110].

Anaerobic digestion is a versatile biological process for organic waste treatment along with biogas production, offering environmental, energy, and resource benefits. Integration of anaerobic digestion with other treatment processes enhances treatment efficiency, energy recovery, and nutrient recycling, contributing to sustainable waste management and resource conservation efforts [111].

TABLE 5.1
Membrane-based Technologies in Wastewater Treatment

Membrane	Pollutants	Activity	References
Membranes with a molecular weight cut-off (MWCO)	Community and commercial sewage	Complete resistance was shown by the membranes towards fouling and the membranes showed high rejection rates towards dyes	[117]
Reverse osmosis and Microfiltration	Macronutrients	Complete elimination of phosphorus followed by 85–99% removal rate for Nitrogen	[119]
Reverse osmosis	Pharmaceuticals	Improved performance	[120]
Microfiltration	Household sewage	Highly efficient in the elimination of macronutrients	[118]
Reverse osmosis	Olive oil mill effluent	24–32 L h^{-1} m^{-2} permeate flux and COD rejection 97.5–99.1%	[117]
Microfiltration	Oily water	90.2% elimination of organic additives	[116]

Source: Adapted by the authors from Ref. [56]

5.3.5 Membrane Filtration Technologies

Membrane filtration technologies play a vital role in the treatment of municipal wastewater, as summarized in Table 5.1. These technologies offer efficient removal of contaminants through physical separation processes. These technologies utilize semi-permeable membranes to selectively separate particles and dissolved substances based on their size, molecular weight, and charge, providing high-quality treated effluent for reuse or discharge into the environment [56].

5.3.5.1 Types of Membrane Filtration

5.3.5.1.1 Microfiltration (MF)

Microfiltration membranes have pore sizes ranging from 0.1 to 10 μm, allowing for the removal of suspended solids, bacteria, and some protozoa from wastewater. MF is commonly used as a pretreatment step to remove larger particles before subsequent membrane processes [112].

5.3.5.1.2 Ultrafiltration (UF)

Ultrafiltration membranes have smaller pore sizes (0.001 to 0.1 μm) than MF membranes, enabling the separation of suspended solids, colloids, viruses, and macromolecules including proteins and polysaccharides. UF is effective in producing high-quality effluent for potable water reuse and industrial applications [113].

5.3.5.1.3 Nanofiltration (NF)

Nanofiltration membranes have even smaller pore sizes (0.001 to 0.01 μm) than UF membranes, allowing for the separation of divalent ions, organic matter, and certain micropollutants while retaining monovalent ions and water molecules. NF is used for the removal of hardness ions, colour, and organic compounds from wastewater [114].

5.3.5.1.4 Reverse Osmosis (RO)

Membranes used in reverse osmosis have the smallest pore sizes (<0.001 μm) among membrane filtration technologies, enabling the removal of nearly all dissolved salts, organic compounds, and micropollutants from wastewater. RO is widely used for desalination, production of high-purity water, and concentration of wastewater streams [115].

5.3.5.2 Challenges and Advancements in Membrane Technology

Despite their numerous benefits, membrane filtration technologies face several challenges, including membrane fouling, scaling, high energy consumption, and membrane material degradation [56]. To address these challenges, significant advancements have been made in membrane materials, module design, fouling mitigation strategies, and process optimization.

5.3.5.2.1 Membrane Materials

Advances in membrane materials, including novel polymers, ceramic membranes, and nanocomposite membranes, have led to improved membrane performance, selectivity, and durability [121].

5.3.5.2.2 Fouling Mitigation

The development of fouling-resistant membranes, surface modifications, and membrane cleaning techniques have helped mitigate fouling and extend membrane lifespan [122].

5.3.5.2.3 Energy Efficiency

Integration of membrane processes with energy recovery systems, such as pressure exchangers, osmotic energy conversion, and solar-powered membranes, has reduced energy consumption and operational costs [58].

5.3.5.2.4 Process Optimization

Advanced process modelling, simulation, and control strategies have optimized membrane system performance, flux stability, and fouling control, enhancing overall treatment efficiency [123].

Conclusively, membrane filtration technologies play a vital role in municipal wastewater treatment, offering efficient removal of contaminants and resource recovery. Developments in the optimization process including choice of material for membranes, component design, and fouling mitigation have improved membrane performance, energy efficiency, and sustainability, paving the way for their widespread application in water reuse, resource recovery, and environmental protection efforts.

5.3.6 Electrocoagulation

Electrocoagulation (EC), as shown in Figure 5.6, is a method of electrochemical water treatment that uses an electric current to extract impurities from wastewater. This process involves the generation of coagulant species through the dissolution of sacrificial electrodes, typically made of aluminum or iron, to destabilize and precipitate pollutants from the wastewater matrix [124].

The electrocoagulation process operates on several electrochemical mechanisms. When electric current is applied to the electrodes immersed in wastewater, metal ions are released from the anode and undergo hydrolysis to form metal hydroxide precipitates. These metal hydroxide species act as coagulants, neutralizing charged particles and contaminants through adsorption and flocculation [125]. Additionally, the release of hydrogen gas (H_2) at the cathode promotes pH adjustment and the formation of hydroxide ions, contributing to the coagulation process [125].

5.3.6.1 Removal of Pollutants Using Electrocoagulation

Electrocoagulation is effective in removing numerous pollutants from wastewater, including suspended solids, colloids, heavy metals, and organic compounds. The coagulation and precipitation of suspended solids and colloidal particles occur due to the destabilization of particles through charge neutralization and their subsequent agglomeration into larger flocs [125]. Metal hydroxide coagulants formed during electrocoagulation also facilitate the removal of heavy metals by complexation and precipitation reactions, leading to their separation from the wastewater matrix [127]. Furthermore, organic pollutants can be adsorbed onto the surface of metal hydroxide

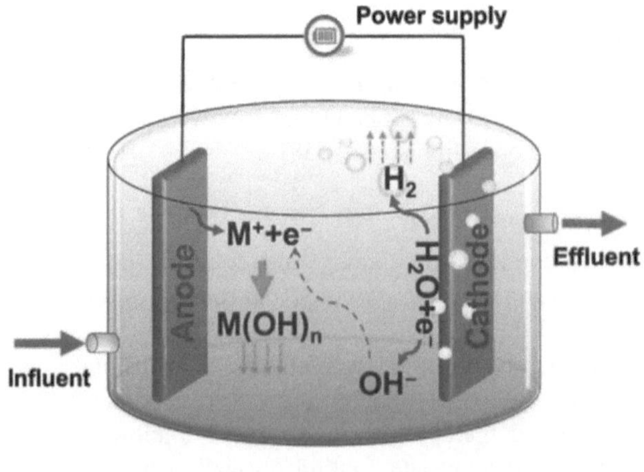

FIGURE 5.6 Schematic diagram of the electrocoagulation process. (Adapted from Ref. [126].)

flocs or undergo electrochemical oxidation, contributing to their removal from the wastewater stream [128].

5.3.6.2 Scale-up and Commercial Applications

Systems for electrocoagulation can be expanded to accommodate different wastewater quantities and flow rates, making them appropriate for both small- and large-scale applications. Commercial electrocoagulation systems generally comprise an electrolytic chamber furnished with consumable electrodes, alongside a power source and control mechanisms to regulate operational settings [129]. These systems can find usage in a wide range of applications, including industrial wastewater treatment, municipal effluent treatment, and emerging applications including decentralized water management and remediation of contaminated groundwater [130].

Commercial applications of electrocoagulation have demonstrated its effectiveness in treating wastewater from various industries, including textiles, electroplating, mining, and food processing, where conventional treatment methods may be inadequate [129]. In municipal wastewater treatment plants, electrocoagulation has been utilized for primary treatment, sludge dewatering, and tertiary treatment to meet discharge standards and achieve water reuse objectives [124]. Additionally, electrocoagulation is increasingly being explored for niche applications such as onsite remediation of contaminated groundwater and treatment of brine and produced water from oil and gas operations [126].

Electrocoagulation is a promising electrochemical water treatment process for removing pollutants from wastewater, offering simplicity, versatility, and effectiveness. Commercial applications of electrocoagulation demonstrate its potential for addressing diverse water treatment challenges and contributing to sustainable wastewater management practices [126].

5.3.7 Resource Recovery from Wastewater

Resource recovery from wastewater involves the extraction and utilization of valuable resources such as nutrients, organic matter, and energy from wastewater streams, contributing to environmental sustainability, circular economy principles, and resource conservation [131].

5.3.7.1 Recovery of Nutrients (Nitrogen, Phosphorus) from Wastewater

Nutrient recovery from wastewater, particularly nitrogen and phosphorus, is essential to mitigate eutrophication and nutrient pollution in receiving water bodies while also providing valuable resources for agriculture and industry [132]. Various technologies have been developed to recover nutrients from wastewater. These include:

5.3.7.1.1 Struvite Precipitation

Struvite ($MgNH_4PO_4{\cdot}6H_2O$) is a crystalline compound that can be precipitated from wastewater containing elevated concentrations of magnesium (Mg^{2+}), ammonium (NH^{4+}), and phosphate (PO_4^{3-}) ions. Struvite precipitation technologies recover phosphorus from wastewater in the form of struvite crystals, which can be used as a slow-release fertilizer in agriculture [133].

5.3.7.1.2 Ammonia Stripping

Ammonia stripping involves the volatilization of ammonia (NH_3) from wastewater by raising the pH and temperature of the wastewater stream, followed by condensation and recovery of ammonia as ammonium sulfate or ammonium nitrate fertilizers [134].

5.3.7.2 Generation of Bioenergy from Wastewater

Wastewater contains organic matter in the form of proteins, lipids, and carbohydrates. A variety of biological and physicochemical processes can transform these organic matter components into bioenergy. Bioenergy generation from wastewater offers dual benefits of renewable energy production and organic waste treatment [135]. Key bioenergy generation pathways from wastewater include:

5.3.7.2.1 Anaerobic Digestion (AD)

Anaerobic digestion biologically converts organic matter present in wastewater into biogas (mainly methane and carbon dioxide) and digestate. Biogas produced from AD can be utilized for the production of heat and electricity or can be used as automotive fuel, contributing to renewable energy production and in mitigation of greenhouse gases [136].

5.3.7.2.2 Microbial Fuel Cells (MFCs)

Microbial fuel cells utilize microorganisms to oxidize and consume organic matter present in wastewater in order to generate electricity through microbial respiration. MFCs offer the potential for direct electricity generation from wastewater while also providing wastewater treatment and nutrient recovery benefits [137].

5.3.7.3 Production of Value-added Products from Wastewater

Wastewater contains various organic compounds, minerals, and metals that can be converted into value-added products through biochemical, chemical, and biotechnological processes [138]. Production of value-added products from wastewater includes:

5.3.7.3.1 Biochemical Conversion

Organic compounds present in wastewater can be converted into bio-based chemicals, biopolymers, and biofuels through fermentation, enzymatic hydrolysis, and chemical synthesis processes. Examples include the production of bioethanol from fermentable sugars and volatile fatty acids derived from wastewater [139].

5.3.7.3.2 Algal Cultivation

Wastewater provides nutrients and CO_2 sources for algal cultivation, enabling the production of biomass, biofuels (e.g., biodiesel, biohydrogen), and high-value products such as omega-3 fatty acids, pigments, and antioxidants from microalgae and cyanobacteria [140].

In conclusion, resource recovery from wastewater presents opportunities for sustainable wastewater management and resource conservation. Technologies for nutrient recovery, bioenergy generation, and production of value-added products offer environmental, economic, and social benefits, playing a role in the shift towards a circular economy and fostering sustainable development.

5.4 FUTURE DIRECTIONS AND CHALLENGES

Advancements in wastewater treatment technologies are continuously evolving to address emerging challenges such as increasing urbanization, population growth, and pollution. These advancements aim to enhance treatment efficiency, resource recovery, and environmental sustainability through the integration of novel approaches and cutting-edge technologies.

5.4.1 Integration of Advanced Oxidation Processes (AOPs) with Membrane Technologies

Advanced oxidation processes (AOPs), that includes ozonation, UV irradiation and photocatalysis are effective for the degradation of recalcitrant organic pollutants and emerging contaminants in wastewater. Integrating AOPs with membrane filtration technologies offers the potential for synergistic treatment effects, where AOPs can pre-treat wastewater to reduce fouling and enhance membrane performance, while membranes can separate and concentrate reaction products for further treatment or resource recovery [79]. This integration can enhance treatment efficacy, decrease energy consumption and enhance the removal of contaminants of emerging concern.

5.4.2 Development of Next-Generation Membrane Materials

Advancements in membrane materials hold great promise for enhancing the efficiency, durability and sustainability of membranous wastewater treatment technologies. Research efforts focus on the development of next-generation membrane materials with improved fouling resistance, selectivity, and permeability, utilizing innovative approaches such as nanotechnology, surface modifications, and biomimetic designs [112]. These advanced membranes exhibit improved separation efficiency, reduced energy consumption, and prolonged service life, contributing to the development of more efficient and cost effective wastewater treatment systems [56].

5.4.3 Implementation of Artificial Intelligence and Machine Learning Techniques

Artificial intelligence (AI) and machine learning (ML) technologies are increasingly employed to optimize wastewater treatment processes, enhance operational efficiency, and predict system performance. AI-based models can analyze complex datasets, sensor measurements, and process variables to optimize process control, identify optimal operating conditions, and detect anomalies or system failures in real-time [138]. Integration of AI and ML techniques into wastewater treatment plants enables adaptive control strategies, predictive maintenance, and decision support systems, leading to improved treatment outcomes, reduced costs, and enhanced sustainability [138].

5.4.4 Exploration of Nature-Based Solutions and Ecological Engineering

Nature-based solutions (NBS) and ecological engineering approaches offer sustainable alternatives for wastewater treatment and ecological restoration, leveraging natural processes and ecosystem services to enhance water quality and habitat biodiversity [39]. NBS such as constructed wetlands, biofiltration systems, and green infrastructure mimic natural wetland ecosystems to remove contaminants, promote nutrient cycling, and provide habitat for wildlife [92]. Ecological engineering principles emphasize the integration of ecological knowledge and engineering design to create multifunctional and resilient wastewater treatment systems that support biodiversity, carbon sequestration, and ecosystem services provision [141].

In conclusion, potential advancements in innovative wastewater treatment technologies offer promising opportunities to address emerging challenges and improve treatment efficiency, resource recovery, and environmental sustainability. Integration of advanced oxidation processes with membrane technologies, development of next-generation membrane materials, implementation of artificial intelligence and machine learning techniques, and exploration of nature-based solutions and ecological engineering approaches are key areas driving innovation in wastewater treatment research and practice.

5.5 CONCLUSION

The management of municipal wastewater is imperative for safeguarding public health and preserving environmental integrity. Over the years, conventional wastewater treatment methods have played a crucial role in mitigating pollution; however, they face limitations in addressing emerging challenges such as the presence of emerging pollutants and the escalating volumes of wastewater due to population growth and urbanization.

To overcome these challenges, the adoption of innovative technologies is essential. Membrane Bioreactors (MBRs) offer enhanced treatment efficiency and footprint reduction compared to conventional activated sludge systems. Advanced Oxidation Processes (AOPs), including ozonation and UV irradiation, provide effective removal of emerging contaminants. Constructed wetlands provide a naturally sustainable method for treating wastewater, while anaerobic digestion not only treats wastewater but also generates biogas for energy recovery. Membrane Filtration Technologies and Electrocoagulation further contribute to pollutant removal and recovery of resources from wastewater. Integration and optimization of these new advancements in wastewater treatment facilities have the potential to greatly improve effectiveness while reducing expenses. Strategies such as combining multiple technologies and implementing optimization techniques are crucial for achieving sustainable wastewater treatment practices. Looking ahead, future advancements in innovative wastewater treatment technologies hold promise for addressing current challenges more effectively. However, it is essential to address issues such as cost-effectiveness, scalability, and regulatory compliance to ensure widespread adoption and implementation of these technologies.

In conclusion, the implementation of innovative treatment technologies for wastewater is vital for achieving sustainable water management practices. Continued research, development, and implementation of these solutions are necessary to meet the growing demands for efficient and environmentally friendly municipal wastewater treatment systems. It is only through collective efforts and commitment that we can ensure the long-term health and sustainability of our water resources.

REFERENCES

[1] Silva JA. Wastewater treatment and reuse for sustainable water resources management: A systematic literature review. *Sustainability* 2023;15:10940. https://doi.org/10.3390/su151410940

[2] Graham KE, Loeb SK, Wolfe MK, Catoe D, Sinnott-Armstrong N, Kim S, et al. SARS-CoV-2 RNA in wastewater settled solids is associated with COVID-19 cases in a large urban sewershed. *Environ Sci Technol* 2021;55:488–98. https://doi.org/10.1021/acs.est.0c06191

[3] Magana-Arachchi DN, Wanigatunge RP. Ubiquitous waterborne pathogens. In Prasad, M.N.V., & Grobelak, A. (Eds.), *Waterborne Pathogens*, Elsevier; 2020, pp. 15–42. https://doi.org/10.1016/B978-0-12-818783-8.00002-5

[4] Bashir I, Lone FA, Bhat RA, Mir SA, Dar ZA, Dar SA. Concerns and threats of contamination on aquatic ecosystems. In: Hakeem, K., Bhat, R., Qadri, H. (Eds), *Bioremediation and Biotechnology*, Springer International Publishing; 2020; 1–26. https://doi.org/10.1007/978-3-030-35691-0_1

[5] Zhang Y, Liang J, Zeng G, Tang W, Lu Y, Luo Y, et al. How climate change and eutrophication interact with microplastic pollution and sediment resuspension in shallow lakes: A review. *Sci Total Environ* 2020;705:135979. https://doi.org/10.1016/j.scitotenv.2019.135979

[6] Obaideen K, Shehata N, Sayed ET, Abdelkareem MA, Mahmoud MS, Olabi AG. The role of wastewater treatment in achieving sustainable development goals (SDGs) and sustainability guideline. *Energy Nexus* 2022;7:100112. https://doi.org/10.1016/j.nexus.2022.100112

[7] Okello C, Tomasello B, Greggio N, Wambiji N, Antonellini M. Impact of population growth and climate change on the freshwater resources of Lamu Island, Kenya. *Water (Basel)* 2015;7:1264–90. https://doi.org/10.3390/w7031264

[8] Ingrao C, Strippoli R, Lagioia G, Huisingh D. Water scarcity in agriculture: An overview of causes, impacts and approaches for reducing the risks. *Heliyon* 2023;9:e18507. https://doi.org/10.1016/j.heliyon.2023.e18507

[9] Schellenberg T, Subramanian V, Ganeshan G, Tompkins D, Pradeep R. Wastewater discharge standards in the evolving context of urban sustainability – The case of India. *Front Environ Sci* 2020;8:30. https://doi.org/10.3389/fenvs.2020.00030

[10] Abubakar IR, Maniruzzaman KM, Dano UL, AlShihri FS, AlShammari MS, Ahmed SMS, et al. Environmental sustainability impacts of solid waste management practices

in the global South. *Int J Environ Res Public Health* 2022;19:12717. https://doi.org/10.3390/ijerph191912717

[11] Ćetković J, Knežević M, Lakić S, Žarković M, Vujadinović R, Živković A, et al. Financial and economic investment evaluation of wastewater treatment plant. *Water (Basel)* 2022;14:122. https://doi.org/10.3390/w14010122

[12] Zarei M. Wastewater resources management for energy recovery from circular economy perspective. *Water – Energy Nexus* 2020;3:170–85. https://doi.org/10.1016/j.wen.2020.11.001

[13] Angelakis AN, Rose JB. *Evolution of Sanitation and Wastewater Technologies Through the Centuries. Water Intelligence Online*, Vol. 13, IWA Publishing; 2014. https://doi.org/10.2166/9781780404851

[14] De Feo G, Antoniou G, Fardin H, El-Gohary F, Zheng X, Reklaityte I, et al. The historical development of sewers worldwide. *Sustainability* 2014;6:3936–74. https://doi.org/10.3390/su6063936

[15] Al Hamedi FH, Kandhan K, Liu Y, Ren M, Jaleel A, Alyafei MAM. Wastewater irrigation: A promising way for future sustainable agriculture and food security in the United Arab Emirates. *Water (Basel)* 2023;15:2284. https://doi.org/10.3390/w15122284

[16] Roman M-D, Sava C, Iluţiu-Varvara D-A, Mare R, Pruteanu L-L, Pică EM, et al. Biological activated sludge from wastewater treatment plant before and during the COVID-19 pandemic. *Int J Environ Res Public Health* 2022;19:11323. https://doi.org/10.3390/ijerph191811323

[17] Keiser DA, Shapiro JS. Consequences of the clean water act and the demand for water quality. *Q J Econ* 2019;134:349–96. https://doi.org/10.1093/qje/qjy019

[18] khalidi-idrissi A, Madinzi A, Anouzla A, Pala A, Mouhir L, Kadmi Y, et al. Recent advances in the biological treatment of wastewater rich in emerging pollutants produced by pharmaceutical industrial discharges. *Int J Environ Sci Technol* 2023;20:11719–40. https://doi.org/10.1007/s13762-023-04867-z

[19] Jain K, Patel AS, Pardhi VP, Flora SJS. Nanotechnology in wastewater management: A new paradigm towards wastewater treatment. *Molecules* 2021;26:1797. https://doi.org/10.3390/molecules26061797

[20] Kookana RS, Drechsel P, Jamwal P, Vanderzalm J. Urbanisation and emerging economies: Issues and potential solutions for water and food security. *Sci Total Environ* 2020;732:139057. https://doi.org/10.1016/j.scitotenv.2020.139057

[21] Varsha M, Senthil Kumar P, Senthil Rathi B. A review on recent trends in the removal of emerging contaminants from aquatic environment using low-cost adsorbents. *Chemosphere* 2022;287:132270. https://doi.org/10.1016/j.chemosphere.2021.132270

[22] Maktabifard M, Al-Hazmi HE, Szulc P, Mousavizadegan M, Xu X, Zaborowska E, et al. Net-zero carbon condition in wastewater treatment plants: A systematic review of mitigation strategies and challenges. *Renew Sustain Energy Rev* 2023;185:113638. https://doi.org/10.1016/j.rser.2023.113638

[23] Beck MB, Cummings RG. Wastewater infrastructure. *Habitat Int* 1996;20:405–20. https://doi.org/10.1016/0197-3975(96)00022-7

[24] Nishat A, Yusuf M, Qadir A, Ezaier Y, Vambol V, Ijaz Khan M, et al. Wastewater treatment: A short assessment on available techniques. *Alex Eng J* 2023;76:505–16. https://doi.org/10.1016/j.aej.2023.06.054

[25] Osuoha JO, Anyanwu BO, Ejileugha C. Pharmaceuticals and personal care products as emerging contaminants: Need for combined treatment strategy. *J Hazard Mater Adv* 2023;9:100206. https://doi.org/10.1016/j.hazadv.2022.100206

[26] Kumar S, Kumar A. Hotspot and trend analysis of forest fires and its relation to climatic factors in the western Himalayas. *Nat Hazard* 2022;114:3529–44. https://doi.org/10.1007/s11069-022-05530-5
[27] Zieliński W, Korzeniewska E, Harnisz M, Drzymała J, Felis E, Bajkacz S. Wastewater treatment plants as a reservoir of integrase and antibiotic resistance genes – An epidemiological threat to workers and environment. *Environ Int* 2021;156:106641. https://doi.org/10.1016/j.envint.2021.106641
[28] Molinos-Senante M, Maziotis A. Evaluation of energy efficiency of wastewater treatment plants: The influence of the technology and aging factors. *Appl Energy* 2022;310:118535. https://doi.org/10.1016/j.apenergy.2022.118535
[29] Gherghel A, Teodosiu C, De Gisi S. A review on wastewater sludge valorisation and its challenges in the context of circular economy. *J Clean Prod* 2019;228:244–63. https://doi.org/10.1016/j.jclepro.2019.04.240
[30] Singh BJ, Chakraborty A, Sehgal R. A systematic review of industrial wastewater management: Evaluating challenges and enablers. *J Environ Manage* 2023;348:119230. https://doi.org/10.1016/j.jenvman.2023.119230
[31] Okoye CO, Okeke ES, Okoye KC, Echude D, Andong FA, Chukwudozie KI, et al. Occurrence and fate of pharmaceuticals, personal care products (PPCPs) and pesticides in African water systems: A need for timely intervention. *Heliyon* 2022;8:e09143. https://doi.org/10.1016/j.heliyon.2022.e09143
[32] Samal K, Mahapatra S, Hibzur Ali M. Pharmaceutical wastewater as Emerging Contaminants (EC): Treatment technologies, impact on environment and human health. *Energy Nexus* 2022;6:100076. https://doi.org/10.1016/j.nexus.2022.100076
[33] Rajasulochana P, Preethy V. Comparison on efficiency of various techniques in treatment of waste and sewage water – A comprehensive review. *Resour-Effic Technol* 2016;2:175–84. https://doi.org/10.1016/j.reffit.2016.09.004
[34] Ebele AJ, Abou-Elwafa Abdallah M, Harrad S. Pharmaceuticals and personal care products (PPCPs) in the freshwater aquatic environment. *Emerg Contam* 2017;3:1–16. https://doi.org/10.1016/j.emcon.2016.12.004
[35] Dhore R, Murthy GS. Per/polyfluoroalkyl substances production, applications and environmental impacts. *Bioresour Technol* 2021;341:125808. https://doi.org/10.1016/j.biortech.2021.125808
[36] Giwa AS, Maurice NJ, Luoyan A, Liu X, Yunlong Y, Hong Z. Advances in sewage sludge application and treatment: Process integration of plasma pyrolysis and anaerobic digestion with the resource recovery. *Heliyon* 2023;9:e19765. https://doi.org/10.1016/j.heliyon.2023.e19765
[37] Bonato M, Corrà F, Bellio M, Guidolin L, Tallandini L, Irato P, et al. PFAS environmental pollution and antioxidant responses: An overview of the impact on human field. *Int J Environ Res Public Health* 2020;17:8020. https://doi.org/10.3390/ijerph17218020
[38] Campanale C, Massarelli C, Savino I, Locaputo V, Uricchio VF. A detailed review study on potential effects of microplastics and additives of concern on human health. *Int J Environ Res Public Health* 2020;17:1212. https://doi.org/10.3390/ijerph17041212
[39] Liu W, Zhang J, Liu H, Guo X, Zhang X, Yao X, et al. A review of the removal of microplastics in global wastewater treatment plants: Characteristics and mechanisms. *Environ Int* 2021;146:106277. https://doi.org/10.1016/j.envint.2020.106277
[40] Joo SH, Liang Y, Kim M, Byun J, Choi H. Microplastics with adsorbed contaminants: Mechanisms and Treatment. *Environ Chall* 2021;3:100042. https://doi.org/10.1016/j.envc.2021.100042

[41] Onu MA, Ayeleru OO, Oboirien B, Olubambi PA. Challenges of wastewater generation and management in Sub-Saharan Africa: A review. *Environ Chall* 2023;11:100686. https://doi.org/10.1016/j.envc.2023.100686
[42] Elena A, Orbeci C, Lazau C, Sfirloaga P, Vlazan P, Bandas C, et al. *Waste Water Treatment Methods. Water Treatment*, InTech; 2013. https://doi.org/10.5772/53755
[43] Irfeey AMM, Alotaibi BA, Najim MMM, Shah AA. Water valuation in urban settings for sustainable water management. *Water (Basel)* 2023;15:3105. https://doi.org/10.3390/w15173105
[44] Casari S, Di Paola M, Banci E, Diallo S, Scarallo L, Renzo S, et al. Changing dietary habits: The impact of urbanization and rising socio-economic status in families from Burkina Faso in Sub-Saharan Africa. *Nutrients* 2022;14:1782. https://doi.org/10.3390/nu14091782
[45] Madias K, Szymkowiak A. Residential sustainable water usage and water management: Systematic review and future research. *Water (Basel)* 2022;14:1027. https://doi.org/10.3390/w14071027
[46] Li C, Liu M, Hu Y, Shi T, Zong M, Walter M. Assessing the impact of urbanization on direct runoff using improved composite CN method in a large urban area. *Int J Environ Res Public Health* 2018;15:775. https://doi.org/10.3390/ijerph15040775.
[47] Ansari A, Dutt D, Kumar V. Revolutionizing urban Waste: Eco-Innovation for Saharanpur's sustainability (Class-II urban Centers). *Waste Manage Bull* 2024;2:184–90. https://doi.org/10.1016/j.wmb.2024.01.006
[48] Capodaglio A. Integrated, decentralized wastewater management for resource recovery in rural and peri-urban areas. *Resources* 2017;6:22. https://doi.org/10.3390/resources6020022
[49] Younas F, Mustafa A, Farooqi ZUR, Wang X, Younas S, Mohy-Ud-Din W, et al. Current and emerging adsorbent technologies for wastewater treatment: Trends, limitations, and environmental implications. *Water (Basel)* 2021;13:215. https://doi.org/10.3390/w13020215
[50] Al-Asheh S, Bagheri M, Aidan A. Membrane bioreactor for wastewater treatment: A review. *Case Stud Chem Environ Eng* 2021;4:100109. https://doi.org/10.1016/j.cscee.2021.100109
[51] Moktadir Md A, Maliha M, Tujjohra F, Munmun SA, Alam Md S, Islam Md A, et al. Treatment of tannery wastewater by different membrane bioreactors: A critical review. *Environ Adv* 2024;15:100478. https://doi.org/10.1016/j.envadv.2023.100478
[52] Iorhemen O, Hamza R, Tay J. Membrane bioreactor (MBR) technology for wastewater treatment and reclamation: Membrane fouling. *Membranes (Basel)* 2016;6:33. https://doi.org/10.3390/membranes6020033
[53] Chattopadhyay I, Banu JR, Usman TMM, Varjani S. Exploring the role of microbial biofilm for industrial effluents treatment. *Bioengineered* 2022;13:6420–40. https://doi.org/10.1080/21655979.2022.2044250
[54] Murshid S, Antonysamy A, Dhakshinamoorthy G, Jayaseelan A, Pugazhendhi A. A review on biofilm-based reactors for wastewater treatment: Recent advancements in biofilm carriers, kinetics, reactors, economics, and future perspectives. *Sci Total Environ* 2023;892:164796. https://doi.org/10.1016/j.scitotenv.2023.164796
[55] Bhattacharyya A, Liu L, Lee K, Miao J. Review of biological processes in a membrane bioreactor (MBR): Effects of wastewater characteristics and operational parameters on biodegradation efficiency when treating industrial oily wastewater. *J Mar Sci Eng* 2022;10:1229. https://doi.org/10.3390/jmse10091229
[56] Obotey Ezugbe E, Rathilal S. Membrane technologies in wastewater treatment: A review. *Membranes (Basel)* 2020;10:89. https://doi.org/10.3390/membranes10050089

[57] Du X, Shi Y, Jegatheesan V, Haq IU. A review on the mechanism, impacts and control methods of membrane fouling in MBR system. *Membranes (Basel)* 2020;10:24. https://doi.org/10.3390/membranes10020024
[58] Xiao T, Zhu Z, Li L, Shi J, Li Z, Zuo X. Membrane fouling and cleaning strategies in microfiltration/ultrafiltration and dynamic membrane. *Sep Purif Technol* 2023;318:123977. https://doi.org/10.1016/j.seppur.2023.123977
[59] Rahman TU, Roy H, Islam Md R, Tahmid M, Fariha A, Mazumder A, et al. The advancement in membrane bioreactor (MBR) technology toward sustainable industrial wastewater management. *Membranes (Basel)* 2023;13:181. https://doi.org/10.3390/membranes13020181
[60] Khan M, Wibowo A, Karim Z, Posoknistakul P, Matsagar B, Wu K, et al. Wastewater treatment using membrane bioreactor technologies: Removal of phenolic contaminants from oil and coal refineries and pharmaceutical industries. *Polymers (Basel)* 2024;16:443. https://doi.org/10.3390/polym16030443
[61] Boano F, Caruso A, Costamagna E, Ridolfi L, Fiore S, Demichelis F, et al. A review of nature-based solutions for greywater treatment: Applications, hydraulic design, and environmental benefits. *Sci Total Environ* 2020;711:134731. https://doi.org/10.1016/j.scitotenv.2019.134731
[62] Nguyen HM, Graber CJ. A critical review of cephalexin and cefadroxil for the treatment of acute uncomplicated lower urinary tract infection in the era of "Bad Bugs, Few Drugs." *Int J Antimicrob Agents* 2020;56:106085. https://doi.org/10.1016/j.ijantimicag.2020.106085
[63] Zhang J, Xiao K, Liu Z, Gao T, Liang S, Huang X. Large-scale membrane bioreactors for industrial wastewater treatment in China: Technical and economic features, driving forces, and perspectives. *Engineering* 2021;7:868–80. https://doi.org/10.1016/j.eng.2020.09.012
[64] Russo T, Alfredo K, Fisher J. Sustainable water management in urban, agricultural, and natural systems. *Water (Basel)* 2014;6:3934–56. https://doi.org/10.3390/w6123934
[65] Yang X, López-Grimau V, Vilaseca M, Crespi M. Treatment of textile wastewater by CAS, MBR, and MBBR: A comparative study from technical, economic, and environmental perspectives. *Water (Basel)* 2020;12:1306. https://doi.org/10.3390/w12051306
[66] Englande AJ, Krenkel P, Shamas J. Wastewater & water reclamation. In: Elias, S. (Eds), *Reference Module in Earth Systems and Environmental Sciences*, Elsevier; 2015; 35 https://doi.org/10.1016/B978-0-12-409548-9.09508-7
[67] Apostolidis N, Hertle C, Young R. Water recycling in Australia. *Water (Basel)* 2011;3:869–81. https://doi.org/10.3390/w3030869
[68] Saravanan A, Deivayanai VC, Kumar PS, Rangasamy G, Hemavathy RV, Harshana T, et al. A detailed review on advanced oxidation process in treatment of wastewater: Mechanism, challenges and future outlook. *Chemosphere* 2022;308:136524. https://doi.org/10.1016/j.chemosphere.2022.136524
[69] Kumari P, Kumar A. Advanced oxidation process: A remediation technique for organic and non-biodegradable pollutant. *Result Surf Interfac* 2023;11:100122. https://doi.org/10.1016/j.rsurfi.2023.100122
[70] Deng Y, Zhao R. Advanced oxidation processes (AOPs) in wastewater treatment. *Curr Pollut Rep* 2015;1:167–76. https://doi.org/10.1007/s40726-015-0015-z
[71] Collin F. Chemical basis of reactive oxygen species reactivity and involvement in neurodegenerative diseases. *Int J Mol Sci* 2019;20:2407. https://doi.org/10.3390/ijms20102407

[72] Badali KM, Zhou S, Aljawhary D, Antiñolo M, Chen WJ, Lok A, et al. Formation of hydroxyl radicals from photolysis of secondary organic aerosol material. *Atmos Chem Phys* 2015;15:7831–40. https://doi.org/10.5194/acp-15-7831-2015

[73] Xu M, Wu C, Zhou Y. Advancements in the Fenton process for wastewater treatment. In Bustillo-Lecompte, C. (Ed), *Advanced Oxidation Processes – Applications, Trends, and Prospects*, IntechOpen; 2020; 61–78. https://doi.org/10.5772/intechopen.90256

[74] Forni C, Facchiano F, Bartoli M, Pieretti S, Facchiano A, D'Arcangelo D, et al. Beneficial role of phytochemicals on oxidative stress and age-related diseases. *Biomed Res Int* 2019;2019:1–16. https://doi.org/10.1155/2019/8748253

[75] Yan M, Lo JC, Edwards JT, Baran PS. Radicals: Reactive intermediates with translational potential. *J Am Chem Soc* 2016;138:12692–714. https://doi.org/10.1021/jacs.6b08856

[76] Pavel M, Anastasescu C, State R-N, Vasile A, Papa F, Balint I. Photocatalytic degradation of organic and inorganic pollutants to harmless end products: Assessment of practical application potential for water and air cleaning. *Catalysts* 2023;13:380. https://doi.org/10.3390/catal13020380

[77] Cuerda-Correa EM, Alexandre-Franco MF, Fernández-González C. Advanced oxidation processes for the removal of antibiotics from water. An overview. *Water (Basel)* 2019;12:102. https://doi.org/10.3390/w12010102

[78] Guengerich FP, Yoshimoto FK. Formation and cleavage of C–C bonds by enzymatic oxidation–reduction reactions. *Chem Rev* 2018;118:6573–655. https://doi.org/10.1021/acs.chemrev.8b00031

[79] Rekhate CV, Srivastava JK. Recent advances in ozone-based advanced oxidation processes for treatment of wastewater – A review. *Chem Eng J Adv* 2020;3:100031. https://doi.org/10.1016/j.ceja.2020.100031

[80] González Y, Gómez G, Moeller-Chávez GE, Vidal G. UV disinfection systems for wastewater treatment: Emphasis on reactivation of microorganisms. *Sustainability* 2023;15:11262. https://doi.org/10.3390/su151411262

[81] Li J, Zhang Q, Chen B, Wang L, Zhu R, Yang J. Hydrogen peroxide formation in water during the VUV/UV irradiation process: Impacts and mechanisms of selected anions. *Environ Res* 2021;195:110751. https://doi.org/10.1016/j.envres.2021.110751

[82] Bar-Niv N, Azaizeh H, Kuc ME, Azerrad S, Haj-Zaroubi M, Menashe O, et al. Advanced oxidation process UV-H_2O_2 combined with biological treatment for the removal and detoxification of phenol. *J Water Process Eng* 2022;48:102923. https://doi.org/10.1016/j.jwpe.2022.102923

[83] Abdullah AM, Garcia-Pinilla MÁ, Pillai SC, O'Shea K. UV and visible light-driven production of hydroxyl radicals by reduced forms of N, F, and P codoped titanium dioxide. *Molecules* 2019;24:2147. https://doi.org/10.3390/molecules24112147

[84] Pandis PK, Kalogirou C, Kanellou E, Vaitsis C, Savvidou MG, Sourkouni G, et al. Key points of advanced oxidation processes (AOPs) for wastewater, organic pollutants and pharmaceutical waste treatment: A mini review. *ChemEngineering* 2022;6:8. https://doi.org/10.3390/chemengineering6010008

[85] Kumar R, Qureshi M, Vishwakarma DK, Al-Ansari N, Kuriqi A, Elbeltagi A, et al. A review on emerging water contaminants and the application of sustainable removal technologies. *Case Stud Chem Environ Eng* 2022;6:100219. https://doi.org/10.1016/j.cscee.2022.100219

[86] Mukherjee J, Lodh BK, Sharma R, Mahata N, Shah MP, Mandal S, et al. Advanced oxidation process for the treatment of industrial wastewater: A review on strategies, mechanisms, bottlenecks and prospects. *Chemosphere* 2023;345:140473. https://doi.org/10.1016/j.chemosphere.2023.140473

[87] Khan ZUH, Gul NS, Sabahat S, Sun J, Tahir K, Shah NS, et al. Removal of organic pollutants through hydroxyl radical-based advanced oxidation processes. *Ecotoxicol Environ Saf* 2023;267:115564. https://doi.org/10.1016/j.ecoenv.2023.115564

[88] Kanakaraju D, Glass BD, Oelgemöller M. Advanced oxidation process-mediated removal of pharmaceuticals from water: A review. *J Environ Manage* 2018;219:189–207. https://doi.org/10.1016/j.jenvman.2018.04.103

[89] Feijoo S, Kamali M, Dewil R. A review of wastewater treatment technologies for the degradation of pharmaceutically active compounds: Carbamazepine as a case study. *Chem Eng J* 2023;455:140589. https://doi.org/10.1016/j.cej.2022.140589

[90] Son H-S, Ko G, Zoh K-D. Kinetics and mechanism of photolysis and TiO_2 photocatalysis of triclosan. *J Hazard Mater* 2009;166:954–60. https://doi.org/10.1016/j.jhazmat.2008.11.107

[91] Zahmatkesh S, Bokhari A, Karimian M, Zahra MMA, Sillanpää M, Panchal H, et al. A comprehensive review of various approaches for treatment of tertiary wastewater with emerging contaminants: What do we know? *Environ Monit Assess* 2022;194:884. https://doi.org/10.1007/s10661-022-10503-z

[92] Vymazal J. Constructed wetlands for wastewater treatment. *Water (Basel)* 2010;2:530–49. https://doi.org/10.3390/w2030530

[93] Zhang BY, Zheng JS, Sharp RG. Phytoremediation in engineered wetlands: Mechanisms and applications. *Procedia Environ Sci* 2010;2:1315–25. https://doi.org/10.1016/j.proenv.2010.10.142

[94] Hassan I, Chowdhury SR, Prihartato PK, Razzak SA. Wastewater treatment using constructed wetland: Current trends and future potential. *Processes* 2021;9:1917. https://doi.org/10.3390/pr9111917

[95] Vymazal J, Zhao Y, Mander Ü. Recent research challenges in constructed wetlands for wastewater treatment: A review. *Ecol Eng* 2021;169:106318. https://doi.org/10.1016/j.ecoleng.2021.106318

[96] Wang J, Long Y, Yu G, Wang G, Zhou Z, Li P, et al. A review on microorganisms in constructed wetlands for typical pollutant removal: Species, function, and diversity. *Front Microbiol* 2022;13:845725. https://doi.org/10.3389/fmicb.2022.845725

[97] Thakur T, Barya M, Dutta J, Mukherjee P, Thakur A, Swamy S, et al. Integrated phytobial remediation of dissolved pollutants from domestic wastewater through constructed wetlands: An interactive macrophyte-microbe-based green and low-cost decontamination technology with prospective resource recovery. *Water (Basel)* 2023;15:3877. https://doi.org/10.3390/w15223877.

[98] Wu H, Fan J, Zhang J, Ngo HH, Guo W, Liang S, et al. Intensified organics and nitrogen removal in the intermittent-aerated constructed wetland using a novel sludge-ceramsite as substrate. *Bioresour Technol* 2016;210:101–7. https://doi.org/10.1016/j.biortech.2016.01.051

[99] Agaton CB, Guila PMC. Ecosystem services valuation of constructed wetland as a nature-based solution to wastewater treatment. *Earth* 2023;4:78–92. https://doi.org/10.3390/earth4010006

[100] Ali Shah F, Mahmood Q, Maroof Shah M, Pervez A, Ahmad Asad S. Microbial ecology of anaerobic digesters: The key players of anaerobiosis. *Sci World J* 2014;2014:1–21. https://doi.org/10.1155/2014/183752

[101] Harirchi S, Wainaina S, Sar T, Nojoumi SA, Parchami M, Parchami M, et al. Microbiological insights into anaerobic digestion for biogas, hydrogen or volatile fatty acids (VFAs): A review. *Bioengineered* 2022;13:6521–57. https://doi.org/10.1080/21655979.2022.2035986

[102] Wainaina S, Lukitawesa, Kumar Awasthi M, Taherzadeh MJ. Bioengineering of anaerobic digestion for volatile fatty acids, hydrogen or methane production: A critical review. *Bioengineered* 2019;10:437–58. https://doi.org/10.1080/21655979.2019.1673937

[103] Zieliński M, Zielińska M, Dębowski M. Application of microwave radiation to biofilm heating during wastewater treatment in trickling filters. *Bioresour Technol* 2013;127:223–30. https://doi.org/10.1016/j.biortech.2012.09.102

[104] Braun M, Schoberth S, Gottschalk G. Enumeration of bacteria forming acetate from H_2 and CO_2 in anaerobic habitats. *Arch Microbiol* 1979;120:201–4. https://doi.org/10.1007/BF00423066

[105] Li Y, Alaimo CP, Kim M, Kado NY, Peppers J, Xue J, et al. Composition and toxicity of biogas produced from different feedstocks in California. *Environ Sci Technol* 2019;53:11569–79. https://doi.org/10.1021/acs.est.9b03003

[106] Kabeyi MJB, Olanrewaju OA. Biogas production and applications in the sustainable energy transition. *J Energy* 2022;2022:1–43. https://doi.org/10.1155/2022/8750221

[107] Karki R, Chuenchart W, Surendra KC, Shrestha S, Raskin L, Sung S, et al. Anaerobic co-digestion: Current status and perspectives. *Bioresour Technol* 2021;330:125001. https://doi.org/10.1016/j.biortech.2021.125001

[108] Mudzanani K, Iyuke SE, Daramola MO. Co-digestion of wastewater treatment sewage sludge with various biowastes: A comparative study for the enhancement of biogas production. *Mater Today Proc* 2022;65:2172–83. https://doi.org/10.1016/j.matpr.2022.05.539

[109] Lorick D, Macura B, Ahlström M, Grimvall A, Harder R. Effectiveness of struvite precipitation and ammonia stripping for recovery of phosphorus and nitrogen from anaerobic digestate: A systematic review. *Environ Evid* 2020;9:27. https://doi.org/10.1186/s13750-020-00211-x

[110] Yu H, Zahidi I, Fai CM, Liang D, Madsen DØ. Mineral waste recycling, sustainable chemical engineering, and circular economy. *Result Eng* 2024;21:101865. https://doi.org/10.1016/j.rineng.2024.101865

[111] Uddin MM, Wright MM. Anaerobic digestion fundamentals, challenges, and technological advances. *Phys Sci Rev* 2023;8:2819–37. https://doi.org/10.1515/psr-2021-0068

[112] Nasir AM, Adam MR, Mohamad Kamal SNEA, Jaafar J, Othman MHD, Ismail AF, et al. A review of the potential of conventional and advanced membrane technology in the removal of pathogens from wastewater. *Sep Purif Technol* 2022;286:120454. https://doi.org/10.1016/j.seppur.2022.120454

[113] Abuwatfa WH, AlSawaftah N, Darwish N, Pitt WG, Husseini GA. A review on membrane fouling prediction using artificial neural networks (ANNs). *Membranes (Basel)* 2023;13:685. https://doi.org/10.3390/membranes13070685

[114] Mahmoud AED, Mostafa E. Nanofiltration membranes for the removal of heavy metals from aqueous solutions: Preparations and applications. *Membranes (Basel)* 2023;13:789. https://doi.org/10.3390/membranes13090789

[115] Yang Z, Zhou Y, Feng Z, Rui X, Zhang T, Zhang Z. A review on reverse osmosis and nanofiltration membranes for water purification. *Polymers (Basel)* 2019;11:1252. https://doi.org/10.3390/polym11081252

[116] Chang Q, Zhou J, Wang Y, Liang J, Zhang X, Cerneaux S, et al. Application of ceramic microfiltration membrane modified by nano-TiO_2 coating in separation of a stable oil-in-water emulsion. *J Memb Sci* 2014;456:128–33. https://doi.org/10.1016/j.memsci.2014.01.029

[117] Nqombolo A, Mpupa A, Moutloali RM, Nomngongo PN. Wastewater treatment using membrane technology. In Yonar, T. (Ed), *Wastewater and Water Quality*, InTech; 2018; 29–40 https://doi.org/10.5772/intechopen.76624

[118] Zuo K, Chen M, Liu F, Xiao K, Zuo J, Cao X, et al. Coupling microfiltration membrane with biocathode microbial desalination cell enhances advanced purification and long-term stability for treatment of domestic wastewater. *J Memb Sci* 2018;547:34–42. https://doi.org/10.1016/j.memsci.2017.10.034

[119] Praveen P, Heng JYP, Loh K-C. Tertiary wastewater treatment in membrane photobioreactor using microalgae: Comparison of forward osmosis & microfiltration. *Bioresour Technol* 2016;222:448–57. https://doi.org/10.1016/j.biortech.2016.09.124

[120] Al-Obaidi M, Kara-Zaitri C, Mujtaba IM. *Wastewater Treatment by Reverse Osmosis Process*. CRC Press; 2020. https://doi.org/10.1201/9781003019343

[121] Hazarika B, Ahmaruzzaman M, Santosh MS, Barceló D, Rtimi S. Advances in polymer-based nanocomposite membranes for water remediation: Preparation methods, critical issues and mechanisms. *J Environ Chem Eng* 2023;11:111401. https://doi.org/10.1016/j.jece.2023.111401

[122] Ahmed MA, Amin S, Mohamed AA. Fouling in reverse osmosis membranes: Monitoring, characterization, mitigation strategies and future directions. *Heliyon* 2023;9:e14908. https://doi.org/10.1016/j.heliyon.2023.e14908

[123] Stoller M, Serrão Mendes R. Advanced control system for membrane processes based on the boundary flux model. *Sep Purif Technol* 2017;175:527–35. https://doi.org/10.1016/j.seppur.2016.09.049

[124] Butler E, Hung Y-T, Yeh RY-L, Suleiman Al Ahmad M. Electrocoagulation in wastewater treatment. *Water (Basel)* 2011;3:495–525. https://doi.org/10.3390/w3020495

[125] Graça NS, Rodrigues AE. The combined implementation of electrocoagulation and adsorption processes for the treatment of wastewaters. *Clean Technol* 2022;4:1020–53. https://doi.org/10.3390/cleantechnol4040063

[126] Mao Y, Cotterill S, Morgan D, Regan S, Zhao Y. Electrocoagulation of peatland runoff: Statistical optimization and economic analysis. *J Water Process Eng* 2022;49:103113. https://doi.org/10.1016/j.jwpe.2022.103113

[127] Kim T, Kim T-K, Zoh K-D. Removal mechanism of heavy metal (Cu, Ni, Zn, and Cr) in the presence of cyanide during electrocoagulation using Fe and Al electrodes. *J Water Process Eng* 2020;33:101109. https://doi.org/10.1016/j.jwpe.2019.101109

[128] Qasem NAA, Mohammed RH, Lawal DU. Removal of heavy metal ions from wastewater: A comprehensive and critical review. *NPJ Clean Water* 2021;4:36. https://doi.org/10.1038/s41545-021-00127-0

[129] Boinpally S, Kolla A, Kainthola J, Kodali R, Vemuri J. A state-of-the-art review of the electrocoagulation technology for wastewater treatment. *Water Cycle* 2023;4:26–36. https://doi.org/10.1016/j.watcyc.2023.01.001

[130] Kalmakhanova MS, Diaz de Tuesta JL, Malakar A, Gomes HT, Snow DD. Wastewater treatment in Central Asia: Treatment alternatives for safe water reuse. *Sustainability* 2023;15:14949. https://doi.org/10.3390/su152014949

[131] Puyol D, Batstone DJ, Hülsen T, Astals S, Peces M, Krömer JO. Resource recovery from wastewater by biological technologies: Opportunities, challenges, and prospects. *Front Microbiol* 2017;7. https://doi.org/10.3389/fmicb.2016.02106

[132] Sengupta S, Nawaz T, Beaudry J. Nitrogen and phosphorus recovery from wastewater. *Curr Pollut Rep* 2015;1:155–66. https://doi.org/10.1007/s40726-015-0013-1

[133] Crutchik D, Garrido JM. Struvite crystallization versus amorphous magnesium and calcium phosphate precipitation during the treatment of a saline industrial wastewater. *Water Sci Technol* 2011;64:2460–7. https://doi.org/10.2166/wst.2011.836
[134] Kinidi L, Tan IAW, Abdul Wahab NB, Tamrin KF Bin, Hipolito CN, Salleh SF. Recent development in ammonia stripping process for industrial wastewater treatment. *Int J Chem Eng* 2018;2018:1–14. https://doi.org/10.1155/2018/3181087
[135] Bhatia RK, Sakhuja D, Mundhe S, Walia A. Renewable energy products through bioremediation of wastewater. *Sustainability* 2020;12:7501. https://doi.org/10.3390/su12187501
[136] Tshemese Z, Deenadayalu N, Linganiso LZ, Chetty M. An overview of biogas production from anaerobic digestion and the possibility of using sugarcane wastewater and municipal solid waste in a South African context. *Appl Syst Innov* 2023;6:13. https://doi.org/10.3390/asi6010013
[137] Roy H, Rahman TU, Tasnim N, Arju J, Rafid Md M, Islam Md R, et al. Microbial fuel cell construction features and application for sustainable wastewater treatment. *Membranes (Basel)* 2023;13:490. https://doi.org/10.3390/membranes13050490
[138] Singh D, Goswami RK, Agrawal K, Chaturvedi V, Verma P. Bio-inspired remediation of wastewater: A contemporary approach for environmental clean-up. *Curr Res Green Sustain Chem* 2022;5:100261. https://doi.org/10.1016/j.crgsc.2022.100261
[139] Boro M, Verma AK, Chettri D, Yata VK, Verma AK. Strategies involved in biofuel production from agro-based lignocellulose biomass. *Environ Technol Innov* 2022;28:102679. https://doi.org/10.1016/j.eti.2022.102679
[140] Srimongkol P, Sangtanoo P, Songserm P, Watsuntorn W, Karnchanatat A. Microalgae-based wastewater treatment for developing economic and environmental sustainability: Current status and future prospects. *Front Bioeng Biotechnol* 2022;10. https://doi.org/10.3389/fbioe.2022.904046
[141] Li P, Li D, Sun X, Chu Z, Xia T, Zheng B. Application of ecological restoration technologies for the improvement of biodiversity and ecosystem in the river. *Water (Basel)* 2022;14:1402. https://doi.org/10.3390/w14091402

6 Recent Technological Developments in Biological Nutrient Removal

Stany Balakumar, Shatakshi Mishra, and Sangeetha Subramanian

6.1 INTRODUCTION

The present era places significant emphasis on wastewater treatment because of the imperative nature of environmental conservation and the sustainable management of resources. One of the most pressing challenges in this subject is the mitigation of excessive nutrients, such as nitrogen and phosphorus, from wastewater before its discharge into natural aquatic ecosystems. The lack of adequate consideration given to the removal of nutrients can initiate a cascade of ecological problems, including the occurrence of harmful algal blooms, reduced levels of oxygen, and the degradation of aquatic ecosystems [1]. Recognizing the pressing nature of this issue, engineers and scientists have diligently endeavored to devise innovative solutions that not only comply with legal mandates but also contribute to the overarching goal of ecological restoration. The phrase "Recent Technological Developments in Biological Nutrient Removal" encapsulates the core aspects of this technical and scientific endeavor. The underlying idea of biological nutrient removal (BNR) is in the remarkable ability of microorganisms to modify and absorb surplus nutrients from wastewater [2]. How human creativity has effectively utilized and enhanced this inherent mechanism has seen significant transformations in recent years. Several factors are converging to facilitate this transition, including heightened environmental awareness, stricter regulatory measures, and rapid technological progress (see Table 6.1).

The initial stage is acknowledging the fundamental importance of nutrient removal in the context of wastewater treatment. Excessive nutrient levels, namely phosphorus and nitrogen, can have detrimental effects on aquatic environments [3]. Fertilizers and similar substances exert comparable effects by promoting the unregulated proliferation of aquatic vegetation, such as algae. As the population of these organisms increases, their decomposition process leads to the depletion of oxygen, leading to the formation of oxygen-deficient areas known as "dead zones" that are inhospitable

DOI: 10.1201/9781003454199-6

for aquatic organisms [3]. The consequences are far-reaching, affecting not only the ecological health of aquatic ecosystems but also the socioeconomic well-being and access to potable water for human populations.

The incorporation of the Biological Nutrient Removal (BNR) process has emerged as an essential element in modern strategies for wastewater treatment, aiming to effectively tackle this pressing issue. Historically, the implementation of BNR techniques has relied on the utilization of various biological reactors. These reactors serve as controlled settings where specific strains of microorganisms are cultivated to facilitate the elimination of nutrients [4]. However, the more recent technologies examined in this compilation have significantly altered the potential and efficacy of BNR techniques. The enhancement of microbial populations represents a significant achievement in BNR. Through extensive exploration of the microbial realm, researchers have unearthed novel species and variants that possess remarkable capabilities in the removal of nutrients. In addition, advancements in synthetic biology and genetic engineering have enabled the targeted customization of microbial consortia for nutrient removal activities, resulting in enhanced accuracy and versatility of BNR processes. Nevertheless, the BNR system encompasses technology that surpasses the limited scope of microbial entities. Contemporary configurations of bioreactors aim to optimize nutrient removal efficiency, minimize energy use, and mitigate adverse environmental impacts. The innovation described has significantly transformed the design of bioreactors. Innovative technologies such as Upflow sludge blanket reactors (UASB)[5], Moving bed biofilm reactors (MBBR) [6], and computational modeling for process optimization [7] have established new benchmarks for performance in biological nutrient removal (BNR).

Moreover, the combination of advanced technologies such as automation and artificial intelligence (AI) has propelled BNR to unprecedented levels of sophistication. To optimize the operational efficiency of biological nutrient removal (BNR) processes, the integration of artificial intelligence (AI) algorithms has been employed. These algorithms serve the purpose of predicting nutrient levels, enhancing process management, and promptly detecting system anomalies [7]. The integration of remote monitoring and automation systems facilitates the efficient operation of biological nutrient removal (BNR) facilities, resulting in reduced operating expenses and decreased reliance on human intervention.

Subsequent sections of the chapter will comprehensively delineate various techniques associated with the effective management of prominent biological nutrients such as nitrogen and phosphorus.

6.2 NITROGEN REMOVAL TECHNIQUES

The process of rapid urbanization has resulted in significant modifications to urban surface characteristics, leading to the reduction in the duration of stormwater runoff and subsequently contributing to the occurrence of urban floods, riverbank erosion, and the destruction of aquatic habitats. Various pollutants, such as suspended particles, organic carbon, nutrients, and bacteria, are also significant contributors to urban stormwater runoff [8]. The utilization of bioretention, which is considered a very effective management strategy, has been employed for pollution treatment (see

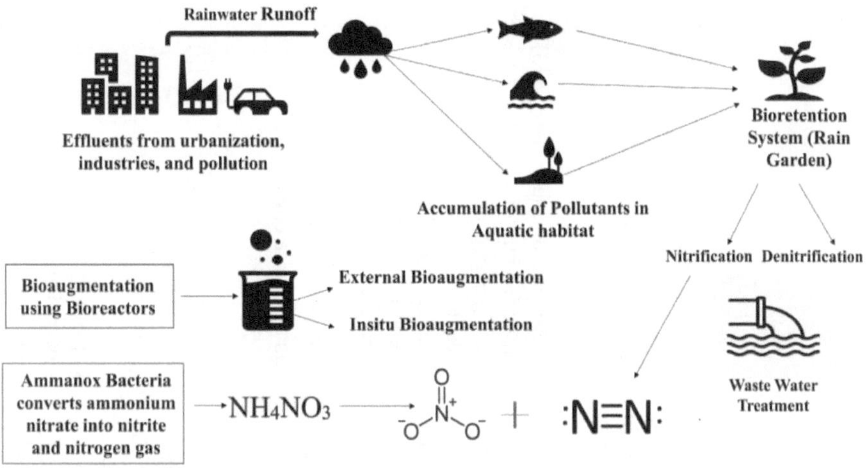

FIGURE 6.1 Techniques involved in Nitrogen removal. (Created by Biorender.com).

Figure 6.1) [9]. However, the successful removal of dissolved pollutants, specifically nitrogen, continues to pose a significant problem, particularly in surface water bodies that are prone to eutrophication [10]. Nitrogen compounds can be more easily converted into forms other than a gaseous state by using nitrogen bio-transforming processes [11]. To accomplish complete nitrogen elimination, these procedures often incorporate denitrification. Nitrogen, nitrous oxide, and dinitrogen are among the gaseous forms produced during the denitrification process from nitrate and nitrite, which are important sources of nitrogen pollution [12].

6.3 ADVANCES IN NITRIFICATION AND DENITRIFICATION

For a considerable period, it was widely accepted that denitrification was highly susceptible to oxygen and only carried out by anaerobic prokaryotes. Hence, it was postulated that denitrification, a biological mechanism responsible for nitrogen removal, could solely transpire in oxygen-deprived settings. The integration of aerobic denitrification, a process that involves conventional nitrification, resulted in increased operational expenses and hindered the widespread use of anaerobic denitrification in bioremediation endeavors. To achieve efficient nitrogen removal, anaerobic denitrification necessitates the establishment of precise operational parameters, including dissolved oxygen levels, pH values, and the presence of organic matter [21]. The bacterium *Paracoccus denitrificans* exhibits the ability to perform aerobic denitrification, a process in which dissimilatory nitrate is converted to dinitrogen in the presence of oxygen. The simultaneous implementation of nitrification and denitrification within a single aerated reactor, while maintaining consistent physicochemical conditions, resulted in a reduction in the size of the bioreactor, the required floor space for infrastructure, and subsequently, in the associated costs of both construction and operation [22]. Over the course of the last decade, there has been a significant focus on a limited number of fungal isolates that can carry out aerobic denitrification.

TABLE 6.1
Techniques Implemented in Biological Nutrient Removal

Biological nutrient removal type	Technology	Description	Advantages	Disadvantages	References
Nitrogen removal techniques	Conventional nitrification and denitrification processes.	Traditional methods involving aerobic and anaerobic stages to remove nitrogen	Effective and widely used. Well-established and robust	High energy and carbon demand, sludge production	[13]
	Anammox process	Anaerobic ammonium oxidation process that directly converts the NH4+ and NO2 to N2 gas.	Low energy and sludge production significantly reduces operational costs.	A slower startup and a specific reactor configuration are required.	[13]
	Bioaugmentation	Introduction of specialized microbial cultures to enhance the nitrogen removal	Improves system performance and faster acclimation to changing conditions.	Potential competition between native and augmented microorganisms.	[14]
	Membrane bioreactors	Integrates a membrane for solid-liquid separation coupled with a biological process	High effluent quality, compact design, improved sludge reduction, and reduced carbon footprints.	Increased operating and maintenance costs.	[7]
	Moving bed biofilm reactors	Uses a mobile biofilm carrier for bioaugmentation and efficient nitrogen removal	Biofilm offers a high biomass concentration, enhanced nutrient removal, and high tolerance to shock loads.	Potential clogging of biofilm carriers.	[15]
	Sequencing batch reactors	Cycles of fill, react, settle, and draw allow for flexible operation and bioaugmentation.	Suitable to small facilities flexible operation and effective nutrient removal.	Requires skilled operations and monitoring	[16]

	Upflow anaerobic sludge blanket	The anaerobic process that can be implemented in the Anammox and bioaugmentation	Reduced sludge generation, low production, and energy demand. suitable for high-strength wastewater.	Slower nutrient removal	[5]
Phosphorous removal techniques	Chemical precipitation	Addition of chemicals usually metal salts to form insoluble phosphorous compounds for removal	Effective for high P removal and rapid phosphorous reduction.	The cost of chemicals is high and sludge disposal is an issue.	[17]
	Biological phosphorous removal	Utilizes microbes to accumulate and remove phosphorous including EBPR and PAOs.	Reduces operational costs, environment friendly, and low chemical use.	Sensitive to environmental changes.	[17]
	Enhanced biological phosphorous removal	A biological process where specific microbes take up and store phosphorous more than their growth.	Reduced chemical usage, energy efficiency, reduced sludge production	Requires precise control and monitoring.	[17]
	Polyphosphate accumulating organisms	Specific microbes capable of accumulating phosphorous as polyphosphate	Efficient phosphorous removal, enhanced biological treatment performance	Sensitive to changes in influent conditions	[18]
	Biological air filter	Combines biological treatment with a solid media filter for phosphorous removal	Reliable phosphorous reduction, effective solids removal, and phosphorous removal	Requires media maintenance	[5]
	Membrane bioreactors	Integrates a membrane for solid-liquid separation with EBPR for efficient phosphorous removal	High-quality effluent reduces the footprint, enhances nutrient removal	High capital and operating costs	[19]

(continued)

TABLE 6.1 (Continued)
Techniques Implemented in Biological Nutrient Removal

Biological nutrient removal type	Technology	Description	Advantages	Disadvantages	References
	Alum and iron coagulation	Addition of aluminum or iron salts to precipitate and remove phosphorous.	Effective in high-strength wastewaters, can be combined with chemical phosphorous removal	Chemical costs are high and sludge generation.	[5]
	Sequencing batch reactors	Cycles, fill, react, settle, and draw combined with EBPR for phosphorous removal,	Flexible operation, improved nutrient removal, suitable for smaller facilities	Skilled operation required	[20]
	Constructed wetlands	Natural or engineered wetlands designed to remove phosphorous through biological process	Low operational and maintenance costs and provides habitat and aesthetic benefits	Limited phosphorous removal capacity.	[3]

These isolates include *Barnettozyma californica K1* [23], *Fusarium solani RADF-77* [24], *Hanseniaspora uvarum* [25], *Sporidiobolus pararoseus* [26], and *Penicillium tropicum* strain *IS0202* [27]. Fungal denitrifiers possess the ability to immobilize germs, while also exhibiting resistance to various environmental conditions, including elevated temperatures and acidic pH levels [28]. Aerobic fungal denitrifiers possess the ability to integrate the advantageous characteristics of fungal denitrifiers into the aerobic denitrification process, hence, promoting the advancement of nitrogen removal technology. Nonetheless, the assessment of individual strains for their purifying capabilities, regardless of whether they are bacterial or fungal denitrifiers, is a laborious process. Furthermore, the utilization of these strains necessitates a particular and consistent environmental context.

6.4 ANAMMOX PROCESSES

Anaerobic ammonium oxidation, or anammox, is a method for removing nitrogen. Because of their autotrophic nature, anammox bacteria do not require organic carbon for growth. The anaerobic reduction of ammonium and nitrite to dinitrogen gas is how they get their energy. Since conventional techniques of wastewater treatment for ammonia-nitrogen removal need high energy and chemical inputs, the anammox process represents a promising new option [29]. Moreover, Anammox systems frequently integrate a small reactor configuration, thereby conserving space, a crucial factor for urban wastewater treatment facilities. In addition, they demonstrate enhanced efficiency in removing nitrogen molecules from effluent.

The Anammox process converts NH_4+ to Nitrogen gas by using Nitrite (NO_2-) as an electron acceptor [30]. Due to technical constraints, the direct application of Anammox to wastewater treatment is not viable. One such constraint is the instability of Nitrite in wastewater [31]. To address this issue, the single-stage Anammox method combines Anammox with partial nitrification in a single reactor. The NH_4+ ion is partially oxidized to NO_2- by ammonia-oxidizing bacteria (AOB) during this process. The NO_2- ion, together with any leftover NH_4+, is then converted into nitrogen gas by anaerobic ammonium-oxidizing bacteria [13]. The efficiency and practicality of the reactor have been enhanced because of recent advancements in reactor design, microbial comprehension, control techniques, and nutrient recovery. Nevertheless, it is imperative to tackle the obstacles associated with the initiation of the process, maintaining a proper balance of nutrients, sensitivity to inhibitors, and scaling up to facilitate the wider use of this technology and enhance its role in promoting environmentally sustainable and effective wastewater treatment systems.

6.5 INTEGRATION OF BIOAUGMENTATION IN NITROGEN REMOVAL

The process of biological nitrification is critical in municipal wastewater treatment plants (WWTPs) for producing high-quality effluent. Ammonia-oxidizing bacteria (AOB) and nitrite-oxidizing bacteria (NOB) aid in the autotrophic process in two processes. Nonetheless, temperature [32], pH levels, dissolved oxygen levels [32], solids retention time (SRT), ammonia and nitrite concentrations, carbon to nitrogen ratio, and the presence of inhibitory compounds can all stymie the nitrification process. The nitrification process can cause severe harm to activated sludge and decrease plant operational performance, resulting in negative ecological repercussions in aquatic habitats. It is critical to undertake techniques aimed at recognizing and decreasing the impact of these inhibitory effects to lessen their influence. The use of bioaugmentation, namely, the introduction of acclimatized nitrifiers to aid in the nitrification process, has considerable promise for alleviating inhibition on a broad scale [15].

Bioaugmentation has been found as a cost-effective technique for improving the efficiency of wastewater treatment in bioreactors. The addition of externally produced nitrifiers from another reactor, known as external bioaugmentation, or the enhancement of nitrification capacities within the aerobic compartment, known as in situ augmentation, are two approaches that can be used to accomplish this [14]. The introduction of an external nitrifiers bioaugmentation technique shows good promise for improving the nitrification process within aerobic reactors running in the mainstream. Advantages include the facilitation of endogenous population growth, the stabilization of biological processes, the supply of streamlined construction procedures, and lower investment and operating expenses. However, one must admit that a significant disadvantage of bioaugmentation is the risk of failure due to poor acclimatization of microorganisms to survive and grow in harsh climatic conditions.

6.6 PHOSPHORUS REMOVAL TECHNIQUES

Phosphorus has the potential to be transmitted to freshwater ecosystems, resulting in the eutrophication of these bodies of water [33], making careful management of phosphorus in agricultural systems essential. The US Environmental Protection Agency (EPA) has established rigorous water quality regulations to control eutrophication. These regulations set maximum allowable levels of total phosphorus (TP) in water systems. A maximum TP concentration of 0.05 mg L21 is allowed in streams that empty into lakes or reservoirs. The limit is even lower within lakes and reservoirs, at just 0.025 mg L21. The EPA recommends that the TP concentration in streams and other flowing waters not drain into lakes or reservoirs not exceed 0.1 mg L21 [34].

Phosphorus removal (P-removal) in activated sludge systems was made possible by the late 1950s introduction of the enhanced biological phosphorus removal (EBPR) system [35]. Numerous EBPR system modifications have subsequently been proposed in the scholarly literature (Figure 6.2) [16]. The EBPR treatment method involves the enrichment of phosphate accumulating organisms (PAOs) within the bacterial community. This enrichment method seeks to increase the accumulation of polyphosphate (poly-P) within PAOs' cells. The accumulation of poly-P in PAOs increases the efficacy of wastewater biological phosphorus removal (bio-P-removal). Due to the strict requirement of cyclic anaerobic, anoxic, and aerobic conditions by the PAOs, the bio-P-removal process from wastewater is more complicated than the removal of nitrogen (N) and chemical oxygen demand (COD).

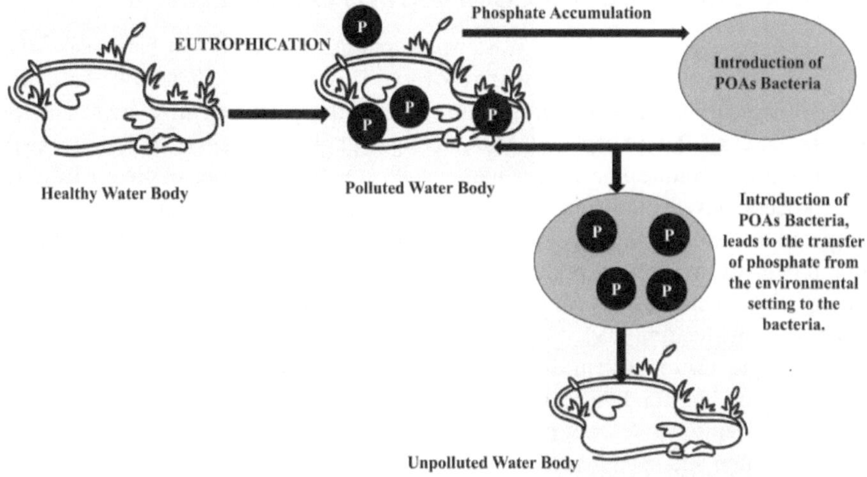

FIGURE 6.2 Techniques involved in Phosphorus removal. (Created by Biorender.com).

6.7 ENHANCED BIOLOGICAL PHOSPHORUS REMOVAL (EBPR)

The procedure of Enhanced Biological Phosphorus Removal (EBPR) in biological wastewater treatment plants (WWTPs) is a cutting-edge technique utilized to effectively eliminate phosphorus from wastewater through the accumulation of intracellular poly-P reserves. The conventional treatment arrangement entails a sequential anaerobic phase before an aerobic phase, with the periodic recycling of sludge through alternating anaerobic and aerobic phases [36]. In the process of enhanced biological phosphorus removal, organic carbon can be effectively eliminated from wastewater in the absence of oxygen, leading to its accumulation as intracellular polyhydroxyalkanoates (PHAs). Simultaneously, intracellular glycogen and poly-P undergo co-degradation, resulting in the excretion of phosphate into the wastewater [17].

Phosphate derived from wastewater has the capability to be accumulated within cells as intracellular poly-P, either in aerobic or denitrifying environments. Additionally, intracellular glycogen reserves are replenished by the oxidation of PHAs. Organisms assimilate net phosphorus, and the removal of phosphorus from wastewater can be efficiently achieved through the disposal of sludge that is high in phosphorus. The microorganisms present in EBPR sludge can be categorized based on their capacity to accumulate phosphorus, distinguishing them as either Phosphorus Accumulating Organisms (PAOs) or non-PAOs. Certain polyphosphate-accumulating organisms (PAOs) can be further categorized as denitrifying PAOs (DPAOs) due to their ability to engage in nitrate respiration. Glycogen-accumulating organisms (GAOs) are recognized for their ability to remove chemical oxygen demand (COD) by recycling carbon like polyphosphate-accumulating organisms (PAOs) [37]. However, GAOs differ in that they aerobically accumulate glycogen instead of polyphosphate.

6.8 POLYPHOSPHATE-ACCUMULATING ORGANISMS

Photo granules are an innovative wastewater treatment method that harnesses solar energy to achieve effective nutrient recovery. This process efficiently eliminates nitrogen and carbon; however, it exhibits comparatively lower conversion rates for phosphorus [18]. Polyphosphate-accumulating organisms (PAOs) are significant in the context of phototrophic cultivation, specifically in the use of photo EBPR[38]. This process has demonstrated enhanced phosphorus removal capabilities when compared to phototrophic systems operating independently. The primary metabolic processes involved in photo EBPR encompass photosynthesis, polyphosphate buildup, chemoheterotrophy for chemical oxygen demand (COD) reduction, nitrification, and denitrification [39]. Nevertheless, the integration of various functionalities within a single reactor system is a considerable challenge owing to the potential competition for nutrients and the inhibitory effects that may arise.

Granules demonstrate both physical and chemical gradients, which contribute to the stratification and spatial heterogeneity of microbial communities [18]. Inside the framework of photo electro-biological phosphorus removal (photo EBPR), photo granules have the potential to serve as habitats for various organisms, facilitating their coexistence inside a single reactor and enabling the achievement of elevated conversion rates essential for efficient wastewater treatment. Nevertheless, the operation of the reactor frequently lacked optimization for the coexistence of phototrophs and PAOs, resulting in unpredictable performance and diminished removal rates in comparison to alternative AGS systems.

To tackle this matter, it is imperative to develop strategies that enable the successful integration of PAOs (polyphosphate-accumulating organisms) into photo granules. Additionally, it is crucial to optimize operational conditions to achieve high conversion rates. Furthermore, a comprehensive understanding of the assembly and roles of the microbial community associated with PAOs is essential. Moreover, light-driven activities are reliant on the inherent diel cycle of solar radiation, potentially resulting in a longer period of anaerobic conditions during the nocturnal hours. The present work incorporated particulate organic aggregates (PAOs) derived from activated granular sludge (AGS) into an existing culture, intending to investigate the assembly and functioning of the microbial community.

6.9 INNOVATIVE MICROBIAL CULTURES AND CONSORTIA

Polyhydroxyalkanoate (PHA) is a bio-polymer that is now being investigated as a potential substitute for plastics derived from petroleum. It is produced as a metabolic by-product by microbes in wastewater treatment facilities (WWTP) that are subjected to high levels of substrate [40]. The initial documentation of polyhydroxyalkanoate (PHA) accumulation in microorganisms engaged in enhanced biological phosphate removal (EBPR) within a wastewater treatment plant (WWTP) was conducted by Wallen and Rohwedder [41]. Nutrient removal in conventional EBPR methods is accomplished by the utilization of a distinct group of microorganisms known as PAOs (Phosphate Accumulating Organisms). These microorganisms operate in a sequential anaerobic-aerobic/anoxic regime. During the anaerobic phase, polyphosphate-accumulating organisms (PAOs) engage in the synthesis of polyhydroxy butyrate (PHB) or polyhydroxy valerate (PHV), while simultaneously releasing phosphate. This release occurs because of the breakdown of the internal polyphosphate pool, which enables the production of adenosine triphosphate (ATP) necessary for cellular energy [42]. During the succeeding electron-rich aerobic/anoxic phase, the stored polyhydroxy butyrate (PHB) is utilized to replenish the phosphate pool, leading to the loss of phosphate in general.

The concurrent accumulation of phosphate and recovery of polyhydroxy butyrate (PHB) is not viable inside a standard EBPR system. The literature extensively covers the recovery of polyhydroxyalkanoates (PHAs) from enhanced biological phosphorus removal (EBPR) enriched sludge, however, most studies do not specifically emphasize nutrient removal. The presence of a varied array of microorganisms could potentially have a role in both the formation of polyhydroxy butyrate (PHB) and the elimination of nutrients. At first, it was hypothesized that nitrate had a suppressive impact on the accumulation of phosphate by polyphosphate-accumulating organisms (PAOs) [43]. However, subsequent research revealed that in anaerobic or anoxic environments, a subset of denitrifiers facilitated the accumulation of phosphate by utilizing intracellular polyhydroxyalkanoates (PHA/PHB) as a source of electrons and nitrate as the final electron acceptor. Dinucleotide phosphate acylphosphatases (DNPAOs) can amass polyhydroxyalkanoates (PHA) without the concomitant liberation of phosphate via glycogen degradation, serving as an alternative energy reservoir.

The efficacy of dynamically enhanced biological phosphorus removal sludge has been verified for the recovery of polyhydroxy butyrate (PHB) under various operational circumstances, including total aerobic and aerobic dynamic feeding, with the inclusion of factors such as hydraulic retention time (HRT), pH, and solids retention time (SRT) [44]. Nevertheless, there exists a dearth of information regarding the conduct of microorganisms and the efficacy of reactors in situations where complete anoxic circumstances are maintained alongside an excess of external substrate. The optimization of carbon concentrations within the reactor is of utmost importance in the development of a sustainable process that effectively balances optimal nutrient removal while simultaneously promoting the formation of PHB.

6.10 GENETIC ENGINEERING IN IMPROVING MICROBIAL PERFORMANCE

The application of genetic engineering holds promise for enhancing microbial efficacy within the Biological Nutrient Removal (BNR) process. Through the manipulation of the genetic composition of microorganisms engaged in biological nutrient removal, scientists and engineers can create microbial strains that exhibit enhanced efficiency, resilience, and adaptability to certain circumstances encountered in wastewater treatment processes. Genetic engineering has the potential to augment food absorption, bolster resistance to stress, minimize the production of byproducts, improve metabolic pathways, regulate competition, amplify biofilm formation, and govern gene expression [45].

The expression of genes involved in byproduct creation pathways can be selectively reduced, whilst genes associated with denitrification processes can be altered to

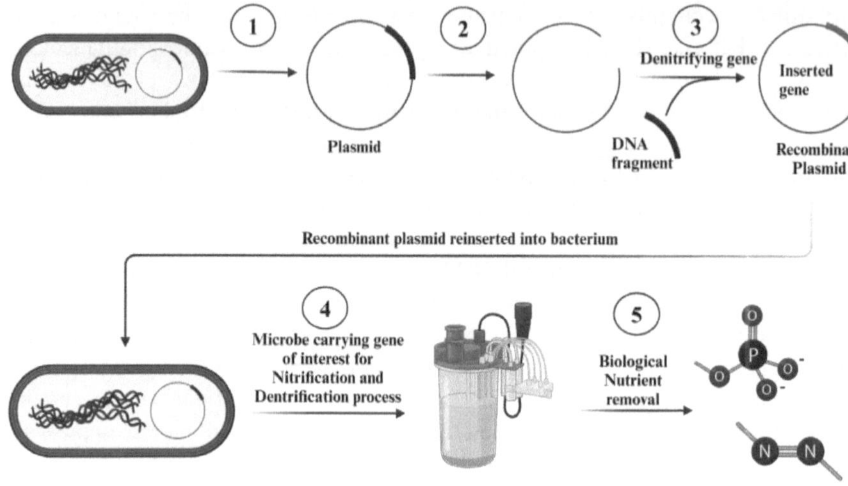

FIGURE 6.3 Genetically improved organism in BNR process. (Created by Biorender.com).

facilitate the complete conversion of nitrate to nitrogen gas. Genetic alterations have the potential to regulate competition among distinct microbial populations within BNR systems. For instance, these modifications can enable phosphorus-accumulating organisms (PAOs) to surpass glycogen-accumulating organisms (GAOs) in terms of using carbon sources.

The enhancement of biofilm development in BNR reactors can be achieved by the modification of genes associated with adhesion and the synthesis of extracellular polymeric substances (EPS) [45]. This genetic modification leads to an improvement in the overall performance of the reactors. The manipulation of genetic switches and promoters enables the regulation of gene expression in response to environmental stimuli, hence facilitating the precise and adaptable management of microbial activity within BNR systems (Figure 6.3).

Genetic engineering has the potential to generate biosensors and reporter strains, which can offer real-time data on the performance of the biological nutrient removal (BNR) system. This capability allows operators to enhance their monitoring of nutrient levels and microbial activity with greater effectiveness. Nevertheless, it is imperative to consider the regulatory and safety considerations associated with the utilization of genetically modified organisms (GMOs) in the context of wastewater treatment [45]. To ensure the safe and responsible utilization of genetically modified organisms (GMOs) in bioenergy with carbon capture and storage (BECCS) systems, it is imperative to thoroughly evaluate ethical and environmental considerations through rigorous risk assessments.

6.11 BIOREACTORS AND PROCESS OPTIMIZATION

Bioreactors are artificially regulated systems in which microorganisms are employed to degrade and eliminate pollutants present in wastewater. Within the realm of biological

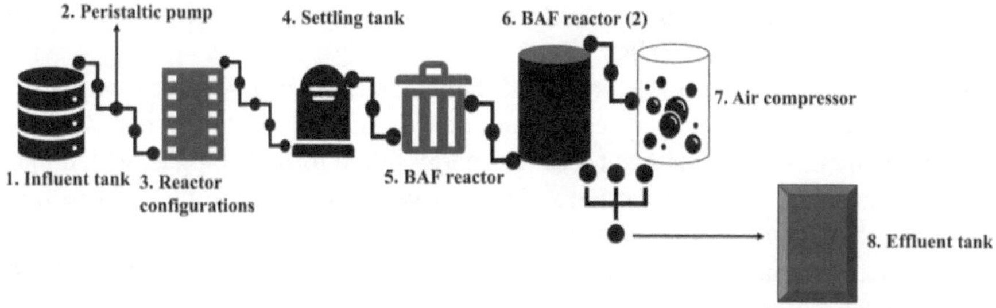

FIGURE 6.4 Biological aerated filter system for wastewater treatment. (Created by Biorender.com).

nutrient removal (BNR), these systems are purposefully designed to cultivate the proliferation and functionality of distinct microbial communities that are accountable for the elimination of nitrogen and phosphorus. The efficacy of BNR bioreactors is contingent upon various aspects, including ambient parameters, microbial activity, and process management measures. Efficient and cost-effective nutrient removal, while adhering to water quality rules, necessitates the optimization of both bioreactor design and treatment processes. The management of operational parameters, such as dissolved oxygen levels, hydraulic retention durations, pH, and influent ratios, plays a crucial role in bioreactor and process optimization. These factors are carefully controlled to provide favorable circumstances that promote the growth and activity of various microbial groups. The field of biotechnology has witnessed significant progress, particularly in the realm of genetic engineering. These advancements have presented novel opportunities for augmenting the efficiency of microbial activity in biological nutrient removal bioreactors. The introduction highlights the significance of BNR in tackling issues related to nutrient pollution. It emphasizes the interdisciplinary character of BNR optimization, which involves the integration of engineering, microbiology, and biotechnology to safeguard water resources and the environment (Figure 6.4).

6.12 UPFLOW ANAEROBIC SLUDGE BLANKET REACTORS (UASB) FOR BNR

The Upflow Anaerobic Sludge Blanket (UASB) reactor is a wastewater treatment technique used for industrial and municipal wastewater anaerobic digestion and organic matter removal. It is well-known for its high-rate treatment capabilities and capacity to efficiently manage high-strength wastewater [5]. Through anaerobic ammonium oxidation (anammox) and denitrification processes, UASB reactors enable partial nitrogen removal.

UASB reactors are often better adapted to remove organic materials than are nitrogen and phosphorus. It can be difficult to achieve total biological nutrient removal in a UASB reactor, and maintain optimal hydraulic and organic loading rates, pH, and temperature management is critical for effective nitrogen removal [5].

In rare circumstances, post-treatment treatments such as aerobic treatment may be required for more thorough BNR.

To supplement existing BNR processes, UASB reactors can be added into larger wastewater treatment systems. These reactors have a smaller footprint than standard aerobic treatment systems and can recover energy in the form of biogas, which may be used to generate heat or electricity. They are very successful in the treatment of high-strength industrial wastewater streams.

UASB reactors have limitations such as restricted nutrient removal, which is less efficient when compared to specialist BNR systems, and temperature sensitivity, which may be difficult in colder locations [5].

6.13 MOVING BED BIOFILM REACTORS (MBBR)

Biofilm reactors are used in biological wastewater treatment to extract organic carbon and nitrogen species. The improved treatment capacity and operational stability provided by the biofilm method as a result of the high active biomass concentration are two of its most salient benefits [46]. Biofilm reactors are suitable for slow-growing microorganisms because microbe immobilization increases the sludge retention time (SRT) [47]. Aeration, liquid recirculation, or mechanical mixing energy are used in moving bed biofilm reactors (MBBRs) to create turbulent energy that keeps biofilms suspended throughout the reactor. This approach can be used to treat wastewater that is aerobic, anoxic, or anaerobic. High-strength effluents, such as those produced by dairy, textile, and wine distilleries, may be treated by anaerobic moving bed biofilm reactors (AMBBRs).

Biofilm carriers (media) play a significant role in AMBBRs. For wastewater treatment, several types of synthetic packing materials with varying geometries, materials, and production procedures have been created. AnoxKaldnes is a well-known manufacturer specializing in biological wastewater treatment. Over the years, the business has produced a variety of various K-type carriers, the majority of which have been employed in MBBR installations.

In the reaction tank, the biofilm carrier can be hung or fastened [48]. Porous foam pads or cylindrical-shaped plastic material are common suspended carriers in AMBBRs. Carriers such as sand or gravel have a low or nil cost [49], whereas artificial supports are expensive and can cause transportation and storage issues in developing nations. Several studies and experiments have been conducted using alternative biofilm carriers, such as pumice stone, porous glass beads, and FLOCOR-RMP in wastewater treatment and biodegradable meal box, licorice (*Glycyrrhiza glabra*), giant reed (*Arundo donax*), and cotton for denitrification of nitrate-contaminated groundwaters [50]. Straw and sisal fiber waste from industry and agriculture are also employed as low-cost biofilm carriers [51].

6.14 CHALLENGES AND FUTURE DIRECTIONS

6.14.1 Emerging Challenges in BNR

Biological nitrogen removal (BNR) is a key course in wastewater treatment that plays a pivotal role in mitigating environmental pollution and related ecological and human health risks. Nitrogen compounds, primarily ammonia and nitrate, are usually present in wastewater, and their effective removal is critical to avoid water pollution. Despite being an engrained method, BNR faces several emerging challenges in the present day. One of the chief encounters in BNR is the growing nitrogen load in wastewater due to population growth, industrialization, and agriculture [52]. Disproportionate nitrogen discharge into natural water bodies can lead to eutrophication, oxygen depletion, and harmful algal blooms, highlighting the urgent need to improve BNR systems that can cope with higher nitrogen concentrations. Additionally, the adaptation of BNR processes to decentralized and small-scale wastewater treatment systems is becoming increasingly crucial. Many regions and communities are seeking cost-effective solutions to address nitrogen removal in localized settings, such as individual households or small treatment plants. This necessitates the development of innovative, compact BNR technologies that can be easily integrated into these systems. Moreover, emerging contaminants in wastewater, such as pharmaceuticals and personal care products, can interfere with conventional BNR processes [52]. These contaminants can affect the microbial communities responsible for nitrogen removal, potentially reducing the efficiency of BNR systems. Research is needed to understand how these emerging pollutants impact BNR and to develop strategies to mitigate their effects. The initiation of resource recovery in wastewater treatment introduces added challenges and opportunities for BNR. While recovering nitrogen from wastewater can be beneficial, it entails advanced BNR processes and technologies to selectively extract nitrogen compounds without negotiating treatment efficiency. Therefore, it is essential to develop and optimize BNR systems that can effectively remove nitrogen compounds from wastewater while addressing these emerging challenges.

6.15 REGULATORY CONSIDERATIONS AND ENVIRONMENTAL IMPACT

Biological Nutrient Removal (BNR) proposes to mitigate the discharge of excessive nutrients, particularly nitrogen and phosphorus, into natural water bodies [53]. While the environmental benefits of BNR are well-established, its successful implementation is contingent upon regulatory considerations and environmental impacts. Regulatory agencies worldwide have imposed stringent standards to limit nutrient discharges in effluents, recognizing the adverse effects of nutrient pollution on aquatic ecosystems [53]. Compliance with these regulations is a primary driver for adopting BNR technologies in wastewater treatment plants. Meeting these standards often necessitates investment in advanced BNR processes and monitoring systems to ensure optimal nutrient removal. However, BNR itself can have environmental impacts. It may generate sludge rich in nutrients, which, if not managed properly, can pose environmental risks. Sustainable disposal or utilization of this sludge, such as agricultural application, is essential to prevent nutrient re-entry into the environment. BNR processes can also indirectly affect the overall energy consumption of treatment plants, potentially leading to increased greenhouse gas emissions. Thus, optimizing BNR to minimize energy consumption and carbon footprint is critical.

6.16 PROMISING DIRECTIONS FOR FURTHER RESEARCH AND INNOVATION

There are several promising avenues for further research and innovation in the field of biological nutrient removal. One such direction involves a more in-depth exploration of advanced microbial ecology, with a focus on optimizing the intricate microbial communities that underpin BNR processes [52]. This approach is expected to yield greater efficiency in nutrient removal. Additionally, resource recovery from removed nutrients aligns with a circular economy approach, and the development of innovative technologies for nutrient reuse is crucial for both efficiency and sustainability. Given the growing demand for decentralized wastewater treatment solutions, compact BNR technologies suitable for small-scale applications are essential. Research should also investigate the impact of emerging contaminants on BNR and ways to mitigate these effects, as pollutants such as pharmaceuticals and microplastics become more prevalent in wastewater. Furthermore, addressing energy efficiency in BNR through innovations in aeration systems, process control, and alternative energy sources will contribute to more sustainable wastewater treatment practices. Lastly, adaptive BNR technologies that can respond to changing influent conditions and pollution events will enhance the resilience and reliability of treatment plants, ensuring compliance with evolving regulatory standards [52]. In summary, these research and innovation directions hold great promise for advancing the field of BNR and improving the sustainability and efficiency of wastewater treatment practices.

6.17 SUSTAINABILITY AND GREEN PRACTICES

The implementation of sustainable approaches to Biological Nutrient Removal (BNR) is of utmost importance in addressing water pollution while minimizing the environmental impact. By incorporating eco-friendly technologies such as low-energy aeration systems and innovative microbial management, BNR processes can significantly decrease their energy consumption [54]. Additionally, the recovery and reuse of extracted nutrients from wastewater align with sustainability principles, effectively transforming waste into a valuable resource. The adoption of decentralized BNR solutions, customized to meet local requirements, promotes efficient and localized treatment, thereby reducing the energy costs associated with transporting wastewater to centralized plants [54]. Sustainable BNR practices not only safeguard ecosystems against nutrient pollution but also contribute to responsible resource management and energy conservation.

6.18 INTEGRATION OF BNR WITH RENEWABLE ENERGY SOURCES

The integration of Biological Nutrient Removal (BNR) with renewable energy sources is an incredibly promising avenue for significantly enhancing the sustainability of wastewater treatment. By cleverly coupling BNR processes with cutting-edge renewable energy technologies such as solar panels, wind turbines, or anaerobic digesters, treatment plants can boldly reduce their reliance on fossil fuels and dramatically decrease their carbon footprint [54]. The energy harnessed from these remarkable sources can power BNR systems, including aeration, pumps, and instrumentation, thereby substantially increasing energy efficiency, and demonstrating exceptional environmental stewardship. This extraordinary synergy not only leads to substantial savings in operational costs but also aligns perfectly with global efforts to transition towards greener and more self-sufficient wastewater treatment solutions, thus making a remarkable contribution to a more sustainable future.

6.19 CARBON FOOTPRINT REDUCTION IN NUTRIENT REMOVAL PROCESSES

Reducing the carbon footprint in nutrient removal processes is a paramount objective in contemporary wastewater treatment. Traditionally, energy-intensive techniques have been utilized in processes such as aeration and chemical dosing for nitrogen and phosphorus removal, resulting in a substantial environmental impact [54]. To address this issue, innovative strategies are being implemented. These strategies encompass optimizing aeration through the utilization of energy-efficient technologies, such as fine-bubble diffusers and process control systems, to minimize energy consumption during biological nutrient removal. Furthermore, resource recovery from the waste stream, such as harnessing biogas generated from sludge digestion to power treatment plants, can further diminish the carbon footprint [54]. By implementing these measures, wastewater treatment facilities can effectively reduce their energy consumption, greenhouse gas emissions, and overall environmental impact while simultaneously adhering to stringent nutrient removal standards.

6.20 CONCLUSION

Recent advances in Biological Nutrient Removal (BNR) technology constitute a big step forward in wastewater treatment and environmental conservation. As we have investigated the various aspects of BNR, these advancements provide novel answers to the pressing problems of nutrient pollution in our water systems. The advancements in nitrogen and phosphorus removal technology, as well as the utilization of specific microbial cultures and consortia, demonstrate our ability to attain greater efficiency and sustainability in nutrient removal procedures. Bioreactors and process optimization methodologies have polished the engineering features of BNR systems, making them more versatile and scalable. Despite these accomplishments, we must stay alert. Emerging issues and changing regulatory requirements involve ongoing research and innovation. The route forward entails not just solving these challenges, but also combining BNR with sustainable practices, renewable energy sources, and carbon-reduction methods. In our quest for excellence, we pave the path for a more ecologically conscious future in which clean water supplies are protected, ecosystems thrive, and the environmental effect of wastewater treatment is reduced. These developments represent a hopeful chapter in our environmental stewardship, promising a brighter and cleaner future for all.

REFERENCES

[1] W. Zuo, Z. Chen, J. Zhang, W. Zhan, H. Yang, L. Li, W. Zhu, and Y. Mao, *Environ Sci Pollut Res Int* **30**, 60920 (2023).

[2] A. Majumder, K. Bhattacharyya, S.C. Kole, and S. Ghosh, *Environ Sci Pollut Res* **20**, 5645 (2013).

[3] X.J. Zuo, H.S. Zhang, and J. Yu, *Chemosphere* **244**, 125626 (2020).

[4] N. Zwart, W. Jonker, R. Ten Broek, J. de Boer, G. Somsen, J. Kool, T. Hamers, C.J. Houtman, and M.H. Lamoree, *Water Res* **168**, 115204 (2020).

[5] X.L. Zou, *Environ Technol (UK)* **36**, 2381 (2015).

[6] A. Sabzali, M. Nikaeen, and B. Bina, *Environ Technol (UK)* **33**, 1803 (2012).

[7] G.T. Daigger and H.X. Littleton, *Water Environ Res* **86**, 245 (2014).

[8] W. Ahmed, K. Hamilton, S. Toze, S. Cook, and D. Page, *Sci Total Environ* **692**, 1304 (2019).

[9] K. Flanagan, P. Branchu, L. Boudahmane, E. Caupos, D. Demare, S. Deshayes, P. Dubois, L. Meffray, C. Partibane, M. Saad, and M.-C. Gromaire, *Water Res* **145**, 562 (2018).

[10] H. Xu, Y. Zhang, X. Zhu, and M. Zheng, *Hydrol Res* **50**, 1062 (2019).

[11] M.M.M. Kuypers, H.K. Marchant, and B. Kartal, *Nat Rev Microbiol* **16**, 263 (2018).

[12] H. Shoun, S. Fushinobu, L. Jiang, S.-W. Kim, and T. Wakagi, *Philos Trans R Soc B Biol Sci* **367**, 1186 (2012).

[13] L. Zuo, H. Yao, H. Li, L. Fan, and F. Jia, *Int J Environ Res Public Health* **17**, 1 (2020).

[14] H.L. Tang and H. Chen, *Bioresour Technol* **190**, 76 (2015).

[15] M. Herrero and D.C. Stuckey, *Chemosphere* **140**, 119 (2015).

[16] Y. Peng and S. Ge, *Bioresour Technol* **102**, 6405 (2011).

[17] A. Oehmen, P.C. Lemos, G. Carvalho, Z. Yuan, J. Keller, L.L. Blackall, and M.A.M. Reis, *Water Res* **41**, 2271 (2007).

[18] L.M. Trebuch, J. Sohier, S. Altenburg, B.O. Oyserman, M. Pronk, M. Janssen, L.E.M. Vet, R.H. Wijffels, and T.V. Fernandes, *Water Res* **235**, 119748 (2023).

[19] J.M.M. Santos, L. Rieger, A.B. Lanham, M. Carvalheira, M.A.M. Reis, and A. Oehmen, *Water Res* **171**, 115373 (2020).

[20] J. Jena, R. Kumar, A. Dixit, S. Pandey, and T. Das, *PLoS One* **10**, 1 (2015).

[21] Y. Yan, H. Lu, J. Zhang, S. Zhu, Y. Wang, Y. Lei, R. Zhang, and L. Song, *Environ Adv* **9**, 100254 (2022).

[22] F. Xie, M. Thiri, and H. Wang, *Bioresour Technol* **319**, 124198 (2021).

[23] J. Fang, S. Liao, S. Zhang, L. Li, S. Tan, W. Li, A. Wang, and J. Ye, *Bioresour Technol* **339**, 125665 (2021).

[24] H.-Y. Cheng, A.-A. Xu, M. Kumar Awasthi, D.-D. Kong, J.-S. Chen, Y.-F. Wang, and P. Xu, *Bioresour Technol* **295**, 122250 (2020).

[25] H. Zhang, Z. Zhao, P. Kang, Y. Wang, J. Feng, J. Jia, and Z. Zhang, *Bioresour Technol* **267**, 569 (2018).

[26] J. Zeng, S. Liao, M. Qiu, M. Chen, J. Ye, J. Zeng, and A. Wang, *Bioresour Technol* **312**, 123593 (2020).

[27] Z. Yao, L. Yang, F. Wang, L. Tian, N. Song, and H. Jiang, *Bioresour Technol* **303**, 122948 (2020).

[28] X. Zuo, W. Xu, S. Wei, S. Jiang, Y. Luo, M. Ling, K. Zhang, Y. Gao, Z. Wang, J. Hu, H.P. Grossart, and Z. Luo, *iScience* **26**, 106824 (2023).

[29] H. Chandel, K. Shyam, N. Kumar, G. Sharma, M. Yadav, S. Murugesan, S. Thakur, and G. Saxena, in *Integrated Environmental Technologies for Wastewater Treatment*

and Sustainable Development, edited by V. Kumar and M. Kumar (Elsevier, 2022), pp. 23–48.
[30] J.G. Kuenen, *Nat Rev Microbiol* **6**, 320 (2008).
[31] L. Zhang, J. Yang, and K. Furukawa, *J Biosci Bioeng* **110**, 441 (2010).
[32] S. Zhang, Y. Wang, W. He, M. Wu, M. Xing, J. Yang, N. Gao, and M. Pan, *Chem Eng J* **236**, 242 (2014).
[33] D.L. Correll, *J Environ Qual* **27**, 261 (1998).
[34] US EPA (U.S. Environmental Protection Agency), EPA – Environmental Protection Agency 395 (1986).
[35] C.M.P. Vazquez, M.C. Wentzel, Y. Comeau, G.A. Ekama, M.C.M. van Loosdrecht, D. Brdjanovic, and A. Oehmen, *Biological Wastewater Treatment – Principles, Modelling and Design*, edited by M. Henze (IWA Publishing, 2008), p. 155.
[36] G.J.F. Smolders, J. van der Meij, M.C.M. van Loosdrecht, and J.J. Heijnen, *Biotechnol Bioeng* **47**, 277 (1995).
[37] R.J. Seviour and S. McIlroy, *J Microbiol* **46**, 115 (2008).
[38] V.C.F. Carvalho, M. Kessler, J.C. Fradinho, A. Oehmen, and M.A.M. Reis, *Sci Total Environ* **793**, 148501 (2021).
[39] A.Y.A. Mohamed, L. Welles, A. Siggins, M.G. Healy, D. Brdjanovic, A.M. Rada-Ariza, and C.M. Lopez-Vazquez, *Water Res* **189**, 116606 (2021).
[40] H. Zou, G.-C. Du, W.-Q. Ruan, and J. Chen, *World J Microbiol Biotechnol* **22**, 701 (2006).
[41] L.L. Wallen and W.K. Rohwedder, *Environ Sci Technol* **8**, 576 (1974).
[42] R.J. Seviour, T. Mino, and M. Onuki, *FEMS Microbiol Rev* **27**, 99 (2003).
[43] J.J. Beun, E.V. Verhoef, M.C.M. Van Loosdrecht, and J.J. Heijnen, *Biotechnol Bioeng* **68**, 496 (2000).
[44] A.S.M. Chua, H. Takabatake, H. Satoh, and T. Mino, *Water Res* **37**, 3602 (2003).
[45] A.K. Dangi, B. Sharma, R.T. Hill, and P. Shukla, *Crit Rev Biotechnol* **39**, 79 (2019).
[46] S. Chen, D. Sun, and J.-S. Chung, *Waste Manage* **28**, 339 (2008).
[47] D. Elenter, K. Milferstedt, W. Zhang, M. Hausner, and E. Morgenroth, *Water Res* **41**, 4657 (2007).
[48] H.S. Kim, J.W. Gellner, J.P. Boltz, R.G. Freudenberg, C.K. Gunsch, and A.J. Schuler, *Water Res* **44**, 1553 (2010).
[49] G. Martínez-Huerta, B. Prendes-Gero, F. Ortega-Fernández, J.M. Mesa-Fernández, *Proceedings of the 2nd International Conference on Environmental and Geological Science and Engineering, EG '09,* p. 44-49 (2009).
[50] M. Kermani, B. Bina, H. Movahedian, M.M. Amin, and M. Nikaeen, *Iran J Biotechnol* **7**, 19 (2009).
[51] J. Andersson and L. Björnsson, *Bioresour Technol* **85**, 51 (2002).
[52] Y. Yang, X. Shi, W. Ballent, and B.K. Mayer, *Water Environ Res* **89**, 2122 (2017).
[53] A. Reyes-Calderón, S. Pérez-Uribe, A.G. Ramos-Delgado, S. Ramalingam, G. Oza, R. Parra-Saldívar, R.A. Ramirez-Mendoza, H.M.N. Iqbal, and A. Sharma, *J Hazard Mater* **423**, 127031 (2022).
[54] H. Zhirong, H. Dwight, and D. Peter, *J Environ Eng* **138**, 307 (2012).

7 Recent Studies, Challenges, and Future Prospects on the Degradation of Phenolic Compounds

Ganesh Swain, Tapas Das, and Ravi Kumar

7.1 INTRODUCTION

Phenolic compounds are classified as aromatic hydrocarbons, like benzene ring structures containing single or multiple hydroxyl groups. The structures of some common phenolic compounds (such as phenol, 4-chlorophenol, catechol, p-cresol, nitrophenol, salicylic acid, etc.) have been represented in Figure 7.1. The high solubility and diffusivity of these compounds have adverse effects on the ecosystems. The US Environmental Protection Agency (USEPA) has listed phenolic compounds as priority pollutants due to their recalcitrant and toxic nature [1,2]. The sources of the phenolic compounds are either natural or anthropogenic activities. These are naturally produced from the decomposition and biosynthesis of plant materials. Under exposure to ultraviolet rays, the phenolic compounds are produced from the amino acids present in the hemicelluloses. In addition, the breakdown of tyrosine in the animal intestinal tract produces phenol and p-cresol [3]. However, human activity plays a major role in generating phenolic compounds in the ecosystem. Due to their versatile nature, phenolic compounds are widely used in various pharmaceuticals, coke ovens, pulp and paper, pesticides, dyes, leather, and paint industries [4,5]. These are also used in the manufacturing of disinfectants, wood preservatives, insecticides, and adhesives. Phenolic compounds are also released to the environment during the incineration of municipal waste, chlorination of the drinking water, smoking, and vehicle exhaust [6]. The concentration of the phenolic compounds in the environment depends upon the concentration of the pollutants in the effluent of the industries. Vashi et al. [7] have reported that the paper and pulp mill effluent contains phenol up to 200 mg/L, whereas, 50–2000 mg/L of phenolic compounds were found in the refinery and coke oven industries effluent.

Therefore, the phenolic compounds should be completely treated before discharging into the environment; otherwise, they will adversely affect human beings

DOI: 10.1201/9781003454199-7

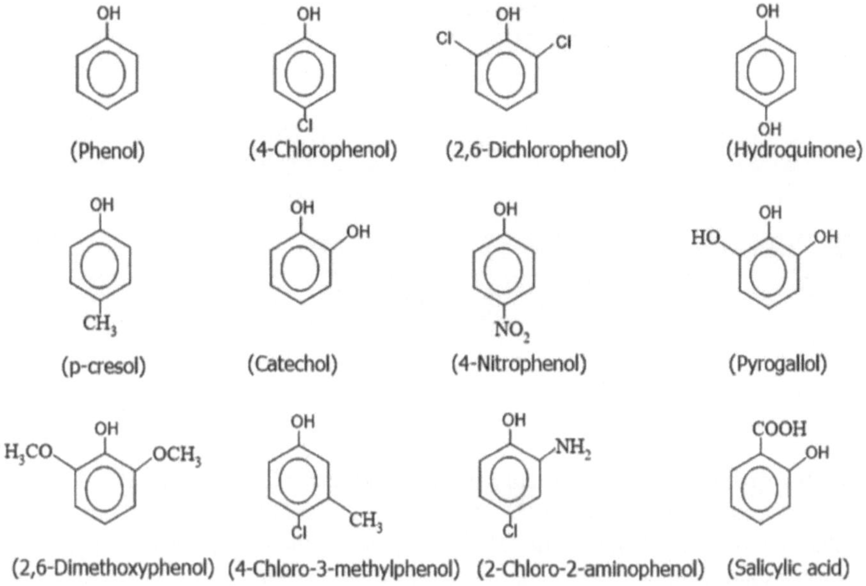

FIGURE 7.1 Chemical structures of phenolic compounds. (Compiled by the authors.)

and aquatic lives. The phenolic compounds undergo oxidation to form more toxic compounds such as quinones and semiquinone radicals, damaging DNA and protein inside the cell membrane [8]. Para-cresol and alkylphenols can induce carcinogenic and teratogenic effects in human beings [6]. The value of median lethal concentration (LC50) was found to be in the range of 1.4–3.88 mg/L for crustaceans, algae, and bacteria [9]. Therefore, the Bureau of Indian Standards (BIS) has set a permissible limit of phenolic concentration in drinking water as 0.002 mg/L [10].

Hence, the degradation of precarious phenolic compounds from wastewater became essential. In this direction, various conventional methods (adsorption, extraction, membrane separation, etc.) were initially studied [11,12]. However, several disadvantages, including the reusability limit of adsorbent, secondary toxic product generation, the high cost of the membrane, and fouling urge the researcher to develop new methodologies for the effective degradation of phenolic compounds. Further, some developed and hybrid methods (i.e., advanced oxidation process, enzymatic treatment, and nanotechnologies) were analyzed on a laboratory scale [13, 14]. However, still, a lot of research is going on regarding the development of a cheap and economical method for the degradation of phenolic pollutants from wastewater. Vashi et al. [7] used a novel technology namely the aerobic granular sludge method for the real phenolic wastewater treatment. The proposed protocol could be able to degrade 74 percent of tannin (a polyphenolic compound) at a concentration of 50 mg/L. Moreover, the concentration level of phenolic compounds was detected in various effluents and found to be 3.0–1200 mg/L (petrochemicals) and 30–1200 mg/L (coke oven plant). A hybrid method comprised of Fenton–biodegradation was utilized for the degradation of high concentration of phenol [15]. They reported that a removal

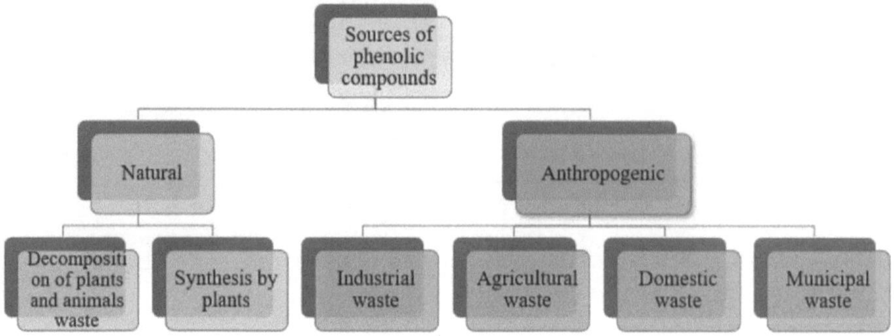

FIGURE 7.2 Schematic representation showing various sources of phenolic compounds. (Compiled by the authors.)

above 99 percent of phenol was achieved using the current hybrid method at an initial phenol concentration of 1000 mg/L. The combined use of electrolysis and anaerobic-aerobic microbial oxidation methods was adopted for the treatment of coal gasification wastewater [16].

The present chapter demonstrates a brief overview of the various sources and adverse impacts of phenolic wastewater on the living system. In addition, a comparative analysis of the various conventional methods, including physical, chemical, and oxidation technologies, were summarized. Moreover, some emerging technologies, which have been preferred in the last few years, were also briefly described in this chapter.

7.2 SOURCES OF PHENOLICS

The phenolic compounds generally originate from two sources such as natural and human-made activities (see Figure 7.2). The natural source releases a very small amount of phenolics, whereas human activities are the major factor for the abundance of phenolic compounds in land and water bodies. The various sources of phenolic compounds are described below.

The decomposition of plant and animal waste are the primitive source of phenol release to the environment. For instance, the hemicellulose present in the wood can produce amino acid, which is a phenolic compound. In addition, the degradation of the biomass produces hydroxybenzene. Phenolics are also found in some fruits, roots, leaves, stems, and vegetables, which are produced by chlorophyll under UV (ultraviolet) exposure [6]. Moreover, some plants release phenolic compounds (such as phenolic acids, flavonoids, and phenylopropanoids) under heavy metal stress, which acts as an oxidizing agent [17]. Thus, phenol accumulates the heavy metals as it has excellent chelating ability and exudes from the plant to the environment. In the meantime, the natural secretion of phenolic compounds has a minimal effect on the ecosystem as they are released in a very small quantity. Therefore, very little attention was given to the control, degradation, or recycling of the natural phenolic sources.

Apart from natural sources, there are certain anthropogenic sources such as manufacturing, agricultural, and municipal waste, which generate phenolics in a higher concentration. Generally, phenolic compounds are categorized into two parts including simple phenols and polyphenols. The polyphenols contain multiple phenolic groups having a higher toxicity compared to simple phenols [18]. Phenol and its derivatives (i.e., chlorophenol, aniline, cresol, etc.) have excellent antimicrobial, flammability, thermal, and bonding properties, which enable the wide use of these aromatic compounds in various chemical, pharmaceutical, wood, textile, dye, petrochemical, and rubber industries [2,19]. It was reported that the steel manufacturing industry uses almost 30–60 m^3 of fresh water for the production of 1.0 tonne of steel. Moreover, the level of p-cresol was found to be in the range of 1200 mg/L in coke oven wastewater and the phenolic compound was successfully degraded using an isolated bacterial species (i.e., *Pseudomonas citronellolis* NS1). The phenolic resin, a well-known thermosetting polymer extensively applied for the manufacturing of bakelite, plywood, automobile, and aerospace industries. Some of the residual phenolic compounds are released into the environment either with untreated wastewater or with the exhaust air. In addition, palm oil mills release a volume of 600–870 kg oil effluent containing various phenolic compounds. The concentration of phenolic compounds was found in the range of 280–680 mg/L [20]. Apart from industrial sources, agricultural waste is also contributing a significant role in the generation of phenolic compounds. It is reported that some phenolics such as 4-chlorophenol, 2,4-dinitrophenol, and catechols are widely used for the manufacturing of pesticides, herbicides, and insecticides [21]. There are currently 22 types of chlorophenolic compounds used in various industries (like dyes, petroleum, and pharmaceuticals), which are highly toxic to the ecosystem [2,22]. Moreover, these pesticides release intermediate phenolic compounds, which contaminate the local aquatic system and land site. In addition, the solid waste generated from the domestic municipal waste treatment plant also contains some amount of phenolic compounds. Hence, these pollutants contaminate and alter the biodiversity of soil and water systems. Therefore, the conventional treatment methods, that is, adsorption, coagulation, membrane separation, extraction, chlorination, biodegradation, photocatalytic oxidation, and so forth were investigated for the degradation of the phenolic pollutants.

7.3 ADVERSE IMPACTS OF PHENOLICS ON THE ECOSYSTEM

The high volatility, as well as the diffusivity of phenolic compounds, can easily reach the living cells. The acute and chronic toxicity of these pollutants aids carcinogenic, mutagenic, and teratogenic effects on humans and aquatic animals. The phenolic compounds, including chlorophenols, aminophenols, chlorocatechols, nitrophenols, methylphenols, and so forth have been reported as highly toxic substances due to their detrimental effect on living animals. Moreover, the phenolic compounds are considered lethal to human beings even at low concentrations (10–26 mg/L). It is reported that the intake of catechol within a range of 50–500 mg/kg of body weight causes death [23]. The short-term contamination of body organs with phenolic compounds causes vomiting, headache, muscle weakness, irregular pulse, convulsion, and so forth. The alteration of DNA characteristics is also a major impact of catechols. Para-cresol, 4-methyl catechol, 4-methoxy phenol, and catechol have a strong affinity to cause cancer in human beings. Bisphenol A alkylphenols damage the endocrine system of human beings, which causes malfunction of mammary glands and potentially delaying the onset of puberty in girls [24]. Chlorophenol poisoning manifests with symptoms like mouth and throat burning, along with necrotic lesions in the oral activity, stomach, and esophagus. Hydroquinone is noted for its detrimental effects on chromosomes [25]. Additionally, it induces abnormal temperature, pulse fluctuation, and muscle weakness. Furthermore, para-cresol and 2,4-dimethyl phenol have carcinogenic effects on human beings [6,8].

Apart from human beings, phenolic compounds are very noxious to aquatic plants and animals. The toxicity of phenolic compounds on different fish embryos (like *Daphnia* sp., rainbow trout, and fathead minnows) was investigated with several phenolic compounds (phenol, catechol, cresols, etc.). It was found that phenol, catechol, and ortho-cresol have LC50 values below 25 mg/L [26]. The body weight and length of these embryos drastically decreased at a very low concentration (less than 3.0 mg/L) of phenolic compounds. A higher adverse impact of OH-phenols (such as hydroquinone, resorcinol, and pyrocatechol) compared to phenol on marine fish (sea bass) was reported in another research article. They found that the fish exposed to OH phenols experienced various metabolic disorders [27]. Moreover, the presence of phenolic compounds in water inhibits the growth of some aquatic plants and fauna. Specifically, phenol is found to be an inhibiting factor for the growth of *cyanobacteria*. Similarly, another researcher reported that the tolerant level of *Azolla filiculoides* hindered above phenol concentration of 1.0 mM and a significant degrowth in the plant was observed at a phenol concentration of more than 2.0 mM [28].

7.4 CONVENTIONAL METHODS FOR PHENOLIC COMPOUNDS TREATMENT

The range concentration of phenolic compounds has been found to be 1.18–2.5 mg/L, which is more than the permissible limit (1.0 mg/L) in rivers and lakes [10]. As is evident from the above, the presence of phenolic compounds in the environment can cause significant deterioration of health and even be fatal to human beings, aquatic cultures, and microorganisms. Therefore, the phenolic concentration in the wastewater should be lower than the permissible limit set by the regulatory authorities. The development and use of various treatment techniques were based on cost-effectiveness, reusability, disposal, sludge generation, and handling operation. Different physical, chemical, advanced oxidations, and biological techniques were extensively studied for their degradation [29].

7.4.1 Physical Method

Generally, the physical method includes conventional techniques, including sedimentation, adsorption, and membrane separation technologies [11]. Among these, sedimentation is a preliminary technique to isolate the larger particles, dirt, and soil from the wastewater. In this process, no physical or chemical reactions are involved during the treatment. Moreover, the inefficacy of this technique from degrading phenolic compounds in the aqueous system makes it unpopular. However, before the treatment of phenolic wastewater, sedimentation process was employed to remove some suspended particles and debris.

The adsorption method involves the molecular force of attraction between the pollutant molecules and the solid adsorbent. The difference in distribution coefficients between the adsorbent and the pollutant phase is the driving force for the separation. This method is simple, easy, and cost-effective. Activated carbons were mainly produced from the carbonization of various natural wastes (plant residue and agricultural waste), polymers (chitosan, sodium alginate), and clay, which have excellent adsorbing properties due to high specific surface area, pore volume, and low cost [30]. The surface of the adsorbate becomes activated by treatment with various agents, including sulfuric acid, phosphoric acid, and nitric acid. The adsorption technique has been widely investigated to remove phenolic compounds and found that high removal efficiency could be achieved within a short retention period. Jain et al. [31] investigated the adsorption ability of cow dung-based activated carbon for phenol removal. They reported that the prepared adsorbent had an excellent regeneration ability having a maximum adsorption capacity of more than 98 percent. Moreover, the enhancement of adsorption ability has been carried out by developing some high-capacity adsorbents, that is, zeolites, montmorillonite, and metal oxide frameworks (MOFs) which ultimately increase the operational costs [32,33]. Nevertheless, higher degradation efficiency, the secondary waste production, is a major limitation of this method.

The membrane-based method is comparatively reliable and effective for removing phenolic compounds due to less power consumption and negligible secondary sludge generation [21]. According to the pore size of the membrane, the above method includes various processes, such as reverse osmosis, microfiltration, ultrafiltration, and nanofiltration. Conidi et al. [34] examined the retention ability of phenolic compounds using ultrafiltration and nanofiltration membranes. They found that the corresponding membranes are able to separate polyphenols and anthocyanins in the order of 84.8 and 90.7 percent, respectively.

Moreover, membrane bioreactor, membrane distillation, emulsion liquid membrane, and pervaporation are widely popular among researchers. In specific, membrane bioreactors (MBRs) are a hybrid system, which utilizes both biological and membrane separation technologies. Boonyaroj et al. [35] reported a two-stage MBR system was able to remove bisphenol A and 4-methyl-2,6-di-tert-butylphenol (BHT) of 65 and 70 percent, respectively. Membrane distillation is one kind of advanced membrane separation technique in which the hydrophobic membrane separates the more volatile liquid from the less volatile solution across the gas-liquid interface. High salt rejection, operational stability even at low temperatures, low operating pressure, and less impact mechanical strength of the membrane are some of the major advantages of this process [12]. The membrane distillation process was able to offer a rejection of salt and organic pollutants of 99.1 percent and 96.2 percent, respectively, for coking wastewater [36]. However, the high cost, biofouling and stability issues of the membrane are the major concerns for the applicability of this technique on a large scale [37].

7.4.2 Chemical and Electrochemical Oxidation Methods

Chemical oxidation methods involve the complete oxidation of the phenolic compounds in the presence of various oxidizing chemicals and catalysts at mild pressure and temperatures. In this destructive method, various reagents and chemicals are used, such as metal cations (Cu^{2+}, Fe^{2+}, Co^{2+}, Mn^{2+}, etc.), ozone, and chlorine dioxide [29]. Among these, ferrate and permanganate oxidation has been extensively studied for phenolic compound degradation due to their superior redox potential (> 1.6 V). Tian et al. [38] explored that ferrate [Fe(VI)] oxidation resulted in 99 percent removal of Bisphenol E within 3.0 min of operation at an initial molar ratio of 10: 1(ferrate: Bisphenol E). In addition, various less-toxic intermediates such as nitrite, bicarbonate, and copper ions were detected during the operation.

However, ozone and hydrogen peroxide are preferred for the treatment of phenolic compounds due to their low cost, less toxic intermediate generation, easy-to-control parameters, and so forth [39]. Despite several advantages, the ozonation method has several demerits, such as low water solubility (which ultimately increases the higher consumption of ozone) and the high cost to produce ozone. Amado-Piña et al. [40] performed a comparative analysis between ozonation and ozonation-hydrogen peroxide oxidation for phenol treatment. They reported that the latter process achieved a phenol removal of more than 2.5 times compared to the ozonation treatment. In addition, the combined process also was able to eliminate the toxic intermediates.

The electrochemical oxidation process is an alternative method, as it requires no additive chemicals and increases the reaction rate [41]. However, this method further adds to the operating cost due to energy and equipment costs [42]. In this method, previously different composite materials, including platinum (Pt), Titanium oxide (TiO_2), RuO_2, SnO_2 (stannic oxide), PbO_2 (lead oxide), IrO_2 (iridium oxide), and boron-doped diamond (BDD) were employed as electrodes. The parameters, such as pH, anode characteristics, types of electrolytes, and current density affect the oxidation of phenolic pollutants. Generally, the electrochemical oxidation process is divided into two categories namely direct and indirect oxidation. The direct oxidation process involves the adsorption of the pollutants on the anode surface whereas the indirect oxidation process involves the generation of intermittent redox species, which hinders the electron movement from electrode to pollutant [13]. Li et al. [43] investigated the oxidation of Bisphenol with BDD and Pt electrodes. They found that a better rate constant was observed in the case of BDD due to its strong oxidizing performance and less generation of chlorinated products. Li et al. [32] analyzed the performance and mechanisms of the electrochemical oxidation of phenol using TiO_2/Ti electrodes. They revealed that the facet-tailored TiO_2/Ti material showed greater ability and fast degradation kinetic of phenol wastewater due to the presence of hydroxyl radical on the electrode surface.

7.4.3 Photocatalytic Oxidation

Photocatalytic degradation is gaining popularity among researchers due to its ability to detoxify pollutants using either solar energy or UV light. In this process, the photocatalyst the free electrons migrate from the valence band to the conduction band, which generates excited electrons (e^-) and holes (h^+). The generated electrons and holes act as reducing and oxidizing agents, respectively, in the photocatalytic process. Further, the generated electron-hole pair reacts with oxygen and water to produce superoxide radicals ($O_2^{\bullet -}$) and hydroxyl radicals ($OH^{\bullet}$), respectively [44]. The photocatalytic reactions involve the following steps:

$$\text{Photocatalyst} + hv \rightarrow h_{VB}^{+} + e_{CB}^{-} \quad (7.1)$$

$$e_{CB}^{-} + O_2 \rightarrow O_2^{\bullet} \quad (7.2)$$

$$h_{VB}^{+} + H2O \rightarrow {}_{\bullet}OH + H^{+} \quad (7.3)$$

$$H_2O_2 + e_{CB}^{-} \rightarrow {}_{\bullet}OH + OH^{-} \quad (7.4)$$

where hv represents the light energy for excitation. Additionally, h_{VB}^{+} and e_{CB}^{-} represent the oxidizing and reducing agents, respectively. In the context of the photocatalytic process, the fundamental stages can be outlined as follows: (a) transfer of pollutants from the aqueous medium to the surface of the photocatalyst; (b) activation of the photocatalyst surface by photon energy, leading to the excitation of electrons from the valence band to the conduction band; (c) adsorption of the pollutants on the photon-activated surface of the photocatalyst; (d) initiation of the photocatalytic reaction on the surface of the photocatalyst and, finally, (d) release of intermediate compounds from the photocatalyst surface and their movement into the bulk solution [45].

The phenolic compounds, especially phenol, react with the hydroxyl group of photocatalysts to produce various intermediates such as hydroquinone, catechol, p-benzoquinone, resorcinol, and so forth. Further, the disintegration of the phenyl group results in malonic acid followed by organic acids (i.e., oxalic acid, acetic acid, 2-hydroxy propanol, 2-hydroxy ethanoic acid and finally converted into CO_2 and H_2O [46]. Kumar et al. [47] have developed a photocatalyst composed of ZnO tetrapod-reduced graphene oxide for the degradation of 4-chlorophenol. They reported that removal of 95 percent was achieved within 180 min of operation at an initial chlorophenol concentration of 20 mg/L. Similarly, the dosing limit, pH, operation time, and phenol concentration were optimized in phenol treatment using TiO_2 nanocatalyst [48]. The study revealed that a maximum phenol removal of 99.48 percent was obtained at corresponding optimal conditions of 1.75 g/L, 5.42, 540 min, and 15.21 mg/L, respectively.

The nature, size, and dosing of the photocatalyst, pH of the wastewater, and intensity of the light have crucial impacts on the photocatalytic process. In this direction, several photocatalysts – TiO_2, ZnO, ZnS, CdS, SiO_2, WO_3, and so forth – are commonly applied for the photodegradation application. Moreover, various novel catalysts, including hydrogels, metal oxide frameworks, lignin, aerogels, sponges, and so forth were utilized for the effective removal of phenolic compounds [49]. Several research works were reported on the account of reusability and regeneration of the photocatalysts. Additionally, the key bottleneck arises due to the broad bandgap and rapid recombination rate of the electron-hole pair, which contributes to lower degradation efficiency. Despite potential degradation ability, the aggregation of semiconductor nanoparticles possess a major challenge resulting in the unavailability of active surface area and impeding light adsorption that induces lower [45].

7.4.4 Biological Method

Besides the various physical, chemical, and advanced oxidation processes (AOP), the biological processes are more favorable due to eco-friendly and no, or less harmful, byproduct formation [5]. Toxic intermediate elimination, ease of operation, and cost-effectiveness are the key advantages of biological methods over other conventional methods [3]. In this method, the degradation of phenolic compounds is carried out by using bacteria, yeast, or fungi (bioremediation), algae (phytoremediation), and earthworms (vermiremediation) [4, 26]. *Pseudomonas*, *Bacillus*, *Arthrobacter*, *Ralstonia,* and *Acinetobacter* sp. are the most popular bacterial species for the degradation of phenolic compounds [6, 42]. Similarly, various fungi (*Amorphoteca* sp., *Neosartorya* sp., *Talaromyces* sp., and *Graphium* sp.), yeast (*Candida* sp., *Yarrowia* sp., and *Pichia* sp.), and algae (*Chlorella* sp., *Phormidium foveolamm,* and *Scenedesmus* sp.) genera were utilized for the mineralization of the phenolic compounds to simpler substances like CO_2 and H_2O [50].

For the practical application of the living cell, bioremediation has been categorized into two methods: free cell system and immobilized cell system [5]. Most of the free cell system is based on a conventional activated sludge process. In this process, the bacterial cell remained in suspended form and degraded the organic matter present in the system CO_2, H_2O, and biomass in the presence of oxygen. Air is supplied through a diffuser into the activated sludge tank to ensure proper microbial growth. However, the major drawbacks of the free cell system (i.e., inability to perform at high phenol concentration, entrainment/washout of the biomass, and poor settle ability) induce the use of an immobilized system. The bacterial cells are immobilized onto the various packing supports to enhance the biodegradation by increasing the resistance against high concentrations of phenolic pollutants, adverse process conditions (pH, temperature, etc.), and a complete washout of biomass from the system [2,50]. Moreover, various packing supports have been used for the immobilization of microorganisms, such as natural polymers (sodium alginate, agar-agar, chitosan, etc.), synthetic polymers (polyvinyl alcohol, low-density polyethylene, polyurethane foam, high-density polyethylene, polypropylene), bio-wastes (sugarcane bagasse, luffa sponge, corncob, and sawdust), and activated carbon.

7.5 EMERGING TECHNOLOGIES FOR THE REMOVAL OF PHENOLIC WASTEWATER

7.5.1 Enzymatic Technology

Enzymes are considered extracellular polymeric substances (i.e., proteins, polysaccharides, nucleic acids), which are secreted from bacterial cells [51]. The enzymes have a catalytic effect during the treatment process. It was found that the enzyme-assisted degradation process provide far better results compared to the bulk cell-assisted treatment of phenol.

Enzymes are very sensitive and induce the reaction under appropriate process conditions. Therefore, specific enzymes have been chosen for the treatment of targeted pollutants. In this direction, peroxidase, a well-known enzyme, has been utilized for the treatment of phenol. The high oxidation capacity of the peroxidase enzymes induces the formation of water-insoluble, high molecular-weight intermediates, which are easy to separate from the wastewater [52]. In addition, the low cost, wide range of substrates, and high susceptibility are major advantages of this enzyme. Besides, laccases and tyrosinases have also been applied for the degradation of phenolic compounds.

For the on-field application, the immobilization of these enzymes on the solid matrix is gaining more popularity as it offers more flexibility in operation, reusability of the solid carrier, and high effectiveness. Zhang and Cai [53] have successfully immobilized peroxidase enzyme on Fe_3O_4/nanotubes and applied the composite for the degradation of phenol. They reported that the immobilized enzyme offered higher reusability and stability. Similarly, the efficacy of the soybean peroxidase immobilized onto calcium alginate beads was investigated in a packed bed bioreactor [54]. They found that the above system could offer a phenol removal of 97 percent at an initial feed flow rate of 5.5. mL/min.

7.5.2 Nanotechnology

Nanomaterials have been widely used recently to separate phenolic compounds from wastewater. In this direction, various nanostructured materials such as metal oxide nanoparticles, carbon nanomaterials, nano-polymers, hybrid nanocomposites, and metal-oxide frameworks (MOFs) have been widely investigated [55]. These materials have high adsorption capacity, surface area-to-volume ratio, and sensitivity. There are generally two ways in which the nanomaterials can be used, that is, (i) nano adsorbents and (ii) nano-assisted membrane filtration [14]. Moreover, the adsorption of the nanomaterials depends upon the active sites and electronic density at the edges.

Carbon is a common material that has been widely investigated to produce carbon-based nanomaterials, including pristine graphene, carbon nanotubes, and carbon nanofibers. These are used to adsorb phenolic compounds due to their excellent adsorption capacity, thermal stability, hydrophobicity, and abundance nature. Saleh et al. [56] prepared a carbon nanotube grafted with a copolymer of acrylic acid and acrylamide for the adsorption of phenol. They reported that the prepared nanomaterial has a high adsorbing capacity and could be used for four consecutive stages. Wang et al. [57] used graphene oxide for the treatment of various phenolic compounds. They reported that the maximum adsorption capacities of 135.6, 191.3, 237.7, 398.6, 420.9, and 585.8 mg/g for phenol, 2-chlorophenol, 4-chlorophenol, 2,4-dichlorophenol, bisphenol-A, and 2,4,6-trichlorophenol, respectively, were achieved by the developed nanomaterial.

However, the membrane-based treatment process has gained more attention for the remediation of phenolic compounds. The development of E-Nano membranes and their application for phenolic wastewater treatment emerged as a promising approach. The nanocomposite structure exhibits high mechanical strength, excellent oil/water separation performance, antifouling properties, and outstanding flux recovery after cleaning [58,59]. However, the E-Nano membranes having low surface roughness and a highly hydrophilic surface have been preferred as they promote enhanced water flux and reduce fouling propensity during the treatment process.

However, the nanomaterials offer several disadvantages during the treatment process, which has anadverse impact on the ecosystem. For instance, nanoparticles come out of the treatment system by leakage, spillage, circulation, or accumulation, which discharge to the environment along with the treated effluent [14,60]. Exposure to the nanoparticles causes damage to internal organs, skin problems, and DNA. Hence, the proper handling of the nanoparticles and ensuring their retention in the treatment system must be accomplished for the development of eco-friendly technology.

7.6 A COMPARATIVE ANALYSIS OF THE VARIOUS TREATMENT TECHNOLOGIES

Each process has its own advantages and disadvantages in terms of cost, simplicity, degradation efficacy, and environmental aspects. Hence, a comparative description of the benefits and drawbacks of the conventional processes are described in Table 7.1.

A brief comparative analysis of the operating cost of various technologies has been represented in Table 7.2. The chemical oxidation technology possesses the second lowest position in terms of operating cost bearing \$7.51/m^3 [29]. However, for a larger scale treatment, this cost would be quite higher than the chemical oxidation process. In the meantime, the adsorption process comes out as an inexpensive technology as the total cost accounted to be \$0.74/m^3 [44]. However, only 30 percent of the total adsorbent can be regenerated during the process. The steep rise in cost is primarily due to the reutilization and safe discarding of the adsorbent. The photocatalytic degradation of phenol costs \$1.69/m^3 whereas the UV-TiO_2 treatment of phenol and m-cresol costs around \$36.38 and \$48.45 per m^3, respectively [44]. The biological processes are somehow different from other technologies as their cost depends upon the biomass produced in the system and the types of wastewater to be treated. It is reported that the annual cost of a biological system to degrade chemical effluent and paint wastewater is \$26,924 and \$11,591, respectively [61]. The operating cost for the production of the Laccase enzyme was reported to be \$6,187/m^3 for the treatment of phenol and catechol [62].

7.7 CONCLUSIONS AND FUTURE PERSPECTIVE

The extensive use and demand of phenolic compounds in various manufacturing sectors worldwide create a large production form of petroleum. However, the noxious and persistent nature of phenolic wastewater raises a critical concern among researchers and induce them to find cost-efficient and eco-friendly technology for degradation purposes. Even though the existing technologies offer satisfactory performances, they still lag in terms of cost, environmental concern, and toxic intermediates generation. Hence, further investigation is required to address these limitations and enhance the effectiveness of phenolic abatement techniques.

The physical technology such as adsorption induces post-treatment of the treated wastewater to regenerate the adsorbent after it becomes saturated. Despite

TABLE 7.1
Summary of Advantages and Disadvantages of the Various Technologies for the Treatment of Phenolic Compounds

Technology	Description	Advantages	Disadvantages
Physical	• Sedimentation • Adsorption • Membrane technology	• Simple • Low-cost, High capacity • High efficiency	• Inefficient • Secondary waste generation • Expensive, high-pressure requirement
Chemical and electrochemical oxidation	Oxidation using • Metal cations • Ozone (O_3) • Hydrogen peroxide (H_2O_2) • Electricity and electrodes	• Less time required • Complete removal • UV/ H_2O_2 system is used in drinking water system • No chemicals required	• High chemical cost • Short half-life, highly toxic • Continuous supply of chemicals is required • High energy and electrode cost
Photocatalytic	Degradation of phenol using UV or visible light and photocatalysts (such as TiO_2, ZnO, ZnS, CdS, etc.)	• Efficient • Complete degradation	• Electron-hole pair recombination • Low usage of visible light
Biological	Phenol treatment by using microorganisms (like bacteria, yeast, fungi, or algae)	• Eco-friendly • Conversion of intermediate pollutants into simpler substances • Low-cost	• Slow process • Depend upon abiotic factors (i.e., pH, temperature, dissolved oxygen, concentration, etc.)
Enzymatic	Degradation of phenol using extracellular polymeric substances	• Consume less time • Degrade the specific pollutant	• High cost of production • High dependency on environmental factors
Nanotechnology	Phenol wastewater degradation by nanomaterials using nanotubes, nanocatalysts, or metal-oxide frameworks	• High adsorption capacity • Low-pressure requirement in nano-filtration	• Human toxicity • Accumulation of nanoparticles in the system • Disposal problem

Source: Compiled by the authors.

TABLE 7.2
The Comparative Analysis of Various Technologies Based on Their Cost

Technology	Description	Cost	Reference
Adsorption	The phenol removal of 85 percent was observed using activated carbon prepared from biomass	\$ 0.74/m^3	[44]
Chemical and electrochemical oxidation	A dosing of 0.025 M H_2O_2 and 13.63 mA/cm^2 current density to degrade 71 percent of phenol	\$ 7.51/m^3	[29]
Photocatalytic	Phenol removal of 81 percent was achieved by TiO_2 photocatalyst under solar light	\$ 1.69/m^3	[44]
Biological	Biodegradation of paint and chemical industries wastewater	\$ 26924 and \$11591 chemical and paint wastewater	[61]
Enzymatic	90 percent of phenol and 3.0 percent of catechol removal occurred using the Laccase enzyme	\$ 6187/m^3 for 10 g/L enzyme production	[62]
Nanotechnology	Capital cost for the production of various nanomaterials	• \$ 2.50 – 1000/g for graphene • \$ 0.1 – 25/g for MWCNT • \$ 25 – 300/g for SWCNT	[63]

Source: Compiled by the authors.

development of an effective and high surface area nano adsorbent, the application resides in the laboratory stage. Hence, the use of cost-effective adsorbents on an industrial scale is essential. Regarding membrane technology, the high cost of the membrane and the rapid fouling tendency make the researcher develop a low-cost membrane for the effective removal of phenol. In the meantime, membrane distillation and pervaporation require additional energy to provide heat to the reactor, which ultimately increases the operational cost. Hence, scaling up the above process on an industrial scale would inevitably demand high energy. Moreover, the effluent also needs further treatment to bring the quality of discharge as per environmental norms. Therefore, in the context of research, it is imperative to approach innovative processes, which address the energy-intensive nature of the membrane-based wastewater treatment method. Additionally, methods like pervaporation and membrane bioreactors are gaining more interest in the removal of phenolic compounds from wastewater. This is primarily attributed to their scalability and the advantage of nearly zero carbon emissions. Similarly, chemical and electrochemical oxidation processes require high chemicals and electricity for the operation. The electrochemical oxidation process is preferred as one of the most efficient methods for wastewater treatment. However, the research work is limited to the treatment of a single pollutant only. Additionally, the real phenolic wastewater will be a mixture of various multi-pollutants. Therefore, future work should emphasize development of efficient electrode materials, illustration of modified mechanism of oxidation, and development of an efficient mass transfer reactor for the electrochemical oxidation process.

The implementation of photocatalysis in a batch reactor using suspended catalysts necessitates an additional step for catalyst recovery. If the catalyst is directly mixed with the pollutant, it would be eliminated together with the treated pollutant, leading to increased costs and the potential for secondary pollution. To address this, catalyst recovery can be achieved through methods such as membrane filtration or absorption. Thus, further research is warranted to develop cost-effective strategies for catalyst retrieval in batch photocatalysis processes.

The biological processes were more like natural processes exhibiting higher degradation ability and complete removal. The immobilization of microorganisms on the solid matrix offers a robust method for the degradation of phenolic compounds. However, sludge generation during the treatment, the influence of environmental factors (like pH, temperature, dissolved oxygen, etc.), and being time-consuming are the major limitations of the biological process. Hence, the biological process could be used as a post or pretreatment method depending upon the chemical oxygen demand concentration in the wastewater.

In the meantime, the enzymatic process can be utilized as an alternative to the biological process as it consumes comparatively less time to degrade the pollutants. However, the high cost of the production of enzymes may be the bottleneck for large-scale applications. The large-scale production of commercial enzymes and their purification for real application is a major challenge. In this direction, recombinant DNA technology would be helpful for effective enzyme production. In the meantime, the substrate inhibition effect and enzyme loss during the operation can be minimized by developing genetically modified enzymes, which will again increase the production cost. Hence, more research work should be emphasized on the high-throughput screening of microbial strains to produce potential enzymes. Besides, the immobilization of enzymes on the various solid supports can avoid the hindrance effect of adverse environmental conditions. To achieve an optimal intramolecular distance between the enzymes, covalent immobilization may be considered as an effective method.

Despite the excellent molecular and thermal properties of nanomaterials, the costly nature and environmental concern are the major issues for real-life applications. Addressing the mentioned challenges substantial research should be emphasized on large-scale applications, safety regulation, energy efficiency, and safe waste disposal. The risk of environmental contamination during the operation is a bigger challenge, which should be addressed. In this direction, the hazardous effect of nanoparticles on living animals needs to be known. Moreover, recycling and appropriate disposal of nanomaterials also should be addressed before real-time application. Additionally, addressing the scalability issues associated with nanotechnology induces extensive research efforts to find viable solutions.

REFERENCES

[1] Lee, CG, Hong, SH, Hong, SG, Choi, JW, Park, SJ. Production of biochar from food waste and its application for phenol removal from aqueous solution. *Wat Air Soil Poll.* 2019; 230: 1–13. https://doi.org/10.1007/s11270-019-4125-x

[2] Swain, G, Sonwani, RK, Giri, B. S, Singh, RS, Jaiswal, RP, Rai, BN. A study of external mass transfer effect on biodegradation of phenol using low-density polyethylene immobilized *Bacillus flexus* GS1 IIT (BHU) in a packed bed bioreactor. *Water Environ J.* 2021; 35(1): 285–294. https://doi.org/10.1111/wej.12626

[3] Putmai, N, Woraruthai, T, Tirapanampai, C, Wongnate, T, Flood, AE. Biodegradation characteristics of mixed phenol and p-cresol contaminants from a swine farm using bacteria immobilized in calcium alginate beads. *Bioresour Technol Rep.* 2023; 23: 101528. https://doi.org/10.1016/j.biteb.2023.101528

[4] Sandhibigraha, S, Mandal, S, Awasthi, M, Kanti BT, Bhunia, B. Optimization of various process parameters for biodegradation of 4-chlorophenol using Taguchi methodology. *Biocatal Agric Biotechnol.* 2020; 24: 101568. https://doi.org/10.1016/j.bcab.2020.101568

[5] Swain, G, Maurya, KL, Sonwani, RK, Singh, RS, Jaiswal, RP, Rai, B. N. Effect of mixing intensity on biodegradation of phenol in a moving bed biofilm reactor: Process optimization and external mass transfer study. *Bioresour Technol.* 2022; 351: 126921. https://doi.org/10.1016/j.biortech.2022.126921

[6] Anku, WW, Mamo, MA, Govender, PP. Phenolic compounds in water: Sources, reactivity, toxicity, and treatment methods. In *Phenolic Compounds-Natural Sources, Importance, and Applications* (pp. 419–443). Edited by Hernández, MS, Tenango, MP, and Mateos, RG. IntechOpen, 2017.

[7] Vashi, H, Iorhemen, OT, Tay, JH. Extensive studies on the treatment of pulp mill wastewater using aerobic granular sludge (AGS) technology. *Chem Eng J.* 2019; 359: 1175–1194. https://doi.org/10.1016/j.cej.2018.11.060

[8] Geng, Y, Liu, X, Yu, Y, Li, W, Mou, Y, Chen, F, Hu, X, Ji, J, Ma, L. From polyphenol to o-quinone: Occurrence, significance, and intervention strategies in foods and health implications. *Compr Rev Food Sci Food Saf.* 2023; 22(4): 3254–3291. https://doi.org/10.1111/1541-4337.13182

[9] Duan, W, Meng, F, Cui, H, Lin, Y, Wang, G, Wu, J. Ecotoxicity of phenol and cresols to aquatic organisms: A review. *Ecotoxicol Environ Saf.* 2018; 157: 441–456. https://doi.org/10.1016/j.ecoenv.2018.03.089

[10] Seth, R, Singh, P, Mohan, M, Singh, R, Aswal, RS. Monitoring of phenolic compounds and surfactants in water of Ganga Canal, Haridwar (India). *Appl Water Sci.* 2013; 3: 717–720. https://doi.org/10.1007/s13201-013-0116-z

[11] Ahmed, SF, Mofijur, M, Nuzhat, S, Chowdhury, AT, Rafa, N, Uddin, MA, Inayat, A, Mahlia, TMI, Ong, HC, Chia, WY, Show, PL. Recent developments in physical, biological, chemical, and hybrid treatment techniques for removing emerging contaminants from wastewater. *J Hazardous Mater.* 2021; 416: 125912. https://doi.org/10.1016/j.jhazmat.2021.125912

[12] Aijaz, MO, Karim, MR, Othman, MHD, Samad, UA. Anti-fouling/wetting electrospun nanofibrous membranes for membrane distillation desalination: A comprehensive review. *Desalination.* 2023; 553: 116475. https://doi.org/10.1016/j.desal.2023.116475

[13] Fu, R, Zhang, PS, Jiang, YX, Sun, L, Sun, XH. Wastewater treatment by anodic oxidation in electrochemical advanced oxidation process: Advance in mechanism, direct and indirect oxidation detection methods. *Chemosphere.* 2022; 311: 136993. https://doi.org/10.1016/j.chemosphere.2022.136993

[14] Younis, SA, Maitlo, HA, Lee, J, Kim, KH. Nanotechnology-based sorption and membrane technologies for the treatment of petroleum-based pollutants in natural ecosystems and wastewater streams. *Adv Colloid Interface Sci.* 2020; 275: 102071. https://doi.org/10.1016/j.cis.2019.102071

[15] Tyagi, M, Kumari, N, Jagadevan, S. A holistic Fenton oxidation-biodegradation system for treatment of phenol from coke oven wastewater: Optimization, toxicity analysis and phylogenetic analysis. *J Water Process Eng.* 2020; 37: 101475. https://doi.org/10.1016/j.jwpe.2020.101475

[16] Huong, DT, Nguyen, VT, Ha, XL, Nguyen Thi, HL, Duong, TT, Nguyen, DC, Nguyen Thi, H. T. Enhanced degradation of phenolic compounds in coal gasification wastewater by methods of microelectrolysis Fe-C and anaerobic-anoxic—oxic moving bed biofilm reactor (A2O-MBBR). *Processes.* 2020; 8(10): 1258. https://doi.org/10.3390/pr8101258

[17] Tuladhar, P, Sasidharan, S, Saudagar, P. Role of phenols and polyphenols in plant defense response to biotic and abiotic stresses. In *Biocontrol Agents and Secondary Metabolites* (pp. 419–441). Edited by Jogaiah, S. Woodhead Publishing, 2021. https://doi.org/10.1016/j.watres.2021.118025

[18] Vardhan, PV, Shukla, LI. Gamma irradiation of medicinally important plants and the enhancement of secondary metabolite production. *Int J Radiat Biol.* 2017; 93(9): 967–979. https://doi.org/10.1080/09553002.2017.1344788

[19] Patti, A, Acierno, D. Towards the sustainability of the plastic industry through biopolymers: Properties and potential applications to the textiles world. *Polymers.* 2022; 14(4): 692. https://doi.org/10.3390/polym14040692

[20] Nur, MA, Garcia, GM, Boelen, P, Buma, AG. Influence of photodegradation on the removal of color and phenolic compounds from palm oil mill effluent by Arthrospira platensis. *J Appl Phycol.* 2021; 33: 901–915. https://doi.org/10.1007/s10811-020-02341-8

[21] Raza, W, Lee, J, Raza, N, Luo, Y, Kim, KH, Yang, J. Removal of phenolic compounds from industrial waste water based on membrane-based technologies. *J Ind Eng Chem.* 2019; 71: 1–18. https://doi.org/10.1016/j.jiec.2018.11.024

[22] Kumar, D, Sharma, C. Reduction of chlorophenols and sludge management from paper industry wastewater using electrocoagulation process. *Sep Sci Technol.* 2020; 55(15): 2844–2854. https://doi.org/10.1080/01496395.2019.1646761

[23] Sundaram, R, Muthu, K, Shanthi, P, Sachdanandam, P. Antioxidant and antihyperlipidemic activities of catechol derivatives and biflavonoid isolated from Semecarpus anacardium seeds. *Toxicol Mech Methods.* 2022; 32(2): 123–131. https://doi.org/10.1080/15376516.2021.1973170

[24] Zulkifli, S, Rahman, AA, Kadir, SHSA, Nor, NSM. Bisphenol A and its effects on the systemic organs of children. *Eur J Pediatr.* 2021; 180(10): 3111–3127. https://doi.org/10.1007/s00431-021-04085-0

[25] Weber, GF, Weber, GF. Molecular inhibitors of growth signals. *Mol Ther Cancer.* 2015; 163–242. https://doi.org/10.1007/978-3-319-13278-5_4

[26] Paisio, CE, Oller, A, Ibaez, SG, Talano, MA, Agostini, E. Bioremediation as a useful biotechnological strategy for the treatment of phenolics: Advances, challenges, and future prospects. In *Bioremediation: Processes, Challenges and Future Prospects.* Edited by Fernández, JBV, and Hernánd, SM. Nova Science Publisher, 2014.

[27] Roche, H, Bogé, G. In vivo effects of phenolic compounds on blood parameters of a marine fish (Dicentrarchus labrax). *Comp Biochem Physiol C Pharmacol, Toxicol Endocrinol.* 2000; 125(3): 345–353. https://doi.org/10.1016/S0742-8413(99)00119-X

[28] Paul, B, Sarkar, A, Roy, S. Appraising the stress responses in Azolla filiculoides elicited by short-term exposure of phenol. *Plant Stress*. 2021; 2: 100032. https://doi.org/10.1016/j.stress.2021.100032
[29] Singh, H, Mishra, BK. Degradation of cyanide, aniline and phenol in pre-treated coke oven wastewater by peroxide assisted electro-oxidation process. *Water Sci Technol*. 2018; 78(10): 2214–2227. https://doi.org/10.2166/wst.2018.503
[30] Zhang, W, Zhang, P, Wang, H, Li, J, Dai, SY. Design of biomass-based renewable materials for environmental remediation. *Trends Biotechnol*. 2022; 40: 1519–1534
[31] Jain, M, Khan, SA, Sahoo, A, Dubey, P, Pant, KK, Ziora, ZM, Blaskovich, MA. Statistical evaluation of cow-dung derived activated biochar for phenol adsorption: Adsorption isotherms, kinetics, and thermodynamic studies. *Bioresour Technol*. 2022; 352: 127030. https://doi.org/10.1016/j.biortech.2022.127030
[32] Li, Y, Fang, Y, Cao, Z, Li, N, Chen, D, Xu, Q, Lu, J. Construction of g-C3N4/PDI@ MOF heterojunctions for the highly efficient visible light-driven degradation of pharmaceutical and phenolic micropollutants. *Appl Catal B Environ*. 2019; 250: 150–162. https://doi.org/10.1016/j.apcatb.2019.03.024
[33] Li, D, Cheng, Y, Zuo, H, Zhang, W, Pan, G, Fu, Y, Wei, Q. Dual-functional biocatalytic membrane containing laccase-embedded metal-organic frameworks for detection and degradation of phenolic pollutant. *J Colloid Interface Sci*. 2021; 603: 771–782. https://doi.org/10.1016/j.jcis.2021.06.155
[34] Conidi, C, Cassano, A, Caiazzo, F, Drioli, E. Separation and purification of phenolic compounds from pomegranate juice by ultrafiltration and nanofiltration membranes. *J Food Eng*. 2017; 195: 1–13. https://doi.org/10.1016/j.jfoodeng.2016.09.017
[35] Boonyaroj, V, Chiemchaisri, C, Chiemchaisri, W, Yamamoto, K. Enhanced biodegradation of phenolic compounds in landfill leachate by enriched nitrifying membrane bioreactor sludge. *J Hazard Mater*. 2017; 323: 311–318. https://doi.org/10.1016/j.jhazmat.2016.06.064
[36] Li, J, Wu, J, Sun, H, Cheng, F, Liu, Y. Advanced treatment of biologically treated coking wastewater by membrane distillation coupled with pre-coagulation. *Desalination*. 2016; 380: 43–51. https://doi.org/10.1016/j.desal.2015.11.020
[37] Said, IA, Chomiak, T, Floyd, J, Li, Q. Sweeping gas membrane distillation (SGMD) for wastewater treatment, concentration, and desalination: A comprehensive review. *Chem Eng Process-Process Intensif*. 2020; 153: 107960. https://doi.org/10.1016/j.cep.2020.107960
[38] Tian, B, Wu, N, Pan, X, Yan, C, Sharma, VK, Qu, R. Ferrate (VI) oxidation of bisphenol E–Kinetics, removal performance, and dihydroxylation mechanism. *Water Res*. 2022; 210: 118025. https://doi.org/10.1016/j.watres.2021.118025
[39] Arifin, MN, Jusoh, R, Abdullah, H, Ainirazali, N, Setiabudi, HD. Recent advances in advanced oxidation processes (AOPs) for the treatment of nitro-and alkyl-phenolic compounds. *Environ Res*. 2023; 229: 115936. https://doi.org/10.1016/j.envres.2023.115936
[40] Amado-Piña, D, Roa-Morales, G, Barrera-Díaz, C, Balderas-Hernandez, P, Romero, R, del Campo, EM, Natividad, R. Synergic effect of ozonation and electrochemical methods on oxidation and toxicity reduction: Phenol degradation. *Fuel*. 2017; 198: 82–90. https://doi.org/10.1016/j.fuel.2016.10.117
[41] Fajardo, AS, Seca, HF, Martins, RC, Corceiro, VN, Freitas, IF, Quinta-Ferreira, ME, Quinta-Ferreira, RM. Electrochemical oxidation of phenolic wastewaters using a batch-stirred reactor with NaCl electrolyte and Ti/RuO2 anodes. *J Electroanal Chem*. 2017; 785: 180–189. https://doi.org/10.1016/j.jelechem.2016.12.033

[42] Singh, T, Bhatiya, AK, Mishra, PK, Srivastava, N. An effective approach for the degradation of phenolic waste: phenols and cresols. In *Abatement of Environmental Pollutants* (pp. 203–243): Elsevier; 2020. https://doi.org/10.1016/B978-0-12-818095-2.00011-4

[43] Li, H, Long, Y, Zhu, X, Tian, Y, Ye, J. Influencing factors and chlorinated byproducts in electrochemical oxidation of bisphenol A with boron-doped diamond anodes. *Electrochim Acta.* 2017; 246: 1121–1130. https://doi.org/10.1016/j.electacta.2017.06.163

[44] Magdy, M, Alalm, MG, El-Etriby, HK. Comparative life cycle assessment of five chemical methods for removal of phenol and its transformation products. *J Clean Prod.* 2021; 291: 125923. https://doi.org/10.1016/j.jclepro.2021.125923

[45] Motamedi, M, Yerushalmi, L, Haghighat, F, Chen, Z. Recent developments in photocatalysis of industrial effluents: A review and example of phenolic compounds degradation. *Chemosphere.* 2022; 296: 133688. https://doi.org/10.1016/j.chemosphere.2022.133688

[46] Pavel, M, Anastasescu, C, State, RN, Vasile, A, Papa, F, Balint, I. Photocatalytic degradation of organic and inorganic pollutants to harmless end products: Assessment of practical application potential for water and air cleaning. *Catalysts.* 2023; 13(2): 380. https://doi.org/10.3390/catal13020380

[47] Kumar, S, Kaushik, RD, Purohit, LP. Novel ZnO tetrapod-reduced graphene oxide nanocomposites for enhanced photocatalytic degradation of phenolic compounds and MB dye. *J Mol Liq.* 2021; 327: 114814. https://doi.org/10.1016/j.molliq.2020.114814

[48] Zulfiqar, M, Samsudin, MFR, Sufian, S. Modelling and optimization of photocatalytic degradation of phenol via TiO2 nanoparticles: An insight into response surface methodology and artificial neural network. *J Photochem Photobiol A Chem.* 2019; 384: 112039. https://doi.org/10.1016/j.jphotochem.2019.112039

[49] Xiao, W, Sun, R, Hu, S, Meng, C, Xie, B, Yi, M, Wu, Y. Recent advances and future perspective on lignocellulose-based materials as adsorbents in diverse water treatment applications. *Int J Biol Macromol.* 2023; 253: 126984. https://doi.org/10.1016/j.ijbiomac.2023.126984

[50] Basak, B, Jeon, BH, Kurade, MB, Saratale, GD, Bhunia, B, Chatterjee, PK, Dey, A. Biodegradation of high-concentration phenol using sugarcane bagasse immobilized Candida tropicalis PHB5 in a packed-bed column reactor. *Ecotoxicol Environ Saf.* 2019; 180: 317–325. https://doi.org/10.1016/j.ecoenv.2019.05.020

[51] Mustafa, G, Zahid, MT, Ihsan, S, Zia, I, Abbas, SZ, Rafatullah, M. Bacterial extracellular polymeric substances for degradation of textile dyes. In *Polymer Technology in Dye-containing Wastewater* (Volume 2, pp. 175–191). Springer Nature; 2022. https://doi.org/10.1007/978-981-19-0886-6_7

[52] Joel, EB, Mafulul, SG, Adamu, HE, Goje, LJ, Tijjani, H, Igunnu, A, Malomo, SO. Peroxidase from waste cabbage (Brassica oleracea capitata L.) exhibits the potential to biodegrade phenol and synthetic dyes from wastewater. *Sci Afr.* 2020; 10: e00608. https://doi.org/10.1016/j.sciaf.2020.e00608

[53] Zhang, C, Cai, X. Immobilization of horseradish peroxidase on Fe3O4/nanotubes composites for biocatalysis-degradation of phenol. *Compos Interfaces.* 2019; 26(5): 379–396. https://doi.org/10.1080/09276440.2018.1504265

[54] Rezvani, F, Azargoshasb, H, Jamialahmadi, O, Hashemi-Najafabadi, S, Mousavi, SM, Shojaosadati, SA. Experimental study and CFD simulation of phenol removal by immobilization of soybean seed coat in a packed-bed bioreactor. *Biochem Eng J.* 2015; 101: 32–43. https://doi.org/10.1016/j.bej.2015.04.019

[55] Pachaiappan, R, Rajendran, S, Ponce, LC. Nanoparticles in industrial wastewater treatment: An overview. In *Industrial Wastewater Treatment: Emerging Technologies for Sustainability* (pp. 133–186). Edited by Karchiyappan, T, Karri, RR, and Dehghani, MH. Springer, 2022. https://doi.org/10.1007/978-3-030-98202-7_6
[56] Saleh, TA, Elsharif, AM, Asiri, S, Mohammed, ARI, Dafalla, H. Synthesis of carbon nanotubes grafted with copolymer of acrylic acid and acrylamide for phenol removal. *Environ Nanotechnol Monit Manag*. 2020; 14: 100302. https://doi.org/10.1016/j.enmm.2020.100302
[57] Wang, W, Gong, Q, Chen, Z, Wang, WD, Huang, Q, Song, S, Chen, J, Wang, X. Adsorption and competition investigation of phenolic compounds on the solid-liquid interface of three-dimensional foam-like graphene oxide. *Chem Eng J*. 2019; 378: 122085. https://doi.org/10.1016/j.cej.2019.122085
[58] Zhuang, J, Dai, J, Ghaffar, SH, Yu, Y, Tian, Q, Fan, M. Development of highly efficient, renewable and durable alginate composite aerogels for oil/water separation. *Surf Coat Technol*. 2020; 388: 125551. https://doi.org/10.1016/j.surfcoat.2020.125551
[59] Cao, W, Ma, W, Lu, T, Jiang, Z, Xiong, R, Huang, C. Multifunctional nanofibrous membranes with sunlight-driven self-cleaning performance for complex oily wastewater remediation. *J Colloid Interface Sci*. 2022; 608: 164–174. https://doi.org/10.1016/j.jcis.2021.09.194
[60] Saravanan, A, Kumar, PS, Jeevanantham, S, Karishma, S, Tajsabreen, B, Yaashikaa, PR, Reshma, B. Effective water/wastewater treatment methodologies for toxic pollutants removal: Processes and applications towards sustainable development. *Chemosphere*. 2021; 280: 130595. https://doi.org/10.1016/j.chemosphere.2021.130595
[61] Sarker, NK, Sarkar, S. A comparative study on cost analysis, efficiency, and process mechanism of effluent treatment plants in Bangladesh. *Environ Qual Manag*. 2018; 27(3): 127–133. https://doi.org/10.1002/tqem.21533
[62] Abejón, R, Belleville, MP, Sanchez-Marcano, J. Design, economic evaluation and optimization of enzymatic membrane reactors for antibiotics degradation in wastewaters. *Sep Purif Technol*. 2015; 156: 183–199. https://doi.org/10.1016/j.seppur.2015.09.072
[63] Adeleye, AS, Conway, JR, Garner, K, Huang, Y, Su, Y, Keller, AA. Engineered nanomaterials for water treatment and remediation: Costs, benefits, and applicability. *Chem Eng J*. 2016; 286: 640–662. https://doi.org/10.1016/j.cej.2015.10.105

8 Innovative Technologies for the Removal of Pollutants in the Chemical Industries

Abhinay Thakur and Ashish Kumar

8.1 INTRODUCTION

8.1.1 The Importance of Sustainability and Environmental Responsibility in Chemical Industries

Sustainability and environmental responsibility have risen to paramount importance within the chemical industry, profoundly impacting both the global ecosystem and human well-being. It underscores the critical significance of these principles in the chemical sector, reflecting the growing recognition that industries, including chemicals, must pivot towards sustainable practices to address pressing environmental challenges [1,2]. The chemical industry plays an integral role in shaping the modern world, supplying products and processes essential for various facets of human life, including pharmaceuticals, agriculture, energy production, and materials manufacturing. However, this industry's historical legacy has been marked by significant environmental degradation, spanning air and water pollution, greenhouse gas emissions, and hazardous substance releases [3, 4]. The contemporary importance of sustainability transcends mere regulatory compliance; it embodies a dedicated commitment to mitigating adverse environmental impacts throughout a product's entire lifecycle, from raw material extraction to disposal. This commitment also encompasses the maximization of resource efficiency, waste reduction, and energy consumption minimization, all of which serve to bolster a company's reputation while simultaneously reducing operational costs and driving innovation [5].

Furthermore, environmental responsibility is inextricably linked with sustainability, underscoring the ethical and moral obligations borne by chemical companies to curtail harm to ecosystems and protect human health. In today's globalized world, where environmental issues disregard national boundaries, chemical industries are increasingly recognizing their role as stewards of the planet. This recognition compels them to adopt a proactive approach aimed at minimizing their environmental footprint. By doing so, they not only ensure their long-term viability in a shifting landscape but also maintain the public's trust and safeguard the well-being of current and

DOI: 10.1201/9781003454199-8

future generations. The synergy between sustainability and environmental responsibility heralds a transformative imperative within the chemical industry, one that seeks to harmonize human progress with ecological preservation.

8.1.2 Overview of Pollution Control in Chemical Processes

Pollution control in chemical processes stands as a cornerstone of responsible and sustainable industry operations. It constitutes a multifaceted domain encompassing the development and application of strategies, technologies, and practices designed to prevent, mitigate, or eliminate the release of pollutants into the environment. These pollutants are diverse and can encompass chemical waste, emissions, and byproducts, all of which possess the potential to inflict harm upon ecosystems and public health [6, 7]. The essence of pollution control lies in its adaptability; it is not a one-size-fits-all endeavor but rather a nuanced and context-specific undertaking. It responds to the intricacies of different industries, processes, and the specific pollutants at play. Pollution control necessitates a two-pronged approach. On one front, preventive measures take center stage, emphasizing the integration of cleaner production methods and waste reduction techniques. This proactive stance seeks to curtail the generation of pollutants at their source, heralding a more sustainable approach to industrial processes [8, 9]. On the other front, end-of-pipe treatments come into play, encompassing techniques such as filtration, scrubbing, and chemical treatment. These measures aim to capture, neutralize, or remove pollutants before they escape into the environment. The section also acknowledges the myriad driving forces behind pollution control efforts. Regulatory requirements often serve as the catalyst, compelling industries to comply with environmental standards and mandates. However, these efforts are not solely dictated by legal mandates. Ethical considerations play a significant role, propelling organizations to take responsibility for their environmental impact and adopt practices that safeguard ecosystems and human health. Beyond the ethical dimension, pollution control holds economic merit. By minimizing waste generation, optimizing resource use, and reducing the burden of environmental liabilities, organizations can achieve tangible cost savings [10, 11]. Moreover, embracing pollution control as a core facet of sustainability can confer a competitive advantage, as it aligns with the growing consumer preference for environmentally responsible products and practices. The implementation of pollution control measures in chemical processes encompasses various specific cost-saving strategies that not only contribute to environmental sustainability but also yield economic benefits across industries.

- Process Optimization: By optimizing industrial processes, companies can enhance efficiency and reduce waste generation, subsequently minimizing the need for costly clean-up and disposal. Process optimization involves adopting advanced technologies and methodologies that improve resource utilization, energy efficiency, and overall production yield. The financial gains from reduced resource consumption and enhanced efficiency have a significant impact on operational costs.
- Waste Minimization and Recycling: Implementing waste minimization strategies, such as adopting cleaner production methods and recycling initiatives,

directly reduces disposal costs. Companies can repurpose byproducts or waste materials within their processes or sell them as raw materials to other industries. This circular approach not only lowers waste management expenses but also contributes to a more sustainable supply chain.

- Energy Efficiency Measures: Pollution control often involves energy-intensive processes. Adopting energy-efficient technologies, such as utilizing renewable energy sources or implementing energy recovery systems, not only reduces carbon emissions but also cuts down on energy costs. The integration of renewable energy, such as solar or wind power, provides a dual advantage by mitigating environmental impact and offering long-term cost savings.
- Advanced Filtration and Treatment Technologies: While introducing advanced filtration systems may require upfront investment, the long-term benefits include reduced maintenance costs and enhanced operational efficiency. These technologies effectively capture and treat pollutants, preventing equipment damage and the need for extensive repairs. Over time, the cost savings from decreased maintenance and increased equipment lifespan outweigh the initial investment.
- Regulatory Compliance Efficiencies: Proactively adhering to environmental regulations ensures that companies avoid costly fines and legal penalties associated with non-compliance. Developing robust compliance strategies not only prevents financial losses but also fosters a positive relationship with regulatory authorities, positioning the company as a responsible and compliant industry player.
- Public Image and Market Advantage: Embracing pollution control as part of sustainable practices enhances a company's public image. As consumers increasingly favor environmentally responsible products and practices, companies that align with these values gain a competitive edge. Positive public perception can translate into increased market share, customer loyalty, and long-term profitability.

In summary, pollution control measures in chemical processes bring about specific cost-saving measures, including process optimization, waste minimization, energy efficiency, advanced filtration technologies, regulatory compliance efficiencies, and a positive public image. These strategies not only contribute to environmental responsibility but also positively impact a company's bottom line across diverse industries.

8.2 ENVIRONMENTAL IMPACT OF CHEMICAL POLLUTION

8.2.1 Global Consequences on Ecosystems

The environmental impact of chemical pollution on ecosystems is a multifaceted and often alarming issue that extends well beyond localized contamination. This section delves into the wide-reaching consequences of chemical pollution on the natural world [12, 13]. Chemical pollutants, when released into the environment, can have devastating effects on ecosystems at various levels. These pollutants can contaminate soil, water bodies, and the atmosphere, leading to direct and indirect harm to flora and fauna [14, 15]. Some key points to elaborate on in this context include the following:

- Biodiversity Loss: One of the most palpable consequences of chemical pollution is the disruption it inflicts upon ecosystems, jeopardizing biodiversity. Chemical pollutants, whether in the form of toxic compounds or altered environmental conditions, poison or transform habitats that countless species call home. This disruption forces many species to grapple with new, inhospitable environments or, in the worst cases, leads to their outright extinction within affected areas. The result is reduced biodiversity, as some species struggle to survive or are entirely eradicated, leading to imbalances in predator-prey relationships and ecological functions.
- Bioaccumulation and Biomagnification: Chemical pollutants, especially persistent organic pollutants and heavy metals, have a knack for infiltrating and accumulating within the tissues of organisms – a phenomenon known as bioaccumulation. However, the ecological drama intensifies as one moves up the food chain. Predators at higher trophic levels ingest prey laden with pollutants, leading to biomagnification. This process means that top-level predators can accumulate exceptionally high concentrations of pollutants in their bodies [14, 16]. These contaminants can cause a slew of health issues, including reproductive problems, weakened immune systems, and organ damage. Alarmingly, humans are not exempt from this phenomenon, as consuming contaminated seafood or other products can introduce these harmful chemicals into our bodies.
- Habitat Destruction: Chemical pollution can unleash direct devastation upon habitats. For instance, acid rain, a consequence of atmospheric chemical pollutants, can corrode forests and aquatic ecosystems. Acid rain's corrosive impact damages plant life disrupts soil chemistry, and acidifies bodies of water, imperiling aquatic species [5]. Moreover, chemical pollutants that infiltrate soil can perturb the delicate balance of essential nutrients necessary for plant growth. This soil contamination, in turn, leads to habitat degradation and the demise of indigenous flora, further compounding the ecological upheaval.
- Altered Ecosystem Function: Beyond the visible disruptions to habitats and species, chemical pollution can precipitate a less conspicuous yet equally pernicious effect: the alteration of fundamental ecosystem functions [4, 17, 18]. Ecosystems rely on intricate processes such as nutrient cycling, pollination, and water purification to maintain their health and resilience. Chemical pollution can disrupt these processes, leading to long-term ecological imbalances. For instance, pollutants can interfere with nutrient cycling by impairing the ability of microorganisms to break down organic matter or by altering the availability of essential nutrients [19, 20]. This, in turn, affects plant growth, which cascades through the food web, impacting herbivores and predators. Likewise, chemical pollutants can impair pollinators, such as bees, through the contamination of nectar and pollen, jeopardizing plant reproduction and agricultural productivity. Furthermore, pollutants in water bodies can impair the natural purification processes, leading to compromised water quality and a host of issues for aquatic ecosystems and human communities reliant on clean water sources.

8.2.2 Human Health Implications

This subsection delves into the critical concern of how chemical pollution affects human health, either directly or indirectly, emphasizing the importance of safeguarding public well-being. Chemical pollutants can pose significant threats to human health, often through exposure to contaminated air, water, soil, or food sources. Some key points to elaborate on include the following:

- Acute and Chronic Health Effects: Chemical pollutants wield a double-edged sword, capable of inflicting both immediate, acute health problems and insidious, long-term chronic conditions. In the short term, exposure to chemical pollutants can result in rapid-onset health issues such as respiratory distress, skin disorders, or acute poisoning [21, 22]. However, the more sinister aspect of chemical pollution lies in its potential to precipitate chronic health conditions. When exposure to these pollutants persists over extended periods, it can culminate in a host of debilitating illnesses, including cancer, neurological disorders, developmental abnormalities, and endocrine disruptions [23, 24]. This dual threat underscores the far-reaching and enduring consequences of chemical pollution for human health.
- Vulnerable Populations: Not all individuals bear the same burden of risk when it comes to chemical pollution. Vulnerable populations, including children, the elderly, and those with pre-existing health conditions, often find themselves disproportionately susceptible to the adverse effects of chemical pollutants. Children, for example, are more vulnerable due to their developing bodies and immune systems, making them susceptible to long-lasting health impacts. Similarly, the elderly may experience exacerbated health effects due to age-related vulnerabilities. Individuals with pre-existing health conditions, such as respiratory or cardiovascular diseases, may see their conditions worsen in the presence of chemical pollutants, further underscoring the importance of safeguarding vulnerable communities [25].
- Occupational Hazards: The chemical industry and occupations entailing exposure to hazardous substances present unique risks. Workers in these sectors confront heightened vulnerabilities due to their proximity to chemicals and industrial processes. Occupational hazards may include chemical spills, accidents, or prolonged exposure to pollutants, posing immediate health risks and long-term health consequences. Consequently, stringent workplace safety measures and regulatory oversight are imperative to protect workers and ensure their well-being [15].
- Community and Environmental Justice: Chemical pollution does not affect all communities equally, amplifying concerns surrounding environmental justice. Marginalized and low-income communities often bear a disproportionate burden of environmental pollution, lacking access to the resources and political influence required to mitigate these impacts [26, 27]. This raises ethical and social equity issues, emphasizing the urgent need for equitable access to a healthy environment. Addressing environmental injustice involves rectifying disparities in exposure, access to healthcare, and the allocation of resources to vulnerable communities [28, 29].

To underscore the significance of human health implications, citing real-world examples and studies that link chemical pollution to specific health problems can be particularly impactful. Additionally, discussing regulatory measures and public health interventions aimed at mitigating these risks can provide a well-rounded perspective on the issue.

8.3 INNOVATIVE POLLUTION MITIGATION APPROACHES

8.3.1 Advanced Filtration Systems

Advanced filtration systems represent a crucial and rapidly evolving facet of innovative pollution mitigation approaches in the chemical industry. These systems offer sophisticated and efficient ways to remove pollutants from various processes and effluents. Here is an elaboration on this topic:

- Applications across the Chemical Industry: The versatility of advanced filtration systems finds applications across diverse domains within the chemical industry [30, 31]. From wastewater treatment to product purification, these systems excel in removing a broad range of contaminants. They are instrumental in eliminating suspended solids, organic contaminants, heavy metals, and even nanoparticles, offering tailored solutions that align with the unique demands of each application.
- Efficiency and Environmental Impact: A hallmark of advanced filtration systems is their exceptional efficiency and selectivity, achieving superior pollutant removal rates. This efficiency not only holds environmental significance by reducing the release of pollutants but also translates into tangible cost savings. By minimizing waste generation and optimizing resource utilization, these systems contribute to economic sustainability.
- Sustainability-Centric Design: Many advanced filtration technologies are meticulously designed with sustainability at the forefront. They often demand less energy and fewer chemicals compared to traditional methods, aligning seamlessly with the principles of environmental responsibility. Reduced resource consumption not only conserves valuable inputs but also diminishes the environmental footprint, making these technologies a pivotal component of sustainable industry practices.
- Overcoming Challenges: Despite their advantages, advanced filtration systems are not without challenges. Issues such as fouling, scaling, and membrane degradation can impede their efficiency over time. However, continuous research and engineering innovation are addressing these limitations, seeking to bolster the longevity and performance of these systems. This commitment to refinement underscores the industry's dedication to pushing the boundaries of pollution control technologies.

In an experiment, Rodriguez et al. [32] developed cost-effective technologies, optimized system design and operation, and treated highly saline-produced water (with total dissolved solids of 120–140 g/L) for hydraulic fracturing. The study involved collecting produced water from a saltwater disposal facility in the Permian Basin, New Mexico. Herein, two treatment methods were compared: chemical coagulation (CC) using ferric chloride and aluminum sulfate as coagulants and electrocoagulation (EC) employing aluminum electrodes for the removal of suspended contaminants. The research investigated the effects of coagulant dose, current density, and hydraulic retention time during EC on turbidity removal. Experimental findings indicated that aluminum sulfate was more efficient and cost-effective than ferric chloride in removing turbidity from produced water. The optimal aluminum dose was achieved with an operating current density of 6.60 mA/cm^2 and a 12-minute contact time during EC treatment, resulting in 74 percent removal of suspended solids and 53–78 percent removal of total organic carbon (TOC). The energy requirement for EC was calculated as 0.36 kWh/m^3 of treated water, with a total operating cost estimated at \$0.44/m^3 of treated water, which was 1.7 or 1.2 times higher than CC using alum or ferric chloride as coagulants, respectively. The higher EC operating cost primarily stemmed from the consumption of aluminum electrode materials due to faradaic reactions and electrode corrosion. The electrocoagulation (EC) system comprised eight parallel aluminium plate electrodes, each measuring 10.2 × 16 × 1.3 cm. Electricity was supplied to these electrodes using a direct current power source, specifically the LH110-3 model from Sorensen in San Diego, California. These electrodes were positioned within a cylindrical reactor with a 2.5 L capacity, constructed from transparent acrylic material. The surface area of the electrodes in contact with the produced water amounted to 379 cm^2, with a 2 cm separation distance between each electrode (refer to Figure 8.1).

EC offers advantages such as shorter retention time, in situ production of coagulants, reduced sludge generation, and high mobility for onsite-produced water treatment. After coagulation, fine particles and other contaminants were further treated in continuous-flow columns packed with various filter media, including

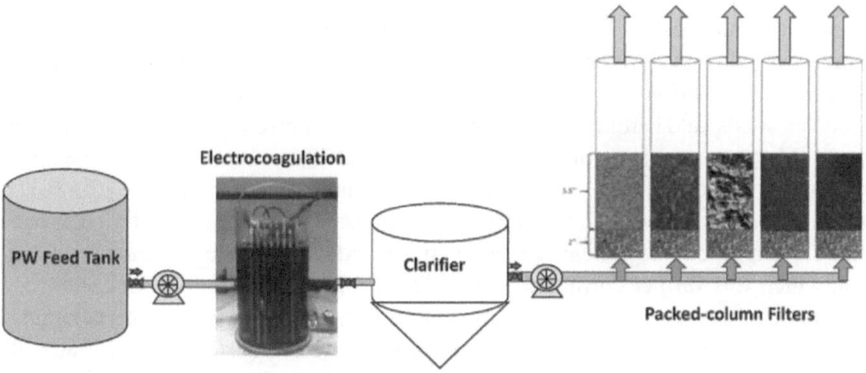

FIGURE 8.1 Schematic representation of the continuous-flow experimental arrangement. The packed columns in use featured a variety of media, arranged from left to right as follows: walnut shell, biochar, pecan shell, new granular activated carbon (GAC), and spent GAC. (Adapted from Ref. [32] under CCBY 4.0.)

agricultural waste products (pecan shell, walnut shell, and biochar) and both new and spent granular activated carbon (GAC). Performance was evaluated based on turbidity, TOC, metals, and electrical conductivity. Biochar and GAC exhibited the highest efficiency in removing turbidity and TOC from produced water. These treatment technologies proved effective in removing suspended constituents and iron, resulting in clean brine suitable for onsite reuse, particularly in hydraulic fracturing operations.

Carra et al. [33] investigated the efficacy of various advanced oxidation processes (AOPs) as tertiary treatment methods for the removal of the pesticide acetamiprid from water. Specifically, we evaluated UV photolysis, UV/TiO_2 photocatalysis, UV/H_2O_2/Fe, and two persulfate-based systems, UV/PS and UV/PS/Fe. The research was conducted using deionized water, synthetic water derived from secondary effluent, and real effluent from a conventional activated sludge treatment plant (CAS effluent). Additionally, the impact of organic matter, such as humic acids, lignin derivatives, and lauryl sulfate, on AOP efficiency was examined. Among all the processes studied, Photo-Fenton at natural pH consistently demonstrated the highest efficiency in all matrices. The persulfate-based systems also exhibited significant potential for the removal of micropollutants. Humic acids were found to diminish the degradation efficiency of UV/H_2O_2/Fe and UV/PS/Fe by competing with radicals. Moreover, at high concentrations, humic acids negatively affected UV photolysis due to their absorbance of UV light. Lignin had an adverse impact on AOPs utilizing iron, as it formed Fe(II)-complexes with high absorbance, effectively acting as a light filter. However, lauryl sulfate, with its chemical structure, did not favor the formation of iron complexes at natural pH, resulting in similar degradation rates for UV/H_2O_2/Fe, UV/H_2O_2, UV/PS, and UV/PS/Fe.

Singh et al. [34] designed a multifunctional three-compartment integrated water filtration system suitable for community usage and validated its effectiveness at both laboratory and pilot scales. The first compartment primarily addresses the removal of flocculating contaminants while improving the taste and odor of drinking water. The second compartment incorporates functionalized sand layers to eliminate metal and microbial contaminants. The third compartment consists of nanoparticle-embedded ceramic candles to further eradicate any remaining contaminants. Furthermore, the analysis of various physicochemical parameters, including pH, turbidity, total dissolved solids, and hardness, before and after filtration revealed stable and improved results. Micro-computed tomography (micro-CT) analysis demonstrated a substantial increase in porosity (~83.11 ± 2.17 %) in baked ceramic candles compared to unbaked candles (26.09 ± 1.33 %). Inductively coupled plasma mass spectrometry (ICP-MS) analysis confirmed a significant reduction in As and Cr levels before and after filtration, which was further corroborated by EDS mapping. Microbial testing indicated a hierarchical decrease in microbial load after each filtration stage at the laboratory scale. This low-cost, multi-stage, gravity-based filtration system demonstrated its effectiveness in removing heavy metal and microbial contaminants from water, producing potable water suitable for daily use. The integrated system's simple design, minimal infrastructure requirements, and promising results suggest its potential for real-time application in supplying safe drinking water to large populations, particularly in rural areas where sophisticated water purification systems may not be feasible.

Mameda et al. [35] focused on the development and evaluation of bifunctional electrocatalytic filter anodes, serving as reactive and separating media, to simultaneously eliminate stubborn dissolved organic compounds and particulate pollutants from actual wastewater effluents. The research reveals that the incorporation of TiO_2 particle interlayers between the Ti fiber support and the upper composite metal oxide catalyst layers aids in reducing filter pores to an effective size range, allowing for the removal of most particulates within the wastewater. The double-sheet design, characterized by a sandwich-structured module with an internal void space for permeation, effectively prevents filter fouling. Moreover, it maintains transmembrane pressure at an exceptionally low level, consistently below 5 kPa, during both batch and continuous flow reactor operations. The study demonstrated significant and concurrent removal of dissolved organics (e.g., chromophores, fluorophores, 1,4-dioxane, chemical oxygen demand, and total organic carbon) and particulate matter (i.e., turbidity). However, removal rates and efficiencies vary depending on the applied electric current density. Decolorization and particulate rejection occur rapidly and immediately, while the degradation of 1,4-dioxane is comparatively slow and time-dependent. The research also proposes potential degradation pathways for 1,4-dioxane during electrocatalysis. Overall, electrochemical filtration technology holds great promise as a component of the next generation of advanced wastewater treatment solutions.

8.3.2 Nanotechnology-based Solutions

Nanotechnology holds immense promise for revolutionizing pollution mitigation in the chemical industry by offering highly precise and efficient approaches to addressing environmental concerns. Here is an elaboration on nanotechnology-based solutions:

- Nanotechnology represents a groundbreaking frontier in pollution control, offering innovative approaches to tackle environmental contaminants at the nanoscale, typically with dimensions less than 100 nanometers [36, 37]. Within this realm, nanoparticles, including nanotubes, nanoparticles, and nanocomposites, emerge as potent tools engineered to possess unique properties, making them ideal candidates for the removal and transformation of pollutants.
- Versatility and Tailored Solutions: Nanotechnology-based solutions are remarkably versatile, providing the flexibility to customize approaches for the specific removal of pollutants. Nanoparticles can be meticulously engineered and functionalized to target a wide array of contaminants, be they heavy metals, organic compounds, or oil droplets present in wastewater or air emissions. This adaptability allows for the creation of tailored solutions that align with the precise requirements of each pollution control application.
- Adsorption and Catalysis: The distinctive features of nanoparticles, including their exceptional surface area-to-volume ratio and unique surface properties, contribute to their remarkable effectiveness in pollution control and remediation. The exceptional surface area-to-volume ratio means that nanoparticles have a large surface area relative to their size, providing more active sites

for interactions with pollutants. Additionally, their unique surface properties enhance their adsorption capacities and catalytic activities. The enhanced adsorption capacities of nanoparticles enable them to efficiently capture pollutants from their surrounding environment. This process involves the adherence of pollutant molecules to the surface of nanoparticles, facilitating their removal from the air, water, or soil. Simultaneously, the catalytic activities of nanoparticles allow them to participate in chemical reactions, transforming captured contaminants into less harmful substances. This dual functionality, encompassing both adsorption and catalysis, makes nanoparticles highly effective in the removal and remediation of pollutants [38]. Nanoparticles excel in both capturing pollutants from their surroundings and catalyzing chemical reactions to transform these contaminants into less harmful substances. This dual functionality renders them highly effective in pollution removal and remediation efforts.

- Emerging Applications: Nanotechnology's influence extends across a multitude of domains within pollution control. It finds applications in water and air purification, where engineered nanoparticles efficiently remove contaminants from these critical resources. Additionally, nanotechnology is employed in soil remediation, aiding in the restoration of polluted terrestrial environments. Furthermore, nanomaterials are being harnessed to create self-cleaning surfaces that resist fouling and contamination, offering innovative solutions in various industries.
- Environmental Concerns and Ongoing Research: While nanotechnology holds immense promise, it also raises environmental concerns related to the use and potential release of engineered nanoparticles. Researchers are diligently investigating these concerns to assess and mitigate any associated risks. Such investigations include studies on the environmental fate and toxicity of nanoparticles, ensuring that their applications are safe and sustainable.

The efficacy of NM-based approaches for pollution mitigation in the chemical industries has been deliberated by several researchers such as Chandio et al. [39] focused on the synthesis of ZnO nanoparticles via a straightforward aqueous chemical growth method. Photocatalysts offer a cost-effective and simple approach to the thermal degradation of chlorophenols (CPs). In this study, ZnO nanostructures were employed as photocatalysts for the simultaneous photodegradation of pentachlorophenol (PCP) and trichlorophenol (TCP) as model dyes under sunlight. The ZnO nanostructures demonstrated remarkable degradation efficiencies of approximately 85–90 percent for PCP and TCP in aqueous solutions, using a minimal catalyst dosage of 240–250 μg, both individually and simultaneously, within a 3-minute exposure to sunlight. The surface characteristics of the ZnO nanostructure were investigated using advanced characterization techniques, specifically the scanning electron microscope (SEM). In Figure 8.2A (magnified at 10,000× resolution) and Figure 8.2B (magnified at 12,000× resolution), we can observe the surface morphology of the ZnO nanostructure. In line with recent research findings, it is noted that ZnO nanoparticles (NPs) exhibit rod-like shapes and aggregations of nanocrystallites. SEM was utilized to assess the morphology and surface roughness

FIGURE 8.2 SEM micrographs and EDS spectra of ZnO nanoparticles. Subfigure (A) presents an SEM image of ZnO particles at a resolution of 10,000×, while Subfigure (B) provides an SEM image at a resolution of 12,000×. Subfigure (C) showcases the EDS spectra, and the inset illustrates the corresponding EDS results. (Adapted from Ref. [39] under CCBY 4.0.)

of these engineered nanostructures. As depicted in Figure 8.2A and 2B, the images clearly illustrate elongated particles with a rod-like nanostructure and highly porous surfaces, rendering them well-suited for catalytic reactions. Furthermore, to confirm the elemental composition and the high degree of purity of the engineered material, SEM was coupled with energy dispersive X-ray (EDX) analysis, as illustrated in Figure 8.2C. The EDX spectra provide conclusive evidence of the presence of zinc and oxygen atoms in high elemental concentrations. Notably, there are no traces of impurities from other elements in the EDX spectrum, underscoring the exceptional purity of the synthesized material.

The superior performance of the ZnO nanostructures in natural degradation under solar irradiation highlights the potential of aqueous chemical growth as a cost-effective method for producing efficient photocatalysts for the photodegradation of various environmental contaminants.

Prajapati et al. [40] aimed to treat wastewater from the distillery industry using both electrochemical and chemical processes with iron as the electrode material. Electrochemical treatment achieved a remarkable 91 percent reduction in chemical oxygen demand and a 71.5 percent reduction in color at pH 7, with an electrode gap of 20 mm, a current density of 26 mA/cm^2, and a reaction time of 70 minutes. In the combined treatment process, the maximum COD reduction reached 97 percent (24.51 mg/dm^3), and the color reduction was 92.6 percent (10.79 PCU) at a mass loading of 1 g/dm3 and pH 7, using ferrous sulfate. This research demonstrated that a combination of electrochemical and chemical treatments, with iron as an electrode material, is highly effective in treating distillery industry wastewater, meeting stringent environmental standards for discharge. Similarly, Saleh et al. [41] incorporated nano-zinc oxide (n-ZnO) onto multi-walled carbon nanotubes (MWCNTs) through a hydrothermal process. Pure n-ZnO was initially synthesized using the pulsed-laser ablation technique, while the MWCNTs were utilized in their as-received state. The resulting MWCNT/ZnO nanocomposites were extensively characterized using various analytical techniques, including X-ray diffraction, field emission scanning electron microscopy, high-resolution transmission electron microscopy, and Fourier transform infrared spectrometry. A structural model was developed to illustrate the arrangement of ZnO on the surface of MWCNTs. These nanocomposites were evaluated for their photocatalytic activity in the degradation of highly toxic cyanide (CN) compounds, utilizing a specialized reactor and pulsed 355 nm UV light generated by the third harmonic of an Nd:YAG laser. The investigation encompassed various experimental parameters, including different laser irradiation durations, incident laser energies, solution pH levels, and the dosage of MWCNT/ZnO nanocomposites. Notably, the study revealed that the MWCNT/ZnO composite exhibited significantly higher photocatalytic activity in CN removal compared to pure n-ZnO and MWCNTs alone. A schematic explanation was provided to elucidate the degradation mechanism facilitated by MWCNT/ZnO. The results indicated that the oxidation process's efficiency is influenced by the pH of the solution, with approximately 90 percent of CN degraded after 20 minutes of UV laser irradiation. This study sheds light on the promising potential of MWCNT/ZnO nanocomposites as efficient catalysts for the removal of toxic cyanide compounds.

Ehmari et al. [42] investigated the photocatalytic degradation of linear alkylbenzene sulfonic acid (LAS) using TiO_2/UV in an aqueous solution. The process was conducted within a continuous reactor equipped with coaxial cylinders. TiO_2 nanoparticles were immobilized on two natural polymer substrates, namely cotton and wool. The interior surface of the polymer textile served as the substrate for nanoparticle deposition. The contaminated solution was passed through the coaxial cylinders. The results indicated that the photocatalytic degradation of LAS was significantly more efficient than UV irradiation alone. Furthermore, it was observed that the wool-based substrate exhibited superior removal efficiency compared to the cotton-based substrate. The system successfully removed approximately 86 percent of pollutants within approximately 100 minutes. The reactor's operation could be tailored to either accommodate recycle flow or extended residence time. Overall, the study revealed that there was a marginal difference between increasing the residence time of the contaminant solution during a single pass through the reactor and employing a recycling flow approach. This research introduces a novel continuous coaxial cylinder reactor for efficient photocatalytic degradation of LAS in wastewater, offering potential solutions for environmental remediation using nanotechnology.

8.3.3 Green Chemistry Principles

Green chemistry principles represent a holistic and sustainable approach to pollution mitigation in the chemical industry. They focus on the design and implementation of chemical processes that minimize or eliminate hazardous substances and waste. Here is an elaboration on this topic:

- Design for Sustainability: At the core of green chemistry lies the principle of designing chemical processes and products with inherent environmental sustainability [43]. This involves crafting processes that are safer, less toxic, and more eco-friendly right from the outset. Key strategies include employing renewable feedstocks instead of finite resources and minimizing the reliance on hazardous chemicals. By focusing on sustainable design, green chemistry paves the way for responsible chemical production that prioritizes the environment and human well-being.
- Atom Economy: Green chemistry champions the maximization of chemical reaction efficiency to minimize waste generation. This principle fosters processes where the majority of reactants become part of the final product, leaving behind minimal byproducts. The result is a reduction in waste disposal requirements and more efficient use of resources, aligning seamlessly with environmental responsibility [17, 44].
- Safer Solvents and Reaction Conditions: A core facet of green chemistry is the promotion of safer solvents and reaction conditions. This entails substituting volatile organic solvents, known for their detrimental effects on human health and the environment, with water-based alternatives or other benign options. By minimizing exposure to harmful chemicals, green chemistry enhances both workplace safety and environmental protection.
- Life Cycle Assessment: Green chemistry transcends traditional boundaries by considering the entire life cycle of chemical products. This comprehensive approach evaluates environmental impacts from production to disposal, ensuring that sustainability considerations are woven into every stage. By addressing potential environmental hotspots across the product's life cycle, green chemistry minimizes adverse effects and strives for holistic sustainability.
- Regulatory Support: Governments and regulatory bodies worldwide are increasingly embracing green chemistry initiatives. They provide incentives and regulations that encourage the adoption of these principles in the chemical industry. These supportive measures create a conducive environment for the development and implementation of sustainable practices, fostering industry-wide transformation.

- Innovation Catalyst: Green chemistry acts as a catalyst for innovation, driving advancements in materials, processes, and technologies. This innovation cycle leads to the creation of more sustainable and environmentally friendly products and practices. As researchers and industries delve into the realm of green chemistry, they unlock novel solutions that propel the chemical industry toward a brighter, more sustainable future.

In an experiment, Almeida et al. [45] produced a biosurfactant from Candida tropicalis UCP0996 using a cost-effective medium formulated with molasses, residual frying oil, and corn steep liquor. They conducted a comprehensive characterization of the biosurfactant and explored its toxicity, formulation, and application in the removal and biodegradation of oil. The surface tension of the medium was significantly reduced to 30.4 mN/m, resulting in the isolation of 4.11 g/L of biosurfactant after 120 hours of fermentation. Their investigations under extreme environmental conditions demonstrated the stability of the biosurfactant. Remarkably, the isolated biosurfactant exhibited no toxicity towards the microcrustacean Artemia salina, and the properties of the formulated biosurfactant remained stable during 120 days of storage. The biosurfactant efficiently removed 66.18 percent of motor oil adsorbed in marine stones and dispersed 70.95 percent of oil in seawater. Furthermore, it enhanced the degradation of motor oil by indigenous seawater microorganisms by 70 percent, showcasing its potential as a commercial additive for oil spill bioremediation.

Rosa-Masegosa et al. [46] presented a review where they discussed the aerobic granular sludge (AGS) which represents a clustered assembly of microbial cells within a three-dimensional matrix, capable of simultaneously removing carbon, nitrogen, phosphorous, and various other pollutants in a single bioreactor under uniform operational conditions. They stated that over the past few decades, the feasibility of implementing AGS technology in wastewater treatment plants (WWTPs) using primarily sequential batch reactors (SBRs) has been extensively investigated. However, the utilization of granular sludge technology through SBRs presents certain drawbacks, including limitations in treating high flow rates and the volume exchange's impact on biomass retention. To address these limitations, continuous flow reactors (CFRs) have emerged as a more competitive alternative. Consequently, numerous research efforts have been directed toward developing various CFR system designs that can efficiently treat urban and industrial wastewater while maintaining granular biomass stability. Despite these endeavors, achieving satisfactory results remains a challenge. Therefore, there is a pressing need for novel technical approaches that can deliver more effective and cost-efficient AGS-CFR systems capable of retaining granular biomass while efficiently treating wastewater. Their review compiles the latest information on diverse technological approaches aimed at establishing AGS-CFR systems and facilitating their upscale integration into real-world plant conditions. They assessed the advantages and disadvantages of these proposals and proposed future trends in the application of aerobic granular systems. In particular, they scrutinize the most significant technical and biological implications of this innovative technology.

Husien et al. [2] explored eco-friendly adsorbents derived from agricultural waste, such as Fava Bean Peels (R–FBP), as potential materials for removing these pollutants.

In this study, R–FBP and their respective physical and chemically activated carbons (P-RFB-AC and C-FBP-AC) were synthesized using H3PO4 acid and characterized through FT-IR and SEM analyses. An optimization process was undertaken to identify the optimal conditions for achieving high removal efficiencies of Methyl Orange (M. Orange) using these materials. The adsorption mechanism was investigated by analyzing the isotherm and kinetics data. The results revealed that physical raw-activated carbon demonstrated the highest removal efficiency of 96.8 percent, surpassing other materials. This outcome was attained through the utilization of an Artificial Neural Network (ANN) combined with the Moth Search Algorithm (MSA), which proved to be the most effective model for achieving superior M. Orange removal efficiency from physical raw fava bean activated carbon. Under specific conditions – namely, 1000 mg/l M. Orange concentration, 2 g/l dose, 15 minutes of contact time, and 120 rpm shaking – the best experimental and predicted removal efficiencies for physical-activated carbon fava bean peels were 96.8 percent, 96.01 percent (as indicated by RSM), and 95.75 percent (predicted by ANN). As for the H_3PO_4 chemical activated carbon from fava bean peels, the highest experimental and predicted removal efficiencies were 94 percent. This study aimed to offer an economical solution for treating industrial wastewater contaminated with anionic M. Orange dye using raw fava bean peels and the activated carbon derived from them, both in physical and chemical forms. The Temkin and Langmuir isotherm models were determined as the best fits for the raw fava bean peels, while the Temkin model was in good agreement with the data from the physical-activated carbon. Both the Temkin and Freundlich models were suitable for fitting the data from the H_3PO_4 chemical-activated carbon. Additionally, the pseudo-second-order kinetics model was identified as the most appropriate for both physically and chemically activated carbons. Future research may explore the potential of the produced activated carbon-based algae for extracting a broader range of contaminants from polluted wastewater. In summary, this study contributes to the development of eco-friendly and cost-effective methods for removing dyes, particularly M. Orange, from industrial effluents. Through the synthesis and characterization of R–FBP and their associated activated carbon materials, the study investigates the adsorption mechanism and identifies optimal conditions for achieving high M. Orange removal efficiencies. The results highlight the superior performance of physical raw-activated carbon in removing M. Orange and shed light on the kinetics of the adsorption process.

Adegoke et al. [3] in their review provided an empirical analysis and methodological insights into the use of sawdust-biomass-based materials for adsorptive pollutant removal from aqueous solutions. The initial sections of the review shed light on the environmental significance of sawdust-biomass-based materials, their potential health and environmental impacts, and the sources, types, compositions, fabrication methods, and configurations of sawdust-based adsorbents. The review also explored the characterization of adsorbents derived from sawdust-biomass materials. Pollutants under consideration in this study encompass endocrine-disrupting chemicals, dyes, pesticides, herbicides, agrochemicals, and heavy metals. A comprehensive evaluation and performance assessment of sawdust-biomass-based adsorbents were presented separately for each pollutant category, allowing for meaningful comparisons within each class. Notably, the adsorption capacity of different sawdust-based adsorbents varies across pollutant categories, with capacities ranging from 10 to 667.9 mg/g (for endocrine-disrupting chemicals and emerging contaminants removal), 69.44–372 mg/g (for pesticides, herbicides, and agrochemicals removal), 3.42–526.3 mg/g (for dyes removal), and 2.87–325 mg/g (for heavy metal removal). The review also delves into the adsorption mechanisms of these materials, and strategies for enhancing the efficiency of sawdust-biomass-based adsorbents, and outlines a potential methodological approach for their engineering applications – an aspect notably lacking in existing literature. Lastly, key observations, conclusions, and identified knowledge gaps pave the way forward for these materials, encouraging further industrial applications and research in the field.

Sahu et al. [4] successfully demonstrated the efficacy of a highly efficient and easily reusable catalyst for on-site water remediation. Their research centers on achieving nearly 100 percent dye adsorption and the effective removal of arsenic (III) from water using a ternary composite composed of ORMOSIL-Fe_3O_4-RGO. They developed a straightforward and low-temperature synthesis method for creating the ORMOSIL-Fe_3O_4-RGO composite, offering a comprehensive solution for water purification challenges. This composite material was specifically employed for the removal of arsenite (III) ions and Rhodamine B dye from water sources, addressing significant water quality issues prevalent in developing/underdeveloped countries, particularly in South Asia and certain American regions. By utilizing Langmuir isotherms, they quantified the adsorption capacity of the ORMOSIL-Fe_3O_4-RGO composite for Rhodamine B, which was found to be approximately 1339 mg/g. This capacity significantly surpasses that of the Fe_3O_4-RGO composite, which stood at around 342 mg/g. Furthermore, their novel composite material exhibited a higher arsenic adsorption capacity, approximately 25 percent greater than that of Fe_3O_4-RGO, as determined by the Langmuir adsorption isotherm.

8.4 INTEGRATION OF ARTIFICIAL INTELLIGENCE AND DATA ANALYTICS

8.4.1 Real-time Pollution Monitoring

- Technological Foundations: The bedrock of real-time pollution monitoring systems is a fusion of cutting-edge technologies. Central to these systems is an array of sensors meticulously designed to detect and measure environmental pollutants. These sensors are often integrated into the broader ecosystem of the Internet of Things (IoT), enabling seamless data collection, transmission, and analysis. Data analytics, powered by artificial intelligence (AI) and machine learning (ML) algorithms, serve as the cognitive engine that processes the vast influx of data in real time. This convergence of technologies empowers industries and regulatory bodies with an unprecedented level of environmental vigilance [42, 47].
- Data Collection and Analysis: Real-time pollution monitoring systems cast a wide net to gather data from various sources. These sources encompass a diverse range of sensors, including air quality sensors, water quality probes, and soil contamination sensors. The data collected from these sensors is transmitted

to a central analytics platform, where AI algorithms come into play. These algorithms diligently analyze the data in real-time, sifting through the digital deluge to discern meaningful patterns and trends. This data-driven approach transforms raw information into actionable insights, providing an invaluable resource for pollution control [48–54].

- Immediate Alerting and Response: A defining feature of real-time pollution monitoring is its capacity to deliver immediate alerts when pollutant levels breach predefined thresholds. AI-driven algorithms exhibit remarkable agility in detecting anomalies, such as sudden spikes in pollutant concentrations. When these anomalies are detected, the system swiftly triggers alerts, notifying environmental managers and regulatory authorities. This rapid response capability empowers stakeholders to initiate timely mitigation measures, mitigating the potential for widespread environmental damage. It also bolsters public safety by providing critical information to communities living in proximity to pollution sources [41].
- Localization and Source Identification: Beyond detecting pollution events, advanced AI models employed in real-time monitoring can pinpoint their precise origins. By cross-referencing data from multiple sensors and harnessing source apportionment modeling, industries can accurately identify the source of pollution incidents. This capability enhances accountability, facilitating the investigation of responsible parties and enabling targeted remediation efforts. It also serves as a valuable tool for industries seeking to proactively address pollution sources and minimize their environmental footprint.
- **Benefits:**
  - Environmental Compliance: At the forefront of the advantages offered by real-time pollution monitoring is its capacity to ensure unwavering environmental compliance. These systems function as vigilant guardians, constantly monitoring pollutant levels to ensure they remain well within regulatory limits. By doing so, they provide industries with the means to preemptively address any deviations that might lead to non-compliance penalties. This proactive approach not only safeguards against financial repercussions but also underscores a commitment to environmental responsibility and adherence to legal obligations.
  - Early Pollution Event Detection: The ability to swiftly detect and respond to pollution events is another cornerstone benefit of real-time monitoring. These systems serve as an environmental early warning system, sounding the alarm when pollutant levels exceed predefined thresholds [41]. This rapid response is pivotal in curbing the potential for long-term environmental damage. By identifying pollution incidents in their infancy, industries can institute timely mitigation measures, averting extensive contamination and costly clean-up efforts. This capability not only safeguards ecosystems and public health but also bolsters an industry's reputation as a responsible steward of the environment [55–57].
  - Data-Driven Decision-Making: The data amassed through real-time pollution monitoring represents a treasure trove of insights for industries. Beyond immediate compliance and pollution event detection, this data

serves as a valuable resource for comprehensive trend analysis. Industries can harness the wealth of information to identify patterns, trends, and areas of concern. This data-driven approach enables informed decision-making and empowers industries to take proactive measures in pollution reduction and process optimization. By leveraging data analytics, industries can fine-tune their practices, minimize environmental impact, and drive operational efficiencies [58].

- **Challenges:**
  - Implementation Costs: Establishing a real-time monitoring system can be capital-intensive, involving investments in sensors, data infrastructure, and AI technology.
  - Data Quality: Ensuring data accuracy and reliability is paramount, as erroneous readings can lead to false alarms or inadequate responses.
  - Integration Complexity: Merging data from various sensors and platforms may pose challenges related to compatibility and data management.

8.4.2 Predictive Maintenance for Pollution Control

Predictive maintenance, underpinned by the formidable capabilities of AI and data analytics, emerges as a formidable defense against equipment failures that could trigger pollution incidents. This section delves into the foundational principles, data sources, and early anomaly detection mechanisms that define predictive maintenance in pollution control, illustrating its pivotal role in safeguarding the environment [35, 40, 59].

- Fundamental Principles: Predictive maintenance is grounded in the fundamental principle of proactively preventing equipment failures that could lead to pollution incidents. This approach hinges on predictive models that harness AI and data analytics to scrutinize a wealth of data sources. These sources include real-time sensor data, historical equipment records, and vast datasets teeming with operational insights. By meticulously analyzing this data, predictive models can forecast when pollution control equipment is likely to require maintenance or replacement. This predictive foresight empowers industries to intervene before equipment failures culminate in environmental harm [60–68].
- Data Sources and Integration: The effectiveness of predictive maintenance is contingent on the diverse and dynamic sources of data it draws upon. IoT sensors, seamlessly integrated into pollution control equipment, serve as frontline data collectors, continuously monitoring the equipment's condition and performance. Historical maintenance records provide crucial insights into the equipment's service history, while operational data offers a real-time snapshot of its ongoing performance. AI models orchestrate the harmonious integration of these disparate data sources, culminating in a holistic view of equipment health. The integration process is the bedrock upon which predictive insights are built [39, 69].

- Early Detection of Anomalies: AI algorithms wield unparalleled prowess in the realm of anomaly detection. By diligently monitoring equipment performance in real-time, these algorithms become attuned to subtle deviations that portend degradation or malfunction. Early signs of equipment anomalies, such as abnormal vibrations, temperature fluctuations, or energy consumption patterns, are swiftly identified and flagged by the AI system. This early detection allows industries to preemptively address equipment issues before they escalate into full-fledged failures. By doing so, predictive maintenance acts as an environmental sentinel, guarding against pollution incidents that could arise from equipment breakdowns.
- **Benefits:**
  - Preventing Unplanned Failures: At its core, predictive maintenance serves as an environmental sentry, preemptively preventing unplanned equipment failures that have the potential to unleash pollution events. By harnessing AI-driven predictive models, industries gain the capability to forecast equipment maintenance needs with remarkable accuracy. These predictive insights enable proactive intervention, averting equipment breakdowns and the environmental incidents they may trigger. This pivotal role in pollution prevention aligns predictive maintenance with the broader mission of environmental stewardship, safeguarding ecosystems and public health.
  - Cost Reduction: The benefits of predictive maintenance extend beyond environmental protection to encompass tangible cost reduction [70]. By optimizing maintenance schedules and resource allocation, industries can drastically curtail maintenance expenses. Unplanned breakdowns, often associated with hefty repair costs and production downtime, are circumvented through proactive maintenance measures. This not only translates into immediate cost savings but also enhances operational efficiency. Predictive maintenance minimizes the expenditure of resources on reactive repairs, redirecting them toward strategic investments that foster sustainability and innovation.
  - Enhanced Equipment Efficiency: Predictive maintenance does not merely serve as a shield against equipment failures; it also acts as a catalyst for equipment efficiency and longevity. By continually monitoring equipment performance and identifying early signs of anomalies or wear and tear, industries can proactively address issues, thus optimizing equipment efficiency. Well-maintained equipment operates at peak performance, consuming fewer resources and extending its lifespan. This synergy between predictive maintenance and equipment efficiency contributes to sustainability efforts, as industries tread the path toward reduced resource consumption and minimized environmental impact [71, 72].
- **Challenges:**
  - Data Quality: Accurate data is essential for predictive models to make reliable forecasts. Poor data quality can lead to inaccurate predictions.
  - Implementation Complexity: Implementing predictive maintenance may require significant changes to existing maintenance practices and the integration of AI technology.
  - Model Updates: Predictive models need periodic updates to adapt to changing conditions and equipment performance.

8.5 SUSTAINABLE PRACTICES IN CHEMICAL PRODUCTION

8.5.1 Adoption of Renewable Energy Sources

- Preventing Unplanned Failures: At its core, predictive maintenance serves as an environmental sentry, preemptively preventing unplanned equipment failures that have the potential to unleash pollution events [73, 74]. By harnessing AI-driven predictive models, industries gain the capability to forecast equipment maintenance needs with remarkable accuracy. These predictive insights enable proactive intervention, averting equipment breakdowns and the environmental incidents they may trigger. This pivotal role in pollution prevention aligns predictive maintenance with the broader mission of environmental stewardship, safeguarding ecosystems and public health.
- Cost Reduction: The benefits of predictive maintenance extend beyond environmental protection to encompass tangible cost reduction. By optimizing maintenance schedules and resource allocation, industries can drastically curtail maintenance expenses. Unplanned breakdowns, often associated with hefty repair costs and production downtime, are circumvented through proactive maintenance measures [75]. This not only translates into immediate cost savings but also enhances operational efficiency. Predictive maintenance minimizes the expenditure of resources on reactive repairs, redirecting them toward strategic investments that foster sustainability and innovation.
- Enhanced Equipment Efficiency: Predictive maintenance doesn't merely serve as a shield against equipment failures; it also acts as a catalyst for equipment efficiency and longevity. By continually monitoring equipment performance and identifying early signs of anomalies or wear and tear, industries can proactively address issues, thus optimizing equipment efficiency. Well-maintained equipment operates at peak performance, consuming fewer resources and extending its lifespan. This synergy between predictive maintenance and equipment efficiency contributes to sustainability efforts, as industries tread the path toward reduced resource consumption and minimized environmental impact.
- **Benefits:**
  - Carbon Emission Reduction: At the forefront of the sustainable chemical production revolution is the significant reduction of carbon emissions. Historically, the chemical industry has been associated with high energy consumption and substantial greenhouse gas emissions, primarily due to its reliance on fossil fuels. The shift toward renewable energy sources such as solar, wind, and biomass fundamentally alters this narrative [34, 76]. These clean energy sources generate electricity with minimal or no carbon emissions, thereby significantly curtailing the industry's carbon footprint. By adopting renewables, the chemical industry becomes an active participant in global climate goals, contributing to the collective effort to combat climate change.
  - Resource Efficiency: Renewable energy sources constitute a cornerstone of resource efficiency in chemical production. Unlike finite fossil fuels,

renewable energy sources are sustainable and inexhaustible. This inherent sustainability translates into reduced environmental impact associated with resource extraction. The elimination of resource scarcity concerns fosters a more resilient and eco-friendly chemical production ecosystem. By tapping into the virtually limitless supply of renewable energy, the industry ensures that energy availability does not become a bottleneck to sustainable growth. This transition to renewable resources aligns with the broader concept of circular economy principles, where resources are conserved, recycled, and utilized efficiently.
  - Enhanced Sustainability Reputation: Beyond environmental benefits, the adoption of renewable energy sources bestows chemical companies with an enhanced sustainability reputation. Industries that prioritize renewable energy are increasingly viewed as environmentally responsible entities. Such companies resonate with eco-conscious customers who seek products and services with minimal environmental impact. Likewise, they attract investors who are keen on supporting businesses committed to sustainability. This enhanced reputation not only enhances the company's brand value but also opens doors to a broader customer base and investment opportunities. It positions the chemical industry as a sustainability trailblazer, poised to thrive in a world where environmental responsibility is paramount.
- **Challenges:**
  - Initial Investment: The adoption of renewable energy sources may require substantial upfront investments in infrastructure and technology.
  - Intermittency: Some renewable sources are intermittent, requiring effective energy storage solutions to ensure continuous production.

8.5.2 Circular Economy Principles for Reduced Environmental Footprint

The integration of circular economy principles into chemical production represents a pivotal shift, one that champions waste reduction, resource conservation, and a minimized environmental footprint. Grounded in fundamental principles like recyclability, material reuse, and waste reduction, this transformation reshapes the chemical industry's approach to sustainability [18, 33].

- Fundamental Principles: The bedrock of circular economy principles in chemical production is an unwavering commitment to minimizing waste and optimizing resource utilization. Designing products with recyclability in mind, promoting material reuse, and slashing waste generation are key tenets. The chemical industry recognizes that the linear "take-make-dispose" model is no longer sustainable. Instead, circular economy principles guide the industry toward a regenerative model where resources are preserved, repurposed, and reimagined throughout their lifecycle.
- Recycling and Upcycling: A central thrust of circular economy adoption in the chemical industry revolves around recycling and upcycling processes. These

strategies minimize waste and extend the lifespan of materials. By recovering and reusing valuable components from waste streams, chemical manufacturers reduce their reliance on virgin materials. This not only curtails waste but also curbs the environmental impact of resource extraction and production. The evolution from a linear to a circular approach heralds a profound transformation in material sourcing and utilization [18, 77, 78].

- Chemical Process Optimization: Circular economy principles in chemical production extend to the very core of the manufacturing process. Optimization becomes paramount, with the overarching goal of maximizing resource efficiency. This optimization spans various dimensions, including minimizing energy consumption, reducing water usage, and implementing catalytic processes that curtail waste generation. Chemical processes are fine-tuned to ensure that inputs are utilized judiciously, with minimal waste as a byproduct. This holistic approach engenders a resource-efficient ecosystem that minimizes the environmental burden.
- Product Life Extension: The concept of product life extension is a cornerstone of circular economy adoption in chemical production. Companies are actively exploring ways to extend the lifespan of their chemical products through refurbishment, maintenance, and remanufacturing. This approach ensures that products remain in use for longer periods, reducing the need for constant replacement and diminishing waste generation. By maximizing the utility of products, chemical manufacturers transition from a culture of disposability to one of sustainability.
- **Benefits:**
    - Waste Reduction: At the core of circular economy practices lies a resolute commitment to waste reduction. The conventional linear model of production and consumption generates copious amounts of waste. In contrast, circular economy principles dictate that waste should be minimized, if not eliminated, from the production process. This shift is especially relevant in the chemical industry, which has historically been associated with waste-intensive processes. By adopting practices such as recycling, upcycling, and waste-to-resource conversion, chemical manufacturers drastically curtail waste generation. This not only mitigates environmental impact but also aligns with global sustainability objectives. Waste reduction signifies a conscious effort to preserve ecosystems, reduce landfill usage, and mitigate the environmental burden of waste disposal [32, 79].
    - Resource Conservation: Circular economy practices champion resource conservation as a fundamental principle. This ethos permeates the entire supply chain, from material sourcing to production and consumption. Reusing materials, minimizing resource extraction, and optimizing resource use become central tenets. In the chemical industry, where resource-intensive processes are common, this approach is especially transformative. By repurposing materials, chemical manufacturers diminish their reliance on virgin resources, preserving them for future generations. This conservationist stance not only reduces the industry's environmental footprint

but also positions it as a responsible steward of resources. It aligns with broader sustainability goals and exemplifies the industry's commitment to safeguarding natural ecosystems and biodiversity.
  - Cost Savings: Circular economy practices bring forth a compelling economic dimension – cost savings. These practices often lead to tangible cost reductions by reducing waste disposal expenses and curtailing resource procurement costs. Waste disposal, especially hazardous waste, can be a significant financial burden for chemical manufacturers. Circular economy strategies substantially diminish waste generation, translating into lower waste disposal expenses. Additionally, by reusing materials and optimizing resource use, the industry trims resource procurement costs [45, 46, 80]. This twofold economic benefit not only enhances the industry's profitability but also positions circular economy adoption as a fiscally prudent choice. It underscores that sustainability and cost-effectiveness can go hand in hand, dispelling the notion that environmentally responsible practices are inherently more expensive.
- **Challenges:**
  - Technological Barriers: Implementing circular economy principles may require innovative technologies and processes, which can be costly to develop and integrate.
  - Behavioral Change: Encouraging consumers to participate in circular economy practices, such as recycling and product refurbishment, can be challenging.

8.6 CONCLUSION

8.6.1 Recap of Key Findings and Insights

Recapping the key findings and insights from our discussion, it is evident that the chemical industry is facing an environmental imperative to shift toward sustainable practices. Historically associated with pollution and resource depletion, the industry is now compelled to embrace sustainability and environmental responsibility as global concerns intensify. This shift is driven by a growing emphasis on sustainability and a recognition of the need for innovation and adaptation. Technological advancements play a pivotal role in this transition, with innovative solutions such as advanced filtration systems, nanotechnology-based approaches, and green chemistry principles leading the way in pollution control and sustainable chemical production. These technologies provide effective means to reduce environmental impact and minimize resource consumption, aligning with the industry's evolving environmental goals. Furthermore, the integration of artificial intelligence (AI) and data analytics is revolutionizing pollution control and equipment maintenance practices. Real-time pollution monitoring enables swift responses to pollution incidents, enhancing environmental protection, while predictive maintenance minimizes equipment failures, reducing environmental risks and costs.

Renewable energy adoption is emerging as a cornerstone of sustainable chemical production. Transitioning from fossil fuels to renewable sources not only mitigates carbon emissions but also contributes significantly to resource conservation, fostering a more sustainable energy ecosystem. Lastly, circular economy principles are gaining prominence in the industry, with practices like recycling, upcycling, and product life extension actively reducing waste generation and resource consumption. These circular approaches not only align with sustainability goals but also promote resource efficiency, marking a transformative step towards more responsible and environmentally conscious chemical production practices. In summary, the chemical industry's journey towards sustainability and environmental responsibility is being driven by technological innovations, AI integration, renewable energy adoption, and circular economy principles, all converging to reshape the industry and reduce its environmental footprint.

8.6.2 Implications for the Future of Chemical Industries in a Sustainable World

The insights garnered from this discussion carry profound implications for the future of chemical industries in a sustainable world. Environmental stewardship emerges as a paramount priority, compelling chemical industries to place a premium on adopting sustainable practices that reduce emissions and minimize resource consumption. This commitment not only serves to mitigate environmental harm but also enhances corporate reputation and competitiveness, as environmentally responsible companies are increasingly favored by consumers and investors alike. Innovation and adaptation will be pivotal in navigating the industry's sustainable future. Chemical companies must continue investing in research and development of new technologies, cultivate a corporate culture that values sustainability, and remain vigilant in staying abreast of evolving regulatory frameworks. These proactive measures will ensure that the industry remains agile and responsive to sustainability goals. Resource efficiency is set to become a driving force in the chemical industry's trajectory. Embracing circular economy principles, optimizing manufacturing processes to reduce resource waste, and curtailing waste generation will be imperative in minimizing the industry's environmental footprint and promoting responsible resource utilization. Global collaboration emerges as a potent strategy for addressing cross-border environmental challenges. Chemical industries should actively engage in collaborative efforts to share best practices, research findings, and cutting-edge technologies. Such cooperation can accelerate progress toward sustainability and foster a global network of environmentally responsible practices.

The landscape of regulatory compliance is evolving rapidly, with governments worldwide enacting stricter environmental regulations. Chemical industries must proactively adapt and comply with these rules, as non-compliance not only risks legal penalties but also undermines the industry's commitment to environmental responsibility. Lastly, consumer expectations are undergoing a significant shift towards sustainability. Chemical companies that prioritize and incorporate sustainable products and practices into their operations will gain a competitive edge by meeting the growing demands of environmentally conscious consumers. In essence, the chemical industry's future in a sustainable world hinges on environmental stewardship, innovation, resource efficiency, global collaboration, regulatory compliance, and meeting the evolving expectations of a sustainability-conscious consumer base.

REFERENCES

[1] Singh A, Goyal V, Singh J, Rawat M. Structural, morphological, optical and photocatalytic properties of green synthesized TiO2 NPs. *Curr Res Green Sustain Chem* 2020; 3: 100033.

[2] Husien S, El-taweel RM, Alrefaey KA *et al.* Experimental investigation of methyl-orange removal using eco-friendly cost-effective materials raw fava bean peels and their formulated physical, and chemically activated carbon. *Curr Res Green Sustain Chem* 2023; 7: 100373.

[3] Adegoke KA, Adesina OO, Okon-Akan OA *et al.* Sawdust-biomass based materials for sequestration of organic and inorganic pollutants and potential for engineering applications. *Curr Res Green Sustain Chem* 2022; 5: 100274.

[4] Sahu TK, Arora S, Banik A, Iyer PK, Qureshi M. Efficient and rapid removal of environmental malignant arsenic(III) and industrial dyes using reusable, recoverable ternary iron oxide – ORMOSIL – reduced graphene oxide composite. *ACS Sustain Chem Eng* 2017; 5: 5912–5921.

[5] Karpudewan M, Ismail Z, Roth WM. Ensuring sustainability of tomorrow through green chemistry integrated with sustainable development concepts (SDCs). *Chem Educ Res Pract* 2012; 13: 120–127.

[6] Al-Bedairy MA, Habeeb Alshamsi HA. Environmentally friendly preparation of zinc oxide, study catalytic performance of photodegradation by sunlight for rhodamine B dye. *Eurasian J Anal Chem* 2018; 13(6): 1–9.

[7] Muinde VM, Onyari JM, Wamalwa B, Wabomba JN. Adsorption of malachite green dye from aqueous solutions using mesoporous chitosan–zinc oxide composite material. *Environ Chem Ecotoxicol* 2020; 2: 115–125.

[8] Mandal A, Bar N, Das SK. Phenol removal from wastewater using low-cost natural bioadsorbent neem (Azadirachta indica) leaves: Adsorption study and MLR modeling. *Sustain Chem Pharm* 2020; 17: 100308.

[9] Qureshi UA, Khatri Z, Ahmed F, Khatri M, Kim IS. Electrospun zein nanofiber as a green and recyclable adsorbent for the removal of reactive black 5 from the aqueous phase. *ACS Sustain Chem Eng* 2017; 5: 4340–4351.

[10] Jain R, Mendiratta S, Kumar L, Srivastava A. Green synthesis of iron nanoparticles using Artocarpus heterophyllus peel extract and their application as a heterogeneous Fenton-like catalyst for the degradation of Fuchsin Basic dye. *Curr Res Green Sustain Chem* 2021; 4: 100086.

[11] Kim H, Watthanaphanit A, Saito N. Simple solution plasma synthesis of hierarchical nanoporous MnO2 for organic dye removal. *ACS Sustain Chem Eng* 2017; 5: 5842–5851.

[12] Lakavat M, Rao LN. Innovative control measures of water pollution-A study on green chemistry. *Am J Mater Sci* 2015; 5: 169–174.

[13] Saravanabhavan S, Aravindhan R, Thanikaivelan P, Rao JR, Nair BU. Green solution for tannery pollution: Effect of enzyme based lime-free unhairing and fibre opening in combination with pickle-free chrome tanning. *Green Chem* 2003; 5: 707–714.

[14] Egbosiuba TC, Abdulkareem AS, Kovo AS *et al.* Adsorption of Cr(VI), Ni(II), Fe(II) and Cd(II) ions by KIAgNPs decorated MWCNTs in a batch and fixed bed process. *Sci Rep* 2021; 11: 1–20.

[15] Sharma RK, Yadav S, Gupta R, Arora G. Synthesis of magnetic nanoparticles using potato extract for dye degradation: A green chemistry experiment. *J Chem Educ* 2019; 96: 3038–3044.
[16] Ramalingam B, Khan MMR, Mondal B, Mandal AB, Das SK. Facile synthesis of silver nanoparticles decorated magnetic-chitosan microsphere for efficient removal of dyes and microbial contaminants. *ACS Sustain Chem Eng* 2015; 3: 2291–2302.
[17] Hlongwane GN, Sekoai PT, Meyyappan M, Moothi K. Simultaneous removal of pollutants from water using nanoparticles: A shift from single pollutant control to multiple pollutant control. *Sci Total Environ* 2019; 656: 808–833.
[18] Abdel-Shafy HI, El-Khateeb MA, Mansour MSM. Treatment of leather industrial wastewater via combined advanced oxidation and membrane filtration. *Water Sci Technol* 2016; 74: 586–594.
[19] Ezung SL, Baruah M, Supong A, Sharma S, Sinha D. Experimental and theoretical insight into the adsorption of 2,4-dichlorophenol on low-cost bamboo sheath activated carbon. *Sustain Chem Pharm* 2022; 26: 100643.
[20] Singh A, Khare P, Verma S *et al.* Pollutant soot for pollutant dye degradation: Soluble graphene nanosheets for visible light induced photodegradation of methylene blue. *ACS Sustain Chem Eng* 2017; 5: 8860–8869.
[21] Ahmed M, Mavukkandy MO, Giwa A *et al.* Recent developments in hazardous pollutants removal from wastewater and water reuse within a circular economy. *NPJ Clean Water* 2022; 5: 1–25.
[22] Pesqueira JFJR, Pereira MFR, Silva AMT. Environmental impact assessment of advanced urban wastewater treatment technologies for the removal of priority substances and contaminants of emerging concern: A review. *J Clean Prod* 2020; 261: 121078 .
[23] Hacioglu T, Tezcaner A, Abbas S, Evis Z. Preparation and characteristics of co-doped hydroxyapatite biomimetic coatings on pretreated ti6al4v alloy. *Surf Rev Lett* 2020; 27(11): 2050012.
[24] Sidhu SS, Singh H, Gepreel MAH. A review on alloy design, biological response, and strengthening of β-titanium alloys as biomaterials. *Mater Sci Eng C* 2021; 121: 111661.
[25] Sarraf M, Rezvani Ghomi E, Alipour S, Ramakrishna S, Liana Sukiman N. A state-of-the-art review of the fabrication and characteristics of titanium and its alloys for biomedical applications. *Bio-Design Manuf* 2022; 5: 371–395.
[26] Kalyoncuoglu UT, Yilmaz B, Gungor S *et al.* Evaluation of the chitosan-coating effectiveness on a dental titanium alloy in terms of microbial and fibroblastic attachment and the effect of aging. *Mater Tehnol* 2015; 49: 925–931.
[27] Kaur M, Singh K. Review on titanium and titanium based alloys as biomaterials for orthopaedic applications. *Mater Sci Eng C* 2019; 102: 844–862.
[28] Dhiman N, Ghosh S, Mishra YK, Tripathi KM. Prospects of nano-carbons as emerging catalysts for enzyme-mimetic applications. *Mater Adv* 2022; 3: 3101–3122.
[29] Kaushik J, Gunture, Tripathi KM, Singh R, Sonkar SK. Thiourea-functionalized graphene aerogel for the aqueous phase sensing of toxic Pb(II) metal ions and H2O2. *Chemosphere* 2022; 287: 132105.
[30] Serhan M, Sprowls M, Jackemeyer D *et al.* Total iron measurement in human serum with a smartphone. *AIChE Annu Meet Conf Proc* 2019; 2019-Novem.
[31] Islam MD, Mahdi MM. Evaluation of micro-pollutants removal from industrial wastewater using conventional and advanced biological treatment processes. *Biodegrad Detoxif Micropollutants Ind Wastewater* 2022; 1: 1–26.
[32] Rodriguez AZ, Wang H, Hu L, Zhang Y, Xu P. Treatment of produced water in the permian basin for hydraulic fracturing: Comparison of diff erent. *Water* 2020; 12(3): 770.

[33] Carra I, Sánchez Pérez JA, Malato S, Autin O, Jefferson B, Jarvis P. Performance of different advanced oxidation processes for tertiary wastewater treatment to remove the pesticide acetamiprid. *J Chem Technol Biotechnol* 2016; 91: 72–81.

[34] Singh P, Andrabi SM, Raina DB, Kumar A. Three staged integrated community-based water filter system for potable water by effective removal of contaminants from ground water. *J Water Process Eng* 2022; 49: 103044.

[35] Mameda N, Park H, Choo KH. Electrochemical filtration process for simultaneous removal of refractory organic and particulate contaminants from wastewater effluents. *Water Res* 2018; 144: 699–708.

[36] Manimozhi V, Saravanathamizhan R, Sivakumar EK, Jaisankar V. Adsorption study of heavy metals removal from wastewater using PVA-nano ferrite composites. *Int J Nanosci Nanotechnol* 2020; 16: 189–200.

[37] Andreozzi R, Caprio V, Insola A, Marotta R. Advanced oxidation processes (AOP) for water purification and recovery. *Catal Today* 1999; 53: 51–59.

[38] Osman M. Waste water treatment in chemical industries: The concept and current technologies. *J Waste Water Treat Anal* 2014; 05: 1–12.

[39] Chandio AD, Pato AH, Channa IA *et al.* Exploring the heterocatalytic proficiencies of ZnO nanostructures in the simultaneous photo-degradation of chlorophenols. *Sustain* 2022; 14(21): 14562.

[40] Prajapati AK, Mehra S, Dewangan T *et al.* Treatment of rice grain based distillery biodigester effluent using iron metal and salt: Chemical oxidation and electro-oxidation combined study in batch mode. *Environ Nanotechnol Monit Manag* 2021; 16: 100585.

[41] Saleh TA, Gondal MA, Drmosh QA. Preparation of a MWCNT/ZnO nanocomposite and its photocatalytic activity for the removal of cyanide from water using a laser. *Nanotechnology* 2010; 21(49): 495705.

[42] Ahmari H, Zeinali Heris S, Khayyat MH. Experimental investigation of new photocatalytic continuous coaxial cylinder reactor for elimination of linear alkylbenzene sulfonic acid from waste water using nanotechnology. *J Mol Liq* 2018; 264: 165–171.

[43] Badmus KO, Tijani JO, Massima E, Petrik L. Treatment of persistent organic pollutants in wastewater using hydrodynamic cavitation in synergy with advanced oxidation process. *Environ Sci Pollut Res* 2018; 25: 7299–7314.

[44] Peng H, Guo J. Removal of chromium from wastewater by membrane filtration, chemical precipitation, ion exchange, adsorption electrocoagulation, electrochemical reduction, electrodialysis, electrodeionization, photocatalysis and nanotechnology: A review. *Environ Chem Lett* 2020; 18: 2055–2068.

[45] Almeida DG, C R De, Meira HM *et al.* Production, characterization and commercial formulation of a biosurfactant from Candida tropicalis UCP0996 and its application in decontamination of petroleum pollutants. *Processes* 2021; 9: 1–18.

[46] Rosa-Masegosa A, Muñoz-Palazon B, Gonzalez-Martinez A, Fenice M, Gorrasi S, Gonzalez-Lopez J. New advances in aerobic granular sludge technology using continuous flow reactors: Engineering and microbiological aspects. *Water (Switzerland)* 2021; 13: 1–20.

[47] Jafarinejad S. Jafarinejad S. A comprehensive study on the application of reverse osmosis (RO) technology for the petroleum industry wastewater treatment. *J Water Environ Nanotechnol* 2017; 2: 243–264.

[48] Thakur A, Kaya S, Kumar A. Recent innovations in nano container-based self-healing coatings in the construction industry. *Curr Nanosci* 2021; 18: 203–216.

[49] Thakur A, Kumar A. Ecotoxicity analysis and risk assessment of nanomaterials for the environmental remediation. *Macromol Symp* 2023; 410: 1–23.

[50] Thakur A, Kumar A, Zhang R. Alcoholic beverage purification applications of activated carbon. In: Verma C, Quraishi MA, editors. *Activated Carbon: Progress and Applications*. The Royal Society of Chemistry, 2023.
[51] Thakur A, Kumar A, Kaya S, Marzouki R, Zhang F, Guo L. Recent advancements in surface modification, characterization and functionalization for enhancing the biocompatibility and corrosion resistance of biomedical implants. *Coatings* 2022; 12: 1459.
[52] Thakur A, Kumar A. Recent advances on rapid detection and remediation of environmental pollutants utilizing nanomaterials-based (bio)sensors. *Sci Total Environ* 2022; 834: 155219.
[53] Thakur A, Sharma S, Ganjoo R, Assad H, Kumar A. Anti-corrosive potential of the sustainable corrosion inhibitors based on biomass waste: A review on preceding and perspective research. *J Phys Conf Ser* 2022; 2267: 012079.
[54] Thakur A, Kumar A, Sharma S, Ganjoo R, Assad H. Computational and experimental studies on the efficiency of Sonchus arvensis as green corrosion inhibitor for mild steel in 0 . 5 M HCl solution. *Mater Today Proc* 2022; 66: 609–621.
[55] Thakur A, Kaya S, Abousalem AS, Kumar A. Experimental, DFT and MC simulation analysis of Vicia Sativa weed aerial extract as sustainable and eco-benign corrosion inhibitor for mild steel in acidic environment. *Sustain Chem Pharm* 2022; 29: 100785.
[56] Thakur A, Kumar A. Corrosion inhibition activity of Pyrazole derivatives on metals and alloys in acidic environment: A review. *J Emerg Technol Innov Res* 2019; 6: 339–343.
[57] Kaya S, Thakur A, Kumar A. The role of in silico / DFT investigations in analyzing dye molecules for enhanced solar cell efficiency and reduced toxicity. *J Mol Graph Model* 2023; 124: 108536.
[58] Villadsen J. Innovative technology to meet the demands of the white biotechnology revolution of chemical production. *Chem Eng Sci* 2007; 62: 6957–6968.
[59] Ramezani M, Enayati M, Ramezani M, Ghorbani A. A study of different strategical views into heavy metal(oid) removal in the environment. *Arab J Geosci* 2021; 14: 1–16.
[60] Thakur A, Kaya S, Abousalem AS *et al.* Computational and experimental studies on the corrosion inhibition performance of an aerial extract of Cnicus Benedictus weed on the acidic corrosion of mild steel. *Process Saf Environ Prot* 2022; 161: 801–818.
[61] Thakur A, Kumar A. Sustainable inhibitors for corrosion mitigation in aggressive corrosive media: A comprehensive study. *J Bio- Tribo-Corrosion* 2021; 7: 1–48.
[62] Thakur A, SAVAŞ K, Kumar A. Recent trends in the characterization and application progress of nano-modified coatings in corrosion mitigation of metals and alloys. *Appl Sci* 2023; 13: 730.
[63] . Sharma, D., Thakur, A., Sharma, M. K., Jakhar, K., Kumar, A., Sharma, A. K., & Hari, O. M. Synthesis, electrochemical, morphological, computational and corrosion inhibition studies of 3-(5-naphthalen-2-yl-[1,3,4]oxadiazol-2-yl)-pyridine against mild steel in 1 M HCl. *Asian J Chem* 2023; 35: 1079–1088.
[64] Dhonchak C, Agnihotri N. Computational insights in the spectrophotometrically 4H-chromen-4-one complex using DFT method. *Biointerface Res Appl Chem* 2023; 13: 357.
[65] Verma C, Thakur A, Ganjoo R, Sharma S, Assad H. Coordination bonding and corrosion inhibition potential of nitrogen-rich heterocycles: Azoles and triazines as specific examples. *Coord Chem Rev* 2023; 488: 215177.
[66] Kumar A, Thakur A. *Encapsulated Nanoparticles in Organic Polymers for Corrosion Inhibition*. Elsevier Inc., 2020.
[67] Thakur A, Kumar A. Recent trends in nanostructured carbon-based electrochemical sensors for the detection and remediation of persistent toxic substances in real- time analysis. *Mater Res Express* 2023; 10: 034001.

[68] Sharma D, Thakur A, Sharma MK *et al.* Effective corrosion inhibition of mild steel using novel 1,3,4-oxadiazole-pyridine hybrids: Synthesis, electrochemical, morphological, and computational insights. *Environ Res* 2023; 234: 116555.
[69] Leson G, Winer AM. Biofiltration: An innovative air pollution control technology for voc emissions. *J Air Waste Manag Assoc* 1991; 41: 1045–1054.
[70] Priya AK, Gnanasekaran L, Rajendran S, Qin J, Vasseghian Y. Occurrences and removal of pharmaceutical and personal care products from aquatic systems using advanced treatment - A review. *Environ Res* 2022; 204: 112298.
[71] Crini G, Cosentino C, Bradu C *et al. Innovative Technologies to Remove Alkylphenols from Wastewater: A Review*. Springer International Publishing, 2022.
[72] dos Santos EV, Bezerra Rocha JH, de Araújo DM, de Moura DC, Martínez-Huitle CA. Decontamination of produced water containing petroleum hydrocarbons by electrochemical methods: A minireview. *Environ Sci Pollut Res* 2014; 21: 8432–8441.
[73] Othmani A, Kadier A, Singh R *et al.* A comprehensive review on green perspectives of electrocoagulation integrated with advanced processes for effective pollutants removal from water environment. *Environ Res* 2022; 215: 114294.
[74] Molinari R, Lavorato C, Argurio P. Application of hybrid membrane processes coupling separation and biological or chemical reaction in advanced wastewater treatment. *Membranes (Basel)* 2020; 10: 1–30.
[75] M DBG, Kadri SUT, Murari SK *et al. Physicochemical–Biotechnological Approaches for Removal of Contaminants from Wastewater*. First Edit. Elsevier Ltd., 2022.
[76] Wayland HA, Boury SN, Chhetri BP *et al.* Advanced cellulosic materials for treatment and detection of industrial contaminants in wastewater. *ChemistrySelect* 2016; 1: 4472–4488.
[77] Vasilachi IC, Asiminicesei DM, Fertu DI, Gavrilescu M. Occurrence and fate of emerging pollutants in water environment and options for their removal. *Water (Switzerland)* 2021; 13: 1–34.
[78] Palani G, Arputhalatha A, Kannan K *et al.* Current trends in the application of nanomaterials for the removal of pollutants from industrial wastewater treatment—A review. *Molecules* 2021; 26(9): 279.
[79] Olivito F, Jagdale P. New technologies to decontaminate pollutants in water: A report about the state of the art. *Toxics* 2022; 10: 16–18.
[80] Shrestha R, Ban S, Devkota S *et al.* Technological trends in heavy metals removal from industrial wastewater: A review. *J Environ Chem Eng* 2021; 9(4): 105688.

9 Advanced Technologies for the Removal of Pollutants in the Pharmaceutical Industries

Nandhi Prabhu, Murugan Rajamehala, Thozhan Gajendran, and Ramalingam Kowsalya

9.1 INTRODUCTION

Development in pharmaceutical research has significantly enhanced the capacity to lead a secure and well-being-oriented lifestyle over an extended period. Nevertheless, these therapeutic constituents have emerged as a novel ecological concern. Polianciuc et al. (2020), Peña et al. et al. (2021) and Khan et al. (2021) have found that the discharge of pharmaceuticals into the natural environment has a cumulative impact on the disruption of aquatic ecosystems and the elevation of public health concerns. Devault et al. (2021)and Archer et al. (2021) have conducted extensive investigations to analyze various substances present in the ecosystem, including antibiotics, endocrine disruptors, personal care products, therapeutic hormones, and other pharmaceuticals. Mukherjee et al. (2021) emphasized the importance of monitoring and assessing the environmental hazards associated with pharmaceuticals in wastewater treatment facilities. This enhances the advancement of therapeutic strategies aimed at mitigating adverse environmental effects (Simon et al., 2021).

Furthermore, the industrial pilot scale techniques employed in pharmaceutical manufacturing have a significant influence on the characteristics of both liquid and solid waste (Bilal et al., 2020). The alterations mentioned have substantial consequences for the public health, environment and the effective process of facilities for wastewater treatment (Wang et al., 2020). Antibiotics demonstrate a targeted and precise effect in eradicating microorganisms when administered at therapeutic levels. The field of environmental antibiotic detection is experiencing expansion and advancement (Kortright et al., 2019; Ayankojo et al., 2020). The presence of antibiotics in water bodies gives rise to two concerns: first, the potential harm these compounds may pose to aquatic organisms and humans when consumed as drinking water; second, the potential for hazardous bacteria to develop resistance to antibiotics

 DOI: 10.1201/9781003454199-9

(Szymańska et al., 2019). Pharmaceutical effluent treatment is traditionally categorized into three stages (Kovalakova et al., 2020). The initial treatment of wastewater entails employing chemical or physicochemical techniques. Secondary methodology generally utilizes biological mechanisms. Finally, tertiary treatment utilizes sophisticated oxidation techniques (Agrawal et al., 2020; Han et al., 2020; Verma and Haritash, 2020; Wang and Zhuan, 2020; Wang and Chen, 2022; Zhu et al., 2021).

Wastewater treatment plants are commonly regarded as the main source of pharmaceutical active compounds in aquatic environments by the discharge of treated water, or in terrestrial environments through the sewage sludge disposal and biological solids waste dumped in landfills. Nevertheless, wastewater treatment plants also serve as a protective measure to control the discharge of pharmaceutical and active chemicals into the environment (Nguyen et al., 2021). Wastewater treatment plants can exert diverse influences on the alteration and breakdown of these substances (Nguyen et al., 2021). Biological processes are generally the most efficient techniques employed in wastewater treatment plants (Rueda-Márquez et al., 2020). Studies have consistently shown that they have a significant impact on the breakdown of different pharmaceuticals and personal care products in wastewater treatment plants. This effect is similar to that of constructed wetlands, membrane bioreactors, conventional activated sludge treatment, and moving bed biofilm reactors (Button et al. 2019; Song et al. 2020; Carolin et al. 2021; Kołecka et al. 2019).

Several advanced anaerobic technologies, including up-flow anaerobic sludge blankets, moving bed biofilm reactors, anaerobic sequencing batch reactors, anaerobic membrane bioreactors,and other hybrid technologies, have proven to be highly effective in the treatment of medicinal waste. In addition, the use of an anaerobic membrane bioreactor has proven to be particularly beneficial since it combines anaerobic processes and membrane filtration, resulting in the creation of an effluent that is completely free from solid waste by retaining all biomass (Shi et al., 2017). Anaerobic membrane bioreactor also facilitates energy generation by efficiently converting a substantial quantity of organic materials found in wastewater into methane. Likewise, it is possible to reduce the size and space needed while producing a significantly higher quality effluent (Anjum et al., 2021). The biological remediation offers greater benefits compared to other chemical techniques. This is because the by-products generated by physicochemical and advanced oxidation treatments have the potential to contaminate water and make it toxic (Lu et al.,2020; Pan et al.,2021). To alleviate the environmental impact of antibiotics, a bioremediation technique is used. This approach involves different approaches such as aerobic, anaerobic, or membrane bioreactors, which leverage microbial transformation activities. These procedures use microbial activity to detoxify antibiotic chemicals found in pharmaceutical effluent, transforming them into less toxic versions.

Aerobic bioremediation is the process of using microorganisms to break down antibiotics in the presence of oxygen. On the other hand, anaerobic bioremediation entails using microbial communities in settings when there is a lack of oxygen (Shah and Shah, 2020). Moreover, membrane bioreactors synergistically integrate physical filtration and biological degradation, hence augmenting the efficacy of antibiotic elimination. In summary, these bioremediation technologies provide efficient methods for tackling antibiotic pollution, hence aiding in the sustainable management of pharmaceutical wastewater and environmental conservation endeavors. Most hybrid treatment techniques are characterized by their high energy consumption and expensive nature (Taoufk et al., 2021). The use of chemicals and higher pH levels in the advanced oxidation process places an additional expensive burden on the process (Tiwari et al.,2020). Tiwari et al. (2020) have devised hybrid treatment techniques that incorporate both anaerobic and aerobic treatment. Priyanka et. al., 2021) have discovered that the combination of membrane bioreactors, reverse osmosis and advanced oxidation process is favored because of its user-friendly characteristics and exceptional performance. Scientific research relies on selecting the suitable approach that effectively fulfils regulatory norms.

The main objective of this study is to thoroughly investigate the existing methods of biological treatment used for pharmaceutical wastewater and assess their efficacy in breaking down and removing different types of pharmaceutical chemicals. Biological treatment methods involve many strategies that utilize microbial activity to decompose organic contaminants found in wastewater. The approaches encompass aerobic and anaerobic processes, along with hybrid systems such as membrane bioreactors. Aerobic treatment entails subjecting the wastewater to conditions abundant in oxygen, where aerobic bacteria break down organic pollutants through metabolism. This procedure commonly takes place in activated sludge systems or aerobic biofilters. Anaerobic treatment happens in the absence of oxygen and relies on anaerobic microbes to break down organic substances through processes like fermentation and methanogenesis. Membrane bioreactors synergistically merge biological treatment with membrane filtration, effectively harnessing the benefits of both processes. This system utilizes microorganisms to break down organic contaminants, with a membrane barrier that separates the treated water from the biomass. As a result, the system produces effluent of superior quality. The effectiveness of these biological treatment methods in decomposing pharmaceutical compounds relies on several factors, including the nature and amount of pollutants, microbial diversity and activity, reactor setup, and operational conditions such as temperature, pH, and hydraulic retention time. Multiple studies have shown that biological therapy systems have the potential to efficiently eliminate a diverse array of pharmacological substances, such as antibiotics, hormones, analgesics, and anticonvulsants. The microorganisms included in these systems undergo metabolic processes to break down medicines into less toxic metabolites. Furthermore, the use of cutting-edge technologies such as bioaugmentation, co-metabolism, and optimization of microbial consortia has demonstrated encouraging outcomes in improving the efficiency of breaking down pharmaceutical chemicals in wastewater. The primary objective of this study is to offer valuable understanding regarding the effectiveness of biological treatment techniques in treating pharmaceutical wastewater. It emphasizes their capacity to serve as environmentally friendly and economically viable options for reducing pharmaceutical contamination in aquatic environments.

9.2 SOURCES OF PHARMACEUTICALS IN WASTEWATER

Pharmaceuticals and various personal care products can affect human health rather than the environment, according to studies conducted by Chami et al. (2021),

Trouillard et al. (2022), and Ghita et al. (2023). Martínez et al. (2021) and Tiwari et al. (2020) noted the ubiquitous presence of these pollutants in the environment, which originate from many sources. Residues of drugs have been identified in several realms of existence (Guo et al., 2020). Their existence and effectiveness generally depend on their origin, the pathway through which they were introduced, and the environmental conditions in which they exist (; Rasheed et al. 2020).

Kar et al. (2020) and Rasheed et al. (2020) discovered a variety of environmental pollutants, including substances that make plastics flexible, anti-inflammatory drugs, preservatives, pharmaceuticals for epilepsy, substances that stimulate the production of solar energy, painkillers, hormones, and toxic metals, in wastewater treatment plants across the world (Gernigon et al., 2020). The occurrence of analgesics and antibacterials in aquatic organisms is primarily ascribed to their substantial consumption (Villemin et al., 2021). Landecker (2021) stated that sulfonamides and tetracyclines are widely utilised antibiotics renowned for their exceptional persistence and propensity to infiltrate the environment. In addition, as stated by Zainab et al. (2020) and Bilal et al. (2020), ibuprofen is commonly found in pharmaceuticals and is recognised as a harmful pollutant in the ecosystem.

Hospitals and pharmaceutical plants are the main sources of pharmaceutical chemicals released into the atmosphere, especially during the COVID-19 pandemic (Bu et al., 2020). Other supplementary sources encompass the disposal of unused prescriptions and the inappropriate management of pharmaceuticals (Kar et al., 2020). Various drugs can be found in the environment, and when evaluating them, characteristics such as ecotoxicity, processing complexity, risk, consumption patterns and biodegradation are considered important (Kairigo et al., 2020 and Kroon et al., 2020). Once leaving the manufacturing core and entering the retail sector, the paths through which the environment is exposed become highly fragmented, posing challenges in tracing and evaluating them. Dey et al. (2019) found that people actively participate in the consumption of many consumer products, including medications. Various goods, such as cosmetics, fungicides, detergents, disinfectants and home cleansers, can immediately enter the aquatic system and introduce pharmaceuticals. Pharmaceutical processing does not promote the reduction of recently emerged chemicals. Without a doubt, the drug is released into the environment in many forms such as the conversion products, metabolites, and original molecule leading to contamination of surface land, marine environment, and groundwater sources. This results in the development of resistance in several detrimental bacterial strains. Prolonged exposure to antibiotics can disturb the diversity of microorganisms and result in harm to the higher trophic levels of the food chain (Dey et al., 2019; Guo et al., 2020; Santos et al., 2020; Bilal et al., 2020;).

9.3 IMPACT OF PHARMACEUTICALS ON LIVING FORMS

Pharmaceutical and personal care products are widely used worldwide. They enter the ecosystem through many methods, potentially endangering different parts of the environment. Active pharmaceutical chemicals can potentially affect aquatic organisms, even at low concentrations (; Niemi et al., 2020; Nguyen et al., 2021; Schwartz et al., 2021). In addition, current regulations have placed greater emphasis on the need to monitor and reduce the presence of specific pharmaceutical and personal care products in wastewater treatment plants, with the aim of protecting water quality (Liang et al., 2021 and Helmreich et. al., 2017). Active pharmaceutical ingredients have been utilized for the treatment of several medical problems, such as cancer, diabetes, goiters, influenza, and COVID-19 (Chaubey et al., 2021;).

The considerable amount of micropollutants found in both land and water environments over the past twenty years. Micropollutants can lead to the emergence of antibiotic resistance in bacteria, which poses a substantial threat to human health (Porika et al., 2021;). The primary worry, as stated by Chaturvedi et al. (2021) is the ongoing release of antibiotics into aquatic ecosystems and soil. Humans frequently come into contact with pharmaceutical residues and personal care items in their daily lives due to the contamination of air, water, or soil. Exposure to pharmaceutical waste can lead to negative health consequences, such as allergies, diseases, and malignancies (Ferreira et al., 2020 and Keerthanan et al., 2021). In a study conducted by Kroon et al. (2020) in Australia, it was found that wastewater included several antibiotics, such as tetracycline, erythromycin, and sulfonamides. Ricky and Shanthakumar (2022) have said that the presence of these compounds in the environment is a result of their accumulation in living organisms and their subsequent introduction into the food chain via sewage systems and agricultural practices. Most studies on the impact of pharmaceutical and personal care products on non-target organisms mostly use acute toxicity data (Liang et al., 2021). Khalil et al. (2020) emphasized as pharmaceuticals have been discovered to have harmful impacts on aquatic ecosystems. These impacts encompass the disturbance of inherent reproduction mechanisms and the eradication of specific organisms. Fish exhibit a heightened level of sensitivity towards pollution as a result of their intimate connection with the environment, which acts as the major conduit for the dissemination of these detrimental compounds. For example, it is expected that diclofenac and gemfibrozil will have a direct negative effect on reproductive organs (Beiras, 2021).

Algae, on the other hand, are photosynthetic creatures that regulate the movement of energy and nutrients in aquatic settings (Hussain et al., 2021). Hence, their behaviors significantly contribute to the conservation of the overall ecosystem. The existence of carbamazepine and diclofenac has been found to cause substantial damage to algae's chloroplasts (Harshkova et al., 2021). Antibiotics are commonly detected in saltwater, groundwater, surface water, and even drinking water at low concentrations. The release of antibiotic molecules or their byproducts into the aquatic ecosystem has profound and enduring ecological impacts, causing pollution and fostering the emergence of antibiotic resistance (Chaturvedi et al., 2021). Shahriar et al. (2021) discovered that the use of wastewater for irrigation leads to a greater accumulation of pharmaceuticals and personal care products in the soil as compared to groundwater.

9.4 IMPACT OF PHARMACEUTICALS ON WASTEWATER TREATMENT PLANTS

The removal of pharmaceutical pollutants is limited by their substantial biological toxicity (; Vasilachi et al., 2021). They actively promote the spread of bacteria and genes that are resistant to antibiotics (Fu et al., 2020;). According to (Zhang et al.,

2022), the growing public concern about antibiotics and other medicines in the environment requires the adoption of efficient techniques to manage and remediate these pollutants. Zhou et al. (2021) have shown that incomplete degradation of antibiotics can result in their persistence as residual chemicals in wastewater treatment plants. These chemicals exert a substantial influence on the efficacy of the treatment plants. The inclusion of phosphorus compounds greatly hinders the decontamination procedures by negatively impacting the functioning of the treatment facilities (Chen et al., 2020). There was a pronounced necessity to effectively purify wastewater containing antibiotics.

Microorganisms employ anaerobic digestion to convert intricate organic matter into methane, which is the primary mechanism for the decomposition of organic matter (Xie et al., 2020). Macrolides and tetracyclines, both antibiotics that target anaerobic fermentation, operate as inhibitors of protein synthesis in bacteria, hence impeding bacterial growth (Kong and Yang, 2021) and Liu et al.(2021) stated that antibiotics have a suppressive impact on bacteria that have the ability to break down substances and produce acid. According to Cheng et al. (2021) sulfonamide antibiotics are found to be a cause of fouling on the membrane during the anaerobic membrane bioreactor process.

9.5 BIOLOGICAL TREATMENT TECHNIQUES FOR PHARMACEUTICAL WASTE

Wastewater can undergo many independent procedures at each step of treatment. Research has been conducted on various effective pharmaceutical waste treatment techniques to remove contaminants. These techniques include primary treatment methods such as flotation systems, primary sedimentation tanks, neutralization tanks, and equalization tanks. Secondary treatment methods involve biological and physicochemical treatment. Tertiary treatment methods include adsorption, membrane filtration, membrane distillation, advanced oxidation processes, solvent extraction, and disinfection.

9.5.1 Anaerobic Treatment of Pharmaceutical Wastewater

The primary mechanism responsible for the breakdown of pharmaceutical chemicals occurs through the biological component in the majority of conventional wastewater treatment plants (Pugazhendi et al., 2022). Due to the presence of a substantial amount of organic matter in these wastewaters, anaerobic techniques have been suggested as sustainable, cost-effective, low-nutrient, low-surface area, and efficient approaches for drug degradation (Khadir et al., 2020).

The primary technologies utilised in anaerobic treatment facilities include of anaerobic digesters, membrane bioreactors, up-flow anaerobic sludge beds (Kong and Shi, 2022), and anaerobic sequencing batch reactors (Chen et al., 2020; Chaturvedi et al., 2022; Huang et al., 2018; Wang et al., 2022; Pugazhendi et al., 2022) (Table 9.1). However, the use of anaerobic technology in antibiotics is hindered by the limited ability to break down and convert refractory organic wastes into minerals, as well as the low efficiency and stability of this process (Kong et al., 2019).

TABLE 9.1
Anaerobic Treatment of Pharmaceutical Wastewater

Type of process	Application	References
Anaerobic Sludge process	Removal of sulfamethoxazole	Zahedi et al., 2022
Anaerobic Sludge process	Removal of clozapine, amitriptyline, trimethoprim, triclocarban and sulfamethoxazole,	Harb et al., 2019
Anaerobic Sludge process	Removal of nitrates and chlorine-substituted contaminants	Dai et al., 2022
Anaerobic Sludge process	Removal of chloramphenicol	Guo et al., 2019
Anaerobic Sludge process	Breakdown of antibiotics	Hu et al., 2020;
Upflow Anaerobic Sludge Blanket (UASB) reactor	Removal of carbamazepine	Moya-Llamas et al., 2021
Upflow Anaerobic Sludge Blanket (UASB) reactor	Treatment of pharmaceutical wastewater	de Oliveira et al., 2020;
Anaerobic digestion process	95% for contaminants treatment of pharmaceutical wastewater	Vistanty and Crisnaningtyas, 2021
UASBbioelectrocatalysis system	Treatment of acetopyrimidine	Sevda et al., 2020
anaerobic digestion process	Breakdown of acetopyrimidine	Wang et al., 2019
Anaerobic digestion process with zero-valent iron and granular activated carbon	Elimination of COD; methane generation	Dai et al., 2022;
Anaerobic digestion process with zero-valent iron and granular activated carbon	Elimination of pharmaceutical intermediates	Dai et al., 2022

Source: Compiled by the authors.

9.5.2 Pharmaceutical Wastewater Treatment by Activated Sludge

The activated sludge process is a widely employed biological technique for wastewater treatment in the pharmaceutical industry. The contaminants undergo a process of transformation, wherein they are converted into gases and digested sludge (Table 9.2). This converted material can be safely released into the environment (Mareai et al., 2020; Quintelas et al., 2020;Hu et al., 2020; Nguyen et al., 2021). Compared to more complex processing techniques, it has a lower initial investment cost and requires fewer ongoing operations. Activated sludge facilities face operational issues such as excessive energy consumption, the generation of substantial quantities of sludge and coloured substances, foaming, and the enlargement of the secondary clarifier. Several crucial aspects influence the performance of a pharmaceutical activated sludge treatment facility (Barati et al., 2022. The

TABLE 9.2
Pharmaceutical Wastewater Treatment by Activated Sludge

Treatment by Activated Sludge	Applications	References
Food in activated sludge processes	Metabolise pharmaceutical substances	Jaén-Gil et al.,2021
Biodegradable feeds to activated sludge processes	Elimination of antibiotics and antibiotic-resistant microorganisms	Cheng et al., 2020
Sequencing bioreactor activated sludge processes	Breakdown of amoxicillin	Zhang et al., 2020;
Antibiotic resistance genes activated sludge processes	Eliminated the amoxicillin	Zhang et al., 2021
Sequential feed-activated sludge approach	Eliminating the specific medicines	Kong et al., 2019
Activated sludge approach	Removal of caffeine, ibuprofen and metronidazole	Kong et. al., 2019
Sequential batch reactor	Treatment of pharmaceutical wastewater	Kołecka et al., 2019
Sequential batch reactor	Treatment of diclofenac	Kołecka et al., 2019
Activated sludge process	Degradation of ibuprofen (IBU)	Alfonso Muniozguren et al., 2021
Activated sludge process	Degradation of and paracetamol	Quintelas et al., 2020
Activated sludge process	Degradation of paracetamol and IBU	Alfonso-Muniozguren et al., 2021;
Activated sludge process	Elimination of Metronidazole	Alfonso Muniozguren et al., 2021
Activated sludge bioreactor	Elimination of acetaminophen	Quintelas et al., 2020

Source: Compiled by the authors.

parameters influencing the process include hydraulic retention time, temperature, pH, dissolved oxygen, organic load, microbial populations, and the presence of hazardous or resistant compounds (Yang et al., 2020andKatam et al., 2021). Modifying the parameters is essential for incorporating wastewater treatment into the pharmaceutical business (Mareai et al., 2020).

9.5.3 Pharmaceutical Wastewater Treatment by Membrane Bioreactor

A membrane bioreactor is a system that integrates a membrane module with a bioreactor. Pharmaceutical wastewater treatment commonly employs this technique as an advanced approach (Table 9.3). This particular bioreactor offers various operational benefits, including reduced sludge production, high-quality wastewater output and improved removal of organic matter, longer sludge residence time, and increased resistance to hazardous compounds. Although MBR offers significant benefits,

TABLE 9.3
Pharmaceutical Wastewater Treatment by Membrane Bioreactor

Type of membrane bioreactor	Application	Reference
Aerobic hollow fibre membrane bioreactor	Removal of pharmaceutical effluent that contains antibiotics	Fakhri et al., 2021
Anaerobic membrane bioreactor	Eliminate erythromycin, sulfamethoxazole and tetracycline from wastewater	Fakhri et al., 2021
Anaerobic membrane bioreactor	inhibits the growth of bacteria that create biofilm.	Aydin and Can, 2020
Bioaugmenter membrane bioreactor	Treat antibiotics like sulfamethoxazole, tetracycline and erythromycin	Aydin and Can, 2020
Anoxic/aerobic membrane bioreactors	Removal of pharmaceutical effluent tetracycline and norfloxacin	Zhang et al., 2020
Aerobic membrane bioreactors	Treatment of wastewater tetracycline	Zhang et al., 2020
Aerobic membrane bioreactors	Removal of antibiotics	Guo et al., 2020
Membrane bioreactors	Antibiotics degraded	Wang et al., 2019
Sequencing membrane bioreactor	Elimination of antibiotics including sulfonamides, tetracyclines, and fluoroquinolones	Xu et al., 2019
Integrated anaerobic permeation membrane bioreactor and membrane distillation system	Eliminating pharmaceuticals and personal care products from wastewater	Arcanjo et al., 2021
Anaerobic membrane bioreactor and membrane distillation system	Eliminating pharmaceuticals and personal care products from wastewater	Wang et al., 2019
Anaerobic membrane bioreactor	Removal of Sulfadiazine and tetracycline	Xu et al., 2019

Source: Compiled by the authors.

researchers have noted that membrane fouling continues to be a significant obstacle in reducing membrane permeability. This phenomenon has a substantial impact on the efficiency and durability of the membrane, leading to higher operational costs for the MBR (Cheng et al., 2021; Liu et al., 2022; Guo et al., 2020; Sengar and Vijayanandan 2022).

9.5.4 Hybrid Treatment of Pharmaceutical Wastewater

Various researchers have developed numerous methods for addressing pharmaceutical wastewater processing through chemical, physical, and biological methods.

Biological wastewater treatment is an environmentally sustainable procedure that generates a small amount of sludge and is economically efficient (Taoufk et al., 2021). The treatment of pharmaceutical waste effluent can be removed effectively by the application of physicochemical hybrid methods, as research conducted by Zamri et al. (2021),and Nidheesh et al. (2021). Nevertheless, these methods suffer from significant drawbacks such as the production of dangerous waste materials, the build-up of membrane fouling during filtration, economic inefficiency, excessive energy usage, higher investment needs and diminished sludge absorption capacity as research conducted by Zamri et al., (2021). . To overcome the constraints of individual techniques, researchers have embraced hybrid approaches, which entail combining physical, chemical, and biological treatment methods. The objective of this integration is to improve the effective elimination of several newly identified pollutants (Saidulu et al., 2021;; Mojiri et al., 2020).

Hybrid systems possess the ability to generate bioenergy, hence aiding in the reduction of operational costs of the system. Multiple treatment methodologies can be utilized to promptly and ecologically effectively eliminate emerging pollutants (Jaén-Gil et al., 2021 and Almaguer et al., 2021). Hybrid systems are widely regarded as the most economically efficient technology and provide improved elimination of pharmaceutical wastewater (Table 9.4)

TABLE 9.4
Hybrid Treatment of Pharmaceutical Wastewater

Hybrid treatment method	Application	Reference
Hybrid systems with biological method and physical or chemical	Effectively eliminate emerging Pharmaceutical pollutants	Pirsaheb et al. (2020)
Biosorption and biological treatment	Removal of drug contaminants	Mojiri et al. (2020)
Activated sludge processes with physical techniques such as ultrafiltration, gamma radiation and reverse osmosis,	Improved elimination of trace organic pollutants	Hu et al. (2020)
Membrane bioreactors, with physical techniques such as UV oxidation, activated carbon and ultrasound ozonation	Broken down multiple medications	Monteoliva-García et al., 2020
Advanced oxidation process	Mineralizing pharmaceuticals	Guo et al., 2020
Combination of chemical and biological treatment approaches	Pharmaceutical wastewater degradation	Angeles et al.,2020
Process Design Objectives (pdos) and biological processes	Pharmaceutical pollutant concentrations can be significantly reduced	Mukimin et al., 2020

(continued)

TABLE 9.4 (Continued)
Hybrid Treatment of Pharmaceutical Wastewater

Hybrid treatment method	Application	Reference
Fenton process with biological treatment	Pharmaceutical wastewater pretreatment technology	Khan et al., 2021;
Advanced oxidation processes (AOP) with microalgae treatment	Drug degradation and the production of non-toxic by-products	Almaguer et al., 2021
Advanced oxidation processes (AOP) with microalgae treatment	Production of biofuels	Samal et al., 2022
UV/H_2O_2 treatment following conventional activated sludge treatment	Eliminating metoprolol and metoprolol acid from hospital wastewater	Jaén-Gil et al., 2021
Integrates activated sludge-based biological treatment with activated carbon-based adsorption	Removal of acetaminophen, caffeine, and ibuprofen	Monteoliva-García et al., 2020
Conventional individual treatment methods	Reduced levels of cod concentration, turbidity, and soluble microbial products	Ferrer-Polonio et al., 2020
Combined activated sludge with a biofilter using slag as a filtering medium	Remove ceftriaxone (CEF) and amoxicillin (AMX)	Çifçi and Meriç, 2022; Cadamuro et al., 2022
ultrafiltration membrane bioreactor (UF-MBR) with UV/H_2O_2 treatment, ozonationand granular activated carbon adsorption	Removing around 87.53% (for AMX) and 93.17% (for CEF), achieve water quality	Pirsaheb et al., 2020
UF-MBR with three additional treatments	Removal drug residues	Mousel et al., 2021
Coagulation, biological method and beta process	Purification of wastewater	Kim et al., 2022
Degradation and detoxification	Effective breakdown and removal of drugs	Changotra et al., 2020
Ozone (O3), activated carbon (AC), and biological treatment techniques	Elimination and detoxification of levosulpiride	Choi et al. (2022)

Source: Compiled by the authors.

9.5.5 Adsorption on Natural Materials

Adsorption on natural materials refers to the process by which substances adhere to and are retained on surfaces of naturally occurring materials. There has been a growing interest in the adsorption of natural materials, such as zeolites. Zeolite active sites enhance the adsorption of metal ions, leading to the creation of both inner and outer-sphere complexes. Outer-sphere complexes are created by inserting a water molecule between the bound ion and the functional group of the adsorbent. In contrast, inner-sphere complexes are formed when there are no water molecules present between the entities involved. A coordination bond will be formed between the heavy metals and the surface functional groups (Zhou and Haynes, 2010; MalamisandKatsou, 2013). For example, Clinoptilolite, a particular type of zeolite, has shown a pronounced affinity for trapping Pb2+, Cd2+, Zn2+, and Cu2+ ions. Moreover, studies have shown that polymeric materials can improve the efficiency of natural clay in removing heavy metals by converting the original clay into a composite material called clay-polymer composite (Şölener et al., 2008). An example involves the creation of a clay-poly (methoxyethyl) acrylamide composite for the purpose of examining its capacity to absorb Pb^{2+} ions. Bentonite, a unique type of clay, has shown remarkable effectiveness (>99%) in removing heavy metals such as Pb^{2+}, Co^{2+}, Zn^{2+}, Ni^{2+}, and Cu^{2+} ions. The metal species displayed adsorption affinities in the following sequence: The order of reduction potential from highest to lowest is $Co^{2+}>Cu^{2+}>Ni^{2+} = Zn^{2+}>Pb^{2+}$ (VhahangweleandMugera, 2015). In another investigation, it was found that the complete removal (100%) of Pb^{2+} from water-based systems can be achieved by using a concentration of 20 g L^{-1} of bentonite (Inglezakis et al., 2007).

9.5.6 Bioremediation

Bioremediation is the process of using living organisms to remove or neutralise pollutants in the environment. Bioremediation is a long-standing and effective process used to break down chlorinated solvents and petroleum hydrocarbons. Typical bioremediation techniques include the use of photocatalysis, biomimetic membranes and ultrafiltration membranes. Bioremediation primarily involves the utilisation of bio stimulation, bioaugmentation, and natural attenuation.

Bio stimulation is the introduction of electron acceptors or nutrients to stimulate the development of microorganisms, hence speeding up and broadening the process of biodegradation of pollutants. Natural attenuation refers to the inherent capacity of specific microorganisms inside a contaminated area to autonomously degrade and remove pollutants. The application of bioremediation resulted in a significant decrease of 86.6 percent in the COD value. A study conducted independently employed a multi-biofilm comprising bacteria and fungi to remediate pharmaceutical effluent. Natural mechanisms have the ability to breakdown most pharmacological compounds, including ibuprofen. Nevertheless, certain compounds, such as trimethoprim and carbamazepine, exhibit resistance to degradation. Furthermore, these biological treatment approaches primarily respond to variations in the environment, such as fluctuations in oxygen levels, temperature, pH levels, and nutrient concentrations. The efficacy and stability of the bioremediation process are greatly influenced by these environmental factors. Therefore, it is crucial to carefully supervise and improve them.

9.5.7 Biofiltration

Biofiltration is the utilization of living organisms to eliminate contaminants from air or water. The biofiltration technique for managing pharmaceutical material waste involves the use of a microbial biofilm linked to a filter media to promote the breakdown of pharmaceutical pollutants by biodegradation via the oxidation process. The medical residue included in the effluent is filtered using a bacterial biofilm. The microbial biofilm utilises pharmaceutical waste products as its primary carbon source and metabolises them using enzyme-mediated metabolic processes.

The waste effluent, after undergoing treatment using biomass, and the carbon dioxide generated through the bacterial oxidation process, are then purified for various uses. Pharmaceutical materials treatment can employ fixed-bed reactors and submerged type biofilters. Fixed-bed types are primarily characterized by the uniform distribution of pharmaceutical waste water throughout the filter matrix. The filter's matrix completely submerges in the submerged arrangement, where it serves as a foundation for the oxidizing bacterial biofilm.

9.6 CHALLENGES IN THE REMOVAL OF PHARMACEUTICAL WASTES

Challenges in the removal of pharmaceutical wastes (PWs) encompass a variety of medical products, such as prescription drugs (both used and unused), antibiotics, hormones, over-the-counter medicines, blood samples, needles, syringes, swabs, histo-pathological samples, faecal specimens, surgical tools, personal care products, and specific radioactive substances. A multitude of refractory chemicals have emerged as new contaminants in the environment, presenting a substantial risk to both public health and environment. Process waste (PW) refers to a wide range of substances, including raw materials, by-products, and end products, that can carry diseases, contain toxins, cause harm, cause genetic damage, and emit radiation. Disposing of PWs poses substantial challenges because of their combination of solid and liquid waste, predominantly produced by healthcare facilities such as hospitals, nursing homes, medical stores, diagnostic laboratories, and similar institutions. Moreover, these variables can contribute to the emergence of antibiotic resistance, which has the capacity to create substantial disruption in the future. Referred to as "medical waste," these provide challenges for both developed and developing nations. Moreover, waste possesses the capacity to be used or diverted for illicit intentions. Passwords are frequently discarded in landfills, which can result in the possible leakage of harmful substances into underground water sources through rivers and lakes, hence inflicting significant harm to both aquatic and terrestrial ecosystems. Any mishandling or improper processing can result in the release of infectious agents and harmful compounds into the environment, which poses a significant risk to both the environment and public health. When evaluating the variety of dangerous compounds, each component requires specific methods of disposal. The user's text is concise and clear. Wastewater treatment plants (WTTs) are acknowledged as major sources of pollution dispersion in the environment. However, it has been shown that the majority of the detrimental components of PW are not adequately eliminated during waste treatment, resulting in environmental contamination. Reportedly, the concentration of persistent contaminants in Indian water is 40 times higher than that observed in Europe and the United States. Other options for disposing of biomedical waste include landfilling, incineration, mechanical treatment, autoclave treatment, microwave treatment, and chemical recycling. Some of these ways can effectively reduce the PWs. Nevertheless, deficient disposalcould potentially have a harmful effect on the ecosystem. Consequently, the development of new waste categories and the drafting of disposal procedures have increased the complexity of remediating polluted water. Furthermore, it is well acknowledged that a substantial amount of PWs produced comprises several compounds with complex and highly stable chemical structures, sometimes including hazardous components. Policymakers, medical professionals, and members of the recycling industry face challenges when it comes to managing PWs.

9.7 PROSPECTIVE ADVANCEMENTS IN PHARMACEUTICAL WASTE DISPOSAL METHODS

The rapid global advancement of modernity has resulted in a substantial rise in the formation of personal care and pharmaceutical products in ecosystems that are both aquatic and terrestrial in nature. Previous methods for handling pharmaceutical contaminants sometimes fail to achieve intended results because they are inefficient in completely breaking down pharmaceuticals and their byproducts, leading to a significant quantity of pharmaceutical residues persisting in the environment. The prospect of pharmaceutical waste elimination from soil and aquatic environments, predominantly by integrated methods, innovative bioremediation techniques, and improved technological solutions are environmentally, economically feasibleand eco-friendly. Previous studies have primarily focused on discovering approaches to improve the effectiveness of removing medicines from the environment. The use of microbial methodologies in conjunction with sustainable energy sources like solar and wind power has the capacity to make a substantial impact on the remediation of pharmaceutical waste. However, its effectiveness may depend on environmental factors. Future pharmaceutical waste removal technologies should focus on the expansion of eco-friendly ways, considering the most recent breakthroughs in technology.

9.8 CONCLUSION

The issue of freshwater contamination is a significant problem in modern times, and the Sustainable Development Goals (SDGs) recognize and tackle this problem in its objectives. The rate of pollution is escalating at a faster pace than the rate of population expansion. This chapter specified the development of pollutants that stem from pharmaceutical sources, which contribute to the pollution in question, emanating from diverse origins. An analysis of scholarly publications was performed on a subset of 2,938 research papers, covering the time period from 1979 to 2020. Hence, we chose the main authors, institutes, countries, publications, and subject areas, as well as the most often used treatment techniques for removing these contaminants from wastewater. The scientific productivity has shown a substantial increase during the previous seven years (2014–2020). Throughout this time frame, a total of 2,057 articles were published, representing 70 percentof the overall amount. Starting from 2014, a considerable amount of scientific research has been linked to the 2030 agenda and the SDGs that were introduced in2015. The exponential growth of the population, widespread industrialization, urbanization, and intensive farming methods are all factors that are adding to the worldwide need for clean and uncontaminated water. In order to remove contaminants from pharmaceutical products disposal, it is necessary to develop new and inventive treatment methods. Emerging technologies, such microbial fuel cells, have been found to have the ability to process pharmaceutical waste materials and their conversion into useful energy. Furthermore, advanced oxidation techniques have demonstrated significant efficacy in the remediation of pharmaceutical wastewater. Recent research has revealed that, while there have been successful technological advancements in the past, in the present circumstances it is crucial to focus on scaling up these technologies. Integrated technology approaches have significant potential and are expected to have a positive future. Nevertheless, the achievement of their goals relies on the unwavering commitment of the present scientific community and on governmental institutions.

REFERENCES

Agrawal, K., Bhatt, A., Bhardwaj, N., Kumar, B., Verma, P. (2020) Integrated approach for the treatment of industrial effluent by physico chemical and microbiological process for sustainable environment. In: Maulin, P., Banerjee, S.A. (eds), *Combined Application of Physico-chemical and Microbiological Processes for Industrial Effluent Treatment*. Springer, pp.119–143.

Alfonso-Muniozguren, P., Serna-Galvis, E.A., Bussemaker, M., Torres-Palma, R.A., Lee, J. (2021) A review on pharmaceuticals removal from waters by single and combined biological, membrane filtration and ultrasound systems. *Ultrason Sonochem* 76: 105656.

Almaguer, M.A., Cruz, Y.R., da Fonseca, F.V. (2021) Combination of advanced oxidation processes and microalgae aiming at recalcitrant wastewater treatment and algal biomass production: A review. *Environ Process* 8: 483–509.

Angeles, F., Luisa, R.A., Mullen, I.J., Huang, C.W., Khunjar, W., Sirotkin, H.I., McElroy, A.E., Aga, D.S. (2020) Assessing pharmaceutical removal and reduction in toxicity provided by advanced wastewater treatment systems. *Environ Sci Water Res Technol*, 6(1): 62–77.

Anjum, F., Khan.I.M., Kim.J., Aslam.M., Blandin.G., Heran.M., Lesage.G. (2021) Trends and progress in ANMBR for domestic wastewater treatment and their impacts on process efficiency and membrane fouling. *Environ. Technol. Innov*. 21: 101204.

Arcanjo, G.S., Ricci, B.C., dos Santos, C.R., Costa, F.C.R., Silva, U.C.M., Mounteer, A.H., Koch, K., da Silva, P.R., Santos, V.L., Amaral, M.C.S. (2021) Effective removal of pharmaceutical compounds and estrogenic activity by a hybrid anaerobic osmotic membrane bioreactor membrane distillation system treating municipal sewage. *Chem Eng J*. 416: 129151.

Archer, E., Volschenk, M., Brocker, L., Wolfaardt, G.M. (2021) A two-year study of emerging micro-pollutants and drugs of abuse in two western cape wastewater treatment works (South Africa). *Chemosphere* 285: 131460.

Ayankojo, A.G., Reut, J., Ciocan, V., Öpik, A., Syritski, V. (2020) Molecularly imprinted polymer-based sensor for electrochemical detection of erythromycin. *Talanta* 209: 120502.

Aydin, S., Can, K. (2020) Pyophage cocktail for the biocontrol of membrane fouling and its effect in aerobic microbial biofilm community during the treatment of antibiotics. *Biores Technol.* 318: 123965.

Barati, B., Zafar, F.F., Wang, S. (2022) Different waste management methods, applications, and limitations. In: Abd El-Fatah A, Wang Q, Huang J (eds) *Waste-to-Energy: Recent Developments and Future Perspectives Towards Circular Economy*. Springer, pp 21–58.

Beiras, R. (2021) Environmental risk assessment of pharmaceutical and personal care products in estuarine and coastal waters. In: Juan C, Álvarez D, Jiménez-Cisneros B (eds) *Pharmaceuticals in Marine and Coastal Environments*. Estuarine and Coastal Sciences Series, vol 1. Elsevier, pp 195–252.

Bilal, M., Mehmood, S., Rasheed, T., Iqbal, H.M.N. (2020) Antibiotics traces in the aquatic environment: Persistence and adverse environmental impact. *Curr. Opin. Environ. Sci. Health Environ. Monit. Assess. Water Energy Food Nexus* 13: 68–74.

Bu, Q., Cao, Y., Gang, Y., He, X., Zhang, H., Sun, J., Yun, M., Cao, Z. (2020) Identifying targets of potential concern by a screening level ecological risk assessment of human use pharmaceuticals in China. *Chemosphere* 246: 125818.

Button, M., Cosway, K., Sui, J., Weber, K. (2019) Impacts and fate of triclosan and sulfamethoxazole in intensifed re-circulating vertical fow constructed wetlands. *Sci. Total Environ.* 649: 1017–1028.

Cadamuro, R.D., Rogovski, P., Pr´a, I.D., Guterres, I.Z., Carmargo, A.F., Bonatto, C., Scapini, T., SpitzaStefanski, F., Fonsecac, T.G., Mass, A.P., Michelon, W., Viancelli, A., da Cruz, A.C.C., Treichel, H., Thaís Silva, I., Fongaro, G. (2022) Chapter 14 – Trends in biofiltration applied to remove pharmaceuticals and personal care products from wastewater. In: Shah, M., Rodriguez-Couto, S., BiswasJ. (eds), *An Innovative Role of Biofiltration in Wastewater Treatment Plants (WWTPs)*. Elsevier, pp. 267–284.

Carolin, C.F., Kumar, P.S., Joshiba, G.J., Kumar, V.V. (2021). Analysis and removal of pharmaceutical residues from wastewater using membrane bioreactors: A review. *Environmental Chemistry Letters*, 19(1): 329–343. https://doi.org/10.1007/s10311-020-01068-9

Chami, K., Feltin, N., Gaffet, E., Lacour, S., Lassus, M., Le Bihan, O., Niaudet, A., Ricaud, M., Nesslany, F. (2021) Les nanomatériaux Manufacture's dansl'environnementpro fessionnel: un aperçu De l'état de l'art. *Arch. Des. Maladies Professionnelles Et De L'environnement* 82(1): 51–68.

Changotra, R., Rajput, H., Guin, J.P., Khader, S.A., Dhir, A. (2020) Technoeconomical evaluation of coupling ionizing radiation and biological treatment process for the remediation of real pharmaceutical wastewater. *J Clean Prod.* 242: 118544.

Chaturvedi, P., Shukla, P., Giri, B.S., Chowdhary, P., Chandra, R., Gupta, P., Pandey, A. (2021) Pharmaceutical and personal care products and antibiotics in environment: A review on emerging contaminants. *Environ. Res.* 194: 110664.

Chaubey, A.K., Patel, M., Mohan, D. (2021) Synthesis of engineered biochar for aqueous pharmaceutical removal. Avril EGU21-9358.

Chen, L., Peijin, Cheng, L., Ye, H.C., Xiangyang, X., Zhu, L. (2020) Biological performance and fouling mitigation in the biochar-amended anaerobic membrane bioreactor (AnMBR) treating pharmaceutical wastewater. *Biores Technol* 302: 122805.

Cheng, D., Ngo, H.H., Guo, W., Chang, S.W., Nguyen, D.D., Nguyen, Q.A., Zhang, J., Liang, S. (2021) Improving sulfonamide antibiotics removal from swine wastewater by supplying a new pomelo peel derived biochar in an anaerobic membrane bioreactor. *Biores Technol.* 319: 124160.

Choi, S., Yoom, H., Son, H., Seo, C., Kim, K., Lee, Y., Kim, Y.M. (2022) Removal efficiency of organic micropollutants in successive wastewater treatment steps in a full-scale wastewater treatment plant: Bench-scale application of tertiary treatment processes to improve removal of organic micropollutants persisting after secondary treatment. *Chemosphere* 288: 132629.

Çifçi, D.İ., Meriç, S. (2022) Chapter 19—A critical review on bio filtration for wastewater treatment: Focus on organic micro pollutants. In: Maulin, S., Rodriguez-Couto, S., Biswas, J. (eds), *An Innovative Role of Bio Filtration in Wastewater Treatment Plants (WWTPs)*. Elsevier, pp.369–388.

Dai, C., Yang, L., Wang, J., Li, D., Zhang, Y., Zhou, X. (2022) Enhancing anaerobic digestion of pharmaceutical industries wastewater with the composite addition of zero valent iron (ZVI) and granular activated carbon (GAC). *Biores Technol* 346: 126566.

de Oliveira, M., Frihling, B.E.F., Velasques, J., Filho, F.J.C.M., Cavalheri, P.S., Migliolo, L. (2020) Pharmaceuticals residues and xenobiotics contaminants: Occurrence, analytical techniques and sustainable alternatives for wastewater treatment. *Sci Total Environ* 705: 135568.

Devault, D.A., Amalric, L., Bristeau, S., Cruz, J., Tapie, N., Karolak, S., Budzinski, H., Lévi, Y. (2021) Removal efficiency of emerging micro pollutants in biofilter wastewater treatment plants in tropical areas. *Environ. Sci. Pollut. Res.* 28(9): 10940–10966.

Dey, S., Bano, F., Malik, A. (2019) 1—Pharmaceuticals and personal care product (PPCP) contamination—a global discharge inventory. In: Majeti, N.V.P., Vithanage, M., Kapley, A. (eds), *Pharmaceuticals and Personal Care Products: Waste Management and Treatment Technology*. Butterworth-Heinemann, pp.1–26.

Fakhri, H., Arabacı, D.N., Ünlü, İ.D., Yangin-Gomec, C., Ovez, S., Aydin, S. (2021) Addition of Trichocladium canadense to an anaerobic membrane bioreactor: Evaluation of the microbial composition and reactor performance. *Biofouling* 37(7): 711–723.

Ferreira, A.R., Guedes, P., Mateus, E.P., Ribeiro, A.B., Couto, N. (2020) Emerging organic contaminants in soil irrigated with effluent: Electrochemical technology as a remediation strategy. *Sci Total Environ* 743: 140544.

Ferrer-Polonio, E., Fernández-Navarro, J., Iborra-Clar, M.-I., Alcaina Miranda, M.-I., Mendoza-Roca, J.A. (2020) Removal of pharmaceutical compounds commonly-found in wastewater through a hybrid biological and adsorption process. *J Environ Manage.* 263: 110368.

Fu, J., Zhang, Q., Huang, B., Fan, N., Jin, R. (2020) A review on anammox process for the treatment of antibiotic-containing wastewater: Linking effects with corresponding mechanisms. *Front Environ Sci Eng* 15(1): 17.

Gernigon, C., Othenin, J., Michaud, A., Vayr, F., Rousseau, V., Montastier, E., Lapeyre-Mestre, M., Niezborala, M., Hérin, F. (2020) Évaluation de la consommationmédicamenteused'une population de travailleurs de la régionToulousaineen 2016. *Archives Des Maladies Professionnelles Et De L'environnement* 81(1): 81.

Ghita, G., Ilie, M., Tudor, G., Arsene, M., Luminarioiu, L., Raischi, M., Petrache, I. (2023). Investigations regarding the contamination with emerging pharmaceutical pollutants in representative rivers and lakes in Romania. In: *E3S Web of Conferences*. EDP Sciences, Vol. 437, p. 02008.

Guo, F., Zhang, J., Yang, X., He, Q., Ao, L., Chen, Y. (2020) Impact of biochar on greenhouse gas emissions from constructed wetlands under various influent chemical oxygen demand to nitrogen ratios. *Bioresour. Technol.* 303: 122908.

Guo, N., Ma, X., Ren, S., Wang, S., Wang, Y. (2019) Mechanisms of metabolic performance enhancement during electrically assisted anaerobic treatment of chloramphenicol wastewater. *Water Res.* 156: 199–207.

Han, Y., Yang, L., Chen Yu, X., Cai, X.Z., Qian, M., Chen, X. et al (2020) Removal of veterinary antibiotics from swine wastewater using anaerobic and aerobic biodegradation. *Sci. Total Environ.* 709: 136094.

Harb, M., Lou, E., Smith, A.L., Stadler, L.B. (2019) Perspectives on the fate of micropollutants in mainstream anaerobic wastewater treatment. *Curr Opin Biotechnol Energy Biotechnol Environ Biotechnol.* 57: 94–100.

Harshkova, D., Majewska, M., Pokora, W., Baścik-Remisiewicz, A., Tułodziecki, S., Aksmann, A. (2021) Diclofenac and atrazine restrict the growth of a synchronous chlamydomonas reinhardtii population via various mechanisms. *Aquat Toxicol* 230: 105698.

Helmreich, Metzger, S.J.I.W.T., Resource, E. (2017) *Recovery Technologies: Impacts on Energy, Environment, Post-treatment for Micropollutants Removal, Innovative Wastewater Treatment & Resource Recovery Technologies: Impacts on Energy, Economy and Environment*, IWA Publishing.

Hu, D., Min, H., Wang, H., Zhao, Y., Cui, Y., Pan, W., Ge, H. et al (2020) Performance of an up-flow anaerobic bio-electrochemical system (UBES) for treating sulfamethoxazole (SMX) antibiotic waste water. *Biores Technol.* 305: 123070.

Huang, B., Wang, H.C., Cui, D., Zhang, B., Chen, Z.B., Wang, A.J. (2018) Treatment of pharmaceutical wastewater containing β-lactams antibiotics by a pilot-scale anaerobic membrane bioreactor (AnMBR).*Chem. Eng. J.* 341: 238–247.

Hussain, M.M., Wang, J., Bibi, I., Shahid, M., Niazi, N.K., Iqbal, J., Mian, I.A. et al (2021) Arsenic speciation and biotransformation pathways in the aquatic ecosystem: The significance of algae. *J Hazard Mater.* 403: 124027.

Jaén-Gil, A., Buttiglieri, G., Benito, A., Mir-Tutusaus, J.A., Gonzalez-Olmos, R., Caminal, G., Barceló, D., Sarrà, M., Rodriguez-Mozaz, S. (2021) Combining biological processes with UV/H2O2 for metoprolol and metoprolol acid removal in hospital wastewater. *Chem Eng J.* 404: 126482.

Kairigo, P., Ngumba, E., Sundberg, L.-R., Gachanja, A., Tuhkanen, T. (2020) Occurrence of antibiotics and risk of antibiotic resistance evolution in selected kenyan wastewaters, surface waters and sediments. *Sci Total Environ.* 720: 137580.

Kar, S., Sanderson, H., Roy, K., Benfenati, E., Leszczynski, J. (2020) Eco toxicological assessment of pharmaceuticals and personal care products using predictive toxicology approaches. *Green Chem.* 22(5): 1458–1516.

Katam, K., Bhattacharyya, D., Soda, S., Shimizu, T. (2021) Effect of hydraulic retention time on the performance of trickling photo bioreactor treating domestic wastewater: Removal of carbon, nutrients, and micropollutants. *J Ind Eng Chem* 102: 351–362.

Keerthanan, S., Jayasinghe, C., Biswas, J.K., Vithanage, M. (2021) Pharmaceutical and personal care products (PPCPs) in the environment: Plant uptake, translocation, bioaccumulation, and human health risks. *Crit Rev Environ Sci Technol.* 51(12): 1221–1258.

Khadir, A., Mollahosseini, A., Negarestani, M., Mardy, A. (2020) Anaerobic biotechnology for the treatment of pharmaceutical compounds and hospital wastewaters. In: Inamuddin, M.I.A., Lichtfouse, E., Asiri, A.M. (eds) *Methods for Bioremediation of Water and Waste Water Pollution. Environmental Chemistry for a Sustainable World.* Springer, pp 61–84.

Khalil, A.M.E., Memon, F.A., Tabish, T.A., Salmon, D., Zhang, S., Butler, D. (2020) Nanostructured porous graphene for efficient removal of emerging contaminants (pharmaceuticals) from water. *Chem Eng J* 398: 125440.

Khan, M. T., Shah, I. A., Ihsanullah, I., Naushad, M., Ali, S., Shah, S. H. A., Mohammad, A. W. (2021). Hospital wastewater as a source of environmental contamination: An overview of management practices, environmental risks, and treatment processes. *Journal of Water Process Engineering*, 41, 101990. https://doi.org/10.1016/j.jwpe.2021.101990

Kim, S., Nam, S.-N., Jang, A., Jang, M., Park, C.M., Son, A., Her, N., Heo, J., Yoon, Y. (2022) Review of adsorption-membrane hybrid systems for water and wastewater treatment. *Chemosphere* 286: 131916.

Kołecka, K., Gajewska, M., Stepnowski, P., Caban, M. (2019) Spatial distribution of pharmaceuticals in conventional wastewater treatment plant with sludge treatment reed beds technology. *Sci. Total Environ.* 647: 149–157.

Kong, L., Shi, X. (2022) Effect of antibiotic mixtures on the characteristics of soluble microbial products and microbial communities in up flow anaerobic sludge blanket. *Chemosphere* 292: 133531.

Kong, Q., Yang, Y. (2021) Recent advances in antibacterial agents. *Bioorg Med Chem Lett* 35: 127799.

Kong, Z., Li, L., Xue, Y., Yang, M., Li, Y.-Y. (2019) Challenges and prospects for the anaerobic treatment of chemical-industrial organic wastewater: A review. *J Clean Prod* 231: 913–927.

Kortright, K.E., Chan, B.K., Kof, J.L., Turner, P.E. (2019) Phage therapy: A renewed approach to combat antibiotic-resistant bacteria. *Cell Host Microbe* 25(2): 219–232.

Kovalakova, P., Cizmas, L., McDonald, T.J., Marsalek, B., Feng, M., Sharma, V.K. (2020) Occurrence and toxicity of antibiotics in the aquatic environment: A review. *Chemosphere* 251: 126351.

Kroon, F.J., Berry, K.L.E., Brinkman, D.L., Kookana, R., Leusch, F.D.L., Melvin, S.D., Neale, P.A. et al (2020) Sources, presence and potential effects of contaminants of emerging concern in the marine environments of the great barrier Reef and Torres Strait, Australia. *Sci Total Environ* 719: 135140.

Landecker, H. (2021) La résistance aux antibiotiques et la biologie de l'Histoire. *Traduit par Lisa Blackman. Revue D'anthropologie Des Connaissances* 15: 15–23.

Liang, J., Meng, H., Zhou, J., Yin, Y., Zhen, H., Zhao, Y., Zhang, K. (2021) Simultaneous occurrence of psychotropic pharmaceuticals in surface water of the megacity shanghai and implication for their ecotoxicological risks. *ACS ES&T Water* 1(4): 825–836.

Lindberg, R.H., Namazkar, S., Lage, S., Östman, M., Gojkovic, Z., Funk, C., Gentili, F.G., Tysklind, M. (2021) Fate of active pharmaceutical ingredients in a northern high-rate algal pond fed with municipal wastewater. *Chemosphere* 271: 129763.

Liu, W., Song, X., Na, Z., Li, G., Luo, W. (2022) Strategies to enhance micropollutant removal from wastewater by membrane bioreactors: Recent advances and future perspectives. *Bioresour Technol.* 344: 126322.

Liu, Y., Li, X., Tan, Z., Yang, C. (2021) Inhibition of tetracycline on anaerobic digestion of swine wastewater. Bioresour Technol 334: 125253.

Lu, Z.-Y., Ma, Y.-L., Zhang, J.-T., Fan, N.-S., Huang, B.-C., Jin, R.-C. (2020) A critical review of antibiotic removal strategies: Performance and mechanisms. *J. Water Process Eng.* 38: 101681.

Malamis, S., Katsou, E. (2013) A review on zinc and nickel adsorption on natural and modified zeolite, bentonite and vermiculite: Examination of process parameters, kinetics and isotherms. *J. Hazard. Mater.* 252–253: 428–461.

Mareai, B.M., Fayed, M., Aly, S.A., Elbarki, W.I. (2020) Performance comparison of phenol removal in pharmaceutical wastewater by activated sludge and extended aeration augmented with activated carbon. *Alex Eng J.* 59(6): 5187–5196.

Martínez-Alcalá, I., Guillén-Navarro, J.M., Lahora, A. (2021) Occurrence and fate of pharmaceuticals in a wastewater treatment plant from southeast of Spain and risk assessment. *J. Environ. Manag.* 279: 111565.

Mojiri, A., Zhou, J., Vakili, M., Van, L.H. (2020) Removal performance and optimisation of pharmaceutical micropollutants from synthetic domestic wastewater by hybrid treatment. *J Contam Hydrol.* 235: 103736.

Monteoliva-García, A., Martín-Pascual, J., Muñío, M.M., Poyatos, J.M. (2020) Effects of carrier addition on water quality and pharmaceutical removal capacity of a membrane bioreactor—advanced oxidation process combined treatment. *Sci Total Environ* 708: 135104.

Mousel, D., Bastian, D., Firk, J., Palmowski, L., Pinnekamp, J. (2021) Removal of pharmaceuticals from wastewater of health care facilities. *Sci Total Environ.* 751: 141310.

Moya-Llamas, M.J., Trapote, A., Prats, D. (2021) Carbamazepine removal from low-strength municipal wastewater using a combined UASB-MBR treatment system. *Water Sci Technol* 83(8): 1920–1931.

Mukherjee, A., Mullick, A., Moulik, S., Roy, A. (2021) Oxidative degradation of emerging micro pollutants induced by rotational hydro dynamic cavitating device: A case study with ciprofoxacin. *J. Environ. Chem. Eng.* 9(4): 105652.

Mukimin, A., Vistanty, H., Zen, N. (2020) Hybrid advanced oxidation process (HAOP) as highly efficient and powerful treatment for complete demineralization of antibiotics. *Sep Purif Technol.* 241: 116728.

Nguyen, P.Y., Carvalho, G., Maria, A.M. (2021) A review of the biotransformation's of priority pharmaceuticals in biological wastewater treatment processes. *Water Res.* 188: 116446.

Nidheesh, P.V., Couras, C., Karim, A.V., Nadais, H. (2021) A review of integrated advanced oxidation processes and biological processes for organic pollutant removal. *Chem Eng Commun* 1: 1–43.

Niemi, L., Taggart, M., Boyd, K., Zhang, Z., Gafney, P.P.J., Pfeger, S., Gibb, S. (2020) Assessing hospital impact on pharmaceutical levels in a rural 'source-to-sink' water system. *Sci. Total Environ.* 737: 139618.

Pan, M., Lyu, T., Zhan, L., Matamoros, V., Angelidaki, I., Cooper, M., Pan, G. (2021) Mitigating antibiotic pollution using cyanobacteria: Removal efficiency, pathways and metabolism. *Water Res.* 190: 116735.

Peña, O. I. G., Zavala, M. Á. L., Ruelas, H. C. (2021). Pharmaceuticals market, consumption trends and disease incidence are not driving the pharmaceutical research on water and wastewater. *International Journal of Environmental Research and Public Health, 18*(5), 2532. https://doi.org/10.3390/ijerph18052532

Pirsaheb, M., Mohamadisorkali, H., Hossaini, H., Hossini, H., Makhdoumi, P. (2020) The hybrid system successfully to consisting of activated sludge and biofilter process from hospital wastewater: Ecotoxicological study. *J Environ Manage.* 276: 111098.

Polianciuc, S.I., Gurzău, A.E., Kiss, B., Ştefan, M.G., Loghin, F. (2020) Anti-biotics in the environment: Causes and consequences. *Med. Pharm. Rep.* 93(3): 231–240.

Porika, M., Ranjit, P., Tippani, R., Reddy, K.V. (2021) Advances in the bioremediation of pharmaceuticals and personal care products (PPCPs): Polluted water and soil. In: Pankaj Kumar A (ed) *Microbial Products for Health, Environment and Agriculture Microorganisms for Sustainability*. Springer, pp 323–358.

Priyanka, M.S., Mohapatra, S., Menon, N.G. (2021) Chapter 23—Hybrid membrane technology: An alternative to industrial wastewater treatment. In: Shah, M.P., Rodriguez-Couto, S. (eds), *Membrane Based Hybrid Processes for Wastewater Treatment*. Elsevier, pp.481–501.

Pugazhendi, A., Jamal, M.T., Al-Mur, B.A., Jeyakumar, R.B. (2022) Bioaugmentation of electrogenic halophiles in the treatment of pharmaceutical industrial wastewater and energy production in microbial fuel cell under saline condition. *Chemosphere* 288: 132515.

Quintelas, C., Mesquita, D.P., Torres, A.M., Costa, I., Ferreira, E.C. (2020) Degradation of widespread pharmaceuticals by activated sludge: Kinetic study, toxicity assessment, and comparison with adsorption processes. *J Water Process Eng* 33: 101061.

Rasheed, T., Bilal, M., Hassan, A.A., Nabeel, F., Bharagava, R.N., Ferreira, L.F.R., Tran, H.N., Hafz, M.N. (2020) Environmental threatening concern and efficient removal of pharmaceutically active compounds using metal-organic frameworks as adsorbents. *Environ. Res.* 185: 109436.

Ricky, R., Shanthakumar, S. (2022) Phycoremediation integrated approach for the removal of pharmaceuticals and personal care products from wastewater—A review. *J Environ Manage* 302: 113998.

Rueda-Márquez, J.J., Palacios-Villarreal, C., Manzano, M., Blanco, E., del Solar, M.R., Levchuk, I. (2020) Photocatalytic degradation of pharmaceutically active compounds (PhACs) in urban wastewater treatment plants effluents under controlled and natural solar irradiation using immobilized TiO2. *Sol. Energy* 208: 480–492.

Saidulu, D., Gupta, B., Gupta, A.K., Ghosal, P.S. (2021) A review on occurrences, eco-toxic effects, and remediation of emerging contaminants from wastewater: Special emphasis on biological treatment based hybrid systems. *J Environ Chem Eng* 9(4): 105282.

Samal, K., Bandyopadhyay, R., Dash, R.R. (2022) Biological treatment of contaminants of emerging concern in wastewater: A review. *J Hazard Toxic Radioact Waste* 26(2): 04022002.

Santos, A., Veiga, F., Figueiras, A. (2020) Dendrimers as pharmaceutical excipients: Synthesis, properties, toxicity and biomedical applications. *Materials* 13(1): 65.

Sathe, S., Chakraborty, I., Dubey, B., Ghangrekar, M. (2022) Microbial fuel cell coupled Fenton oxidation for the cathodic degradation of emerging contaminants from wastewater: Applications and challenges. *Environ. Res.* 204: 112135.

Schwartz, H., Marushka, L., Chan, H.M., Batal, M., Sadik, T., Ing, A., Fediuk, K., Tikhonov, C. (2021) Pharmaceuticals in source waters of 95 first nations in Canada. *Can. J. Public Health* 112(1): 133–153.

Sengar, A., Vijayanandan, A. (2022) Effects of pharmaceuticals on membrane bioreactor: Review on membrane fouling mechanisms and fouling control strategies. *Sci Total Environ* 808: 152132.

Sevda, S., Garlapati, V.K., Naha, S., Sharma, M., Ray, S.G., Sreekrishnan, T.R., Goswami, P. (2020) Biosensing capabilities of bioelectrochemical systems towards sustainable water streams: Technological implications and future prospects. *J Biosci Bioeng* 129(6): 647–656.

Shah, Shah, M. (2020) Characterisation and bioremediation of wastewater: A review exploring bioremediation as a sustainable technique for pharmaceutical wastewater. *Groundwater Sustain. Dev.* 11: 100383.

Shahriar, A., Tan, J., Sharma, P., Hanigan, D., Verburg, P., Pagilla, K., Yang, Y. (2021) Modeling the fate and human health impacts of pharmaceuticals and personal care products in reclaimed wastewater irrigation for agriculture. *Environ Pollut.* 276: 116532.

Shi, X., Leong, K.Y., Ng, H.Y. (2017) Anaerobic treatment of pharmaceutical wastewater: A critical review. *Biores. Technol.* 245: 1238–1244.

Simon, M., Kumar, A., Garg, A. (2021) Biological treatment of pharmaceuticals and personal care products (PPCPs) before discharging to environment. In: Kumar, P., Bharagava, G.R.M. (eds), *Fate and Transport of Subsurface Pollutants. Microorganisms for Sustainability*. Springer, pp. 259–282.

Şölener, M., Tunali, S., Özcan, A. S., Özcan, A. Gedikbey, T. (2008) Adsorption characteristics of lead(II) ions onto the clay/poly(methoxyethyl)acrylamide (PMEA) composite from aqueous solution. *Desalination* 223: 308–322.

Song, Z., Zhang, X., Sun, F., Ngo, H.H., Guo, W., Wen, H., Li, C., Zhang, Z. (2020) Specific microbial diversity and functional gene (AOBAmoA) analysis of a sponge-based aerobic nitrifying moving bed biofilm reactor exposed to typical pharmaceuticals. *Sci. Total Environ.* 742: 140660.

Szymańska, U., Wiergowski, M., Sołtyszewski, I., Kuzemko, J., Wiergowska, G., Woźniak, M.K. (2019) Presence of antibiotics in the aquatic environment in Europe and their analytical monitoring: Recent trends and perspectives. *Micro. Chem. J.* 147: 729–740.

Taoufk, N., Boumya, W., Achak, M., Sillanpää, M., Barka, N. (2021) Comparative overview of advanced oxidation processes and biological approaches for the removal pharmaceuticals. *J. Environ. Manag.* 288: 112404.

Tiwari, B., Drogui, P., Tyagi, R.D. (2020) Chapter 18—Removal of emerging micro-pollutants from pharmaceutical industry wastewater. In: Sunita, V., Pandey, A., Tyagi, R.D., Ngo, H.H., Larroche, C. (eds), *Current Developments in Biotechnology and Bioengineering*. Elsevier, pp.457–480.

Trouillard, A., Sabri, A., Prot-Labarthe, S., Storme, T., Bourrat, E., Soulairol, I. (2022) Exposition pédiatrique aux perturbateurs endocriniens et substances cancérogčnes, mutagčnes ou reprotoxiques par des formes pharmaceutiques destinées ā la voie cutanée: Dispositions réglementaires en France, en Europe, et état de l'art des connaissances scientifiques. *Ann Pharmaceut Franjais* 80(4): 426–439.

Vasilachi, I.C., Asiminicesei, D.M., Fertu, D.I., Gavrilescu, M. (2021) Occurrence and fate of emerging pollutants in water environment and options for their removal. *Water* 13(2): 181.

Verma, M., Haritash, A.K. (2020) Review of advanced oxidation processes (AOPs) for treatment of pharmaceutical wastewater. *Adv. Environ. Res.* 9(1): 1–17.

Vhahangwele, M., Mugera, G.W. (2015) The potential of ball-milled South African bentonite clay for attenuation of heavy metals from acidic wastewaters: Simultaneous sorption of Co 2+, Cu 2+, Ni 2+, Pb^{2+}, and Zn $^{2+}$ ions. *J. Environ. Chem. Eng.* 3: 2416–2425.

Villemin, C., Abel, F., Dispersyn, G., Cotty-Eslous, M., Marchand, S. (2021) La réalitévirtu ellecommeantidouleur: une revue systématique de la littérature. *Douleur Et Analgésie* 34(3): 163–170.

Vistanty, H., Crisnaningtyas, F. (2021) Integration of upflow anaerobic sludge blanket and constructed wetlands for pharmaceutical wastewater treatment. *IOP Conf Ser Earth Environ Sci* 623(1): 012082.

Wang, D., Li, Y., Zhuang, H., Yi, X., Yang, F., Han, H. (2019) Direct current triggering enhanced anaerobic treatment of acetyl pyrimidine containing wastewater in up-flow anaerobic sludge blanket coupled with bioelectrocatalytic system. *Chemosphere* 231: 457–467.

Wang, J., Chen, X. (2022) Removal of antibiotic resistance genes (ARGs) in various wastewater treatment processes: An overview. *Crit. Rev. Environ. Sci. Technol.* 52(4): 571–630.

Wang, J., Man, Y., Ruan, W., Tam, N.F.-Y., Tao, R., Yin, L., Yang, Y., Dai, Y., Tai, Y. (2022) The effect of rhizosphere and the plant species on the degradation of sulfonamides in model constructed wetlands treating synthetic domestic wastewater. *Chemosphere* 288(Pt 2): 132487.

Wang, J., Zhuan, R. (2020) Degradation of antibiotics by advanced oxidation processes: An overview. *Sci. Total Environ.* 701: 135023.

Wang, M., Chen, H., Liu, S., Xiao, L. (2021b) Removal of pathogen and antibiotic resistance genes from waste activated sludge by different pre-treatment approaches. *Sci. Total Environ.* 763: 143014.

Xie, S., Li, X., Wang, C., Kulandaivelu, J., Jiang, G. (2020) Enhanced anaerobic digestion of primary sludge with additives: Performance and mechanisms. *Biores Technol* 316: 123970.

Xu, R., Zhiyong, W., Zhou, Z., Meng, F. (2019) Removal of sulfadiazine and tetracycline in membrane bioreactors: Linking pathway to microbial community shift. *Environ Technol.* 40(2): 134–143.

Yadav, B., Pandey, A.K., Kumar, L.R., Kaur, R., Yellapu, S.K., Sellamuthu, B., Drogui, P. (2020) Introduction to wastewater microbiology: Special emphasis on hospital wastewater. In:Tyagi, R.D., Sellamuthu, B., Tiwari, B., Yan, S., Drogui, P., Zhang, X., Pandey, A. (eds), *Current Developments in Biotechnology and Bioengineering*. Elsevier, pp.1–41.

Yang, Y., Wang, L., Xiang, F., Zhao, L., Qiao, Z. (2020) Activated sludge microbial community and treatment performance of wastewater treatment plants in industrial and municipal zones. *Int J Environ Res Public Health* 17(2): 436.

Zahedi, S., Gros, M., Casabella, O., Petrovic, M., Balcazar, J.L., Pijuan, M. (2022) Occurrence of veterinary drugs and resistance genes during anaerobic digestion of poultry and cattle manures. *Sci Total Environ* 822: 153477.

Zainab, S.M., Junaid, M., Nan, X., Malik, R.N. (2020) Antibiotics and anti-biotic resistant genes (ARGs) in groundwater: A global review on dissemination, sources, interactions, environmental and human health risks. *Water Res.* 187: 116455.

Zamri, M.F.M.A., Bahru, R., Pramanik, S.K., Fattah, I.M.R. (2021) Treatment strategies for enhancing the removal of endocrine-disrupting chemicals in water and wastewater systems. *J Water Process Eng* 41: 102017.

Zhang, C., Dong, S., Chen, C., Zhang, Q., Zhou, D. (2020) Co-substrate addition accelerated amoxicillin degradation and detoxification by up-regulating degradation related enzymes and promoting cell resistance. *J Hazard Mater* 394: 122574.

Zhang, R., Xuming, X., Jia, D., Lyu, Y., Jingrun, H., Chen, Q., Sun, W. (2022) Sediments alleviate the inhibition effects of antibiotics on denitrification: Functional gene, microbial community, and antibiotic resistance gene analysis. *Sci Total Environ* 804: 150092.

Zhou, J., Ding, J., Wan, H., Guan, G. (2021) Boosting photocatalytic degradation of antibiotic wastewater by synergy effect of heterojunction and phosphorus doping. *J Colloid Interface Sci* 582: 961–968.

Zhou, Y.-F., Haynes, R.J.(2010)Sorption of heavy metals by inorganic and organic components of solid wastes: Significance to use of wastes as low-cost adsorbents and immobilizing agents. *Crit. Rev. Environ. Sci. Technol.* 40: 909–977.

Zhu, T.-T., Su, Z.-x., Lai, W.-X., Zhang, Y.-B., Liu, Y.-W. (2021) Insights into the fate and removal of antibiotics and antibiotic resistance genes using biological wastewater treatment technology. *Sci. Total Environ.* 776: 145906.

10 The Fundamental of Membrane as a Treatment Method for Wastewater Including Both Potential and Challenges for Reuse and Recovery

Momina and Kafeel Ahmad

10.1 INTRODUCTION

Human activities heavily rely on water resources. As the global population grows, significant amounts of wastewater are generated daily, spanning domestic, industrial, and agricultural sectors. Unfortunately, the available freshwater sources are not being replenished sufficiently to keep up with the escalating demands, resulting in an inequitable allocation among different sectors. Effective wastewater treatment has emerged as a critical issue due to the potential adverse impacts on both the environment and human well-being resulting from inadequate wastewater management. Inadequate wastewater treatment can result in consequences such as oxygen depletion, diminished biodiversity in both aquatic and terrestrial ecosystems, and the closure of beaches [1]. During recent years, membrane technology has experienced substantial advancements and adoption within wastewater treatment strategies. This progress can be attributed to the advantages the technology presents in the realm of water and wastewater treatment. Notably, membrane technology has witnessed remarkable growth in the past few decades, characterized by reduced equipment size, energy demands, and capital expenses. Consequently, this technology holds promising prospects for enhancing wastewater treatment processes. It has the potential to overcome the gap of economic viability and sustainability [2]. It boasts features such as minimal or zero chemical usage, environmentally friendly attributes, and convenient accessibility.

DOI: 10.1201/9781003454199-10

The membrane-based processes used for wastewater treatment encompass microfiltration (MF), ultrafiltration (UF), electrodialysis (ED), nanofiltration (NF), and reverse osmosis (RO). These membranes serve as barriers that allow water to pass through while preventing undesirable contaminants from also passing through, resembling the role of cell walls in the human body [3]. This innovative method has emerged as a leading and effective strategy in wastewater treatment, contributing significantly to its safe disposal into the environment. Although membrane technology is not novel, its application remains dynamic due to the diverse and intricate characteristics of wastewater. Membrane technology offers advantages such as minimal or no chemical usage, which is in line with environmentally conscious characteristics. This not only diminishes the environmental effects of using chemicals but also promotes the overall sustainability of wastewater treatment methods [4]. Opportunities for improvement persist in terms of efficiency, spatial demands, energy consumption, permeate quality, and technical expertise. One of the main challenges connected with membrane-based procedures is fouling. The buildup of unwanted substances on the surface of the membrane decreases effectiveness, leading to a need for frequent cleaning and replacement. Ongoing advancements focus on refining membrane modules and elements to mitigate membrane fouling, a significant challenge in membrane processes. Moreover, there is a robust exploration, development, and adoption of hybrid methodologies that combine multiple membrane processes or integrate additional technologies [5]. These integrated approaches enhance wastewater treatment capabilities across a wide range of facilities. Advancements have led to a decrease in energy needs, but the issue of energy consumption still persists. Finding an equilibrium between attaining optimum treatment effectiveness and reducing energy usage is essential for the enduring viability of membrane-based wastewater treatment systems [6]. This chapter provides a thorough examination of different membrane technologies used in wastewater treatment, aiming to address and alleviate the difficulties associated with membrane-based technologies. The chapter focuses on identifying and addressing the challenges associated with membrane technologies for wastewater treatment, such as membrane fouling, technical expertise requirements, and energy consumption concerns.

10.2 DIFFERENT TECHNOLOGIES OF MEMBRANE-BASED WASTEWATER TREATMENT

Different types of membrane had been employed by researchers in various types of application of wastewater treatment such as organic membrane and inorganic membrane based on their material composition. Organic membranes are primarily composed of synthetic organic polymers and are commonly employed in pressure-driven separation methods like UF, MF, RO, and NF. On the other hand, inorganic membranes consist of materials such as ceramics, zeolites, silica, or metals. They exhibit remarkable chemical and thermal stability and find applications in various industries, including ultrafiltration, hydrogen separation, and MF. The movement of substances/contaminants from water can occur by different mechanism such as pressure driven forces, non-pressure driven forces, equilibrium/non-equilibrium driven forces. All these types of membrane are discussed in this chapter along with their applications in water and wastewater treatment for reuse and recovery.

10.2.1 Pressure-driven Membranes

Pressure-driven membrane technologies are widely employed in wastewater treatment, spanning from pretreatment to posttreatment stages. These methods utilize hydraulic pressure to achieve separation of permeate and retention phases for efficient membrane-based separation processes. The solute concentration in the permeate phase is lower when compared to that in the retention and feed solutions. The overall operational expenses of the system are influenced by the pressure applied. This membrane includes MF,UF, NF and RO. Membranes are classified based on their pore sizes, which determine their specific types. Tese include RO (with pores of 0.0001 – 0.001 μm), UF (0.01-0.1 μm), MF (0.1–5 μm), and NF (0.001 – 0.01 μm) membranes. Some other features of these membrane are depicted in Figure 10.1.

10.2.1.1 Ultrafiltration (UF)

UF is a widely accepted method for treating wastewater. It is capable of effectively removing large molecules and maintains a good flow rate in the permeate. UF membranes typically have pores ranging from 0.005 to 0.1 μm, and their primary mechanism of action is size-based exclusion. However, factors such as the shape of the molecules, their charge, and the fluid dynamics can also significantly impact how efficiently UF membranes reject substances [7]. The UF process is extensively utilized for the direct treatment of real textile wastewater. This method effectively produces high-quality permeate that can be readily employed in washing and dyeing operations [8,9]. In particular, UF membranes characterized by their compact structure and a low molecular weight cut-off (1–7 kDa) demonstrate exceptional capabilities in the separation of direct and reactive dyes, all the while permitting the passage of salt ions. This capability facilitates the recovery of dyes [10,11]. A commercial UF membrane

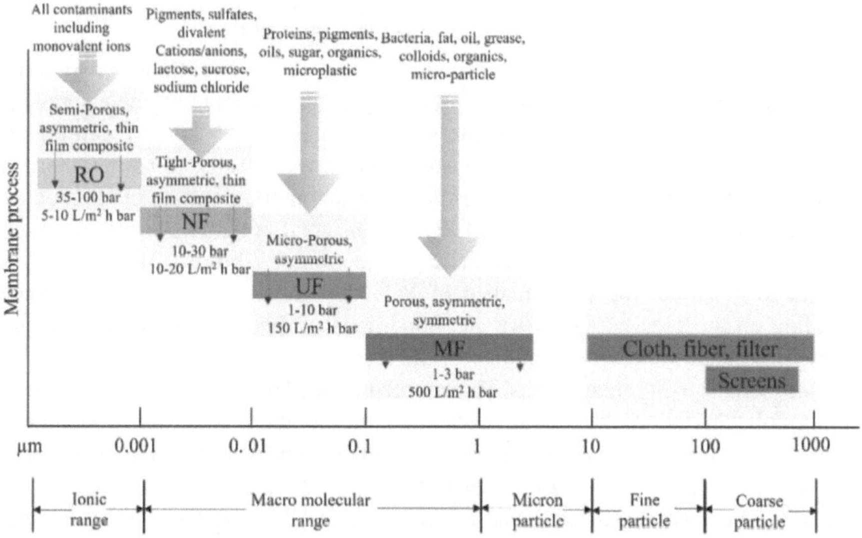

FIGURE 10.1 Features of pressure-driven membrane processes. (Compiled by authors.)

with a low molecular weight of 2 kDa was able to achieve impressive removal rates of over 80 percent for COD, total organic carbon (TOC), color, and divalent ions when applied to the treatment of real textile wastewater. The resulting permeate was suitable for use in washing dyed knitted fabrics [12]. For the separation of dyes from highly saline textile wastewater, hydrolyzed polyacrylonitrile (HPAN) UF membranes were fabricated using a co-deposition of polydopamine/polyethylenimine (PDA/PEI) with MWCO options ranging from 1 to 5 kDa. The UF membrane with a molecular weight of 1.7 kDa demonstrated efficient rejection of various reactive dyes within the molecular weight range less than 1.7 kDa. Remarkably, it maintained a robust permeation rate for inorganic salts, surpassing 97.0 percent (Ye et al. 2020). Fully sponge-like structure UF membrane was formulated using Polyethersulfone (PES) and sulfonated polysulfone (SPSf) with 11 percent adipic acid as a pore-forming agent. The membrane consists of negatively charged particles on the surface, attributable to the exposed anionic sulfonic acid groups. The membrane shows 69.5–100 percent of dye rejection under conditions of low ionic strength [14]. Recent discoveries have revealed that the rejection of dye molecules, particularly those with molecular weights surpassing 696.7 Da, is not solely governed by the Donnan effect. Instead, it is significantly impacted by the aggregation of these molecules. This aggregation phenomenon arises from the hydrophobic attraction between the aromatic rings present in neighbouring dye molecules. Consequently, the enlargement of dye clusters enhances the efficiency of the dye rejection process [15]. Numerous research endeavours have been dedicated to enhancing the efficacy of textile effluent treatment through the incorporation of extra functionalities or by customizing the surface charges of UF membranes such as fungal bioactive UF membrane for textile effluent treatment [16], Polyvinyl chloride (PVC) based UF membrane for purification of the actual oilfield produced water [17], carbon dot based UF membrane for textile wastewater [18], and so forth.

10.2.1.2 Nanofiltration (NF) and Reverse Osmosis (RO)

Nanofiltration (NF) and reverse osmosis (RO) possess similar operational characteristics and membrane structures, yet they are discernable by their pore size and their respective abilities to eliminate specific substances. While NF can effectively eliminate natural organic matter, viruses, and divalent salts, RO surpasses this by being able to remove a broader spectrum of contaminants, resulting in the production of highly pure water. Both NF and RO exhibit great potential for the treatment of saline wastewater due to their remarkable rejection capabilities. However, it is essential to note that the performance of RO and NF membranes may fluctuate depending on the specific chemistry and conditions of the saline wastewater being treated. Advanced filtration technology in the realm of NF is evolving with the aim of crafting high-quality membranes. Various techniques are being employed, such as grafting polymerization, interfacial polymerization, the integration of nanoparticles, and more. The primary objective of these methodologies is to produce membranes that exhibit outstanding separation capabilities, increased productivity, and selectivity, enhanced thermal stability, and resistance to fouling. Nonetheless, there is a pressing need for substantial enhancements in NF membranes. These improvements should encompass increased rejection and separation efficiency, fortified antifouling properties, prolonged membrane lifespan, and heightened resistance to chemical agents [19]. Recent advancements in the modifications of membrane involves the introduction of additives into the thin film layer, like nanoparticles, leading to formation of thin-film nanocomposite (TFC) membranes. Contrarily, grafting polymerization involves attaching monomers covalently to the primary polymer chain and causing them to polymerize as extra side chains. Grafting is a popular technique for adding different functional groups to a polymer matrix. Techniques including electron beam irradiation, UV/photo-grafting, layer-by-layer methods and plasma treatment, comes under grafting polymerization.

Mogan, initially proposed and developed composite membrane using interfacial polycondensation method for the formation of thin layer polymeric surface substrate [20]. However, until Cadotte and his research team discovered that they could produce a series of composite membranes with remarkably high flux by using interfacial crosslinking of piperazine with a mixture of trimesoyl chloride and isophthaloyl chloride, this approach did not gain attention in industrial manufacturing [21]. These membranes not only achieved high flux production but also demonstrated a strong rejection of sulphate ions, however, lower selectivity for chloride ions. This method has been widely recognized as interfacial polymerization and has had a significant impact on the field of membrane science and technology. It signifies a substantial advancement towards the production of high-flux, low-pressure membranes capable of effectively removing salts while maintaining their performance. With this concept, different TFC membranes have been synthesized and used for wastewater treatment and re-use. However, it is important to note that this aspect has not been thoroughly explored, as existing literature indicates a predominant focus on the development of the top thin active layer, in both current and past research efforts. TFC membrane configuration comprising a slender selective layer with a thickness ranging from 100 nm to 200 nm, positioned atop an UF or MF substrate with a thickness of 50 μm to 100 μm as shown in Figure 10.2. To reinforce the support layer, a non-woven polyester fabric can also be incorporated. The interfacial polymerization technique, which uses amine- and acyl-chloride monomers, is often used to create the selective layer utilising polyamide (PA). Monomers of diamine and acid chloride are dissolved in distinct, immiscible solvents. These two monomers are then brought into contact with the membrane surface, initiating a polymerization reaction that results in the formation of a thin PA film. Numerous papers and reviews have been published thus far regarding PA-based TFN membranes. These publications extensively discuss the influence of various nanomaterials on the physicochemical properties of TFN membranes and explore how alterations in the intrinsic characteristics of membranes can impact their performance [22–24]. One of the modifications involves customizing the surface charge of the selective layer by carefully selecting appropriate monomers or by introducing nanomaterials with the desired surface charges [25,26], modification of support layers for providing thermal stability especially for high temperature wastewater [27].

Surface modification techniques are commonly employed to enhance the surface properties of TFC-NF membranes. These techniques include plasma-induced polymerization, graft polymerization, ion implantation, and surface coating. The objective of applying coating polymers is to enhance the hydrophilicity of the membranes and

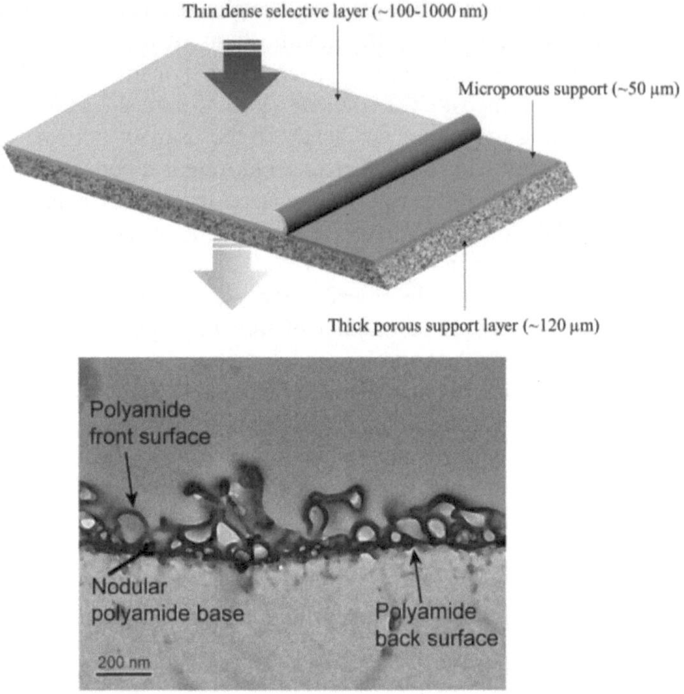

FIGURE 10.2 Schematic structure of TFC membrane and FESEM images of PA-based TFC membrane. (Adapted from Ref [58].)

reduce surface roughness. Coating can be absorbed onto the surface, covalently attached to the surface polymer, or cross-linked with itself. Moreover, it can introduce either positive or negative charges to the membrane surface. This process can be used to tailor the membrane's chemical properties, improve selectivity, increase fouling resistance, or introduce specific functionalities, ultimately optimizing their suitability for various applications such as water purification, gas separation, or medical filtration, thereby expanding their versatility and efficiency in diverse industrial and scientific contexts. Graft polymerization, involves the covalent attachment of new functional groups to the substrate's surface [28]. Ultraviolet (UV) radiation method is used by researchers for generation of free radicals for grafting, which improves the fouling resistance and permeability due to enhancement in surface hydrophilicity, surface roughness, and charges on the surface of membranes [29]. However, opposite charges contaminants can be easily attracted on the surface of membrane through electrostatic attraction which ultimately increases the fouling of membrane Thus, modification of membrane by chemical or physical means and incorporation of nanomaterials such as carbon dots (CDs), carbon nanotubes (MWCNTs, SWNTs, HNTs) has demonstrated the ability to create a preferred flow pathway and enhance intermolecular hydrogen bonding [18,30]. This leads to an increase in chlorine resistance and improved resistance to fouling in the membranes.

10.2.1.3 Forward Osmosis (FO)

FO stands apart from pressure-driven processes because it does not require external pressure to operate. Instead, it relies on the osmotic pressure difference generated by varying concentrations between a draw solution (high concentration) and feed solution (less concentration) [31] . FO has gained increasing attention within the water treatment community because it consumes less energy compared to RO. Additionally, FO process can achieve high water recovery rates [32]. These types of membranes are mostly used for saline wastewater treatment. In contrast to various pressure-driven membranes, it is important to note that in a FO membrane, both its active layer and support layer play crucial roles in determining its performance. The active side of membrane is typically facing the feed solution, while the support side is facing the draw solution. This presents a unique challenge in the manufacturing of FO membranes, as it requires precise formulation and modification to achieve the desired characteristics on both sides. The inclusion of nanomaterials in the synthesis of FO membranes has yielded promising results, enhancing the overall membrane properties. A wide array of nanomaterials, including organic, inorganic, and biomimetic substances, have been extensively integrated into the fabrication process of FO membranes [33]. However, the commercialization of these synthesized membrane has not been up to mark due to a lack of compatibility, stability, and feasibility of these FO membranes.

The operational principles of the FO process present certain challenges when used as a standalone unit operation in various processes. Therefore, to fully exploit the capabilities of the FO process, it is often integrated or combined with other advanced treatment processes. For instance, brine is fed into a high-pressure chamber in the Pressure Retarded Osmosis (PRO) process, which is isolated from a nearby feed chamber by a semi-permeable membrane. With this configuration, water may be dragged into the high-pressure chamber under conditions of high pressure and concentration. The feedwater introduced into this system can be sourced from rivers or municipal wastewater effluent, but it can lead to fouling of membrane. In this case FO can be applied as pretreatment process for PRO to reduce the fouling issues, and it requires a lower energy input, resulting in a substantial increase in overall efficiency [34]. Some researchers study the hybrid system of RO-FO process and concluded the reduced energy consumption. The hybrid system significantly lowered energy requirements in comparison to single RO system from 2.5 kWh/m^3 to just 1.5 kWh/m^3. Moreover, this hybrid approach effectively mitigated fouling issues, leading to a substantial increase in water flux [35]. In a different research study, a multi-stage flash desalination unit was integrated with FO pre-treatment process exhibited a remarkable 32 percent improvement in water recovery. [36]. One prominent use of pre-treatment involves combining membrane distillation (MD) with FO to control the waste from a water/oil separator. This hybrid process demonstrated exceptional removal efficiency, exceeding 93 percent, for various contaminants including organics, inorganics, and oil. The permeates that were produced had a good grade, making them acceptable for reusing in industrial operations. Importantly, the FO streams were highly concentrated, ranging from 77 percent to 84 percent, resulting in significantly reduced disposal volumes.

Additionally, it led to a reduction in the concentrations of magnesium, calcium, and sulfate ions in the feed solution, which enhanced water permeability and flux. These hybrid systems offer sustainability and applicability due to their independence from chemical additives and external water sources. Moreover, they exhibited no issues related to membrane wetting or bacterial contamination. However, there were certain technical difficulties that had to be overcome, such as internal concentration polarisation brought on by monovalent ions and a decrease in the forward osmosis flux due to the scaling of $CaSiO_3$ on the support layer. Additionally, the presence of $CaSO_4$ and NaCl crystals on the active layer further contributed to a reduction in forward osmosis flux. Furthermore, the permeate was susceptible to the presence of organic nanoparticles and volatile fatty acids [37].

Considering the present status of membrane technology and water treatment processes, it is evident that there is no single solution capable of addressing all the complex morphological issues associated with water and wastewater treatment. Therefore, an integrated approach involving multiple membrane technologies, including forward and reverse osmosis, is essential for an effective water and wastewater treatment strategy. The pace at which forward osmosis becomes a prominent part of these solutions hinges on the results of techno-economic analyses that favour cost-effective and straightforward membrane production methods.

10.2.2 Concentration-driven Membranes

Pervaporation (PV) is a separation method used to remove tiny amounts of volatile substances from liquid mixtures. It does this by allowing these substances to turn into vapor and pass through a membrane, which can be either porous or nonporous. This process combines two actions: membrane permeation and evaporation, to separate the different components within the liquid mixture. PV is applied in various industries, including those dealing with hydrocarbons and volatile organic compounds. This technique is also used in biotechnology, desalination, food, medicine, and other industrial fields. The main driving force behind this technique is the difference in concentration within the mixture. It operates based on a mechanism called solution-diffusion, which results in the formation of vapor as the components move through the membrane. To remove the vapor produced during the process, either low pressure is applied, or an inert substance is introduced in the later stages. One practical example of pervaporation is its use in separating ethanol from water mixtures [38]. Several techniques can be employed in the creation of PV membranes. These methods encompass hollow fiber spinning, solution casting, interface polymerization, solution coating, and membrane modification.

The Gas Separation (GS) process operates on a principle similar to PV. The feed is first sorbed into the membrane, then permeates through the membrane, and finally, desorption occurs at the low-pressure side. Selectivity plays a crucial role in the GS process . Gaseous molecules are transported through the membrane via the solution diffusion mechanism. It involves the use of polymeric membrane which are asymmetric and homogenous to separate gaseous mixtures from polar vapours. However, this membrane possesses certain challenges in high temperature bases applications like in natural gas treatment, petrochemical industries, heavy hydrocarbon separation, and so forth. These challenges include achieving high selectivity in gas mixtures with closely similar components, maintaining structural integrity and separation efficiency under high pressure and temperature, which iscontinuous concern that affects its long-term stability, and addressing fouling caused by sulphur compounds, heavy hydrocarbons, and particulates. The choice of membrane materials must align with the chemical composition of the gas streams, scale up complexity due to higher cost and lower energy efficiency. Addressing these multifaceted challenges in gas separation membranes is vital to optimizing petrochemical processes, improve efficiency, and reduce environmental impacts in this critical industry.

10.2.3 Electrically Driven Membranes

Electrodialysis (ED) is a membrane process that operates under the influence of an electric field. In this process, ions with opposite charges in the mixture continually move through ion exchange membranes due to the applied electrical current. ED has shown promise in the separation and fractionation of organics and salts. For the recovery of salt ions from textile wastewater containing Remazol Brilliant Blue R dye, bipolar membrane ED has been investigated. This process generates acid and base products. A linear relationship between dye concentration and the final diluted salt concentration was observed. To predict the desalination rate, a new parameter called critical salt concentration was introduced as shown in Figure 10.3 (a) [39]. The seta potential of the dye was regulated to control the fouling of ion exchange membrane which resulted in 74.4 percent of desalination rate from a mixture of Na_2SO_4 and dye. Within this process, approximately 72.2 percent of sodium ions and 67.9 percent of sulfate ions were transformed into sodium hydroxide and sulfuric acid, respectively. In response to the common fouling issues associated with an ion exchange membrane, an innovative ED system was proposed for textile wastewater treatment and dye recovery as shown in Figure 10.3 (b). This system utilized a loose NF membrane as an alternative to the traditional ion exchange membrane [40]. This approach resulted in rejection of dyes with higher efficiency and allows the rapid transport of anions through nano-channels of NF membrane. This configuration outperformed commercial ion-exchange membranes with astounding results of 98.9 percent salt rejection and 99.4 percent dye recovery. Moreover, minimal fouling was observed over five ED cycles in comparison to ion-exchange membranes. This enhancement was credited to the NF membrane's inherent negative charge, which successfully stopped dye species from entering and settling within the membrane's pore structure.

Lately, there has been a significant focus on wastewater treatment using reverse electrodialysis (RED). This process allows the conversion of the salinity gradient into electrical energy, generating cell potentials during treatment. Simultaneously, pollutants are broken down in the anodic and cathodic compartments of the system. Utilizing wastewater with varying salinity levels to drive electrochemical treatment offers several advantages, particularly considering the constant production of saline water from diverse sources. Another intriguing use in this direction is the use of RED stacks for autonomous wastewater treatment [41]. In a RED stack with 60 membrane pairs, a salinity gradient was established by using two synthetic wastewaters

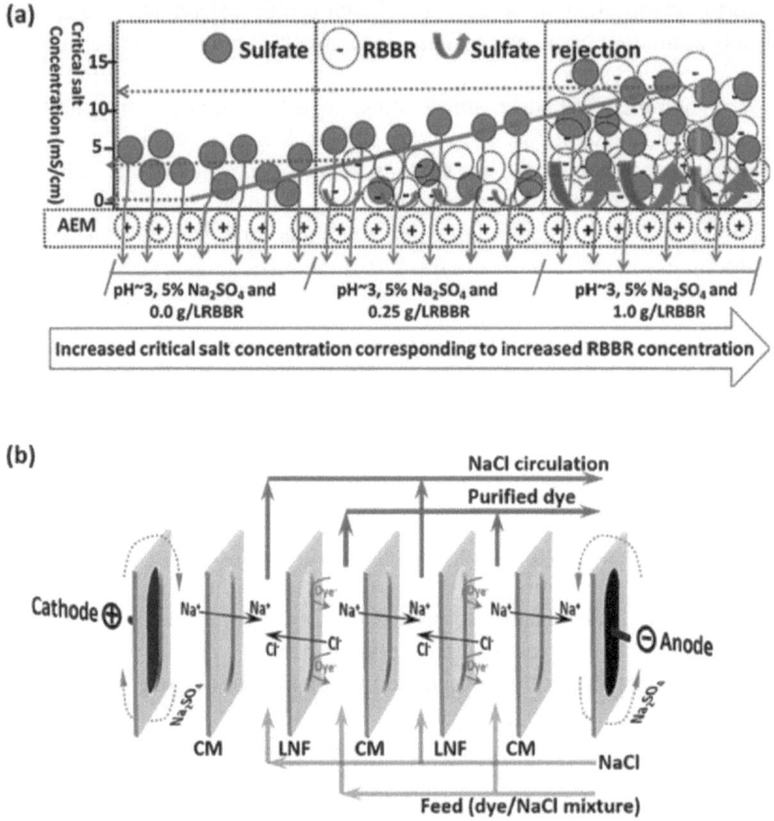

FIGURE 10.3 (a) The diagram illustrates the linear relationship between critical salt concentration and RBBR dye concentration. (Adapted from Ref [39].) (b) This schematic shows the configuration of an ED assembly featuring a cation exchange membrane and a loose NF membrane for the separation of dye and salt (CM: cation exchange membrane, LNF: loose NF membrane). (Adapted from Ref [40].)

of different salt concentrations and a model organic contaminant. The low-salinity wastewater was sent to the cathodic compartment, while high-salinity wastewater was directed to the anodic compartment. This configuration resulted in a higher rate of organic decomposition, which was maintained by adjusting the volume ratio of the two wastewater streams. As a result, an impressive 70 percent per hour of total organic carbon removal was achieved, primarily due to the generation of chlorine through electrochemical processes.

An innovative electrodialysis membrane configuration was proposed for treating tannery wastewater, incorporating a double anti-fouling membrane design. This design aimed to protect an electrodialysis anion exchange membrane with a UF membrane, guarding against organic foulants originating from animal skins which have high molecular weight and are negatively charged [42]. Organic materials like proteins and fat were concentrated in the dilate compartment while 72.3 percent of sulphide, 67 percent of chloride, and 62.5 percent of calcium ions, respectively, were removed from tannery effluents. the ED method managed to reach a total demineralization efficiency of 56 percent even with just one desalination compartment and no membrane fouling. Table 10.1 summarizes the application of various membrane and their performance in wastewater treatment, recovery and reuse.

TABLE 10.1
Studies Conducted on a Variety of Membranes for the Purpose of Treating, Recovering, and Reusing Wastewater

Membrane type	Type of wastewater	Rejection Efficiency of pollutants	Reference
Commercial flat sheet UF membrane	Textile wastewater	COD = 79.8% TOC = 82.4% Turbidity = 99.3% Sulfate = 93.1% Calcium = 98.6%	[59]
Polydopamine (PDA)/ polyethylenimine (PEI) triggered by ammonium persulphate UF membrane	Textile wastewater	Dye >98.2% Salt >97.0%	[13]
UF	Textile wastewater	Dye = 97.47% Salt = 99.84%	[15]
PES/SPSf tight UF membrane	Textile wastewater	Dye = 69.4–100% Salt = 10%	[60]
2,2′-benzidinedisulfonic acid (BDSA) and trimesoyl chloride (TMC) NF membrane	Textile wastewater	Dye >99% Flux recovery = 95.6%	[61]
poly(vinylidene fluoride) (PVDF) NF membrane	Shale gas flowback and produced water	Turbidity: 99.9% COD: 94.2% Sulfate: 91.7% Calcium: 72.8%	[62]
Polyvinylidene fluoride (PVDF)/ styrene-maleic anhydride (SMA) copolymer NF membranes	Textile wastewater	Congo red = 99.9% Methyl blue = 96.7% Salt = 1.7%	[63]
Polyether amine (PEA) and trimesoyl chloride NF membrane	Textile wastewater	Methly blue = 95.4% NaCl =16.4%	[64]
Commercial TFC RO membrane	Agriculture wastewater	BW30 for MCPA >95% SW30 for MCPA >95%	[65]
RO membrane	Tannery wastewater	Chromium >99.5%	[66]
PA TFC membrane with Polysulfone+SPEEK/ polyetherimide co-cast Substrate (FO membrane)	Shale gas produced water	Water recovery rate of 53.2%	[67]

(continued)

TABLE 10.1 (Continued)
Studies Conducted on a Variety of Membranes for the Purpose of Treating, Recovering, and Reusing Wastewater

Membrane type	Type of wastewater	Rejection Efficiency of pollutants	Reference
APTES-SiO_2 TFC membrane	Aquaculture wastewater	Water recovery rate of 47%	[68]
Fluorinated polyoxadiazole hollow fiber MD membrane	Shale gas produced water	Salt >99.9%	[69]
Commercial hydrophobic MD membrane	Textile wastewater	COD: 90% Colour: 94%	[70]
SGMD membrane	Textile wastewater	Dye: 100%	[71]
Bipolar Membrane Electrodialysis (BMED)	Textile wastewater	Desalination rate: 74.4%	[39]
Loose NF-based ED	Textile wastewater	Desalination efficiency = 98.9% Dye rejection = 99.4%	[40]
ED-UF membrane	Tannery wastewater	Demineralization efficiency = 56 ± 1.25%	[42]
ED-UF membrane	Textile wastewater	TDS = 94.2%	[72]

10.3 CHALLENGES IN CURRENT MEMBRANE-BASED WASTEWATER TREATMENT PROCESSES AND ITS AREA OF IMPROVEMENT

Fouling poses a persistent challenge when it comes to treating wastewater using membrane-based technologies. In essence, fouling occurs as unwelcome substances, termed foulants, accumulate on a membrane's surface, consequently reducing its flow rate or flux [43]. In certain scenarios, these foulants can obstruct the membrane's tiny pores, resulting in internal fouling [44]. The fouling issue can be classified into three primary categories: organic, inorganic and biofouling, based on the nature of the foulants present. Inorganic fouling, often referred to as scaling, typically transpires when the saturation point of the inorganic ions in the water surpasses. This occurrence triggers nucleation, followed by the growth of crystals and the settling of inorganic ions within the membrane's pores or on its surface [45]. Several factors can influence the fouling process, including the inherent properties of the membrane and the feedwater's composition as well as the operational factors such as temperature and pH [46]. Saline wastewater may harbor diverse foulants, resulting in the development of intricate layers of fouling. Biofouling, a phenomenon where the micro-organisms are attached on the surface of membrane, is often intensified by wastewater's salinity. The saline environment can stimulate bacterial respiration and encourage the aggregation of cells, leading to the production of a greater quantity of extracellular polymeric substances (EPS) through cellular secretion [47]. The elevated EPS concentration can elevate the viscosity of the wastewater and contribute to the irreversible settlement of a cake-like layer on the surface of the membrane. This layer is notoriously difficult to remove and can severely impede membrane performance [48,49]

A commonly employed strategy for fouling mitigation involves implementing effective pretreatment procedures before initiating the membrane-based treatment process. Pretreatment processes encompass a diverse array of technologies selected based on the specific wastewater characteristics, and they can be categorized as either membrane or non-membrane processes. Within membrane-based pretreatment, MF and UF are frequently employed owing to their effectiveness in eliminating larger suspended particles that could otherwise cause severe fouling in tighter membranes like NF and RO. Some researchers used MF and UF membrane former to NF and RO for saline wastewater treatment and found the reduction in fouling from lipids [50]. While minor decline in the permeate flux was observed, this decrease was primarily caused to the precipitation of calcium and phosphorus. However, it was possible to maintain the flux at a stable level. An anti-scalant was introduced into the produced water after it had undergone UF treatment, just before it entered the RO-AGMD pilot setup [51]. Their findings indicated that MD operation at a low permeate flux of 1.4 LMH, coupled with the anti-scalant's effects, proved to be effective in managing scaling issues. However, it is crucial to note that the amount of anti-scalant used must be carefully adjusted to prevent the formation of additional fouling, which can occur when an excessive amount of anti-scalant is introduced. Some studies also coagulation process before membrane filtration, for example, poly aluminium chloride is used as chemical coagulant for pre-treatment of textile effluents [52]. In their research, they implemented a pre-treatment system consisting of coagulation, sand filtration, and magnetic ion exchange. This pre-treatment system proved to be more effective in mitigating fouling in RO compared to the pre-treatment system involving coagulation, sand filtration, and UF. The reason behind this observation is that the fouling layer formed on the RO membrane was more compact. In the case of the coagulation-sand filter-UF system, organic molecules which are having lower molecular weight might easily penetrate the UF membrane and deposit on the valleys of the RO membrane, leading to fouling. This underscores the importance of integrating numerous processes with the core membrane filtration to minimize the issues like fouling. Furthermore, the inclusion of coagulant such as polyaluminium chloride before Membrane Bioreactor (MBR) treatment has resulted in effective reduction in fouling in the bioreactor during the textile wastewater treatment [47].

These insights can be applied to the MD process, prompting researchers to explore enhancements in feed water quality by integrating appropriate pretreatment steps. The impact of pretreatment before MD on the overall system performance, remains an area that has not been extensively explored. Therefore, it is crucial to assess both the practicality of incorporating a pretreatment stage and the holistic benefits it can bring, which include cost savings related to cleaning frequency and membrane replacement expenses. Moreover, membrane modification using graft polymerization and surface coatings to improve hydrophilicity and smoothness can also enhance membrane performance and reduces fouling. Innovative configurations of membrane systems have also been suggested to mitigate fouling challenges for the treatment of saline wastewater. For example, modification to the ED system by incorporating a UF membrane onto the anionic membrane was used for leather-processing wastewater [53]. This UF membrane allowed the passage of anions like sulphide, NaCl, and amino acids due to their lower molecular weight while effectively preventing peptide and protein fouling on the anionic membrane. As a result of this modified ED system, the removal of 56 percent of conductivity, 62.5 percent of calcium, 72.3 percent of sulphide, 67 percent of chlorides, and 10.4 percent of COD was achieved. Additionally, multiple pre-treatment steps were included to enhance the process. Another study by Hayes and Saverin [54]aimed to address fouling caused by the high calcium content (4000 mg/L) in flowback water from shale gas hydraulic fracturing, which also contained 30,000 mg/L of NaCl. This elevated the content of calcium, which led to severe fouling within the electrode cell. To mitigate this, they proposed a cleaner approach involving the addition of HCl to the electrolyte and the use of a monovalent-selective membrane, resulting in a reduction of up to 70 percent in calcium flux.

Concentration polarization, particularly the internal type, has consistently been a key challenge in the performance of FO processes. This phenomenon often leads to discrepancies between the expected and actual performance of FO systems. Given the central role that internal concentration polarization plays in FO processes, it is crucial to explore strategies for mitigating its effects. These strategies should include improvements to the inherent properties of FO membranes and the implementation of external measures aimed at preventing and mitigating concentration polarization and fouling issues. The implementation of an air sparging configuration has proven effective in providing aeration, which serves to minimize concentration polarization and membrane fouling [55]. Gao et al., [56] took advantage of biogas as the source of sparging gas in an anaerobic MBR system for the treatment of pulping wastewater. They observed that increasing the biogas sparging rate helped reduce the deposition and the accumulation of sludge on and within the membrane module. This, in turn, improved membrane flux and the long-term stability of the operation when treating pulping wastewater. In another study, a combination of activated carbon scrubbing, biogas sparging, and liquid circulation was employed to additionally enhance the control of fouling in an anaerobic MBR system for the treatment of seafood processed wastewater [57]. Their findings revealed that this configuration for the MBR furnished with biogas circulation has led to the formation of a thinner cake layer on the membrane surface after 18 days of operation compared to the configuration without biogas sparging, which experienced fouling within the first 11 days of operation. Additionally, fouling due to inorganic and organic contaminants like phosphate and struvite was effectively reduced by 90 percent at an 8-hour hydraulic retention time. This discovery holds promise for MBR development as it could lead to reduced cleaning frequency and lower maintenance costs.

The evolution of pressure-driven membrane technology has primarily revolved around exploring innovative additives and polymer blends to enhance the performance of the membrane in saline wastewater treatment. However, a common challenge encountered in these endeavors is the delicate balance between maximizing permeability and maintaining selectivity. Achieving this equilibrium remains a significant objective. In contrast, FO, MD, and ED have garnered increasing attention for their applications in treatment of saline wastewater and resource recovery. While the potential of these procedures has been substantiated, much of the research has concentrated on assessing their suitability for treating saline wastewater and recovering valuable nutrients. There remains a pressing need for further research and optimization in membrane development and system design concerning FO, MD, and ED. Goh et al., [7] presents an overview of the characteristics, current technological readiness, and areas that require enhancement within the major membrane procedures used for the treatment of saline wastewater.

While the effective filtration and retrieval performance of membrane determine its practical applicability, cost-effectiveness is also one of the most critical factors influencing their commercial viability. Costs can vary significantly based on factors such as site-specific conditions, infrastructure requirements, and the need for final brine treatment. As a result, variations in cost estimations are common when comparing results from different sources. Numerous economic analyses have been conducted in the field of saline wastewater treatment and resource recovery. These studies have delivered valuable insights into the economic feasibility of implementing or enhancing various treatment processes in terms of optimization. However, many of these analyses have limitations. For instance, most studies primarily focus on operational expenses, particularly energy costs. Capital costs have been considered in some cases, but these calculations are often based on simplified flow models and small-scale capacities. Estimations of expenditures are also hindered by limited available data and discrepancies between datasets. Additionally, some economic evaluations fail to adequately account for factors like membrane fouling, which can significantly impact operating costs. Given the importance of realistic economic analyses, it is crucial to expand research efforts in this area in parallel with technological developments [7]. Figure 10.4 provides an overview of the features, scalability, and areas needing improvement for membrane-based technologies in saline wastewater treatment.

10.4 CONCLUSION

In summary, this chapter has delved into membrane-based wastewater treatment methods, showcasing their potential as potent tools for wastewater treatment, reuse, and recovery. We explored various membrane processes, spanning pressure-driven, concentration-driven, and electric-driven techniques, highlighting their advantages such as robust contaminant removal, reduced chemical usage, and resource reclamation. These technologies have found applications across diverse sectors, including municipal wastewater treatment, industrial processes, and desalination.

Nonetheless, challenges accompany these benefits, with fouling, scaling, concentration polarization, and membrane wear being prominent issues. Effective pretreatment methods, membrane enhancements, and system optimization are crucial to address these hurdles. Additionally, the economic feasibility of membrane technologies is pivotal for their widespread adoption. While membrane development has progressed, the permeability-selectivity trade-off remains a challenge that necessitates ongoing research and innovation.

In future, the integration of hybrid approaches and the exploration of novel materials and techniques hold great potential for improving performance and cost-effectiveness

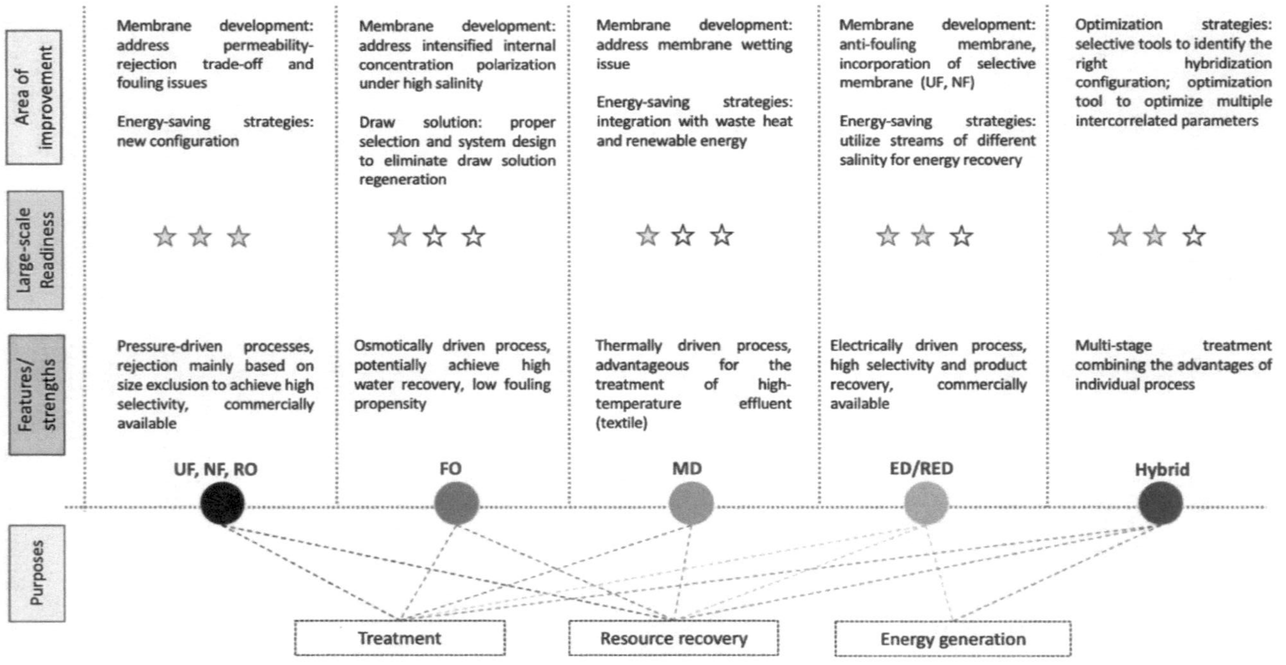

FIGURE 10.4 Features large-scale readiness and area of improvement of membrane-based technologies for saline wastewater treatment. (Adapted from Ref [7].)

in membrane-based wastewater treatment. With the world grappling with water scarcity and environmental concerns, membrane processes are poised to assume a pivotal role in sustainable wastewater management and resource recovery.

REFERENCES

1. Obaideen K, Shehata N, Sayed ET, Abdelkareem MA, Mahmoud MS, Olabi AG. The role of wastewater treatment in achieving sustainable development goals (SDGs) and sustainability guideline. *Energy Nexus*. 2022;7:100112.
2. Ezugbe EO, Rathilal S. Membrane technologies in wastewater treatment: A review. *Membranes (Basel)*. 2020;10(5):89.
3. Nqombolo A, Mpupa A, Moutloali RM, Nomngongo PN, et al. Wastewater treatment using membrane technology. In: *Wastewater and Water Quality*. IntechOpen; Editor: Taner Yonar, 2018.
4. Abdullah N, Yusof N, Lau WJ, Jaafar J, Ismail AF. Recent trends of heavy metal removal from water/wastewater by membrane technologies. *J Ind Eng Chem*. 2019;76:17–38.
5. Shoshaa R, Ashfaq MY, Al-Ghouti MA. Recent developments in ultrafiltration membrane technology for the removal of potentially toxic elements, and enhanced antifouling performance: A review. *Environ Technol Innov*. 2023;31:103162.
6. Das S, O'Connell MG, Xu H, Bernstein R, Kim JH, Sankhala K, et al. Assessing advances in anti-fouling membranes to improve process economics and sustainability of water treatment. *ACS ES&T Eng*. 2022;2(11):2159–73.
7. Goh PS, Wong KC, Ismail AF. Membrane technology: A versatile tool for saline wastewater treatment and resource recovery. *Desalination*. 2022;521:115377.
8. Buscio V, Crespi M, Gutiérrez-Bouzán C. Application of PVDF ultrafiltration membranes to treat and reuse textile wastewater. *Desalin Water Treat*. 2015;57(18):8090–6.
9. Buscio V, Marín MJ, Crespi M, Gutiérrez-Bouzán C. Reuse of textile wastewater after homogenization–decantation treatment coupled to PVDF ultrafiltration membranes. *Chem Eng J*. 2014;265:122–8.
10. Han G, Feng Y, Chung TS, Weber M, Maletzko C. Phase inversion directly induced tight ultrafiltration (UF) hollow fiber membranes for effective removal of textile dyes. *Environ Sci Technol*. 2017;51(24):14254–61.
11. Lin J, Ye W, Baltaru MC, Pan Tang Y, Bernstein NJ, Gao P, et al. Tight ultrafiltration membranes for enhanced separation of dyes and Na_2SO_4 during textile wastewater treatment. *J Memb Sci*. 2016;514:217–28.
12. Ćurić I, Dolar D, Karadakić K. Textile wastewater reusability in knitted fabric washing process using UF membrane technology. *J Clean Prod*. 2021;299:126899.
13. Ye W, Ye K, Lin F, Liu H, Jiang M, Wang J, et al. Enhanced fractionation of dye/salt mixtures by tight ultrafiltration membranes via fast bio-inspired co-deposition for sustainable textile wastewater management. *Chem Eng J*. 2020;379:122321.
14. Hu M, Yang S, Liu X, Tao R, Cui Z, Matindi C, et al. Selective separation of dye and salt by PES/SPSf tight ultrafiltration membrane: Roles of size sieving and charge effect. *Sep Purif Technol*. 2021;266:118587.
15. Jiang M, Ye K, Deng J, Lin J, Ye W, Zhao S, et al. Conventional ultrafiltration as effective strategy for dye/salt fractionation in textile wastewater treatment. *Environ Sci Technol*. 2018;52(18):10698–708.
16. Isik Z, Arikan EB, Bouras HD, Dizge N. Bioactive ultrafiltration membrane manufactured from *Aspergillus carbonarius* M333 filamentous fungi for treatment of real textile wastewater. *Bioresour Technol Rep*. 2019;5:212–9.

17. Ahmad T, Guria C, Mandal A. Optimal synthesis and operation of low-cost polyvinyl chloride/bentonite ultrafiltration membranes for the purification of oilfield produced water. *J Memb Sci*. 2018;564:859–77.
18. Heng ZW, Tan YY, Chong WC, Mahmoudi E, Mohammad AW, Teoh HC, et al. Preparation of a novel polysulfone membrane by incorporated with carbon dots grafted silica from rice husk for dye removal. *J Water Process Eng*. 2021;40:101805.
19. Van DBB, Mäntt M, Mänttäri M, Nyström M, Nyström N. Drawbacks of applying nanofiltration and how to avoid them: A review. *Sep Purif Technol*. 2008;63:251–63.
20. Morgan PW. *Condensation Polymers: By Interfacial and Solution Methods*. Vol. 10. Interscience; 1965, pp. 19–64.
21. John EC, Robert JP. *Research on In Situ-Formed Condensation Polymer for Reverse Osmosis Membranes: Final Report*. United states: MRI-North Star Division; 1978, 44 pp.
22. Yang Z, Ma XH, Tang Y. Recent development of novel membranes for desalination. *Desalination*. 2017;434:37–59.
23. Esfahani MR, Aktij SA, Dabaghian Z, Firouzjaei MD, Rahimpour A, Eke J, et al. Nanocomposite membranes for water separation and purification: Fabrication, modification, and applications. *Sep Purif Technol*. 2019;213:465–99.
24. Lau WJ, Gray S, Matsuura T, Emadzadeh D, Paul Chen J, Ismail AF. A review on polyamide thin film nanocomposite (TFN) membranes: History, applications, challenges and approaches. *Water Res*. 2015;80:306–24.
25. Kang D, Shao H, Chen G, Dong X, Qin S. Fabrication of highly permeable PVDF loose nanofiltration composite membranes for the effective separation of dye/salt mixtures. *J Memb Sci*. 2021;621:118951.
26. Metecan A, Cihanoğlu A, Alsoy Altinkaya S. A positively charged loose nanofiltration membrane fabricated through complexing of alginate and polyethyleneimine with metal ions on the polyamideimide support for dye desalination. *Chem Eng J*. 2021;416:128946.
27. Karami P, Khorshidi B, McGregor M, Peichel JT, Soares JBP, Sadrzadeh M. Thermally stable thin film composite polymeric membranes for water treatment: A review. *J Clean Prod*. 2020;250:119447.
28. Korolkov IV, Gorin YG, Yeszhanov AB, Kozlovskiy AL, Zdorovets MV, Str I. Preparation of PET track-etched membranes for membrane distillation by photo-induced graft polymerization. *Mater Chem Phys*. 2017;205:55–63.
29. Vatanpour V, Zoqi N. Surface modification of commercial seawater reverse osmosis membranes by grafting of hydrophilic monomer blended with carboxylated multiwalled carbon nanotubes. *Appl Surf Sci*. 2017;396:1478–89.
30. Zargar M, Jin B, Dai S. An integrated statistic and systematic approach to study correlation of synthesis condition and desalination performance of thin film composite membranes. *Desalination*. 2016;394:138–47.
31. Shaffer DL, Werber JR, Jaramillo H, Lin S, Elimelech M. Forward osmosis: Where are we now? *Desalination*. 2015;356:271–84.
32. Ang WL, Wahab Mohammad A, Johnson D, Hilal N. Forward osmosis research trends in desalination and wastewater treatment: A review of research trends over the past decade. *J Water Process Eng*. 2019;31:100886.
33. Wang J, Liu X. Forward osmosis technology for water treatment: Recent advances and future perspectives. *J Clean Prod*. 2021;280:124354.
34. Jamil S, Jeong S, Vigneswaran S. Application of pressure assisted forward osmosis for water purification and reuse of reverse osmosis concentrate from a water reclamation plant. *Sep Purif Technol*. 2016;171:182–90.

35. Yangali-Quintanilla V, Li Z, Valladares R, Li Q, Amy G. Indirect desalination of Red Sea water with forward osmosis and low pressure reverse osmosis for water reuse. *Desalination*. 2011;280(1–3):160–6.
36. Altaee A, Mabrouk A, Bourouni K, Palenzuela P. Forward osmosis pretreatment of seawater to thermal desalination: High temperature FO-MSF/MED hybrid system. *Desalination*. 2014;339(1):18–25.
37. Saqib NM, Alamoudi T, Soukane S, Soo Son H, Jin Y, Medina SC, et al. Performance and implications of forward osmosis–membrane distillation hybrid system for simultaneous treatment of different real produced water streams. *Chem Eng J*. 2022;450:138479.
38. Ravanchi MT, Kaghazchi T, Kargari A. Application of membrane separation processes in petrochemical industry: A review. *Desalination*. 2009;235:199–244.
39. Berkessa YW, Lang Q, Yan B, Kuang S, Mao D, Shu L, et al. Anion exchange membrane organic fouling and mitigation in salt valorization process from high salinity textile wastewater by bipolar membrane electrodialysis. *Desalination*. 2019;465:94–103.
40. Ye W, Liu R, Chen X, Chen Q, Lin J, Lin X, et al. Loose nanofiltration-based electrodialysis for highly efficient textile wastewater treatment. *J Memb Sci*. 2020;608:118182.
41. Ma P, Hao X, Galia A, Scialdone O. Development of a process for the treatment of synthetic wastewater without energy inputs using the salinity gradient of wastewaters and a reverse electrodialysis stack. *Chemosphere*. 2020;248:125994.
42. Tamersit S, Bouhidel KE, Zidani Z. Investigation of electrodialysis anti-fouling configuration for desalting and treating tannery unhairing wastewater: Feasibility of by-products recovery and water recycling. *J Environ Manage*. 2018;207:334–40.
43. Ho KC, Teow YH, Mohammad AW, Ang WL, Lee PH. Development of graphene oxide (GO)/multi-walled carbon nanotubes (MWCNTs) nanocomposite conductive membranes for electrically enhanced fouling mitigation. *J Memb Sci*. 2018;552:189–201.
44. Lee WJ, Ng ZC, Hubadillah SK, Goh PS, Lau WJ, Othman MHD, et al. Fouling mitigation in forward osmosis and membrane distillation for desalination. *Desalination*. 2020;480:114338.
45. Jiang S, Li Y, Ladewig BP. A review of reverse osmosis membrane fouling and control strategies. *Sci Total Environ*. 2017;595:567–83.
46. Manawi Y, Kochkodan V, Hussein MA, Khaleel MA, Khraisheh M, Hilal N. Can carbon-based nanomaterials revolutionize membrane fabrication for water treatment and desalination? *Desalination*. 2016;391:69–88.
47. Tan X, Acquah I, Liu H, Li W, Tan S. A critical review on saline wastewater treatment by membrane bioreactor (MBR) from a microbial perspective. *Chemosphere*. 2019;220:1150–62.
48. Capodici M, Cosenza A, Di Bella G, Di Trapani D, Viviani G, Mannina G. High salinity wastewater treatment by membrane bioreactors. In *Current Developments in Biotechnology and Bioengineering. Advanced Membrane Separation Processes for Sustainable Water and Wastewater Management – Case Studies and Sustainability Analysis*. Editor: Ashok Pandey, Sangeeta Negi and Carlos Ricardo Soccol. Publisher. Elsevier; 2020; pp. 177–204.
49. Liu L, Xiao Z, Liu Y, Li X, Yin H, Volkov A, et al. Understanding the fouling/scaling resistance of superhydrophobic/ omniphobic membranes in membrane distillation. *Desalination*. 2021;499:114864.
50. Abdelkader S, Gross F, Winter D, Went J, Koschikowski J, Geissen SU, et al. Application of direct contact membrane distillation for saline dairy effluent treatment: Performance and fouling analysis. *Environ Sci Pollut Res*. 2019;26(19):18979–92.

51. Duong HC, Chivas AR, Nelemans B, Duke M, Gray S, Cath TY, et al. Treatment of RO brine from CSG produced water by spiral-wound air gap membrane distillation – A pilot study. *Desalination*. 2014;366:121–9.
52. Yin Z, Yang C, Long C, Li A. Effect of integrated pretreatment technologies on RO membrane fouling for treating textile secondary effluent: Laboratory and pilot-scale experiments. *Chem Eng J*. 2018;332:109–17.
53. Tamersit S, Bouhidel KE, Zidani Z. Investigation of electrodialysis anti-fouling configuration for desalting and treating tannery unhairing wastewater: Feasibility of by-products recovery and water recycling. *J Environ Manage*. 2017;207:334–40.
54. Hayes TD, Severin BF. Electrodialysis of highly concentrated brines: Effects of calcium. *Sep Purif Technol*. 2017;175:443–53.
55. Bagheri M, Ahmad SM. Critical review of fouling mitigation strategies in membrane bioreactors treating water and wastewater. *Bioresour Technol*. 2018;258:318–34.
56. Gao W, Han M, Xu C, Liao B, Hong Y, Cumin J, et al. Performance of submerged anaerobic membrane bioreactor for thermomechanical pulping wastewater treatment. *J Water Process Eng*. 2016;13:70–8.
57. Chaiprapat S, Thongsai A, Charnnok B, Khongnakorn W, Bae J. Influences of liquid, solid, and gas media circulation in anaerobic membrane bioreactor (AnMBR) as a post treatment alternative of aerobic system in seafood industry. *J Memb Sci*. 2016;509:116–24.
58. Pacheco F, Sougrat R, Reinhard M, Leckie JO, Pinnau I. 3D visualization of the internal nanostructure of polyamide thin films in RO membranes. *J Memb Sci*. 2016;501:33–44.
59. Ćurić I, Dolar D, Karadakić K. Textile wastewater reusability in knitted fabric washing process using UF membrane technology. *J Clean Prod*. 2021;299:126899.
60. Hu M, Yang S, Liu X, Tao R, Cui Z, Matindi C, et al. Selective separation of dye and salt by PES/SPSf tight ultrafiltration membrane: Roles of size sieving and charge effect. *Sep Purif Technol*. 2021;266:118587.
61. Li M, Yao Y, Zhang W, Zheng J, Zhang X, Wang L. Fractionation and concentration of high-salinity textile wastewater using an ultra-permeable sulfonated thin-film composite. *Environ Sci Technol*. 2017;51(16):9252–60.
62. Chang H, Liu B, Yang B, Yang X, Guo C, He Q, et al. An integrated coagulation-ultrafiltration-nanofiltration process for internal reuse of shale gas flowback and produced water. *Sep Purif Technol*. 2019;211:310–21.
63. Kang D, Shao H, Chen G, Dong X, Qin S. Fabrication of highly permeable PVDF loose nanofiltration composite membranes for the effective separation of dye/salt mixtures. *J Memb Sci*. 2021;621:118951.
64. Mi YF, Wang N, Qi Q, Yu B, Peng XD, Cao ZH. A loose polyamide nanofiltration membrane prepared by polyether amine interfacial polymerization for dye desalination. *Sep Purif Technol*. 2020;248:117079.
65. Zhang J, Weston G, Yang X, Gray S, Duke M. Removal of herbicide 2-methyl-4-chlorophenoxyacetic acid (MCPA) from saline industrial wastewater by reverse osmosis and nanofiltration. *Desalination*. 2020;496:114691.
66. Mohammed K, Sahu O. Recovery of chromium from tannery industry waste water by membrane separation technology: Health and engineering aspects. *Sci Afr*. 2019;4:e00096.
67. Chen G, Liu R, Shon HK, Wang Y, Song J, Li XM, et al. Open porous hydrophilic supported thin-film composite forward osmosis membrane via co-casting for treatment of high-salinity wastewater. *Desalination*. 2017;405:76–84.

68. Nguyen TQ, Tung KL, Lin YL, Dong CD, Chen CW, Wu CH. Modifying thin-film composite forward osmosis membranes using various SiO_2 nanoparticles for aquaculture wastewater recovery. *Chemosphere*. 2021;281:130796.
69. Xu J, Srivatsa Bettahalli NM, Chisca S, Khalid MK, Ghaffour N, Vilagines R, et al. Polyoxadiazole hollow fibers for produced water treatment by direct contact membrane distillation. *Desalination*. 2018;432:32–9.
70. Li F, Huang J, Xia Q, Lou M, Yang B, Tian Q, et al. Direct contact membrane distillation for the treatment of industrial dyeing wastewater and characteristic pollutants. *Sep Purif Technol*. 2018;195:83–91.
71. Mousavi SA, Arab Aboosadi Z, Mansourizadeh A, Honarvar B. Surface modified porous polyetherimide hollow fiber membrane for sweeping gas membrane distillation of dyeing wastewater. *Colloids Surf A Physicochem Eng Asp*. 2021;610:125439.
72. Lafi R, Gzara L, Lajimi RH, Hafiane A. Treatment of textile wastewater by a hybrid ultrafiltration/electrodialysis process. *Chem Eng Process – Process Intensif*. 2018;132:105–13.

11 The Use of Membrane Bioreactors in the Treatment of Low-strength Wastewater Applications

Ahmed Albahnasawi, Murat Eyvaz, Motasem Y.D. Alazaiza, Nurullah Özdoğan, Ercan Gurbulak, Sahar Alhout, and Ebubekir Yuksel

11.1 INTRODUCTION

Industrial activities have seen a rise because of the increase in global population. This phenomenon has presented a dual challenge: it exacerbates environmental issues while simultaneously spurring an unprecedented demand for water resources [1]. Given the current state of our water resources, there is an urgent imperative to explore innovative solutions that can revolutionize water cycle management in both public and industrial domains. Furthermore, it is crucial to recognize and harness the intrinsic value of water by integrating sustainable techniques into the water cycle [2]. In this context, the recovery and reuse of wastewater emerge as invaluable resources. Advanced technologies offer a promising avenue to achieve this goal [3]. Among the alternative technologies for wastewater treatment, one stands out: the MBR, which combines biological processes with membrane filtration, resulting in a highly efficient wastewater treatment system. Within the bioreactor tank, biomass degradation occurs, while the membrane module helps the separation of treated wastewater from microorganisms (Figure 11.1). Over the past two decades, MBR technology has garnered considerable attention, thanks to its proven ability to produce high-quality effluent. It has now matured into a cornerstone of wastewater treatment [4] .

In fact, the MBR market is poised for remarkable growth, with a projected Compound Annual Growth Rate (CAGR) of 22.4 percent [5]. According to a recent report from Business Communication Company (BCC), the MBR market reached a size of three billion USD in 2019 and is predicted to surge to 4.2 billion USD by 2024, exhibiting a robust CAGR of 7 percent. The Asia-Pacific region (APAC), a

 DOI: 10.1201/9781003454199-11

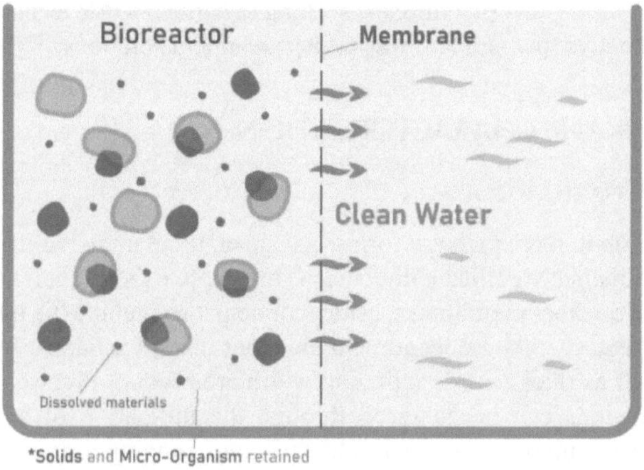

FIGURE 11.1 Membrane bioreactor schematic. (Compiled by the authors.)

significant segment of the MBR market, is particularly noteworthy, highlighting the highest market growth globally[6] .

Conventional Activated Sludge (CAS) processes typically entail a two-step approach. In the first step, wastewater undergoes treatment within an aeration tank, where active microorganisms (referred to as activated sludge) play a central role in the purification process. Subsequently, sedimentation tan is used to separate the activated sludge and treated water, often referred to as a secondary clarifier. It is worth noting that achieving complete separation of activated sludge from the treated effluent in CAS systems is a persistent challenge, with a part of the lighter fraction often accompanying the discharged treated water. In contrast, MBR technology presents a changing solution by incorporating membranes with varying pore sizes. This innovation allows MBRs to achieve a significantly improved separation of activated sludge from the treated effluent [7].

This chapter provides a comprehensive overview of the application of MBRs in treating low-strength wastewater. Beginning with an introduction, the chapter sets the stage by outlining the significance of MBR technology in addressing the challenges posed by low-strength wastewater streams. Section 2 delves into membrane characterization, offering insights into the various membrane types and their properties crucial for efficient wastewater treatment. Section 3 explores different types of MBR systems and novel configurations, highlighting advancements in technology that enhance performance and reliability. Moving forward, Section 4 focuses on the biological performance of MBRs, examining the intricate interplay between microorganisms and membranes to achieve high-quality effluent standards. Finally, Section 5 presents recommendations and future prospects, offering valuable insights into potential advancements and areas for further research to optimize MBR technology for low-strength wastewater treatment. Through a structured approach encompassing introductory concepts, membrane characterization, system types, biological performance evaluation, and future outlooks, this chapter serves as a comprehensive resource for researchers, engineers, and practitioners seeking to harness MBR technology for sustainable and effective low-strength wastewater treatment solutions.

11.2 MEMBRANE CHARACTERIZATION

11.2.1 Membrane Modules

Membranes come in three primary forms: flat sheet, tubular, and hollow fiber. Each type possesses distinct significant dimensions that dictate their functionality [8]. For tubular and hollow fiber membranes, critical dimensions include the inner diameter, and outer diameter membrane length. On the other hand, for flat sheet membranes, parameters such as thickness, length, and width are crucial. Precise measurement of these dimensions can be achieved through the utilization of advanced tools including calipers, micrometers, and field emission scanning electron microscopes (FE-SEM) [9]. The dimensions of a membrane can be accurately measured using a FE-SEM due to its high-resolution imaging capabilities. FE-SEM employs a concentrated electron beam to examine the surface of the sample, delivering intricate images with magnifications ranging from tens to hundreds of thousands of times. To measure the dimensions of a membrane with FE-SEM, the sample is prepared and mounted onto a specialized holder or stage within the microscope. The electron beam scans across the surface of the membrane, producing high-resolution images that reveal fine structural details. The software associated with the FE-SEM typically includes measurement tools that allow users to precisely determine dimensions such as thickness, pore size, and surface roughness. These tools often utilize features like edge detection algorithms or manual measurement tools to accurately determine dimensions based on the captured images. Flat sheet membranes offer several advantages, including straightforward cleaning and replacement procedures, making them highly efficient for commercial applications [10]. However, their primary drawback lies in their relatively high cost, prompting ongoing efforts by researchers to develop more cost-effective alternatives. Hollow fiber membranes, on the other hand, offer distinct advantages, including resilience. cost-effective fabrication processes, and high surface area per unit volume in the face of challenging operating conditions [11]. Nonetheless, they are susceptible to fouling, needing frequent cleaning, which can be a drawback [12]. Tubular membranes also present enticing features, including robust mechanical strength, low fouling rates, extended operational lifespans, and ease of maintenance [13]. However, their low packing density and higher capital and operational costs have limited their widespread adoption on a large scale [14].

11.2.2 Pore Size

Two fundamental categories of membranes employed in wastewater treatment are porous and non-porous [15]. Porous membranes rely on size exclusion mechanisms, such as Nanofiltration (NF), Ultrafiltration (UF), and Microfiltration (MF) to separate particles. MF membranes typically have pore sizes ranging from 0.1 to 10 micrometers (μm). These relatively large pores enable MF membranes to effectively remove suspended solids, bacteria, and large colloidal particles from water, making it suitable for applications such as clarification and pre-treatment. UF membranes feature smaller pore sizes compared to MF, typically ranging from 0.001 to 0.1 micrometers (μm). UF membranes are capable of removing smaller particles such as viruses, proteins, macromolecules, and finer colloidal particles, providing a higher level of filtration and purification. NF membranes have even smaller pore sizes, typically ranging from 0.001 to 0.01 micrometers (μm). NF membranes are effective in removing divalent ions, small organic molecules, and certain dissolved solids while allowing smaller ions and molecules to pass through, making it suitable for applications such as desalination, softening, and color removal. In contrast, non-porous membranes operEnd NFed on principles like solubility or diffusivity to effect particle separation, as seen in Tight-End NF and Reverse Osmosis (RO) [16]. RO membranes have the smallest pore sizes among these membrane processes, typically ranging from 0.0001 to 0.001 micrometers (μm). RO membranes operate by applying pressure to overcome osmotic pressure, effectively removing dissolved salts, ions, organic compounds, and other contaminants from water, producing high-quality permeate suitable for various applications including drinking water purification, wastewater treatment, and industrial processes. The interaction between the size of particles in wastewater and the pore size of the membrane plays a pivotal role in membrane fouling. When membrane pore size increases, finer particles gain easier access to the membrane, increasing the likelihood of pore blockage mechanisms [17]. Conversely, smaller pore sizes tend to result in the collection of smaller particles on the membrane surface, often forming a layer that can be removed using methods such as air scouring. Understanding the pore size distribution of a membrane is essential for accurately assessing its performance and suitability for specific applications. While Field Emission Scanning Electron Microscope (FE-SEM) images of the membrane surface can provide insight into pore size, obtaining a comprehensive pore size distribution for the entire membrane surface proves challenging due to the limited fraction of the membrane that these images represent. Therefore, pore size distribution is typically calculated based on bulk information using methods such as bubble point, particle rejection, and polymer rejection [18]. These approaches provide a more holistic understanding of membrane characteristics, facilitating their effective use in wastewater treatment processes [19].

11.2.3 Hydrophilicity

The propensity of a membrane to foul is significantly influenced by its hydrophilicity, a measure of its affinity for water. A membrane is considered hydrophobic if air within its pores cannot easily be displaced by water. Many polymer-based membranes fall into this hydrophobic category, which presents a major drawback: a heightened susceptibility to fouling. Within the framework of MBR, microorganisms present in the system produce biomass as activated sludge floc and organic matter, which typically possess inherent hydrophobic properties. Consequently, as a result, the biomass tends to adhere strongly to the membrane surface, forming a fouling layer that diminishes the membrane's efficiency. To address this issue, during membrane fabrication, hydrophilic materials can be introduced to enhance the membrane's hydrophilicity [20].

Additionally, wetting agents are employed to help the penetration of water into the membrane pores, aiding in the enhancement of water permeation. However hydraulic pressure becomes indispensable to expel air from the pores when wetting agents are not used. The minimum hydraulic pressure required to expel air from the pores of the membrane depends on several factors. Firstly, the average pore diameter of the membrane plays a crucial role. Smaller pores may require higher pressure to effectively remove air compared to larger pores. Secondly, the surface tension of the water also influences the required hydraulic pressure. Water with higher surface tension tends to trap air bubbles more tenaciously within the pores, necessitating greater pressure to dislodge them. Additionally, the contact angle, which refers to the angle formed between the water-air interface and the membrane surface, impacts the ease of air removal. A higher contact angle indicates less favorable wetting of the membrane surface by water, making it more challenging to expel air bubbles. Therefore, when designing and operating MBR systems, it is essential to consider these factors to ensure adequate hydraulic pressure is applied for effective air removal from the membrane pores, thus optimizing system performance and efficiency [20].

11.2.4 Electric Charge (Zeta Potential)

Membrane surface characteristics play a crucial role in determining fouling propensity, with factors such as electric charge and hydrophilicity significantly influencing fouling behavior. In the case of MF and UF membranes, frequently employed in MBRs, the main foulants are organic substances, typically carrying a negative surface charge. Consequently, selecting a membrane with a higher negative charge on its surface can significantly reduce fouling potential [21]. At the surface of the membrane, the zeta potential serves as a crucial parameter for quantifying the electric charge. The streaming potential method is currently the most advanced technique utilized for measuring zeta potential. This technique assesses the potential produced when electrolytes move between two materials submerged in an aqueous solution. Other approaches used to measure zeta potential could include electrophoresis, electroosmosis, and attributes associated with sedimentation potential [21]. By understanding and controlling these electric charge dynamics, it is possible to design and operate membranes more effectively in MBR systems, reducing fouling-related issues.

11.2.5 Surface Roughness

A rough surface provides a greater contact area, leading to stronger interactions between foulants and the membrane. As a result, membranes with rougher surfaces tend to experience higher levels of fouling due to increased interaction opportunities [22]. Technological tools like the Atomic Force Microscope (AFM), a type of Scanning Probe Microscope (SPM), enable the measurement of membrane surface roughness [23]. AFM analysis has revealed that membrane surface roughness tends to be more significant in membranes with larger pore sizes compared to those with smaller pore sizes. Moreover, there is a direct correlation between fouling rate and surface roughness: an increase in surface roughness corresponds to an increase in fouling rate [23]. Understanding and managing membrane surface roughness is crucial for mitigating fouling-related challenges in various applications.

11.3 MBR AND NOVEL CONFIGURATIONS

11.3.1 Aerobic and Anaerobic Membrane Bioreactor (AnMBR)

Aerobic treatment technologies have been instrumental in wastewater and industrial effluent treatment for over a century. However, they come with notable drawbacks, including a massive energy need for the aeration, substantial sludge generation, the emission of greenhouse gases like nitrous oxide (N_2O), a large physical footprint, and significant maintenance costs. Achieving a reduction in the energy required is crucial for broader adoption of aerobic MBR technology. Aeration control strategies within the aeration tanks play a pivotal role in achieving this reduction in overall energy consumption. Recent studies in full-scale MBRs have shown reductions of 20 percent in aeration and 4 percent in energy consumption rates by implementing ammonia-N-based aeration control strategies [24]. It is worth noting that decreasing airflow rates to reduce energy consumption may affect greenhouse gas. Understanding the correlation between operational conditions and direct/indirect greenhouse gas emissions is crucial for reducing the environmental impact, particularly concerning emissions stemming from incomplete nitrification. Successful reductions in the operational costs of MBR plants by 1317 percent have been reported, depending on influent dynamics, through the application of closed-loop aeration to maintain constant dissolved oxygen concentrations within the aerobic reactor, as opposed to open-loop systems [25]. Effluents with biodegradable Chemical Oxygen Demand (COD) levels below 1000 mg/L are typically treated using aerobic processes., while highly polluted effluents with BOD exceeding 4000 mg/L often undergo anaerobic treatment. Organic matter decomposition in wastewater through anaerobic treatment is a process that not only treats wastewater but also produces renewable fuel (biogas). Additionally, it recovers nutrients that can be used in agriculture by generating fertilizers [26]. Another MBR configuration, an alternative approach, combines membrane filtration with anaerobic digestion treatment to address specific challenges encountered in the MBR process. In this configuration, the decomposition of organic matter into methane-rich biogas reduces the energy requirements for wastewater treatment. Additionally, nutrient recovery becomes feasible by converting nutrients into chemically available forms through precipitation. Despite these advantages, challenges such as membrane stability, fouling, resource dilution, and salinity buildup impede the widespread adoption of this approach [27]. Membrane models and anaerobic bioreactor constitute the two core components of an AnMBR plant. Among the bioreactor configurations, completely stirred tank reactor (CSTR), up-flow anaerobic sludge blanket (UASB), and anaerobic fluidized bed bioreactor (AFBR) are the most common for AnMBRs, with CSTR being the most often used due to its simplicity in operation and construction [28]. Recently, the external submerged configuration has garnered attention in pilot-scale applications and demonstrates promise for extensive adoption in domestic wastewater treatment. Advance research indicates lower energy requirements for pilot-scale external submerged AnMBRs, often utilizing hollow fiber membranes, compared to laboratory-scale AnMBRs and aerobic MBRs, with reported values ranging from 0.04 to 135 kWh [29].

11.3.2 Anaerobic Fluidized Membrane Bioreactor (AFMBR)

Although AnMBRs have demonstrated promising results in the removal of antibiotics and offer various advantages, the persistent issue of membrane fouling has hindered their widespread adoption. To address this challenge and simultaneously enhance antibiotic removal from wastewater, a novel solution has emerged in the form of Anaerobic Fluidized Bed Membrane Bioreactors (AFMB). AFMB stands for an innovative fusion of membrane technology, liquid circulation, and particle injection. It delivers several distinct advantages over traditional AnMBRs:

Enhanced Antibiotic Removal: AFMB promotes the accelerated growth of antibiotic-degrading bacteria, leading to increased efficiency in antibiotic removal.

Reduced Concentration of EPSs and SMPs: This technology results in lower concentrations of Extracellular Polymeric Substances (EPSs) and Soluble Microbial Products (SMPs), contributing to overall improved performance.

Stable Sludge with Larger Particle Size: AFMB fosters the development of a more stable sludge characterized by larger particle sizes, a crucial factor in mitigating fouling occurrences. However, achieving a comprehensive understanding of the performance of this innovative model requires further investigation. Notably, the incorporation of carriers into the AFMBR has been proven to substantially reduce energy requirements compared to traditional AnMBRs. Energy consumption for AFMBRs typically falls within the range of 0.039–0.13 kWh/m^3, while conventional AnMBRs can consume energy ranging from 0.25–7.3 kWh/m^3 [30]. This underscores the potential of AFMB as a more energy-efficient and effective solution for removing antibiotics and addressing fouling challenges in wastewater treatment.

11.3.3 Membrane Photobioreactor (MPBR)

An MPBR typically includes various submerged MF or UF membranes, including flat sheet or hollow fibers, integrated with Photobioreactors (PBRs). This innovative configuration presents numerous advantages such as efficient microalgae separation, system stability, and enhanced effluent quality. Nevertheless, a notable challenge in advancing this setup lies in its limited efficacy in treating primary raw domestic wastewater, which frequently contains elevated levels of organic matter. This is primarily due to the lower concentration of organic matter in such wastewater, which serves as a vital food source for microalgae used in treatment of wastewater [31]. In the realm of wastewater treatment, researchers have delved into the investigation of MPBRs, aiming to understand their performance in depth. Studies have been conducted to scrutinize the influence of various parameters, such as Hydraulic Retention Time (HRT) and organic loading rate (OLR), on the effectiveness of

MPBRs in treating wastewater. Optimal conditions for MPBR performance were found to include an OLR of up to 0.014 kg/m^3/day and a 2-day HRT, which effectively maintained a fouling frequency of 0.016 day-1 during the treatment of domestic wastewater. Remarkably, this system operates independently of external aeration sources, proficiently producing microalgal biomass while concurrently eliminating organics and nutrients [31]. Introducing an electric field within the MBR has emerged as a promising strategy to combat fouling issues that have long hindered the widespread use of MBRs. Recent developments suggest that this approach effectively mitigates the deposition of foulants on the membrane surface, leading to a reduction in fouling incidents [32]. Consequently, In recent years, a novel configuration has attracted considerable attention, involving the integration of Microbial Fuel Cells (MFCs) with MBRs. This configuration harnesses the electricity generated from wastewater treatment by MFCs to effectively control fouling tendencies in MBRs. This innovative approach not only addresses fouling issues but also demonstrates the potential for sustainable and energy-efficient wastewater treatment processes [32]. Unlike the direct application of electricity within MBRs, this new configuration reduces the energy requirements for the process and simultaneously does not compromise bacterial performance due to the presence of high electric fields.

Researchers have delved into the fouling characteristics of sludge in MFC-MBRs, employing hollow fiber membranes and utilizing the extended Derjaguin-Landau-Verwey-Overbeek (XDLVO) model for fouling analysis. The findings revealed a decrease in the free energy of adhesion between Soluble Microbial Products (SMPs) and both clean and fouled membranes in MFC-MBRs compared to control systems (C-MBRs). Furthermore, it was deduced that SMPs encounter heightened energy barriers for adsorption onto the membrane surface in MFC-MBRs, effectively inhibiting fouling processes. This configuration also exhibits reduced hydrophobicity and a diminished negative surface charge of sludge flocs, further enhancing its advantages [33]. Numerous studies have evaluated the performance of different MBR configurations in treating various types of wastewaters, whether synthetic or industrial effluents. These assessments have taken place at different scales, ranging from laboratory to pilot applications.

11.4 PERFORMANCE AND REMOVAL MECHANISMS IN MEMBRANE BIOREACTORS

11.4.1 Microbiological Aspects

Within the biochemical phase of wastewater treatment, microbes play a pivotal role in the removal of organic carbon and nutrients. These microorganisms thrive within intricate three-dimensional structures known as flocs, where they are intertwined by extracellular polymeric substances (EPS). The formation of flocs is of paramount importance in the context of activated sludge wastewater treatment. This floc structure serves not only to adsorb soluble substrates but also to capture colloidal matter and macromolecules present in the wastewater [34]. Activated sludge harbors a highly diverse microbial community, encompassing eukaryotes (protozoa, nematodes, rotifers), prokaryotes (bacteria), and viruses. Among these, bacteria predominate, playing a central role in the degradation of contaminants [35]. In the realm of MBR technology, which amalgamates biochemical processes and sludge separation into a unified step, a perpetual cycle ensues. This process operates through a continuous cycle wherein new sludge is continually produced as organic materials are consumed. Simultaneously, a portion of this sludge undergoes decay through a process known as endogenous respiration. Endogenous respiration encompasses various sides of biomass loss and energy requisites unrelated to growth. This includes processes such as maintenance, decay, lysis, endogenous respiration, predation, and death. Notably, endogenous respiration can occur in both anoxic and aerobic environments, although it continues at a slower pace under anoxic conditions, particularly due to reduced predation activity among protozoa [36].

Maintaining a high sludge age, which equates to an elevated sludge concentration, can be used to stimulate endogenous respiration among a microbial population in an MBR. The availability of substrate determines the amount of energy that bacteria can use. There is a theoretical possibility of having a situation where the energy supplied equals the demands for maintenance if the sludge retention time (SRT) is extended, hence increasing biomass content. According to this theory, which was developed by Pirt, maintenance energy is the bare minimum of bioenergetics needed for sludge endogenous respiration. Microorganisms prioritize meeting their maintenance energy needs over producing additional biomass. This understanding underscores the importance of SRT management in optimizing MBR performance. Consequently, when nutrient supplies dwindle, external substrate primarily sustains essential bacterial functions, without a significant change in bacterial population. Furthermore, as biomass concentration rises, sludge loading decreases, as reflected in the food-to-microorganism (F/M) ratio (g COD/g TSS/day) [37] .

As the sludge loading diminishes to a certain threshold, excess sludge production dwindles, and in some cases, it ceases altogether [38]. Various investigations into the application of MBR technology in wastewater treatment have shown the possibility of achieving zero sludge production at different F/M ratios. These outcomes are contingent on the specific feed compositions that dictate the growth of microbial populations [39]. However, a best biomass concentration, linked to the SRT, is pivotal for the effective operation of an MBR. High sludge ages occurs when biomass retention results in a population that grows slowly, characterized by. In such conditions, cell dormancy and death contribute to a reduction in population viability [40]. Several factors contribute to these phenomena. MBR sludge behaves as a non-Newtonian fluid, leading to exponential increases in sludge viscosity as the mixed-liquor suspended solids (MLSS) concentration rises. This leads to restrictions in mass transfer for oxygen and substrate, resulting in increased aeration costs and significant membrane fouling [41]. On the contrary, a decrease in Mixed Liquor Suspended Solids (MLSS) concentrations offers a larger specific surface area for substrate uptake and enzyme production, thereby enhancing enzymatic activity. As a result, when operating at low Sludge Retention Time (SRT), the system exhibits a quicker response to the degradation of xenobiotic waste. Furthermore, there is an increased likelihood of genetic mutation and microbial adaptation to fluctuating organic loads [42]. While high sludge ages offer certain advantages, they also present challenges. One such challenge is the potential limitation of substrate solubility. To address this issue, researchers have proposed modifying sludge morphology, which includes encouraging the growth of non-flocculating microorganisms. This approach aims to improve removal performance, as evidenced by recent studies. Furthermore, the association between the chemical components of EPS and the surface characteristics and structure of biological flocs in activated sludge has been brought to light by current study. It is interesting to note that operating circumstances might have a substantial impact on these qualities. Findings in this field, nevertheless, are not necessarily reliable. For instance, numerous studies have found that SRTs lower floc size, possibly as a result of decreased generation of EPS, which is essential for floc development. However, some experts recommend that a minimal rate of sludge waste be maintained in order to guarantee an ideal range of sludge concentrations in an MBR [43]. In the event that the reactor's sludge is not eliminated, the buildup of inorganic compounds becomes problematic. These non-biodegradable substances may be created by the microbiological process or may come from the influent. Although protozoan activity might not entirely break down bacterial cell walls, it can help in the creation of inert material. Remarkably, inert materials are not completely inert; depending on the SRT, slow-growing microorganisms may be responsible for their breakdown. Apart from the prolonged SRTapproach, one may also increase the pace of sludge deterioration in MBRs by causing some of the sludge to disintegrate. The most popular method for accomplishing this is sludge lysis, which is the death and fragmentation of cells that results in the bacterial elimination. Autochthonous substrates produced by this mechanism are recycled in microbial metabolism and add to organic loading. Interestingly, there is no discernible difference between the biomass growth on these substrates and the growth on the original organic substrate [43]. One of the most important steps in cell lysis is the breakdown of the cell wall, which can be sped up using chemical or physical therapies. Through heat treatment, researchers have successfully increased endogenous metabolism in an MBR by causing cell lysis and death. In their study, biomass was extracted and subjected to various temperatures (50, 70, and 90°C), and the resulting hydrolysates were recycled back into the bioreactor. This innovative approach resulted in an improvement in endogenous metabolism through cryptic growth, particularly when both the Hydraulic Retention Time (HRT) and SRT were maintained at extremely low levels (2 and 10 hours, respectively [44,45]. A variety of techniques, including alkaline treatment, ozone-induced biodegradation, and ultrasonic disintegration, can be used to promote cryptic development in an MBR. According to experimental results, ozone treatment in particular can significantly reduce the formation of surplus sludge without having a negative impact on the processes of biological mineralization and nitrification [46,47]. Cultivating controlled predators is a further tactic that has been effective in lowering the amount of extra sludge produced in bench and pilot-scale reactors. Protozoa and metazoa make up the majority of these predators; they feed on bacteria without interfering with the breakdown of their substrate. A portion of the energy transferred from lower to higher trophic levels is wasted as a result of ineffective biomass conversion. This energy loss is maximised under ideal circumstances, resulting in the least amount of biomass produced overall [48,49].

Environmental factors, such as temperature and pH, exert a significant influence on microbial growth and act as limiting factors in MBR systems. Temperature not only governs the treatment rate but also impacts the composition of bacterial communities. For instance, fewer strict aerobic bacteria are found at lower temperatures, suggesting less oxygen transmission. This is partially because mixed liquor has a lower viscosity at these temperatures. For maximum MBR function, temperatures should ideally be between 15 and 25°C. Below 10°C, treatment effectiveness starts to decline. Autotrophic metabolism is negatively impacted by pH levels outside of the ideal range of 7.2–8.5. Notwithstanding progress, the effective functioning of MBR systems is still up for discussion since there is a dearth of thorough data on the establishment of microbial community structures inside the reactor [50–52]. The characteristics of sludge morphology, including the presence of dispersed bacteria, a reduced quantity of large filamentous bacteria, and floc densification, play a pivotal role in removal efficiencies while also influencing sludge filterability and fouling mechanisms [53,54]

In scenarios of high organic loading, particularly observed in low Sludge Retention Time (SRT) conditions, challenges such as foaming and sludge bulking may become more pronounced. Furthermore, there is still more study to be done on how membrane separation, as opposed to settling separation, alters the structure of sludge [55]. MF and UF membranes are effective in retaining dispersed bacteria, colloidal matter, and supra-colloidal material. Because of this, an MBR's biological habitat might differ greatly from an ordinary activated sludge system's. One theory is that bacteria in MBRs compete fiercely with one another for inflowing substrates at high sludge ages. The microbiology and physiology of MBRs, however, remain present many mysteries and difficulties despite advances, underscoring the need for more study and comprehension in this field.

11.4.2 Phosphorus Removal and Nitrification/Denitrification

The excessive use of agricultural pesticides and fertilizers, coupled with the discharge of inadequately treated municipal and industrial wastewater, has resulted in elevated concentrations of nitrogen and phosphate in both groundwater and surface water. This discharge often leads to the overgrowth of certain weed species, causing eutrophication in lakes and other water bodies [56]. Eutrophication, characterized by the excessive growth of algae, is primarily driven by the presence of nitrates and phosphates, which act as rate-limiting factors in this process. The effective removal of these compounds is vital for successful wastewater treatment. Additionally, gastric cancer is a significant health concern associated with water contaminated with nitrates and nitrites. This contamination has been linked to various health issues, including methemoglobinemia in infants, commonly referred to as blue baby disease. Additionally, these compounds have the potential to induce DNA mutations, further exacerbating health risks [57]. An essential aerobic process called biological nitrification involves changing ammonia into nitrite (NO_2) and then nitrate (NO_3). After nitrification, anoxic denitrification – a process that converts nitrate to nitrogen gas (N_2) can be used to extract nitrogen from wastewater. Because nitrifying bacteria have a poor cell output and a sluggish growth rate, nitrification frequently presents a bottleneck in the biological nitrogen removal process. Making sure that the net accumulation rate of biomass (and therefore the net removal rate of biomass from the system) stays lower than the growth rate of nitrifying bacteria is a crucial prerequisite for successful nitrification [58–60]. Extended SRTs in MBRs increase the activated sludge's capacity to nitrate by preventing nitrifying bacteria from being washed out of the bioreactor. Furthermore, faster-growing heterotrophic bacteria, which are better rivals for ammonia nitrogen (NH3-N), are less likely to threaten nitrifiers. Extensive research has demonstrated that MBRs can serve as high-rate nitrification technologies suitable for wastewater containing high concentrations of ammonia nitrogen [61,62]. On the contrary, denitrification, the process of removing nitrate, requires anoxic conditions. Typically, to enhance denitrification, an anoxic tank is added upstream from the aerated tank. Anoxic conditions can also be induced by operating MBRs in an intermittent aeration mode, even for submerged MBRs that necessitate continuous bubbling. Ammonium is mostly nitrified to nitrate during the aerobic phase (aeration) in this intermittent aeration MBR setup. Then, during the anoxic phase (non-aeration), deposited nitrate is fully denitrified together with concurrent phosphorus (P) absorption. By discarding sludge after the aerobic stage, when the biomass contains a high concentration of polyphosphates (polyP), the net removal of phosphorus is accomplished [63,64]. Phosphorus (P) in wastewater primarily exists as phosphates, including, condensed phosphates, orthophosphates, and organic phosphate fractions. Removing P from wastewater can be accomplished through adsorption, precipitation, or a biological process known as luxury uptake [65]. It is interesting to note that cell development and metabolism only use a little portion of the TSS mass in the mixed liquor – roughly 1 to 2 percent. In certain circumstances, such an appropriate pH and the presence of iron or calcium ions, are necessary for precipitation and adsorption to successfully remove P. A critical phase in wastewater treatment facilities (WWTPs) is the luxury absorption of P. This process involves introducing an anaerobic phase before the aerobic phase in the wastewater treatment process and recycling sludge through both phases. The purpose is to promote the growth of phosphate accumulating microorganisms (PAOs) through a competitive mechanism. PAOs excel in anaerobic uptake of lower fatty acids, ensuring there are no fatty acids left when transitioning to the aerobic phase. The term "Enhanced Biological Phosphorus Removal" (EBPR) refers to this complete process. By running MBR treatment units in intermittent aeration mode, the EBPR process may be integrated into the system. Furthermore, PAOs (which are usually bigger than 0.5 μm) can be physically retained to improve phosphorus removal within an MBR. An MF membrane (0.2 μm) that acts as a physical barrier to retain the PAOs inside the reactor aids in this retention. Consequently, this setup ensures there is enough biomass present for the EBPR mechanism to operate effectively [66,67].

By combining nitrification, denitrification, phosphorus absorption, and phosphorus release processes in one reactor, intermittently aerated MBRs have the ability to remove nitrogen and phosphorus simultaneously. However, because of nitrate inhibition, intermittent aeration presents difficulties for efficient phosphorus removal, particularly at higher removal levels. In the anaerobic stage, nitrate prevents phosphate release, while in the aerobic stage, it hinders phosphate absorption. Furthermore, nitrate is effectively used by denitrification as a last electron acceptor for the development of non-polyP heterotrophs. As a result, less substrate is available for polyP organisms, which lowers the amount of phosphorus removed. Not all phosphate-accumulating bacteria can use nitrate as an electron acceptor, despite some research suggesting that polyP species can engage in denitrification [68,69]. Because of partial denitrification, intermittently aerated MBRs have difficulty stabilising nitrogen removal, particularly when treating rural home sewage. Sufficient oxygen concentration must be maintained for stable nitrification. The most common ammonium-oxidizing bacteria (AOB) and nitrate-oxidizing bacteria (NOB) in MBRs are identified as Nitrosomonas and Nitrospira-like bacteria, respectively. Under aerobic circumstances, these bacteria are very productive, but under anoxic environments, they have difficulty storing or using their substrate. Stress and metabolic damage can be brought on by oxygen deprivation, and NOB is especially vulnerable to these effects. Achieving consistent and effective nitrogen removal in intermittently aerated MBRs requires careful management of oxygen levels to support the activity of both AOB and NOB bacteria. Researchers have extensively studied the impact of SRT on MBR performance concerning phosphorus removal and, nitrification, denitrification. Some findings suggest that at extremely low SRTs (2 days), there is a decrease in nitrification rates, likely due to the partial loss of nitrifying microorganisms. On the other hand, it has been observed that increasing the sludge concentration, indicated by the sludge retention time, leads to a decrease in the population of nitrifiers. Prolonged SRTs have also been linked to negative influences on nitrification performance, explained by hindered oxygen and substrate transfer resulting from elevated Mixed Liquor Suspended Solids (MLSS) concentrations. Similar trends regarding the impact of SSRT on phosphorus removal have been documented, demonstrating decreased phosphorus removal with longer SRTs. This reduction can be attributed to competitive conditions arising between Phosphate-Accumulating Organisms (GAOs) and Polymeric Acid Oxidising Organisms (PAOs) in SRTs longer than 20 days. These results highlight how crucial it is to keep a balanced SRT in place to avoid nitrifier washout and to lessen the negative consequences of too extended SRTs, such decreased mass transfer from inadequate aeration and competition with GAOs [70,71]. However, it is worth noting that MBR systems typically achieve over 90 percent removal of NH3-N (ammonia-nitrogen), almost independently of the SRT.

Several other critical factors play a significant role in nitrogen removal, including temperature, alkalinity, and the organic and nitrogen loads, often expressed as the Carbon-to-Nitrogen (C/N) ratio, which is essentially the ratio of biological oxygen demand (BOD) to total nitrogen (TN). The C/N ratio needs to be sufficiently high to facilitate denitrification effectively. However, the success of this ratio depends on the presence of readily degradable organic matter, such as volatile fatty acids. Temperature also plays a crucial role, with a recommended upper limit of 40°C to ensure adequate nitrification. Proper temperature control can lead to a nitrification rate exceeding 99 percent, even when the inflow TN concentration varies [72]. Firstly, aeration facilitates the thorough removal of all particles, which typically contain up to 0.1 mg of phosphorus per mg of total solids (TS). Secondly, aeration in MBRs aids in preventing phosphate release that might occur under anoxic conditions. Furthermore, there is a growing interest in utilizing MBR technology for phosphorus recycling. This interest is driven by the anticipation that the phosphorus content of sludge will rise as the Sludge SRT is extended [73,74].

11.4.3 Suspended Solids and Organic Matter

Understanding COD removal processes in the context of combining activated sludge with real wastewater is still quite limited. Microbial reactions to the dynamic conditions within a real wastewater treatment system may deviate from a mere escalation in cell numbers, encompassing diverse substrate-removal mechanisms like sorption, accumulation, and storage [75,76].

Several studies have investigated the effectiveness of both MBRs and Conventional Activated Sludge (CAS) processes while operating under comparable conditions. Consistently, these studies consistently show that MBRs outperform other methods in terms of removing COD, ammonia-nitrogen (NH3-N), and suspended solids (SS). MBR effluents have a lower organic carbon concentration compared to CAS procedures due to many contributing variables. These characteristics include increased retention times and reduced floc diameters. Furthermore, the membrane's ability to fully retain particulate materials is of great importance. The membrane selectively blocks a significant amount of soluble organic molecules and colloids, which improves their removal by promoting higher levels of lysis activity in the reactor. This is driven by the increasing concentrations of these chemicals. Increasing the solid retention times (SRTs) results in longer sludge ages, which in turn enables a more comprehensive breakdown of biodegradable organic compounds present in the influent water. This also enables microbial adaptability to chemicals that are less susceptible to biodegradation. As a result, biomass has the ability to adapt to the wastewater environment without being limited to microbes that grow quickly and generate flocs [77,78].

The improved efficiency of membrane bioreactors (MBRs) operated at prolonged solid retention times (SRTs) can be attributed to many variables. The existence of distributed bacteria is a vital component that gives an edge in competing among the entire population. This advantage is especially significant when the concentration of substrate becomes very low, which is usually found at low F/M ratios and high sludge ages. In such circumstances, the bioreactor exhibits smaller flocs. The decrease in floc size enhances the transfer of both oxygen and carbon, resulting in elevated rates of removal and improved capacity to adjust to changes in the quality and quantity of the influent [79,80]. UF and MF membranes play a crucial role in collecting all suspended solids (SS) in the reactor due to their small pore size. Hence, non-biodegradable organic components are removed by separating them from the particles, which are then released together with the sludge [81]. Aeration flow stands as a critical factor influencing the biochemical processes of BOD and COD removal. It is imperative to provide the right amount of oxygen to microorganisms to meet their three specific demands:

> Carbonaceous BOD refers to the process of transforming carbon-based organic matter in wastewater into cellular tissue and a variety of end products. Nitrogenous BOD: In this process, ammoniacal nitrogen is oxidised to produce nitrite, which is then transformed to nitrate by nitrification. The inorganic chemical oxygen demand (COD) process involves the oxidation of reduced inorganic chemicals present in the wastewater.
>
> (Ref. [82])

The properties of biomass, such as soluble microbial products (SMP) and extracellular polymeric substances (EPS), have a significant effect on the rate at which oxygen is transferred and, consequently, on the pace at which organic matter is removed. These molecules are also often acknowledged as major substances that cause fouling in membranes. SMP, which is soluble and found in the liquid phase, and EPS, which is attached to cells and forms part of the solid phase, have important functions. Oxygen needs to pass through the liquid layer that surrounds the flocs and then move through the floc matrix (EPS) in order to reach the active spots on the bacterial cell membrane. The amounts of EPS vary in response to alterations in the microbial state and operating parameters of the bioreactor. In intermittently aerated membrane bioreactors (MBRs), the levels of extracellular polymeric substances (EPS) grow in direct correlation with the duration of non-aeration [83].

11.5 RECOMMENDATIONS AND FUTURE PROSPECTIVE

Future MBR system design and operation should focus on optimizing aeration strategies to achieve efficient oxygen transfer, particularly considering the influence of biomass characteristics like SMP and EPS. This optimization can enhance organics removal and reduce fouling potential. The relationship between HRT and biomass concentration should be carefully balanced to ensure complete organic matter oxidation while avoiding overloading conditions. Further research is needed to determine the optimal HRT for various wastewater types. Since membrane fouling stays a challenge, efforts should be directed toward continuous monitoring and mitigation strategies. Enhanced fouling control mechanisms can improve MBR efficiency and extend membrane life. Understanding the dynamics of microbial communities during nitrification within MBRs is crucial. Research should focus on characterizing and manipulating these communities to optimize treatment processes. To fully harness the potential of MBRs, future research should prioritize enhanced process design and modeling, considering the contributions of different mechanisms to overall COD removal. This will aid in developing more efficient and reliable MBR systems. The development of advanced membrane materials with improved fouling resistance and separation efficiency holds promise for enhancing MBR performance. Integration of intelligent process control systems, such as artificial intelligence and machine learning, can help optimize MBR operation in real time, making it more adaptive to varying influent conditions. Future MBR systems should explore opportunities for resource recovery, including phosphorus recycling from sludge, which aligns with sustainability goals. Decentralized MBR systems may become more prevalent, providing efficient wastewater treatment solutions for smaller communities and industries. Combining MBR technology with other advanced treatment processes, such as forward osmosis or membrane distillation, can offer novel solutions for specific water treatment challenges.

11.6 CONCLUSION

MBR technology has emerged as a highly effective and versatile approach for wastewater treatment, showing remarkable capabilities in organic pollutant and suspended solid removal. This paper reviews key aspects of MBR operation, focusing on factors influencing its performance and potential future directions. MBRs benefit from longer HRTs, enabling complete organic matter oxidation and reducing the risk of biomass washout. Optimizing aeration strategies to enhance oxygen transfer efficiency remains an essential research area, as it significantly influences organics removal and membrane fouling. Membrane fouling, a persistent challenge, needs ongoing research in monitoring and fouling mitigation strategies. Understanding microbial community dynamics during nitrification within MBRs is crucial for efficient treatment. Balancing HRT and biomass concentration is vital to achieving best removal rates while avoiding overloading conditions. Prospects include advanced membrane materials, intelligent process control, resource recovery, decentralized MBR systems, and hybrid treatment approaches. These developments aim to improve MBR efficiency, reliability, and sustainability. In conclusion, MBR technology continues to evolve, offering a promising solution for wastewater treatment and resource recovery needs. Ongoing research and innovation will further enhance its capabilities, making it an adaptable and indispensable tool in the field of wastewater treatment.

REFERENCES

[1] Albahnasawi A, Yüksel E, Gürbulak E, et al. Fate of aromatic amines through decolorization of real textile wastewater under anoxic-aerobic membrane bioreactor. *J Environ Chem Eng* [Internet]. 2020;8:104226. Available from: https://doi.org/10.1016/j.jece.2020.104226

[2] Albahnasawi A, Yüksel E, Eyvaz M, et al. Performances of anoxic-aerobic membrane bioreactors for the treatment of real textile wastewater. *Global Nest J.* 2020;22:22–27.

[3] Puig S, Hasan SW, Naddeo V. Valuing water into the circular economy: The role of new technology and innovative solutions. *Case Stud Chem Environ Eng* [Internet]. 2022 [cited 2023 September 28];5:100203–100213. Available from: www.iris.unisa.it/handle/11386/4807164

[4] Marrot B, Barrios-Martinez A, Moulin P, et al. Industrial wastewater treatment in a membrane bioreactor: A review. *Environ Prog* [Internet]. 2004 [cited 2023 September 28];23:59–68. Available from: https://onlinelibrary.wiley.com/doi/full/10.1002/ep.10001

[5] Krzeminski P, Leverette L, Malamis S, et al. Membrane bioreactors – A review on recent developments in energy reduction, fouling control, novel configurations, LCA and market prospects. *J Memb Sci.* 2017;527:207–227.

[6] The Membrane Bioreactor Market, Industry Size Forecast [Internet] [cited 2023 September 28]. Available from: www.marketsandmarkets.com/Market-Reports/membrane-bioreactor-market-484.html

[7] Judd S. The status of membrane bioreactor technology. *Trends Biotechnol.* 2008;26:109–116.

[8] Giwa A, Dindi A, Kujawa J. Membrane bioreactors and electrochemical processes for treatment of wastewaters containing heavy metal ions, organics, micropollutants and dyes: Recent developments. *J Hazard Mater* [Internet]. 2019;370:172–195. https://doi.org/10.1016/j.jhazmat.2018.06.025

[9] Huong Le TX, Dumée LF, Lacour S, et al. Hybrid graphene-decorated metal hollow fibre membrane reactors for efficient electro-Fenton – Filtration co-processes. *J Memb*

Sci [Internet]. 2019;587:117182. Available from: https://doi.org/10.1016/j.memsci.2019.117182

[10] Mogheir Y, Albahnasawi A. Use of nanofiltration for nitrate removal from Gaza Strip groundwater. *J Eng Res Technol*. 2014;1:32–39.

[11] Mason S, Ewert J, Ratsey H, et al. Flat sheet membrane bioreactors operational experiences – A New Zealand perspective, Canterbury, New Zealand [Internet]. 2010;18. Available from: http://scholar.google.com/scholar?hl=en&btnG=Search&q=intitle:FLAT+SHEET+MEMBRANE+BIOREACTOR+OPERATIONAL+EXPERIENCES+–+A+NEW+ZEALAND+PERSPECTIVE#0

[12] Liao Y, Bokhary A, Maleki E, et al. A review of membrane fouling and its control in algal-related membrane processes. *Bioresour Technol*. 2018;264:343–358.

[13] Ghouil B, Harabi A, Bouzerara F, et al. Development and characterization of tubular composite ceramic membranes using natural alumino-silicates for microfiltration applications. *Mater Charact* [Internet]. 2015;103:18–27. Available from: http://dx.doi.org/10.1016/j.matchar.2015.03.009

[14] Lim M, Patureau D, Heran M, et al. Removal of organic micropollutants in anaerobic membrane bioreactors in wastewater treatment: Critical review. *Environ Sci (Camb)*. 2020;6:1230–1243.

[15] El-Aswar EI, Ramadan H, Elkik H, et al. A comprehensive review on preparation, functionalization and recent applications of nanofiber membranes in wastewater treatment. *J Environ Manage*. 2022;301:113908.

[16] Li C, Sun W, Lu Z, et al. Ceramic nanocomposite membranes and membrane fouling: A review. *Water Res* [Internet]. 2020 [cited 2020 July 25];175:115674. Available from: www.sciencedirect.com/science/article/pii/S0043135420302104?casa_token=K5ZfK-LdCeAAAAAA:FeMLg6mtPa04XXFgUoQav79x6Pcg-jP7axSA5M7V7hwTq2KrsO6MZkA9wsvJwAkCP0E7wgSQlpU

[17] Van den Brink P, Vergeldt F, Van As H, et al. Potential of mechanical cleaning of membranes from a membrane bioreactor. *J Memb Sci* [Internet]. 2013;429:259–267. Available from: http://dx.doi.org/10.1016/j.memsci.2012.11.061

[18] Mogheir Y, Albahnasawi A, Abuhabib A. Optimal use of nanofiltration for nitrate removal from Gaza Strip municipal wells. *IUG J Nat Eng Stud*. 2014;22:45–60.

[19] Chang IS, Lee CH, Ahn KH. Membrane filtration characteristics in membrane-coupled activated sludge system: The effect of floc structure on membrane fouling. *Sep Sci Technol*. 1999;34:1743–1758.

[20] Wang Z, Chen Y, Sun X, et al. Mechanism of pore wetting in membrane distillation with alcohol vs. surfactant. *J Memb Sci*. 2018;559:183–195.

[21] Drexler ILC, Yeh DH. Membrane applications for microalgae cultivation and harvesting: A review. *Rev Environ Sci Biotechnol* [Internet]. 2014 [cited 2023 September 28];13:487–504. Available from: https://link.springer.com/article/10.1007/s11157-014-9350-6

[22] Chang H, Qu F, Liu B, et al. Hydraulic irreversibility of ultrafiltration membrane fouling by humic acid: Effects of membrane properties and backwash water composition. *J Memb Sci*. 2015;493:723–733.

[23] Iorhemen OT, Hamza RA, Tay JH. Membrane bioreactor (MBR) technology for wastewater treatment and reclamation: Membrane fouling. *Membranes* [Internet]. 2016 [cited 2023 September 28];6:33. Available from: www.mdpi.com/2077-0375/6/2/33/htm

[24] Sun J, Liang P, Yan X, et al. Reducing aeration energy consumption in a large-scale membrane bioreactor: Process simulation and engineering application. *Water Res*. 2016;93:205–213.

[25] Maere T, Verrecht B, Moerenhout S, et al. BSM-MBR: A benchmark simulation model to compare control and operational strategies for membrane bioreactors. *Water Res.* 2011;45:2181–2190.
[26] Chan YJ, Chong MF, Law CL, et al. A review on anaerobic–aerobic treatment of industrial and municipal wastewater. *Chem Eng J.* 2009;155:1–18.
[27] Maaz M, Yasin M, Aslam M, et al. Anaerobic membrane bioreactors for wastewater treatment: Novel configurations, fouling control and energy considerations. *Bioresour Technol.* 2019;283:358–372.
[28] Chen C, Guo W, Ngo HH, et al. Challenges in biogas production from anaerobic membrane bioreactors. *Renew Energy.* 2016;98:120–134.
[29] Shin C, Bae J. Current status of the pilot-scale anaerobic membrane bioreactor treatments of domestic wastewaters: A critical review. *Bioresour Technol.* 2018;247:1038–1046.
[30] Guo W, Cheng D, Ngo HH, et al. *Anaerobic Membrane Bioreactors for Antibiotic Wastewater Treatment. Current Developments in Biotechnology and Bioengineering: Advanced Membrane Separation Processes for Sustainable Water and Wastewater Management , Anaerobic Membrane Bioreactor Processes and Technologies.* Elsevier, 2020;219–239.
[31] Ashadullah AKM, Shafiquzzaman M, Haider H, et al. Wastewater treatment by microalgal membrane bioreactor: Evaluating the effect of organic loading rate and hydraulic residence time. *J Environ Manage.* 2021;278:111548.
[32] Yin X, Li X, Hua Z, et al. The growth process of the cake layer and membrane fouling alleviation mechanism in a MBR assisted with the self-generated electric field. *Water Res.* 2020;171:115452.
[33] Li H, Xing Y, Cao T, et al. Evaluation of the fouling potential of sludge in a membrane bioreactor integrated with microbial fuel cell. *Chemosphere.* 2021;262:128405.
[34] Liwarska-Bizukojc E, Bizukojc M. Digital image analysis to estimate the influence of sodium dodecyl sulphate on activated sludge flocs. *Process Biochem.* 2005;40:2067–2072.
[35] Deowan SA, Bouhadjar SI, Hoinkis J. Membrane bioreactors for water treatment. In *Advances in Membrane Technologies for Water Treatment: Materials, Processes and Applications.* Woodhead Publishing, 2015;155–184.
[36] Gujer W, Henze M, Mino T, et al. Activated sludge model No. 3. *Water Sci Technol.* 1999;39:183–193.
[37] Radjenović J, Matošić M, Mijatović I, et al. Membrane bioreactor (MBR) as an advanced wastewater treatment technology. In *Emerging Contaminants from Industrial and Municipal Waste*, edited by Barceló D, Petrovic M. Springer, 2008;37–101. Available from: https://link.springer.com/chapter/10.1007/978-3-540-79210-9_2
[38] Ognier S, Wisniewski C, Grasmick A. Membrane bioreactor fouling in sub-critical filtration conditions: A local critical flux concept. *J Memb Sci.* 2004;229:171–177.
[39] Yoon SH, Kim HS, Chung YC. Effect of acidity consumption/production on the pH of aeration tank during the biodegradation of acetic acid/epichlorohydrin. *Water Res.* 2002;36:2695–2702.
[40] Macomber J, Cicek N, Suidan MT, et al. Biological kinetic data evaluation of an activated sludge system coupled with an ultrafiltration membrane. *J Environ Eng* [Internet]. 2005 [cited 2023 September 28];131:579–586. Available from: https://ascelibrary.org/doi/abs/10.1061/%28ASCE%290733-9372%282005%29131%3A4%28579%29
[41] Wei Y, Van Houten RT, Borger AR, et al. Minimization of excess sludge production for biological wastewater treatment. *Water Res.* 2003;37:4453–4467.

[42] Cicek N, Macomber J, Davel J, et al. Effect of solids retention time on the performance and biological characteristics of a membrane bioreactor. *Water Sci Technol.* 2001;43:43–50.
[43] Iorhemen OT, Hamza RA, Tay JH. Membrane bioreactor (MBR) technology for wastewater treatment and reclamation: Membrane fouling. *Membranes (Basel).* 2016;6:13–16.
[44] Shin C, Tilmans SH, Chen F, et al. Temperate climate energy-positive anaerobic secondary treatment of domestic wastewater at pilot-scale. *Water Res.* 2021;204:117598.
[45] Al-Mardeai S, Elnajjar E, Hashaikeh R, et al. Membrane bioreactors: A promising approach to enhanced enzymatic hydrolysis of cellulose. *Catalysts* [Internet]. 2022 [cited 2023 Sep 29];12:1121. Available from: www.mdpi.com/2073-4344/12/10/1121/htm
[46] Peralta-Hernández JM, Brillas E. A critical review over the removal of paracetamol (acetaminophen) from synthetic waters and real wastewaters by direct, hybrid catalytic, and sequential ozonation processes. *Chemosphere.* 2023;313:137411.
[47] Khan NA, Khan AH, Ahmed S, et al. Efficient removal of ibuprofen and ofloxacin pharmaceuticals using biofilm reactors for hospital wastewater treatment. *Chemosphere.* 2022;298:134243.
[48] Jiang J, Zhou Z, Huang J, et al. Repurposing hydrolysis acidification tank in municipal wastewater treatment plants for sludge reduction and biological nutrient removal. *Chem Eng J.* 2020;396:125327.
[49] Xiao X, Guo H, Ma F, et al. New insights into mycelial pellets for aerobic sludge granulation in membrane bioreactor: Bio-functional interactions among metazoans, microbial communities and protein expression. *Water Res.* 2023;228:119361.
[50] Cai Y, Ben T, Zaidi AA, et al. Nitrogen removal augmentation of ship sewage by an innovative aerobic–anaerobic micro-sludge MBR technology. *Process Biochem.* 2019;82:123–134.
[51] Moya-Llamas MJ, Pacazocchi MG, Trapote A. Respirometric tests in a combined UASB-MBR system treating wastewater containing emerging contaminants at different OLRs and temperatures: Biokinetic analysis. *J Environ Manage.* 2023;345:118643.
[52] Feng Z, Li T, Lin Y, et al. Microbial communities and interactions in full-scale A^2/O and MBR wastewater treatment plants. *J Water Process Eng.* 2022;46:102660.
[53] Campo R, Capodici M, Corsino F, et al. Membrane bioreactors sludge: From production to disposal. In *Current Developments in Biotechnology and Bioengineering: Advanced Membrane Separation Processes for Sustainable Water and Wastewater Management – Case Studies and Sustainability Analysis,* edited by Mannina G, Pandey A, Larroche C, Ng H, Ngo H. Elsevier, 2020;323–351.
[54] Sun L, Ma J, Li L, et al. Exploring the essential factors of performance improvement in sludge membrane bioreactor technology coupled with symbiotic algae. *Water Res.* 2020;181:115843.
[55] Vo TDH, Bui XT, Dang BT, et al. Influence of organic loading rates on treatment performance of membrane bioreactor treating tannery wastewater. *Environ Technol Innov.* 2021;24:101810.
[56] Qin B, Zhang Y, Zhu G, et al. Eutrophication control of large shallow lakes in China. *Sci Total Environ.* 2023;881:163494.
[57] Kornijów R. Eutrophication and derivative concepts. Origins, compatibility and unresolved issues. *Ecohydrol Hydrobiol.* 2024;24:289–298.
[58] Bozal-Leorri A, Corrochano-Monsalve M, Arregui LM, et al. Evaluation of a crop rotation with biological inhibition potential to avoid N_2O emissions in comparison with synthetic nitrification inhibition. *J Environ Sci.* 2023;127:222–233.

[59] Wang X, Bai J, Wang C, et al. Two newly-identified biological nitrification inhibitors in *Suaeda salsa*: Synthetic pathways and influencing mechanisms. *Chem Eng J.* 2023;454:140172.
[60] Yue X, Liu H, Wei H, et al. Reactive and microbial inhibitory mechanisms depicting the panoramic view of pH stress effect on common biological nitrification. *Water Res.* 2023;231:119660.
[61] Lagum AA, Elektorowicz M. Modification of nitrifying microbial community via DC electrical field application. *J Environ Chem Eng.* 2022;10:107743.
[62] Wang J, Tian Y, Wei J, et al. Impacts of dibutyl phthalate on biological municipal wastewater treatment in a pilot-scale A^2/O-MBR system. *Chemosphere.* 2022;308:136559.
[63] Corsino SF, de Oliveira TS, Di Trapani D, et al. Simultaneous sludge minimization, biological phosphorous removal and membrane fouling mitigation in a novel plant layout for MBR. *J Environ Manage.* 2020;259:109826.
[64] Li S, Guo Y, Zhang X, et al. Advanced nitrogen and phosphorus removal by the symbiosis of PAOs, DPAOs and DGAOs in a pilot-scale A^2O/A+MBR process with a low C/N ratio of influent. *Water Res.* 2023;229:119459.
[65] Lv N, Li X. Phosphorus removal from wastewater using Ca-modified attapulgite: Fixed-bed column performance and breakthrough curves analysis. *J Environ Manage.* 2023;328:116905.
[66] Hei S, Xu H, Liu Y, et al. Redox environment inducing strategy for enhancing biological phosphorus removal in a full-scale municipal wastewater treatment plant. *J Clean Prod.* 2022;376:134237.
[67] Sarvajith M, Nancharaiah Y V. De novo granulation of sewage-borne microorganisms: A proof of concept on cultivating aerobic granular sludge without activated sludge and effective enhanced biological phosphorus removal. *Environ Res.* 2023;224:115500.
[68] Photiou P, Poulizou M, Vyrides I. Recovery of phosphates from anaerobic MBR effluent using columns of eggshell and seagrass residues and their final use as a fertilizer. *Sustain Chem Pharm.* 2023;33:101039.
[69] Dai H, Sun Y, Wan D, et al. Simultaneous denitrification and phosphorus removal: A review on the functional strains and activated sludge processes. *Sci Total Environ.* 2022;835:155409.
[70] Faridizad G, Abdollahzadeh Sharghi E, Bonakdarpour B. The use of membrane bioreactors in high rate activated sludge processes: How and why sludge retention time affects membrane fouling. *J Water Process Eng.* 2022;47:102807.
[71] Do KU, Chu XQ. Performances of membrane bioreactor technology for treating domestic wastewater operated at different sludge retention time. In *Development in Wastewater Treatment Research and Processes: Removal of Emerging Contaminants from Wastewater through Bio-Nanotechnology*, edited by Shah M, Rodriguez-Couto S, Biswas J. Elsevier, 2022;107–122.
[72] Zhang Y, Hu W, Lin L, et al. Enhanced treatment of sludge drying condensate by A/O-MBR process: Microbial activity and community structure. *Chemosphere.* 2023;340:139911.
[73] He C, Wang K, Wang W, et al. Chemically enhanced high-loaded membrane bioreactor (CE-HLMBR) for A-stage municipal wastewater treatment: Pilot-scale experiments and practical feasibility evaluation. *Sep Purif Technol.* 2023;307:122853.
[74] Radmehr S, Rissanen T, Kallioinen-Mänttäri M, et al. Reducing mechanical aeration in membrane bioreactors by inoculation of algal cells into activated sludge biomass. *J Water Process Eng.* 2022;49:103047.

[75] Huang L, Li Z, Wang G, et al. A^2O–MBR–BAF–O_3 process for treating high organic wastewater with high ammonia nitrogen. *Biochem Eng J*. 2022;186:108574.
[76] Gharibian S, Hazrati H. Towards practical integration of MBR with electrochemical AOP: Improved biodegradability of real pharmaceutical wastewater and fouling mitigation. *Water Res*. 2022;218:118478.
[77] Tang K, Xie J, Pan Y, et al. The optimization and regulation of energy consumption for MBR process: A critical review. *J Environ Chem Eng*. 2022;10:108406.
[78] Najafi Chaleshtori S, Shamskilani M, Babaei A, et al. Municipal wastewater treatment and fouling in microalgal-activated sludge membrane bioreactor: Cultivation in raw and treated wastewater. *J Water Process Eng*. 2022;49:103069.
[79] Abdullah R, Astira D, Widyanto AR, et al. Recent development of mixed matrix membrane as a membrane bioreactor for wastewater treatment: A review. *Case Stud Chem Environ Eng*. 2023;8:100485.
[80] Nguyen PT, Phuc Hanh Tran D, Le LT, et al. Characterization of reciprocation membrane bioreactor on treatment performance, energy consumption and membrane fouling. *Bioresour Technol*. 2023;381:129146.
[81] Hou H, Mengting Z, Duan L, et al. Removal performance and biodegradation mechanism of sulfonamides antibiotic contained wastewater by IFAS-MBR bioreactor. *J Mol Liq*. 2022;367:120572.
[82] Hang YD. Determination of oxygen demand. In *Food Analysis*, edited by Nielsen S.S. Springer, 2017;503–507. https://doi.org/10.1007/978-3-319-45776-5_28
[83] Mannina G, Ni BJ, Makinia J, et al. Biological processes modelling for MBR systems: A review of the state-of-the-art focusing on SMP and EPS. *Water Res*. 2023;242:120275.

12 Innovative Technologies for the Treatment of Industrial Wastewater

Larissa Capeletti Romani, Mayana Cardoso Raimundi, Vitoria Dassoler Longo, Aline Frumi Camargo, and Helen Treichel

12.1 INTRODUCTION

Industrial effluents, liquid waste from industrial production processes, represent a significant environmental challenge when released into the environment without proper treatment. These effluents contain a range of pollutants, including toxic substances that can wreak havoc on aquatic ecosystems, disrupt the food chain, and reduce biodiversity. Furthermore, they can contaminate vital drinking water sources, risking public health.

In this chapter, we embark on a journey through the universe of innovative technologies for treating industrial effluents, primarily focusing on bioprocesses. These cutting-edge approaches have emerged as effective and sustainable solutions to address the pressing issue of industrial pollution. By harnessing the power of biology, bioprocesses offer a means of removing pollutants from wastewater and recovering valuable resources from these waste streams.

Bioremediation, which uses microorganisms to decompose and transform organic compounds in effluents into less harmful substances, is at the heart of our research. These microbial agents, which may occur naturally or be intentionally introduced into treatment systems, play a vital role in remediation. The field of stabilization ponds, economical treatment systems that depend on a consortium of aerobic and anaerobic microorganisms, is also deepened. These tiny but diligent workers work harmoniously to break down wastewater organic matter and nutrients through biological processes.

Activated sludge reactors, another intriguing facet of our discussion, harness the power of aerobic microorganisms to degrade organic matter in industrial effluents. Furthermore, activated sludge, composed of a dynamic mix of microorganisms and biomass, thrives in aerated environments that promote robust biological activity.

In this sense, this chapter aims to highlight the importance of investing in innovative bioprocesses. In doing so, it is possible to mitigate the adverse impacts of

DOI: 10.1201/9781003454199-12

industrial pollution and ensure the health of aquatic ecosystems and the availability of high-quality water for future generations.

12.2 BIOREMEDIATION

In the current context, the use of bioremediation to treat industrial effluents is part of a time of search for sustainable and ecologically responsible approaches. The history of this practice dates back to the end of the twentieth century when awareness about the impacts of industrial pollution on ecosystems gained prominence. Initially, the focus was on applying microorganisms capable of selectively degrading polluting compounds, aiming to reduce toxicity levels and the permanence of contaminants present in effluents (Figure 12.1) (Saravanan et al., 2023).

However, the evolution of molecular biology techniques and understanding of metabolic processes has led to improving bioremediation strategies. Furthermore, the use of customized bioreactors and the integration of biotechnological methods have further increased the effectiveness of this process. Currently, the approach has gained relevance as a valuable tool in treating industrial effluents by promoting the effective and sustainable degradation of pollutants, contributing to environmental preservation, and mitigating impacts associated with productive activities (Choudhary et al., 2022).

In the industrial context, lead release is estimated to be responsible for approximately 10 percent of environmental heavy metal contamination. When it reacts with

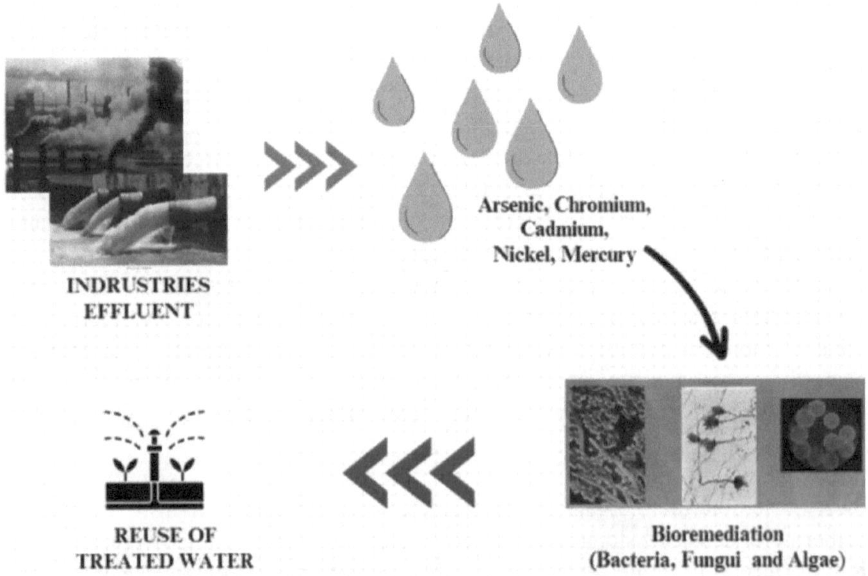

FIGURE 12.1 Diagram of the bioremediation steps carried out using different microorganisms (compiled by the authors).

the air, it forms several harmful chemical compounds. Therefore, reducing this type of contamination, especially in water, is essential to mitigate the formation of these toxic products (Tufail et al., 2022). Thus, biosorption, a microbial treatment procedure, has stood out concerning traditional approaches for removing metals in aqueous solutions. While conventional methods such as chemical oxidation/reduction, precipitation, and ion exchange have disadvantages such as inefficiency, high prices, and waste generation, biosorption stands out with high metal binding efficiency, selectivity, and absence of chemical waste (Singh et al., 2023).

According to recent research, the ability of bacteria to absorb heavy metals from contaminated environments is emerging as a promising bioremediation strategy. In addition, the resistance and ability to remove lead have been proven and demonstrated, through different tests, by bacterial strains such as *B. cereus* and *Bacillus* sp., for example, revealing their promising application in the bioremediation of polluted environments (Idrees et al., 2023).

However, the scientists' recommendation for future studies to further explore the mechanisms underlying the absorption and homeostasis of heavy metals by these strains, including lead, is notable, with the aim of further improving the effectiveness of this remediation method.

Increasing research on bioremediation is being undertaken to develop forward osmosis (FO) processes as a cost-effective alternative to energy-intensive water vacuum suction equipment in submerged membrane bioreactor applications (MBR). Perceived as a spontaneous process driven by water without external pressures applied, FO has been progressively implemented on laboratory scales (Lee et al., 2019). However, the prospect of realizing the use of FO processes in overhauling currently existing energy-intensive osmotic MBR processes is discussed, with research needs highlighted. Likewise, the extent of FO water recovery and the selection of extraction solutes can influence the composition of preconcentrated wastewater regarding organic matter retention and salinity accumulation. The incompatibility between high salinity and anaerobic microorganisms represents the most prominent challenge of integrating FO with anaerobic treatment.

Furthermore, water recovery also determines the accumulation of dissolved solutes in wastewater (Ansari et al., 2018). To overcome this challenge, it is recommended, according to research in the area, that the sludge retention time be reduced and the selectivity of FO membranes improved to optimize extraction solutions better. Therefore, considering all technical and economic restrictions, the need to explore the use of FO more comprehensively and work with other water/wastewater treatment processes is highlighted. Still, in manufacturing, it is estimated that more than 10 million tons of phenolic compounds are abandoned into the environment by the petrochemical, pharmaceutical, leather, textile, and agrochemical industries. Furthermore, it is believed that industrial methodologies for producing paints, paper, cellulose, and pesticides are also primarily responsible for the discharge of phenolic compounds into the environment (Blandin et al., 2018).

Various physicochemical technologies have been employed in treating industrial wastewater, including distillation, nanofiltration, reverse osmosis, chemical, electrochemical and catalytic oxidation, solvent extraction, ozonization, and adsorption. Although these techniques have many advantages, they have disadvantages,

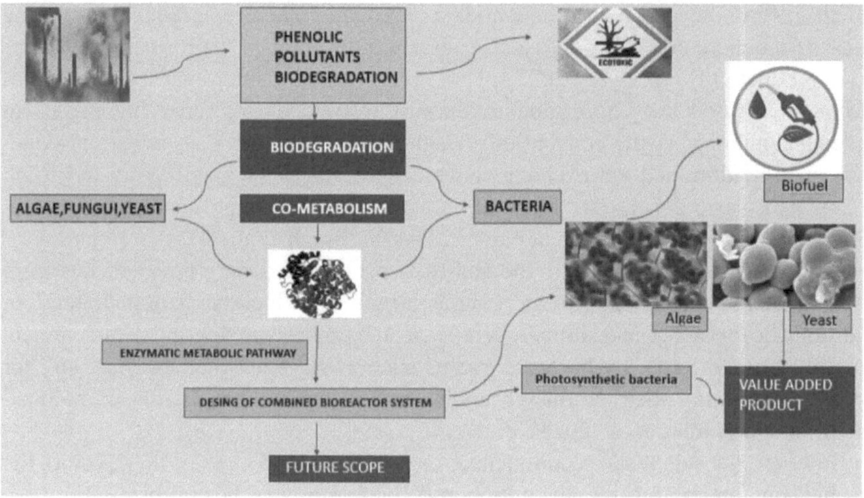

FIGURE 12.2 Diagram of biodegradative processes and their corresponding products using mixed populations (compiled by the authors).

such as high operational and maintenance costs, pretreatment requirements, fouling, and pH level restrictions. However, in recent years, there has been a shift towards using environmentally compatible technologies for wastewater treatment to reduce operating costs and their environmental impact. According to scientific literature, species belonging to the genera *Pseudomonas* and *Bacillus* have been capable of degrading phenol through the meta-cleavage pathway since these organisms have a wide range of catabolic enzymes involved in the phenol degradation process (Figure 12.2) (Panigrahy et al., 2022). Furthermore, the use of microbial consortia in the remediation of phenolic compounds is promising, as mixed populations increase tolerance to toxic pollutants and degradation efficiency due to cooperation between microorganisms that secrete metabolites and enzymes.

Currently, studies also show the importance of moving carbon-based bed biofilm reactors (MBBRs) in that they provide a protected surface where different species of microorganisms can coalesce. Combining activated carbon with activated sludge offers many benefits, such as adsorption of pollutants and superior shock resistance (Bibi et al.,2023). According to a study, two MBBRs were operated using different lignite-activated coke (LAC) and activated carbon (AC) carriers to estimate phenol removal. In this system, phenol was used as the primary carbon source in the first three stages of the treatment, with an initial attenuation in degradation noted; however, the performance stabilized probably due to the formation of biofilm and the adsorption capabilities of the carriers used reaching up to 96 percent phenol removal (Bibi et al., 2023). It was also witnessed that LAC-based MBBR had more efficient performance when phenol levels were increased; however, both reactors had a similar tolerance limit for phenols (around 450 mg/L). LAC-based MBBR also demonstrated improved shock loading at higher ammonia levels, leading to more effective extraction in phenols (88.68% versus 94.61%) when compared to AC-based MBBR. Greater impact tolerance was built due to LAC's resilient adsorption capabilities.

Furthermore, the slime flakes improved compactness, stability, size, and size distribution, enhancing resistance against high-concentration shocks and toxicity. In LAC-based MBBR, phenol degradation was dominated by practical cooperation between microorganisms. The facultative anaerobes *Cloacibacterium* and *Hydrogenophaga* contributed to the cleavage of the phenol ring and increased denitrification. The predatory bacterium *Bdellovibrio* vigorously participated in nitrogen fixation and biomass conversion by converting complex biopolymers into extracellular substances, leading to more compacted flocs. As for the anaerobes in the biofilm, *Tissierella* showed tolerance to higher ammonia levels and stimulated methanogens. For Archaea, the *Thaumarchaeota* proportion of LAC was twice that of AC, specifically for *Nitrososphaera*, which are very effective nitrifiers. Due to the adaptability of Comamonas belonging to *Burkholderiales*, these species were considered extremely important as they were capable of phenol biodegradation, nitrification, and denitrification.

Furthermore, they demonstrated interspecific cooperation with other bacteria. These species also produced biopolymers and created flakes to protect against toxicants and predators (Zheng et al., 2019). For Tran et al. (2013), it can be concluded from these studies that co-metabolism is extremely important for the complete degradation of phenolic compounds and achieving higher toxicity tolerances. Electrobioremediation also offers a promising treatment to eliminate persistent pollutants from groundwater, as it stimulates biological dechlorination activity, uses renewable electricity to carry out the process, and avoids the presence of chemicals (Zeppili et al., 2024). In recent studies, a microbial electrolysis cell configuration consisting of two chambers performed reductive and oxidative tetrachloroethane (TeCA) degradation. By polarizing the cathode chamber with graphite beads at a potential of -650 mV relative to the standard hydrogen electrode and using a metal oxide (MMO) mixture to generate oxygen, the necessary reductive and oxidative environment conditions were established for the effective removal of TeCA (Lai et al.,2021). Continuous experiments were conducted using two different feed solutions: a mineral solution optimized for dechlorination by microorganisms and synthetic groundwater containing sulfate and nitrate anions to investigate possible secondary reactions. The bioelectrochemical process has effectively reduced TeCA in the cathode chamber, producing a mixture containing trans-dichloroethylene, vinyl chloride, and ethylene. These products were subsequently subjected to oxidation in the anode chamber, obtaining removal efficiencies of $37 \pm 2\%$, $100 \pm 4\%$, and $100 \pm 5\%$, respectively. The addition of synthetic groundwater containing nitrate and sulfate caused reductions of these ions in the cathode chamber, resulting in a 17 percent decrease in the rate of reductive dechlorination.

Furthermore, additional chlorinated byproducts were observed in the cathode chamber effluent, including cis-dichloroethylene and 1,2-dichloroethane (1,2-DCA) (Lai et al., 2015). It is essential to highlight that, despite reducing the reductive dechlorination rate during operation with synthetic groundwater, the aerobic dechlorinating microorganisms in the anode chamber eliminated vinyl chloride and

1,2-DCA. This study represents the first evidence of a sequential bioelectrochemical process that combines oxidation and reduction for TeCA mineralization, simulating contaminated groundwater conditions (Zeppili et al., 2024).

12.3 STABILIZATION PONDS

With the increasing demand for aquaculture products, water reuse will likely gain greater adoption in global aquaculture activities. In this context, wastewater stabilization ponds (WSP) represent sustainable and cost-effective approaches to effectively treat nutrients and several pollutants of emerging concern present in effluents. The present study analyzed the accumulation of selected medicines from different therapeutic categories by two main fish species in aquaculture, *Cyprinus carpio* and *Sander lucioperca*, with varying food preferences. Their everyday prey items were also examined to investigate the occurrence of specific compound accumulation. Of the 66 chosen pharmaceuticals and their metabolites, 49 were positively identified in water and sediment samples collected from the representative WSP. During a six-month study that simulated typical aquaculture operations, 14 pharmaceuticals and their metabolites were identified (in more than 50% of samples) in at least one fish tissue obtained from WSP. Notable differences in species-specific bioaccumulation factors (BAFs) and also between different organs were observed among the evaluated compounds. Although muscle tissues consistently accumulated lower levels of the target compounds, several substances showed elevated concentrations in both fish species' brains, liver, and kidney tissues, exemplified by sertraline (Grabicova et al., 2020).

Globally, pollution is a serious health threat (Landrigan et al., 2017). Chemical contamination of aquatic ecosystems by emerging contaminants, such as perfluoroalkyl substances, human and veterinary pharmaceuticals, personal care products, new pesticides, metabolites, and transformation products, has aroused increasing interest due to frequent detection and worldwide dissemination (aus der Beek et al., 2016; Fairbairn et al., 2016; Roberts et al., 2016). One prominent emerging contaminant category is human pharmaceuticals (Fekadu et al., 2019; Klatte et al., 2017; Kummerer, 2010; Vazquez-Roig et al., 2012; Verlicchi et al., 2012). Residues of pharmaceutical products and their metabolites, which are regularly found in relatively low concentrations (ng/L to μg/L) in aquatic environments in developed countries, can cause adverse effects on non-target organisms, including changes in behavior and bioenergetic parameters (Saaristo et al., 2018; Sehonova et al., 2018). Figure 12.3 shows a flowchart of variables that contribute to wastewater composition.

Many pharmaceuticals are hydrophobic and can therefore accumulate in aquatic organisms (Du et al., 2016; Steinbach et al., 2016). The magnitude of this accumulation can vary between trophic levels, with organisms at lower levels accumulating more than those at higher levels (Zenker et al., 2014). Incomplete removal of emerging contaminants, including pharmaceuticals, during wastewater treatment processes results in their discharge into water bodies (Yang et al., 2017). Wastewater stabilization ponds (WSP) emerge as a low-cost option to reduce nutrients and micropollutants in treated effluents, with their removal efficiency varying depending on the design and environmental conditions (Figure 12.4) (Gruchlik et al., 2018). In

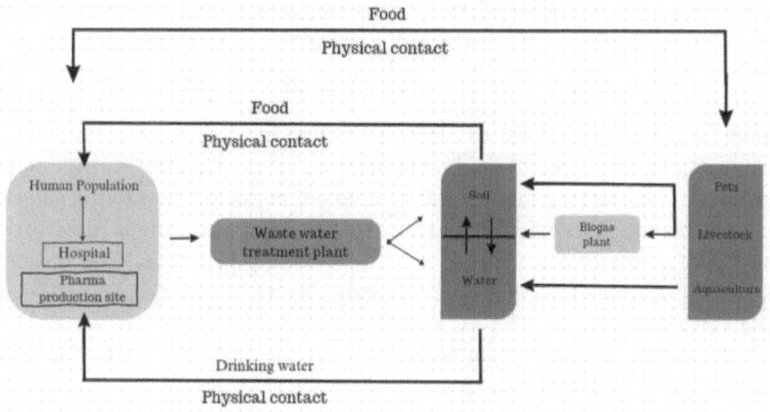

FIGURE 12.3 Flowchart of variables that contribute to wastewater composition (compiled by the authors).

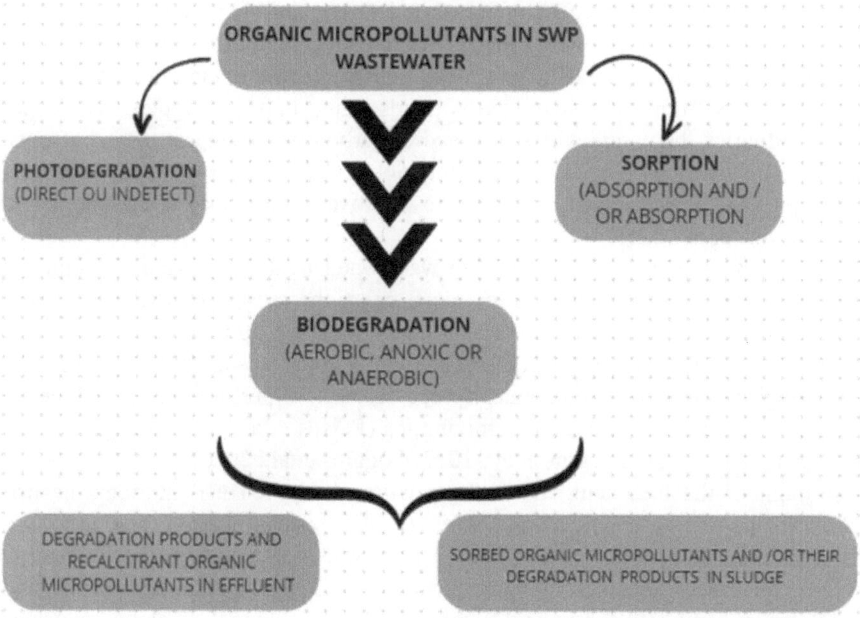

FIGURE 12.4 Flowchart of the biodegradation stages of organic micropollutants in wastewater stabilization ponds (compiled by the authors).

water reuse, using WSP for aquaculture is common, but the relationship between this practice and the bioaccumulation of emerging contaminants remains poorly understood (Brooks and Conkle, 2019). This study focused on exposure and bioaccumulation of pharmaceuticals in two species of fish raised in WSP over six months (Koba et al., 2018).

Obtaining accurate estimates of gaseous emissions and carbon capture in wastewater treatment is crucial to guide treatment facilities' planning, operation, and design, especially considering greenhouse gas reduction goals. This study examined the relationship between biological activity, water dynamics, and carbon-based greenhouse gas (GHG) release involving diffusion and boiling processes. The objective was to provide guidelines for more accurate assessments of emissions in waste stabilization ponds (WSPs).

The lakes showed stratification during the day and mixing during the night. Buoyancy flow contributed significantly (from 40% to 75%) to turbulence in the water column during nocturnal cooling events, resulting in considerable increases in surface CO_2 and CH_4 concentrations. The onset of stratification and the proliferation of phytoplankton at the surface, associated with an elevated pH and fluctuations in CO_2 partial pressure, led to an overall decrease in CO_2 emission.

Boiling was responsible for a substantial portion (from 40% to 99%) of total CH_4 emissions and up to 95 percent of integrated GHG emissions during wastewater treatment (in terms of CO_2 equivalent). Hydrodynamic conditions, diurnal variation, and boiling effect are crucial factors to be considered for reliable assessments of GHG emissions from waste stabilization ponds. This study contributes to a deeper understanding of the complexities involved in carbon processing at these hotspots. The role of waste stabilization ponds in the global GHG budget will likely increase with population growth unless measures to improve performance are implemented (Bartosiewicz et al., 2021).

Waste stabilization ponds (WSPs) are wastewater treatment systems recognized for their simplicity, efficiency, and low cost, being the most commonly used technology in the world. In Australia, these systems dominate 60 percent of wastewater treatment plants (Coggins et al., 2019). The treatment process in WSPs involves natural biological, physical, and chemical processes, whose performance is influenced by water dynamics and microbial communities (Passos et al., 2016; Coggins et al., 2017; Dias et al., 2017; Coggins et al., 2017; Coggins et al., 2020).

Throughout the treatment, the transformation of carbon and nutrients occurs, leading to the release of greenhouse gases (GHG), such as carbon dioxide (CO_2), methane (CH_4), and nitrous oxide (N_2O). These systems can contribute significantly to GHG emissions due to their high organic matter load, dynamic mixing regime, and shallow water depth (Holgerson and Raymond, 2016). Despite their global relevance in wastewater treatment, few studies have directly investigated GHG emissions from WSPs, considering the controls involved (Silva et al., 2012; Glaz et al., 2016; Pan et al., 2016). CH_4, a potent GHG, is generated during the microbial decomposition of organic matter, with its evasion into the atmosphere occurring through diffusion and boiling. These processes vary depending on local conditions, such as oxygen levels and temperature. Water stratification and mixing affect the distribution of CH_4 in the water column, resulting in temporal and spatial variations in emissions. Furthermore, turbulence generated by wind and buoyancy flow influences GHG evasion rates.

Accuracy in measuring GHG emissions is crucial and depends on the sampling method and gas exchange model choice. Both modeling and floating chamber approaches are used to estimate these emissions, with the modeling providing high-resolution estimates. However, validation of the models is still necessary. Accurately quantifying GHG fluxes requires consideration of the physical and biological factors involved. This study investigated the transformation of sedimentary phosphorus in a wastewater stabilization pond, focusing on adsorption and release under different pH and redox potential (Eh) conditions. The binding capacity of different phosphorus fractions varied under different pHs: iron-phosphorus (Fe-P) and aluminum-phosphorus (Al-P) showed higher critical power at pH 7–8.

In comparison, organic phosphorus (OP alk) and calcium-phosphorus (Ca-P) had greater capacity at pH 5 and 9, respectively. Redox potential played a significant role, where the change from Fe(II) to Fe(III) with increasing Eh affected the adsorption and release of sedimentary phosphorus. No specific threshold in Eh was observed for phosphorus adsorption, indicating a continuous process (Peng et al., 2007). In this study, the authors evaluated the efficiency of a waste stabilization pond (WSP) system in reducing contaminants in hospital wastewater in a rural area of South Africa. WSPs are widely used due to their simplicity and low cost. We carried out monthly samplings for six months and monitored physicochemical and microbiological parameters. In some months, microbial parameters in the effluent, such as Escherichia coli and enterococci, exceeded influent levels.

Furthermore, pathogen removal efficiency was low (<1 log reduction). The effluent's chemical oxygen demand (COD) levels exceeded discharge guidelines, although some reductions were observed in certain months. The WSP system showed no trend toward contaminant reduction except for zinc. It was found that the WSP Siloé system was not working correctly, releasing poor-quality effluents into the river. The WSP system was recommended to be expanded to handle the additional wastewater load and sludge removal under best practices. Furthermore, we highlight the importance of continuous system monitoring to ensure compliance with regulatory guidelines (Edokpayi et al., 2021).

The global scarcity of freshwater, worsened in arid and semi-arid regions, results from population growth, water contamination, and climate variability (Edokpayi et al., 2020; Ahmed et al., 2020). Contaminated water for multiple purposes is ordinary in rural areas of developing countries due to a lack of sustainable access to clean water (Olasoji et al., 2019; Edokpayi et al., 2018). The continued exploitation of surface water for domestic use is essential, as 785 million people lack access to basic drinking water (Bessong et al., 2009). The microbiological and physicochemical quality of surface waters is of concern due to contamination resulting from human activities (Bessong et al., 2009).

Untreated wastewater is a significant source of surface water pollution (Edokpayi et al., 2020). Wastewater management is challenging in developing countries due to urbanization, industrialization, and lack of knowledge (Edokpayi et al., 2018). Lack of wastewater treatment infrastructure expansion often leads to inadequate Treatment, causing environmental damage and health problems (Olukanni and Ducost, 2011). Pollutants such as nitrates and phosphates from inadequately treated wastewater can cause eutrophication and health problems (Naidoo and Olaniran, 2014).

Wastewater management is crucial to prevent environmental degradation and health risks, but the high cost of treatment plants is challenging for developing countries. Wastewater stabilization ponds (WSPs) are low-cost, simple treatment alternatives to operate and maintain. Therefore, well-managed WSPs can produce quality effluents. However, many WSP systems do not meet discharge standards, highlighting the importance of proper management (Edokpayi et al., 2017).

This study evaluated the efficiency of the WSP Siloé system in removing chemical and microbiological contaminants from hospital wastewater in South Africa. It was concluded that many WSP systems do not meet discharge standards due to poor management and degraded infrastructure (Edokpayi et al., 2017). Removing chemical pollutants is vital due to their persistence and impact on human health and aquatic life (Edokpayi et al., 2017).

12.4 ACTIVATED SLUDGE

Activated sludge (AS) represents most of the waste generated in wastewater processing and produced in wastewater treatment plants. The wastewater and sludge treatment concept is limited to the treatment and available resources. Furthermore, activated sludge and wastewater can become valuable raw materials for producing sustainable products, boosting the circular economy. The activated sludge process is equivalent to the highest cost in effluent treatment (Chen et al., 2023), based on forced aeration to stimulate the dissolution and dissemination of oxygen, enabling microbial growth.

Sedimentation of sludge under the effect of gravity is one of the most used methods in the effluent treatment process. Still, the reflux of sludge from the secondary tank to the primary reaction zone means that microorganisms always remain in an alternating environment, resulting in a limited enrichment of these microbes (Chen et al., 2023). The separation and control of microorganisms in activated sludge can significantly impact the energy efficiency of the Treatment, considering gravity, the internal environment, and particle size. Economic and energy benefits can be achieved through functional distribution and different bioreactors (Chen et al., 2023).

To develop technologies that are considered promising, persulfate has been widely researched for waste-activated sludge (WAS). Persulfate oxidation appears more advantageous for alkaline fermentation than the Fenton method, a consolidated technology with disadvantages such as treatment costs and difficulties operating hydrogen peroxide (Xiao, 2023). The application of persulfate is not yet compact due to the lack of certainty resulting from the irregular results when applied in wastewater treatment. Still, its oxidation is considered adequate for the degradation of pollutants in WAS (Wang et al., 2016)

In 2019, the largest sludge producers were the European Union and China, around 55 and 39 million tons, respectively (Ma et al., 2021; Wei et al., 2020). Much of this sludge is incinerated or disposed of in landfills. In 2020, China produced more than 60 Mt/a of sewage sludge, and it is estimated that in 2025, this will increase to 90 Mt/

a, resulting in high treatment costs, as the sludge contains polluting and dangerous compounds (Quan et al., 2022).

The ecological method of transforming sludge into biochar through pyrolysis emerged in the last decade and has been used in agriculture and energy. WAS is considered a renewable energy source as it has a notable amount of lipids, carbohydrates, and proteins (Pezoti et al. 2016). Another ecological approach is to use sludge to replace nutrients in the soil (Kor-Bicakci et al., 2019); the soil increases cation exchange capacity, soil quality, humic acid, and water retention capacity (Awasthi et al., 2022).

To combine the features of suspended sludge and attached biofilm, the integrated fixed film activated sludge system (IFAS) realizes effective organic carbon removal and enhanced biological nitrogen removal (Machat et al., 2019). It has been recognized as a flexible solution for wastewater treatment and has emerged as a cutting-edge treatment technology. IFAS also expands treatment ability and decreases the dangers of slow-growing microorganism decline (Waqas et al., 2020). It was recently recognized as an adaptable and compact resource for treating municipal and industrial wastewater and providing excellent nitrogen removal due to its nitrifiers that promote system compaction (Bai et al., 2015).

Sulfonamides (SAs) have been widely detected in drinking water sources and deserve further studies on their removal in water treatment processes. Sulfadiazine (SDZ), sulfamethoxazole (SMX), and sulfathiazole (STZ) were investigated using fixed film-activated sludge membrane bioreactors (IFAS-MBRs). Membrane bioreactor (MBR) is an advanced treatment technology widely used in wastewater treatment. However, membrane fouling mitigation remains challenging (Hou et al., 2022).

Upflow anaerobic sludge bed (UASB), as one of the well-developed anaerobic digestion techniques, can anaerobically treat a variety of industrial wastewater due to ease of operation, sludge granulation, and high resistance to shock load, being universally used (Muralikrishna and Manickam, 2017).

The UASB anaerobic reactor can promote excellent quality treated effluent for reuse when used as the first stage of partial degradation of MO and ECs and supplemented with MBR. The recirculation between the two reactors and the high sludge retention times -SRT led to a significant increase in the bacterial diversity developed in the reactors (Kim et al., 2012).

Aerobic granular sludge (AGS) has attracted increasing attention in the last two decades. This technology offers several advantages over conventional biological systems, such as high biomass concentration, long sludge retention time, small reactor area, and low rate of sludge discharge relative to traditional biological systems (Figure 12.5) (Cai et al., 2021; Hang et al., 2021; Sguanci et al., 2019). Furthermore, aerobic granular sludge is more compact and enormous, compresses strongly, and forms a cake on the surface (Wang et al., 2023).

AGS deals with granular-activated sludge formed by the self-condensation of microorganisms under aerobic conditions). Spontaneously the organisms form and accumulate in dense structures (Nancharaiah and Sarvajith, 2019; Wang et al., 2018). Aerobic granular sludge (AGS) technology has been applied to some types of real industrial effluents, such as wineries (Ghosh and Chakraborty, 2019), dyes

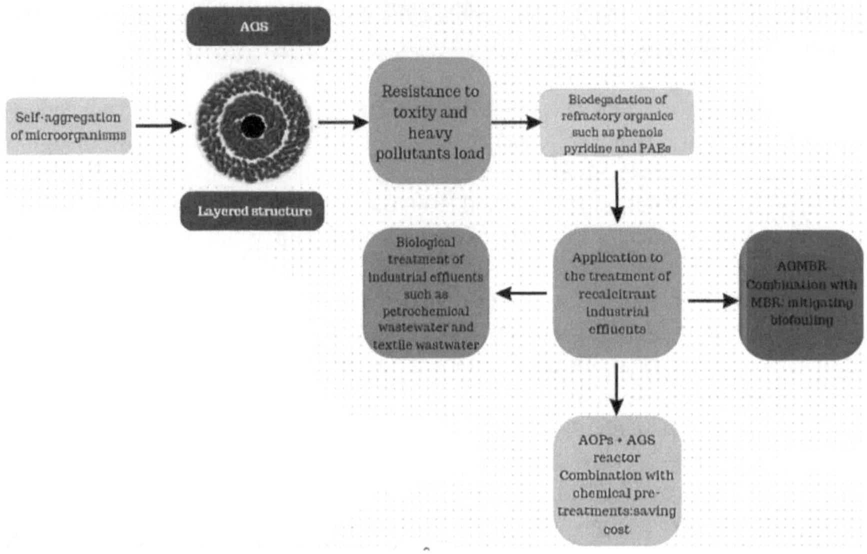

FIGURE 12.5 Representative diagram of AGS applications in the recalcitration of industrial effluents (compiled by the authors).

(Liu et al., 2020), slaughterhouses (Aziz et al., 2022) and pharmacists (Jiang et al., 2021). AGS has been considered a promising process for treating industrial effluents when layered with distributed microorganisms and combined with extracellular polymeric substance (EPS). Increasing tolerance against shock loads and toxic environments commonly found in industrial wastewater (Muñoz-Palazon et al., 2019).

Combining AGS with membrane bioreactors (MBR) would make up the AGS-MBR system's new bioreactor configuration. This system improves effluent quality and reduces membrane fouling problems (Zhang et al., 2023). According to Suja (2015) and Winkler (2012), the high density of aerobic granular sludge (AGS) promotes more efficient sediment ability, resulting in more compact effluent treatment plants (ETEs) compared to CAS due to the shorter time and space required to separate the liquid-solid phases (Nancharaiah et al., 2018; Bassin et al., 2019).

Recently, research has been carried out on developing AGS continuous flow reactors (CFRs) as they offer more advantages than SBRs. A CFR is more accessible to build, operate, and maintain, and the challenge is to find a suitable design for AGS-CFR that simultaneously allows the retention of granular biomass and promotes the washout of filamentous bacteria and fungi (Muñoz-Palazon et al., 2019). Anaerobic digestion is applied in WAS for its economic and environmental benefits. It stabilizes sludge, causing a small portion of solids to be discarded. Using the methane generated in heat or electricity after conversion is also possible. In contrast, anaerobic digestion is generally limited by low methane generation efficiency (Leite et al., 2023). Figure 12.6 presents a flowchart demonstrating the origin and destination of wastewater and its reimplementation in the environment.

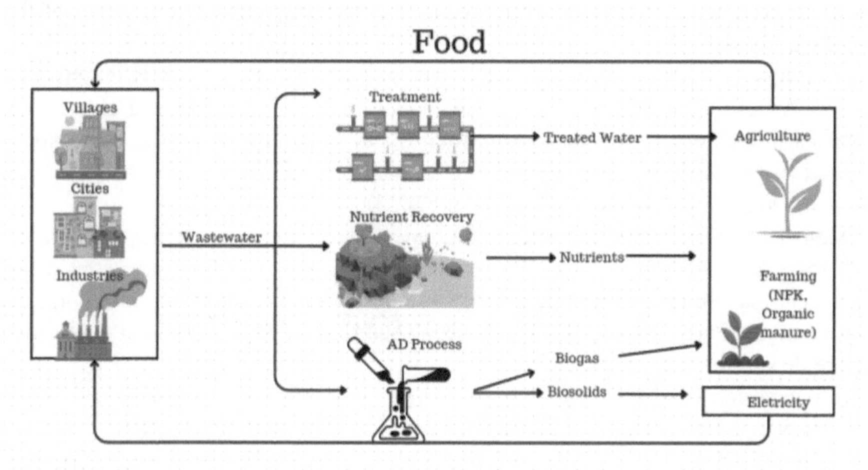

FIGURE 12.6 Flowchart demonstrating the origin and destination of wastewater and its reimplementation in the environment (compiled by the authors).

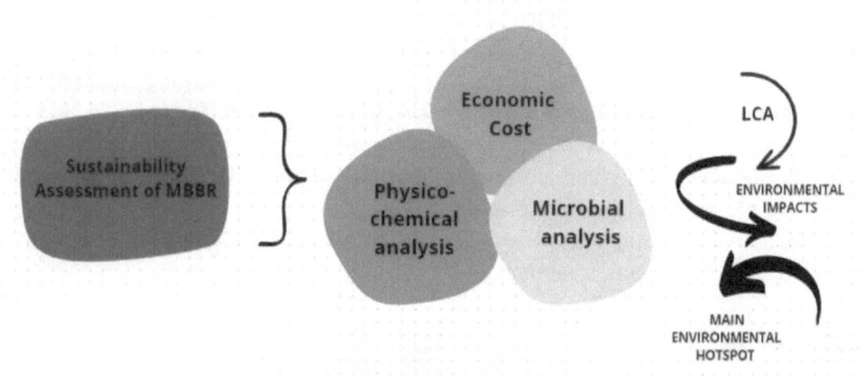

FIGURE 12.7 Representative diagram about the economic and sustainable advantages of the MBBR process (compiled by the authors).

Running MBBR offers several benefits, such as easy operation, high efficiency, and maintenance (Yi et al., 2020). Large-scale production and commercialization showed that MBBR could substitute CAS. MBBR technology can be implemented by adding carriers to the currently activated sludge system without the need for additional construction (Singh et al., 2018), being suitable for treating domestic and industrial wastewater with a recalcitrant character (Leyva-Díaz et al., 2016; Tang et al., 2021) and reuse of treated wastewater (Leyva-Díaz,2020). Furthermore, studies investigate the possibility of combining UASB with MBBR to update the effluent treatment process simply and economically (Liu et al., 2020). Figure 12.7

shows a representative diagram of the economic and sustainable advantages of the MBBR process.

Currently, research is investigating the consequences of metal ions in activated sludge. According to Zhang, Fe 3+ can reduce the zeta potential and change the particles prone to forming flakes, improving sludge flocculation. The same study showed that the low concentration of Mg 2+ changed the sludge microstructure and improved sludge flocculation (Zhang et al., 2023).

Many technologies are proposed to improve sludge dewatering, remove pollutants, and recover resources. Advanced oxidation processes (AOPs) are considered flashy and can destroy the structure of organic chemicals and mineralize organic contaminants. Furthermore, AOPs have been used for sludge conditioning to improve sludge dewatering (Xu et al., 2020).

In recent decades, this technology based on potassium permanganate has gained prominence. Still, it needs to be improved to improve $KMnO_4$ oxidation, identify dominant reactive species, and effectiveness in wastewater decontamination and sludge treatment (Wang et al., 2023).

12.5 CONCLUSION

Research continues to explore new technologies such as forward osmosis, phenolic compound bioremediation, electro-bioremediation, and other innovative strategies demonstrating the potential to treat persistent pollutants and protect the environment. Meanwhile, aquaculture faces challenges in bioaccumulating emerging contaminants and greenhouse gas emissions in wastewater stabilization pond systems.

In the context of activated sludge treatment, transforming this waste into valuable resources, such as biochar and nutrient replacement in the soil, highlights the search for sustainable approaches. Furthermore, advanced technologies such as integrated fixed film activated sludge (IFAS) and aerobic granular sludge (AGS) offer effective alternatives for wastewater treatment.

REFERENCES

Ahmed, Toqeer, Mohammad Zounemat-Kermani, Miklas Scholz. 2020. "Climate change, water quality and water-related challenges: A review with focus on Pakistan". *International Journal of Environmental Research and Public Health* 17(22): 8518. https://doi.org/10.3390/ijerph17228518

Ansari, Ashley J., Faisal I Hai, William E. E. Price, Huu H. Ngo, Wenshan Guo, Long D. Nghiem. 2018. "Assessing the integration of forward osmosis and anaerobic digestion for simultaneous wastewater treatment and resource recovery". *Bioresource Technology* 260: 221–226. https://doi.org/10.1016/j.biortech.2018.03.120

Awasthi, Mukesh K., Ekta Singh, Parameswaran Binod, Raveendran Sindhu, Surendra Sarsaiya, Aman Kumar, Hongyu Chen, Yumin Duan, Ashok Pandey, Sunil Kumar, Mohammad J. Taherzadeh, Ji Li, Zengqiang Zhang. 2022. "Biotechnological strategies for bio-transforming biosolid into resources toward circular bio-economy: A review". *Renewable and Sustainable Energy Reviews* 156: 111987. https://doi.org/10.1016/j.rser.2021.111987

Aziz, Asad, Hasan Rameez, Ashish Sengar, Diwakar Sharma, Izharul H. Farooqi, Farrukh Basheer. 2022. "Biogas production and nutrients removal from slaughterhouse wastewater using integrated anaerobic and aerobic granular intermittent SBRs – Bioreactors stability and microbial dynamics". *Science of the Total Environment* 848: 157575. https://doi.org/10.1016/j.scitotenv.2022.157575

Bai, Yang, Yaobin Zhang, Xie Quan, Shuo Chen. 2015. "Enhancing nitrogen removal efficiency and reducing nitrate liquor recirculation ratio by improving simultaneous nitrification and denitrification in integrated fixed-film activated sludge (IFAS) process". *Water Science and Technology* 73: 827–834. https://doi.org/10.2166/wst.2015.558

Bartosiewicz, Maciej, Liah X. Coggins, Patricia Glaz, Alicia Cortés, Sebastien Bourget, Elke S. Reichwaldt, Sally MacIntyre, Anas Ghadouani, Isabelle Laurion. 2021. "Integrated approach towards quantifying carbon dioxide and methane release from waste stabilization ponds". *Water Research* 202: 117389. https://doi.org/10.1016/j.watres.2021.117389

Bassin, J.P., D.C. Tavares, R.C. Borges, M. Dezotti. 2019. "Development of aerobic granular sludge under tropical climate conditions: The key role of inoculum adaptation under reduced sludge washout for stable granulation". *Journal of Environmental Management* 230: 168–182. https://doi.org/10.1016/j.jenvman.2018.09.072

Beek, Tim Aus der, Frank-Andreas Weber, Axel Bergmann, Silke Hickmann, Ina Ebert, Arne Hein, Anette Küster. 2016. "Pharmaceuticals in the environment–Global occurrences and perspectives". *Environmental Toxicology and Chemistry*, 35: 823–835. https://doi.org/10.1002/etc.3339

Bessong, Pascal O., John O. Odiyo, Justice N. Musekene, Abera Tessema. 2009. "Spatial distribution of diarrhoea and microbial quality of domestic water during an outbreak of diarrhoea in the Tshikuwi Community in Venda, South Africa". *Journal of Health, Population and Nutrition* 27: 652–659. https://doi.org/10.3329/jhpn.v27i5.3642

Bibi, Amina, Shazia Bibi, Mohammed Abu-Dieyeh, Mohammad A. Al-Ghouti. 2023. "Towards sustainable physiochemical and biological techniques for the remediation of phenol from wastewater: A review on current applications and removal mechanisms". *Journal of Cleaner Production* 417: 137810. https://doi.org/10.1016/j.jclepro.2023.137810

Blandin, Gaetan, Clément Gautier, Marc S. Toran, Hector Monclús, Ignasi Rodriguez-Roda, Joaquim Comas. 2018. "Retrofitting membrane bioreactor (MBR) into osmotic membrane bioreactor (OMBR): A pilot scale study". *Chemical Engineering Journal* 339: 267–277. https://doi.org/10.1016/j.cej.2018.01.103

Brooks, Bryan W., Jeremy L. Conkle. 2019. "Commentary: Perspectives on aquaculture, urbanization and water quality". *Comparative Biochemistry and Physiology Part C: Toxicology & Pharmacology* 217: 1–4. https://doi.org/10.1016/j.cbpc.2018.11.014

Cai, Fangrui, Lirong Lei, Youming Li, Yuancai Chen. 2021. "A review of aerobic granular sludge (AGS) treating recalcitrant wastewater: Refractory organics removal mechanism, application and prospect". *Science of the Total Environment* 782: 146852. https://doi.org/10.1016/j.scitotenv.2021.146852

Chen, Ben, Gengrui Wei, Tao Zhang, Haizhen Wu, Chaofei Wu, Acong Chen, Heng Zhang, Xianghong Guan, Yuan Ren, Chunhua Feng, Guanglei Qiu, Chaohai Wei. 2023. "Flow drag force contributes high bio-treatment efficiency in a circulating fluidized bed reactor: Mechanism of selective separation of functional decayed sludge". *Chemical Engineering Journal* 454: 140448. https://doi.org/10.1016/j.cej.2022.140448

Choudhary, Khushboo, Vivekanand Vivekanand, Nidhi Pareek. 2022. "Recent advancements in microbial bioremediation of industrial effluents: challenges and future outlook". In *Microbial Biodegradation and Bioremediation (Second Edition).* Edited by Surajit Das and Hirak Ranjan Dash. Elsevier: 293–303. https://doi.org/10.1016/B978-0-323-85455-9.00022-9.

Coggins, Liah X., Irma Larma, Amy Hinchliffe, Ruben Props, Anas Ghadouani. 2020. "Flow cytometry for rapid characterisation of microbial community dynamics in waste stabilisation ponds". *Water Research* 169: 115243. https://doi.org/10.1016/j.watres.2019.115243

Coggins, Liah X., Marco Ghisalberti, Anas Ghadouani. 2017. "Sludge accumulation and distribution impact the hydraulic performance in waste stabilisation ponds". *Water Research* 110: 354–365. https://doi.org/10.1016/j.watres.2016.11.031

Coggins, Liah X., Nicholas D. Crosbie, Anas Ghadouani. 2019. "The small, the big, and the beautiful: Emerging challenges and opportunities for waste stabilization ponds in Australia". *Wiley Interdisciplinary Reviews* 6: 1383. https://doi.org/10.1002/wat2.1383

Dias, D.F.C., R.G. Passos, M. von Sperling. 2017. "A review of bacterial indicator disinfection mechanisms in waste stabilisation ponds". *Reviews in Environmental Science and Bio/Technology* 16: 517–539. https://doi.org/10.1007/s11157-017-9433-2

Du, Bowen, Samuel P. Haddad, Andreas Luek, W. Casan Scott, Gavin N. Saari, S. Rebekah Burket, Christopher S. Breed, Martin Kelly, Linda Broach, Joseph B. Rasmussen, C. Kevin Chambliss, Bryan W. Brooks. 2016. "Bioaccumulation of human pharmaceuticals in fish across habitats of a tidally influenced urban bayou". *Environmental Toxicology and Chemistry* 35: 966–974. https://doi.org/10.1002/etc.3221

Edokpayi, Joshua N., Abimbola M. Enitan-Folami, Adeyemi O. Adeeyo, Olatunde S. Durowoju, Ademola O. Jegede, John O. Odiyo. 2020. "Recent trends and national policies for water provision and wastewater treatment in South Africa". In *Water Conservation and Wastewater Treatment in BRICS Nations.* Edited by Pardeep Singh, Yulia Milshina, Kangming Tian, Deepak Gusain, João Paulo Bassin. Elsevier: 187–211. https://doi.org/10.1016/B978-0-12-818339-7.00009-6

Edokpayi, Joshua N., Elizabeth T. Rogawski, David M. Kahler, Courtney L. Hill, Catherine Reynolds, Emanuel Nyathi, James A. Smith, John O. Odiyo, Amidou Samie, Pascal Bessong, Rebecca Dillingham. 2018. "Challenges to sustainable safe drinking water: A case study of water quality and use across seasons in rural communities in Limpopo Province, South Africa". *Water* 10:159. https://doi.org/10.3390/w10020159

Edokpayi, Joshua N., John O. Odiyo, Elizabeth O. Popoola, Titus A.M. Msagati. 2021. "Evaluation of contaminants removal by waste stabilization ponds: A case study of Siloam WSPs in Vhembe District, South Africa". *Heliyon* 7: e06207. https://doi.org/10.1016/j.heliyon.2021.e06207

Edokpayi, Joshua N., John O. Odiyo and Olatunde S. Durowoju. 2017. "Impact of wastewater on surface water quality in developing countries: A case study of South Africa." In *Water Quality.* Edited by: Izharul Haq, Ajay Kalamdhad and Ashok Pandey. Elsevier: 247–260. https://doi.org/10.1016/B978-0-323-91902-9.00013-4

Fairbairn, David J., M. Ekrem Karpuzcu, William A. Arnold, Brian L. Barber, Elizabeth F. Kaufenberg, William C. Koskinen, Paige J. Novak, Pamela J. Rice, Deborah L. Swackhamer. 2016. "Sources and transport of contaminants of emerging concern: A two-year study of occurrence and spatiotemporal variation in a mixed land use watershed". *Science of the Total Environment* 551–552: 605–613. https://doi.org/10.1016/j.scitotenv.2016.02.056

Fekadu, Samuel, Esayas Alemayehu, Raf Dewil, Bart Van der Bruggen. 2019. "Pharmaceuticals in freshwater aquatic environments: A comparison of the African and European challenge". *Science of the Total Environment* 654: 324–337. https://doi.org/10.1016/j.scitotenv.2018.11.072

Ghosh, Sayanti, Saswati Chakraborty. 2019. "Influence of inoculum variation on formation and stability of aerobic granules in oily wastewater treatment". *Journal of Environmental Management* 248: 109239. https://doi.org/10.1016/j.jenvman.2019.07.010

Glaz, Patricia, Maciej Bartosiewicz, Isabelle Laurion, Elke S. Reichwaldt, Roxane Maranger, Anas Ghadouani. 2016. "Greenhouse gas emissions from waste stabilisation ponds in Western Australia and Quebec (Canada)". *Water Research* 101: 64–74. https://doi.org/10.1016/j.watres.2016.05.060

Grabicová, Kateřina, Roman Grabic, Ganna Fedorova, Andrea Vojs Staňová, Martin Bláha, Tomáš Randák, Bryan W. Brooks, Vladimír Žlábek. 2020. "Pharmaceutical bioaccumulation by fish during tertiary treatment in a wastewater stabilization pond". *Environmental Pollution* 267: 115593. https://doi.org/10.1016/j.envpol.2020.115593

Gruchlik, Yolanta, Kathryn Linge, Cynthia Joll. 2018. Removal of organic micropollutants in waste stabilisation ponds: A review. *Journal of Environmental Management* 206: 202–214. https://doi.org/10.1016/j.jenvman.2017.10.020

Hang, Zhenyu, Zhihua Li, Yuanmo Zhu, Jing Zhang, Fan Yang, Tianyu Zhang. 2021. "A new approach to valuate and improve the stability of aerobic sludge systems based on maintenance coefficient". *Journal of Environmental Management* 296: 113192. https://doi.org/10.1016/j.jenvman.2021.113192

Holgerson, Meredith A., Peter A. Raymond. 2016. "Large contribution to inland water CO_2 and CH_4 emissions from very small ponds". *Nature Geoscience* 9: 222–226. https://doi.org/10.1038/ngeo2654

Hou, Huanhuan, Zhu Mengting, Liang Duan, Yang Zhao, Zhichao Zhang, Meichen Yao, Beihai Zhou, Haiya Zhang, Slawomir W. Hermanowicz. 2022. "Removal performance and biodegradation mechanism of sulfonamides antibiotic contained wastewater by IFAS-MBR bioreactor". *Journal of Molecular Liquids* 367: 120572. https://doi.org/10.1016/j.molliq.2022.120572

Idrees, Muhammad, Shakir Ali, Abdul Rehman, Syed Zajif Hussain, Dilara Abbas Bukhari. 2023. "Uptake of lead by bacteria isolated from industrial effluents and their potential use in bioremediation of wastewater". *Saudi Journal of Biological Sciences* 30: 103740. https://doi.org/10.1016/j.sjbs.2023.103740

Jiang, Yu, Xueqing Shi, How Yong Ng. 2021. "Aerobic granular sludge systems for treating hypersaline pharmaceutical wastewater: Start-up, long-term performances and metabolic function". *Journal of Hazardous Materials* 412:125229. https://doi.org/10.1016/j.jhazmat.2021.125229

Kor-Bicakci, Gokce, Emine Ubay-Cokgor, Cigdem Eskicioglu. 2019. "Effect of dewatered sludge microwave pretreatment temperature and duration on net energy generation and biosolids quality from anaerobic digestion". *Energy* 186: 782–795. https://doi.org/10.1016/j.energy.2018.11.103

Kim, Young Mo, Dong-Hyun Chon, Hee-Sik Kimm, Chul Park. 2012. "Investigation of bacterial community in activated sludge with an anaerobic side-stream reactor (ASSR) to decrease the generation of excess sludge". *Water Research* 46: 4292– 4300. https://doi.org/10.1016/j.watres.2012.04.040

Klatte, Stephanie, Hans-Christian Schaefer, Maximilian Hempel. 2017. "Pharmaceuticals in the environment – A short review on options to minimize the exposure of humans, animals and ecosystems". *Sustainable Chemistry and Pharmacy* 5: 61–66. https://doi.org/10.1016/j.scp.2016.07.001

Koba, Olga, Katerina Grabicova, Daniel Cerveny, Jan Turek, Jitka Kolarova, Tomas Randak, Vladimir Zlabek, Roman Grabic. 2018. "Transport of pharmaceuticals and their metabolites between water and sediments as a further potential exposure for aquatic organisms". *Journal of Hazardous Materials* 342: 401–407. https://doi.org/10.1016/j.jhazmat.2017.08.039

Kummerer, Klaus. 2010. "Pharmaceuticals in the environment". *Annual Review of Environment and Resources* 35: 57–75. https://doi.org/10.1146/annurev-environ-052809-161223

Lai, Agnese, Maria L. Astolfi, Veronica Bertelli, Valeria G. Agostinelli, Marco Zeppilli, Mauro Majone. 2021. "Chromate fate and effect in bioelectrochemical systems for remediation of chlorinated solvents". *New Biotechnology* 60: 27–35. https://doi.org/10.1016/j.nbt.2020.06.006

Lai, Agnese, Roberta Verdini, Federico Aulenta, Mauro Majone. 2015. "Influence of nitrate and sulfate reduction in the bioelectrochemically assisted dechlorination of cis-DCE". *Chemosphere* 125: 147–154. https://doi.org/10.1016/j.chemosphere.2014.12.023

Landrigan, Philip J., Richard Fuller, Nereus J. R. Acosta, Olusoji Adeyi, Robert Arnold, Niladri (Nil) Basu, et al. 2017. "The *Lancet* Commission on pollution and health". The Lancet Commissions 391: 462–512. https://doi.org/10.1016/S0140-6736(17)32345-0

Lee, Duu-Jong, Meng-Huan Hsieh. 2019. "Forward osmosis membrane processes for wastewater bioremediation: Research needs". *Bioresource Technology* 290: 121795. https://doi.org/10.1016/j.biortech.2019.121795

Leite, R.M. Wanderli, Bruna S. Magnus, Barbara A.B. Moraes, Mario T. Kato, Lourdinha Florencio, Rejane H.R. Costa, Paulo B. Filho. "Mesophilic anaerobic digestion of waste activated sludge in an intermittent mixing reactor: Effect of hydraulic retention time and organic loading rate". *Journal of Environmental Management* 338: 117839. Epub 2023 March 30. PMID: 37003223. doi: 10.1016/j.jenvman.2023.117839

Leyva-Díaz, J.C., J. Martín-Pascual, J.M. Poyatos. 2016. "Moving bed biofilm reactor to treat wastewater". *International Journal of Environmental Science and Technology* 14: 881–910. https://doi.org/10.1007/s13762-016-1169-y

Liu, Yonghong, Ning Wang, Yuxi Wei, Kang Dang, Miaojie Li, Yaozhong Li, Qinghui Li, Ruihua Mu. 2020. "Pilot study on the upgrading configuration of UASB-MBBR with two carriers: Treatment effect, sludge reduction and functional microbial identification". *Process Biochemistry* 99: 211–221. https://doi.org/10.1016/j.procbio.2020.09.007

Ma, Yongfei, Ming Li, Ping Li, Lie Yang, Li Wu, Feng Gao, Xuebin Qi, and Zulin Zhang. 2021. "Hydrothermal synthesis of magnetic sludge biochar for tetracycline and ciprofloxacin adsorptive removal". *Bioresource Technology* 319. https://doi.org/10.1016/j.biortech.2020.124199

Machat, Hela, Chedly Boudokhane, Nicolas Roche, and Hatem Dhaouadi. 2019. "Effects of C/N Ratio and DO concentration on carbon and nitrogen removals in a hybrid biological reactor". *Biochemical Engineering Journal* 151: 107313. doi:https://doi.org/10.1016/j.bej.2019.107313

Muñoz-Palazon, Barbara, Alejandro Rodriguez-Sanchez, Miguel Hurtado-Martinez, Ines Manuel de Castro, Belén Juarez-Jimenez, Alejandro Gonzalez-Martinez, Jesus Gonzalez-Lopez. 2019. "Performance and microbial community structure of an aerobic granular sludge system at different phenolic acid concentrations." *Journal of Hazardous Materials* 376: 58–67. https://doi.org/10.1016/j.jhazmat.2019.05.015

Muralikrishna, Iyyanki, Valli Manickam. 2017. "Industrial wastewater treatment technologies, recycling, and reuse". In *Environmental Management*. Edited by Iyyanki V. Muralikrishna and Valli Manickam. Elsevier: 295–336. http://dx.doi.org/10.1016/B978-0-12-811989-1.00013-0

Naidoo, Shalinee, Ademola O. Olaniran. 2014. "Treated wastewater effluent as a source of microbial pollution of surface water resources". *International Journal of Environmental Research and Public Health*. 11: 249–270. https://doi.org/10.3390/ijerph110100249

Nancharaiah, Yarlagadda V., Manjunath Sarvajith. 2019. "Aerobic granular sludge process: A fast growing biological treatment for sustainable wastewater treatment". *Current Opinion in Environmental Science & Health* 12: 57–65. https://doi.org/10.1016/j.coesh.2019.09.011

Nancharaiah, Y.V., G. Kiran Kumar Reddy. 2018. "Aerobic granular sludge technology: Mechanisms of granulation and biotechnological applications". *Bioresource Technology* 247: 1128–1143. https://doi.org/10.1016/j.biortech.2017.09.131

Olasoji, Samuel O., Nather O. Oyewole, Bayode Abiola, Joshua N. Edokpayi. 2019. "Water quality assessment of surface and groundwater sources using a water quality index method: A case study of a peri-urban town in Southwest, Nigeria". *Environments* 6: 23. https://doi.org/10.3390/environments6020023

Olukanni, David O., Joel J. Ducoste. 2011. "Optimization of waste stabilization pond design for developing nations using computational fluid dynamics". *Ecological Engineering* 37: 1878–1888. https://doi.org/10.1016/j.ecoleng.2011.06.003

Pan, Yuting, Liu Ye, Ben van den Akker, Ramon Ganigué Pagès, Ronald S. Musenze, Zhiguo Yuan. 2016. "Sludge-drying lagoons: A potential significant methane source in wastewater treatment plants". *Environmental Science & Technology* 50: 1368–1375. https://doi.org/10.1021/acs.est.5b04844

Panigrahy, Namita, Ankita Priyadarshini, Mitali Madhusmita Sahoo, Akshaya Kumar Verma, Achlesh Daverey, Naresh Kumar Sahoo. 2022. "A comprehensive review on eco-toxicity and biodegradation of phenolics: Recent progress and future outlook." *Environmental Technology & Innovation* 27: 102423. https://doi.org/10.1016/j.eti.2022.102423

Passos, Ricardo G., Daniel F.C. Dias, Marcos von Sperling. 2016. "Review of practical aspects for modelling of stabilization ponds using Computational Fluid Dynamics". *Environmental Technology Reviews* 5: 78–102. https://doi.org/10.1080/21622515.2016.1251499

Peng, Jian-Feng, Bao-Zhen Wang, Yong-Hui Song, Peng Yuan, Zhenhua Liu. 2007. "Adsorption and release of phosphorus in the surface sediment of a wastewater stabilization pond". *Ecological Engineering* 31: 92–97. https://doi.org/10.1016/j.ecoleng.2007.06.005

Pezoti, Osvaldo, André L. Cazetta, Karen C. Bedin, Lucas S. Souza, Alessandro C. Martins, Taís L. Silva, Oscar O. Santos Júnior, Jesuí V. Visentainer, Vitor C. Almeida. 2016. "NaOH-activated carbon of high surface area produced from guava seeds as a high-efficiency adsorbent for amoxicillin removal: Kinetic, isotherm and thermodynamic studies". *Chemical Engineering Journal* 288: 778–788. https://doi.org/10.1016/j.cej.2015.12.042

Quan, Le M., Hesam Kamyab, Ali Yuzir, Veeramuthu Ashokkumar, Seyed E. Hosseini, Balamuralikrishnan Balasubramanian, Irina Kirpichnikova. 2022. "Review of the application of gasification and combustion technology and waste-to-energy technologies in sewage sludge treatment". *Fuel* 316: 123199. https://doi.org/10.1016/j.fuel.2022.123199

Roberts, Jenna, Anupama Kumar, Jun Du, Christopher Hepplewhite, David J. Ellis, Andrew G. Christy, Sara G. Beavis. 2016. "Pharmaceuticals and personal care products (PPCPs) in Australia's largest inland sewage treatment plant, and its contribution to a major Australian river during high and low flow". *Science of the Total Environment* 541: 1625–1637. https://doi.org/10.1016/j.scitotenv.2015.03.145

Saaristo, Minna, Tomas Brodin, Sigal Balshine, Michael G. Bertram, Bryan W. Brooks, Sean M. Ehlman, Erin S. McCallum, et al. 2018. "Direct and indirect effects of chemical contaminants on the behaviour, ecology and evolution of wildlife." *Proceedings of the Royal Society B* 285: 20181297. https://doi.org/10.1098/rspb.2018.1297

Saravanan, A., Ponnusamy Senthil Kumar, Pham Anh Duc, Gayathri Rangasamy. 2023. "Strategies for microbial bioremediation of environmental pollutants from industrial wastewater: A sustainable approach". *Chemosphere* 313: 137323.https://doi.org/10.1016/j.chemosphere.2022.137323

Sehonova, Pavla, Zdenka Svobodova, Petra Dolezelova, Petra Vosmerova, Caterina Faggio. 2018. "Effects of waterborne antidepressants on non-target animals living in the aquatic environment: A review". *Science of the Total Environment* 631–632: 789–794. https://doi.org/10.1016/j.scitotenv.2018.03.076

Sguanci, S., C. Lubello, S. Caffaz, T . Lotti. 2019. "Long-term stability of aerobic granular sludge for the treatment of very low-strength real domestic wastewater". *Journal of Cleaner Production* 222: 882–890. https://doi.org/10.1016/j.jclepro.2019.03.061

Silva, Joao P., José L.Ruiz, Miguel R.Peña, Henk Lubberding, Huub Gijzen. 2012. "Influence of photoperiod on carbon dioxide and methane emissions from two pilot-scale stabilization ponds". *Water Science and Technology* 66(9): 1930–40. doi: 10.2166/wst.2012.396

Singh, Anju, Sheetal Jaisingh Kamble, Megha Sawant, Yogita Chakravarthy, Absar Kazmi, Enrique Aymerich, Markus Starkl et al. 2018. "Technical, hygiene, economic, and life cycle assessment of full-scale moving bed biofilm reactors for wastewater treatment in India". *Environmental Science and Pollution Research* 25: 2552–2569. https://doi.org/10.1007/s11356-017-0605-y

Singh, Anshu, Izharul Haq. 2023. "Chapter 14 – Bioremediation approaches for the removal of emerging contaminants from industrial wastewater". In *Current Developments in Biotechnology and Bioengineering*. Edited by Izharul Haq, Ajay Kalamdhad and Ashok Pandey. Elsevier: 247–260. https://doi.org/10.1016/j.biortech.2015.07.045

Steinbach, Christoph, Roman Grabic, Ganna Fedorova, Olga Koba, Oksana Golovko, Katerina Grabicova, Hana K. Kroupova. 2016. "Bioconcentration, metabolism and half-life time of the human therapeutic drug diltiazem in rainbow trout Oncorhynchus mykiss". *Chemosphere* 144: 154–159. https://doi.org/10.1016/j.chemosphere.2015.08.038

Suja, E., Y.V. Nancharaiah, T.V. Krishna Mohan, V.P. Venugopalan. 2015. "Denitrification accelerates granular sludge formation in sequencing batch reactors". *Bioresource Technology* 196: 28–34.

Tang, Kai, Peter Rosborg, Emma S. Rasmussen, Adam Hambly, Michael Madsen, Niels M. Jensen, Aviaja A. Hansen, et al. 2021. "Impact of intermittent feeding on polishing of micropollutants by moving bed biofilm reactors (MBBR)". *Journal of Hazardous Materials* 403: 123536. https://doi.org/10.1016/j.jhazmat.2020.123536

Tran, Ngoc H., Taro Urase, Huu H. Ngo, Jiangyong Hu, Say L. Ong. 2013. "Insight into metabolic and cometabolic activities of autotrophic and heterotrophic microorganisms in the biodegradation of emerging trace organic contaminants". *Bioresource Technology* 146: 721–731. https://doi.org/10.1016/j.biortech.2013.07.083

Tufail, Muhammad A., Jawaria Iltaf, Tahreem Zaheer, Leeza Tariq, Muhammad Bilal Amir, Rida Fátima, Ayesha Asbat, et al. 2022. "Recent advances in bioremediation of heavy metals and persistent organic pollutants: A review". *Science of the Total Environment* 850: 157961. https://doi.org/10.1016/j.scitotenv.2022.157961

Vazquez-Roig, Pablo, Vicente Andreu, Cristina Blasco, Yolanda Picó. 2012. "Risk assessment on the presence of pharmaceuticals in sediments, soils and waters of the Pego-Oliva Marshlands (Valencia, eastern Spain)". *Science of the Total Environment* 440: 24–32. https://doi.org/10.1016/j.scitotenv.2012.08.036

Verlicchi, P., M. Al Aukidy, E. Zambello. 2012. "Occurrence of pharmaceutical compounds in urban wastewater: Removal, mass load and environmental risk after a secondary treatment – A review". *Science of the Total Environment* 429: 123–155. https://doi.org/10.1016/j.scitotenv.2012.04.028

Wang, Li, Xiang Liu, Duu-Jong Lee, Joo-Hwa Tay, Yi Zhang, Chun-Li Wan, Xiao-Feng Chen. 2018. "Recent advances on biosorption by aerobic granular sludge". *Journal of Hazardous Materials* 357: 253–270. https://doi.org/10.1016/j.jhazmat.2018.06.010

Wang, Yufen, Kaixin Zheng, Haixiao Guo, Lixin Tian, Yanying He, Xiaomin Wang, Tingting Zhu, Peizhe Sun, Yiwen Liu. 2023. "Potassium permanganate-based advanced oxidation processes for wastewater decontamination and sludge treatment: A review". *Chemical Engineering Journal* 452: 139529. https://doi.org/10.1016/j.cej.2022.139529

Wang, Yuxian, Zhimin Ao, Hongqi Sun, Xiaoguang Duan, Shaobin Wang. 2016. "Activation of peroxymonosulfate by carbonaceous oxygen groups: Experimental and density functional theory calculations". *Applied Catalysis B: Environmental* 198: 295–302. https://doi.org/10.1016/j.apcatb.2016.05.075

Waqas, Sharjeel, Muhammad Roil Bilad, Zakaria Man, Yusuf Wibisono, Juhana Jaafar, Teuku Meurah Indra Mahlia, Asim Laeeq Khan, Muhammad Aslam. 2020. "Recent progress in integrated fixed-film activated sludge process for wastewater treatment: A review". *Journal of Environmental Management* 268: 110718. https://doi.org/10.1016/j.jenvman.2020.110718

Wei, Liangliang, Xinhui Xia, Fengyi Zhu, Qiaoyang Li, Mao Xue, Jianju Li, Bo Sun, Junqiu Jiang, Qingliang Zhao. 2020. "Dewatering efficiency of sewage sludge during Fe^{2+}-activated persulfate oxidation: Effect of hydrophobic/hydrophilic properties of sludge EPS". *Water Research* 181: 115903. https://doi.org/10.1016/j.watres.2020.115903

William J. Cosgrove, Daniel P. Loucks. 2015. "Water management: Current and future challenges and research directions". *Water Resources Research* 51: 4823–4839. https://doi.org/10.1002/2014WR016869

Winkler, M.-K.H, J.P. Bassin, R. Kleerebezem, R.G.J.M. van der Lans, M.C.M. van Loosdrecht. 2012. "Temperature and salt effects on settling velocity in granular sludge technology". *Water Research* 46: 5445–5451. https://doi.org/10.1016/j.watres.2012.07.022

Xiao, Jun, Dandan He, Yuhang Ye, Bin Yang, Abing Duan, Dongbo Wang. 2023. "Recent progress in persulfate to improve waste activated sludge treatment: Principles, challenges and perspectives". *Chemical Engineering Journal* 469: 143596. https://doi.org/10.1016/j.cej.2023.143956

Xu, Qiuxiang, Qi-Su Huang, Wei Wei, Jiang Sun, Xiaohu Dai, Bing-Jie Ni. 2020. "Improving the treatment of waste activated sludge using calcium peroxide". *Water Research* 187: 116440. https://doi.org/10.1016/j.watres.2020.116440

Yang, Yi, Yong Sik Ok, Ki-Hyun Kim, Eilhann E. Kwon, Yiu Fai Tsang. 2017. "Occurrences and removal of pharmaceuticals and personal care products (PPCPs) in drinking water and water/sewage treatment plants: A review". *Science of the Total Environment* 596–597: 303–320. https://doi.org/10.1016/j.scitotenv.2017.04.102

Yi, Huan, Minfang Li, Xiuqin Huo, Guangming Zeng, Cui Lai, Danlian Huang, Ziwen An, et al. 2020. "Recent development of advanced biotechnology for wastewater treatment". *Critical Reviews in Biotechnology* 40: 99–118. https://doi.org/10.1080/07388551.2019.1682964

Zenker, Armin, Maria Rita Cicero, Francesca Prestinaci, Paola Bottoni, Mario Carere. 2014. "Bioaccumulation and biomagnification potential of pharmaceuticals with a focus to the aquatic environment". *Journal of Environmental Management* 133: 378–387. https://doi.org/10.1016/j.jenvman.2013.12.017

Zeppilli, Marco, Hafsa Yaqoubi, Edoardo Dell'Armi, Agnese Lai,Mustapha Belfaquir, Laura Lorini, Marco Petrangeli Papini. 2024. "Tetrachloroethane (TeCA) removal through sequential graphite-mixed metal oxide electrodes in a bioelectrochemical reactor".

Environmental Science and Ecotechnology 17: 100309. https://doi.org/10.1016/j.ese.2023.100309

Zhang, Huanlong, Yue Xu, Liying Bin, Ping Li, Fenglian Fu, Shaosong Huang, Bing Tang. 2023. "Rapid granulation of aerobic granular sludge in an integrated moving bed biofilm reactor-membrane bioreactor: Effects and mechanism of *Streptomyces cellulosus* microspheres". *Bioresource Technology Reports* 22: 101412. https://doi.org/10.1016/j.biteb.2023.101412

Zheng, Mengqi, Hao Zhu, Yuxing Hanb, Xu Chunyan, Zhengwen Zhang, Hangjun Han. 2019. "Comparative investigation on carbon-based moving bed biofilm reactor (MBBR) for synchronous removal of phenols and ammonia in treating coal pyrolysis wastewater at pilot-scale". *Bioresource Technology* 288: 121590. https://doi.org/10.1016/j.biortech.2019.121590

13 Remediation Strategies Based on Nanotechnology for the Removal of Heavy Metal Contaminants from Wastewater

Parvindar M. Sah, Harshala S. Naik, and Rajesh W. Raut

13.1 INTRODUCTION

Wastewater contamination by heavy metals poses a substantial threat to both the environment and human health (Briffa et al., 2020; Roy, 2014), primarily from industrial activities, mining operations, and improper electronic waste disposal (Doyle et al., 2020). Key contributors such as lead, cadmium, mercury, chromium, and arsenic can cause detrimental effects on ecosystems and living organisms, leading to health issues like neurological disorders, organ damage, and cancer (Hashem, 2014; M.S. Islam et al., 2023). Thus, the imperative for safeguarding human and environmental well-being necessitates the efficient removal of heavy metals from wastewater (Borji et al., 2020; Cheriyamundath & Vavilala, 2021; Pandey, 2018).

While traditional methods like precipitation (Xiaowang Liu et al., 2009), ion exchange (Al-Enezi et al., 2004), adsorption (I. Ali & Gupta, 2007), membrane filtration (Xiang et al., 2022), and chemical coagulation (Charerntanyarak, 1999; Sakhi et al., 2019), have been employed, and they exhibit inherent limitations. Precipitation techniques (Xuyang Liu et al., 2015) are ineffective for low concentrations of heavy metals and generate sludge requiring proper disposal. Ion exchange and adsorption processes, while selective, often pose cost-prohibitive challenges and necessitate adsorbent regeneration. Membrane filtration (Xiang et al., 2022), can be expensive and susceptible to fouling, while chemical coagulation may introduce secondary pollutants. Therefore, the pursuit of alternative remediation strategies that are more efficient, cost-effective, and sustainable becomes imperative (Rizwan et al., 2014).

DOI: 10.1201/9781003454199-13

This chapter focuses on the significance of nanotechnology-based remediation strategies for heavy metal removal, leveraging the unique properties exhibited by nanomaterials at the nanoscale (Kiruba Daniel et al., 2014). Various nanomaterials, including nanoscale zero-valent iron (nZVI) particles, carbon nanotubes (CNTs), graphene oxide (GO), silver nanoparticles (AgNPs), magnetic nanoparticles (MNPs), and titanium dioxide (TiO2) nanoparticles for photocatalysis, show considerable potential in heavy metal remediation. Each nanomaterial possesses distinctive characteristics suitable for heavy metal removal(X. Zhao et al., 2016).

While nanotechnology-based strategies provide significant advantages, they are not without limitations. Factors such as pH, dissolved oxygen, and the presence of other contaminants can influence the reactivity and efficiency of these nanomaterials (Borji et al., 2020; Cheriyamundath & Vavilala, 2021; Pandey, 2018). Agglomeration and settling of particles may necessitate additional measures for effective deployment. Additionally, the potential release of nanomaterials into the environment raises concerns about their long-term impact, requiring careful consideration (Ibrahim et al., 2016).

Nevertheless, nanotechnology-based remediation strategies offer promising solutions for the efficient and sustainable removal of heavy metals from wastewater (Sharma & Sharma, 2022). The utilization of nanoscale zero-valent iron particles, carbon nanotubes, graphene oxide, silver nanoparticles, magnetic nanoparticles, and titanium dioxide nanoparticles holds great potential for overcoming the limitations of traditional methods. Further research and development are essential to optimize these strategies, address their limitations, and ensure their safe implementation on a larger scale (Yang et al., 2019). Our objectives in this chapter are twofold: first, to evaluate the effectiveness of nanomaterials, including nZVI, CNTs, GO, AgNPs, MNPs, and TiO_2 nanoparticles, for efficient heavy metal removal; and second, by uncovering both advantages and limitations, to call for additional research aiming to optimize these strategies for safe and sustainable large-scale implementation.

13.1.1 Traditional Methods of Heavy Metal Removal and Their Limitations

It should be emphasized that conventional approaches, though commonly employed, frequently face challenges related to selectivity, removal efficiency, operational expenses, and waste handling. Technological advancements have given rise to more efficient and sustainable alternatives, including membrane filtration, electrochemical methods, and advanced oxidation processes (Chadha et al., 2022; Liosis et al., 2021; Xiang et al., 2022; Yang et al., 2019). These alternatives can overcome some of the limitations associated with traditional methods and offer improved effectiveness in removing heavy metals from wastewater, as outlined in Table 13.1, which compares the traditional approach along with its advantages.

13.1.2 Importance of Nanotechnology-based Remediation Strategies

Nanotechnology-driven remediation approaches have become crucial in combatting environmental pollution, especially in instances of heavy metal contamination. These

TABLE 13.1
Traditional Way of Heavy Metal Removal: Its Advantages and Limitation

Traditional Methods	Advantages	Ref.	Limitation
Chemical Precipitation: Addition of chemicals to 8wastewater to form insoluble precipitates with heavy metal ions, which are then separated through sedimentation.	**Effectiveness**: Varies for different heavy metals. **pH Dependence:** Efficiency is highly dependent on wastewater pH, requiring pH adjustment and maintenance.	(Charerntanyarak, 1999)	Some metals may not readily form insoluble precipitates, limiting removal efficiency. **Sludge Disposal:** Separated precipitates form sludge that needs proper disposal, posing environmental challenges.
Ion Exchange: Exchange of heavy metal ions in wastewater with ions of a different charge on an exchange resin material.	Ion exchange is including efficient removal of specific ions, regeneration capability for extended use.	(Al-Enezi et al., 2004)	**Regeneration:** Exchange resin becomes saturated and requires regeneration or replacement, adding to operational costs.
Adsorption: Utilization of adsorbent materials, such as activated carbon or zeolites, to attract and bind heavy metal ions from wastewater.	Adsorbent selection includes tailored removal of specific contaminants, cost-effectiveness, and versatility in addressing diverse water treatment challenges.	(Al-Rifai et al., 2018)	**Saturation and Regeneration**: Adsorbents become saturated and require regeneration or replacement. Regeneration may not always be feasible or cost-effective. **Competing Ions:** The presence of other ions can compete with heavy metals for adsorption sites, reducing overall removal efficiency.
Precipitative Softening: Addition of lime or other alkaline substances raises w88astewater pH, promoting the precipitation of heavy metal hydroxides.	Precipitative softening offers including the reduction of water hardness, extended appliance lifespan, lower maintenance costs, enhanced water quality, increased energy efficiency.	(Charerntanyarak, 1999; Masarwa et al., 1997; Pohl, 2020)	**Sludge Generation**: Like chemical precipitation, precipitative softening generates sludge requiring proper handling and disposal. **Metal Solubility:** Some heavy metals may not precipitate effectively under typical operating conditions, reducing removal efficiency.

Source: Compiled by authors.

approaches capitalize on the distinctive properties of nanomaterials to augment the efficiency and efficacy of remediation processes (Yang et al., 2019).

Here are some key reasons highlighting the importance of nanotechnology-based remediation strategies:

a. **Enhanced Removal Efficiency:** Nanomaterials, characterized by a high surface area-to-volume ratio, facilitate heightened interaction with contaminants. This increased surface reactivity enables more efficient processes such as adsorption, absorption, and catalytic degradation of heavy metals found in wastewater. Nanoparticles and nanocomposites, as types of nanomaterials, prove effective in binding and immobilizing heavy metal ions, leading to superior removal efficiency when compared to conventional methods (Borji et al., 2020).
b. **Selectivity and Specificity:** Nanotechnology provides the benefit of customizing the properties of nanomaterials to display precise adsorption or catalytic characteristics for heavy metals. Through surface functionalization or modification of nanomaterials, it becomes possible to selectively bind and eliminate particular heavy metals, facilitating targeted remediation. This selectivity minimizes interference from other co-existing contaminants, thereby enhancing the overall efficiency of the remediation process (Jjagwe et al., 2021).
c. **Versatility and Adaptability:** Nanotechnology offers a versatile foundation for developing a variety of remediation approaches. Nanomaterials can be tailored and engineered to meet the requirements of different remediation techniques (Park et al., 2018), including adsorption (White et al., 2018), membrane filtration (Abu-Dalo et al., 2021), photocatalysis (Ghasemzadeh et al., 2014), and electrochemical methods Samarghandi et al., 2020) This adaptability allows for the application of nanotechnology-based strategies in various wastewater sources, ensuring efficient removal of heavy metals from industrial effluents, agricultural runoff, and urban wastewater (Yang et al., 2019).
d. **Minimized Environmental Footprint:** Remediation strategies employing nanotechnology can present eco-friendly alternatives when compared to conventional methods. The synthesis of nanomaterials can be achieved through greener and sustainable approaches, minimizing the use of hazardous chemicals, and reducing energy consumption. Additionally, the heightened efficiency of nanomaterials in removing heavy metals leads to decreased waste generation and reduced sludge volume, thereby mitigating the environmental impact linked to waste disposal (Singh et al., 2018).
e. **Remediation in Complex Systems:** Nanotechnology-based approaches have shown promise in addressing heavy metal contamination in complex systems, such as groundwater and soil. Nanoparticles can penetrate and reach subsurface environments, facilitating in situ remediation of contaminated groundwater and soil matrices. The small size and high mobility of nanoparticles allow them to access hard-to-reach areas, effectively treating localized sources of heavy metal contamination (Yang et al., 2019).

f. **Potential for Remediation Monitoring:** Nanotechnology provides possibilities for the real-time monitoring and evaluation of remediation processes. Using functionalized nanomaterials acting as sensing probes, heavy metal concentrations can be detected and quantified in situ. This capability facilitates the ongoing assessment of remediation effectiveness, ensuring timely interventions and adjustments as needed. (Liosis et al., 2021; Manager & Minerals, 2016; K. K. Yadav et al., 2017; Krishna Kumar Yadav et al., 2017).

Although nanotechnology-based remediation strategies possess significant potential, it is crucial to acknowledge associated challenges, including long-term environmental implications, potential nanomaterial toxicity, and cost considerations. Ensuring the safe and responsible application of nanotechnology for heavy metal remediation requires diligent risk assessment and regulation (Yang et al., 2019).

13.2 NANOTECHNOLOGY-BASED REMEDIATION STRATEGIES

Nanotechnology-based remediation strategies present promising solutions for addressing heavy metal contamination. Nanoscale zero-valent iron (nZVI) particles demonstrate high reactivity and a significant surface area, allowing for effective adsorption (Xu et al., 2022), reduction, and precipitation of heavy metals (Pohl, 2020) Carbon nanotubes (CNTs) (Mallakpour & Khadem, 2018) and graphene oxide (GO) (Gontrani et al., 2021) exhibit high adsorption capacities and customizable properties, facilitating efficient heavy metal removal through adsorption, complexation, and electrostatic interactions (Kumar et al., 2021). Silver nanoparticles (AgNPs) possess unique properties and can remove heavy metals through adsorption (M. A. Islam et al., 2021), redox reactions, and complex formation . Magnetic nanoparticles (MNPs) (J. Chen et al., 2015) enable the efficient separation and recovery of heavy metals due to their magnetic properties. Photocatalysis with titanium dioxide (TiO_2) (J. Chen et al., 2015) nanoparticles leverages their high photocatalytic activity to generate reactive oxygen species and degrade heavy metal pollutants (Koul & Taak, 2018) (Figure 13.1).

These nanotechnology-based strategies offer significant advantages, including enhanced removal efficiency, selectivity, and promoting environmental remediation. However, they also come with limitations such as potential challenges related to aggregation, nanoparticle release, scalability, and environmental and toxicity concerns are mentioned in Table 13.2. By addressing their limitations and ensuring responsible deployment, these strategies can play a vital role in safeguarding ecosystems and human health from the harmful effects of heavy metal pollution (K. K. Yadav et al., 2017).

13.2.1 Nanoscale Zero-valent Iron (nZVI) Particles

Nanoscale zero-valent iron (nZVI) consists of an Fe (0) core and Fe oxide outer shell, offering a unique combination of properties for heavy metal removal. The Fe (0) core possesses reducing capabilities, while the outer shell acts as a reaction site for chemisorption and electrostatic interactions. Various mechanisms, including

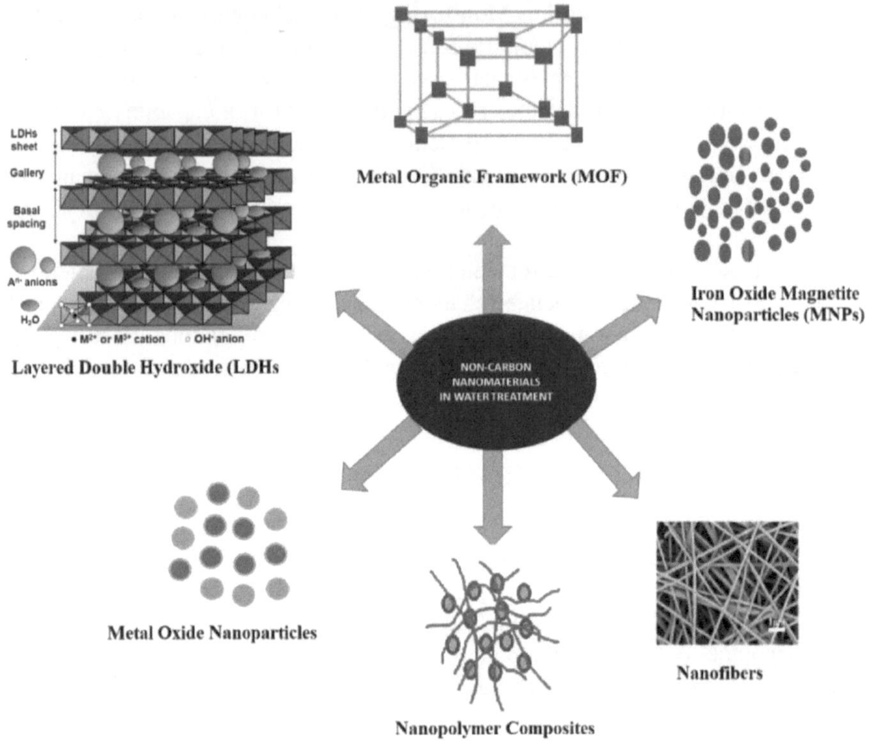

FIGURE 13.1 Non-carbon nanomaterial used for waste water treatment. (Image under creative common licence CC by 4. Reproduced with permission (Baby et al., 2022) Copyright © 2022 MDPI.

reduction, absorption, precipitation, and mineralization, contribute to the removal of heavy metals from aqueous solutions using nZVI (Tarekegn et al., 2021).

In water, nZVI exhibits ligand-like properties. At low pH, the iron oxide shell carries a positive charge, enabling the attraction of anions such as phosphates and sulphates. On the other hand, at high pH, the shell becomes negatively charged, facilitating the attraction of cations like metal ions. As a result, heavy metals in solution are sequestered onto the solid surface of nZVI through these reactions (Tarekegn et al., 2021).

Nanoscale zero-valent iron (nZVI) particles refer to extremely small particles of iron with a size range typically between 1 and 100 nanometres. These particles are chemically pure forms of iron in a zero-valent state, meaning they lack any charge or oxidation state. Due to their nanoscale size, nZVI particles possess unique properties and characteristics that make them highly effective in various applications, particularly in environmental remediation (Puthukkara P et al., 2021).

The small size of nZVI particles provides them with a large surface area-to-volume ratio, which enhances their reactivity and allows for increased contact with contaminants. This high reactivity is attributed to the increased number of

TABLE 13.2
Nanotechnology-based Remediation Strategies for Heavy Metal Removal and Its Characteristics, Application, and Limitation

Nanomaterials	Characterisation	Application	Limitation	Reference
Modified nZVI	High surface area and reactivity	Adsorption and chemical reduction of heavy metal ions.	pH, agglomeration, potential environmental effects	(Ken & Sinha, 2020)
Graphene-Based Nanocomposite	High surface area and excellent adsorption Capacities.	Selective removal of heavy metals from wastewater.	pH sensitivity, potential environmental impact	(Krishna et al., 2023)
Nanoengineered CNTs	Excellent adsorption Capacities, Selective heavy metals removal	Wastewater treatment and removal of heavy metals contaminants.	Cost challenges, Adsorbent regeneration	(Yang et al., 2019)
Metals-organic Frameworks (MOFs)	High surface area and tunable structure.	Adsorption and immobilisation of heavy metal ions	Limited Selectivity, potential toxicity	(Shi et al., 2022)
Nanofibers	High surface area and Flexibility.	Adsorption and filtration of heavy metals in wastewater	Prone to fouling, potential environmental impact	(Salehi et al., 2021)
Nanocomposites	Versatile combinations of different nanomaterials	Enhanced heavy metals removal through synergistic effects	Complexity in synthesis, cost considerations	(Ayalew et al., 2022)
Hydrogels	Porus structure, high water content.	Adsorption and removal of heavy metal ions in water	Limited mechanicals in strengths, potentials leaching	(C. Zhao et al., 2023)
Nanofiltration Membranes	Porus structure, effective Filteration.	Filtration and separation of heavy metal ions from water	Susceptible to fouling, may require frequent maintenance	(Covaliu-Mierlă et al., 2023)

Source: Compiled by authors

surface-active sites available for chemical reactions to occur. As a result, nZVI particles have been extensively studied and utilized for the removal and degradation of various contaminants, including heavy metals, chlorinated compounds, pesticides, and organic pollutants (Puthukkara P et al., 2021).

Nanoscale zero-valent iron (nZVI) particles exhibit diverse types characterized by synthesis methods, surface alterations, and particle structure. The classifications encompass bare nZVI, unmodified iron particles formed via chemical reduction or electrochemical processes; stabilized nZVI, surface-modified to enhance stability and dispersion using substances like surfactants or polymers; polymer-encapsulated nZVI, iron particles shielded by polymer coatings to prevent oxidation and regulate ion release; core-shell nZVI, with iron cores enveloped by materials such as metal oxides for heightened reactivity; supported nZVI, immobilized on solid materials to augment stability, reactivity, and separation; and magnetic nZVI, featuring magnetic properties enabled by surface modifications like magnetite or maghemite, allowing manipulation via external fields for improved recovery and reuse in remediation applications (Liosis et al., 2021).

One of the key advantages of nZVI particle-based remediation is their high reactivity and effectiveness across a wide range of contaminants. They can be applied in both in-situ and ex-situ remediation approaches, allowing for flexibility in addressing contaminated sites. Moreover, nZVI particles can be easily injected into subsurface environments or incorporated into permeable reactive barriers to treat groundwater contaminated with heavy metals (Tarekegn et al., 2021).

However, nZVI particle-based remediation also has limitations. The particles may exhibit aggregation or agglomeration, reducing their mobility and dispersibility in the treatment system. This can limit their effectiveness in reaching the target contaminants. Additionally, post-treatment waste management is an important consideration, as the spent nZVI particles may need to be properly handled and disposed of to prevent any potential release of iron or residual contaminants into the environment (Park et al., 2018).

13.2.2 Carbon Nanotubes (CNTs) and Graphene Oxide (GO)

The use of carbon nanotubes (CNTs) and graphene nanoparticles for the removal of heavy metals from water is a cutting-edge and promising approach in the field of water purification. Both nanomaterials possess unique properties that make them highly effective for heavy metal remediation, and ongoing research is demonstrating their potential in addressing one of the most pressing environmental concerns (Hong et al., 2020). Carbon nanotubes, particularly multi-walled carbon nanotubes (MWCNTs) and single-walled carbon nanotubes (SWCNTs), have garnered attention for their exceptional surface area and tunable properties (Chadha et al., 2022). Researchers have successfully functionalized these nanotubes to enhance heavy metal adsorption, showcasing their adaptability in tackling a variety of heavy metal contaminants (Hasdi, 2020). The use of MWCNTs functionalized with amine groups to remove lead (Pb) and cadmium (Cd) is a noteworthy example of their practicality (Chadha et al., 2022). CNTs can be capable of acting as promising support for composite adsorbents because of the tunable surface chemistry and adjustable pore size. CNTs can act as pollutant removal scaffolds. For example, CNT iron oxide for europium adsorption, chromium adsorption using carbon nanoparticles. Additionally, their incorporation into composites has expanded their potential for widespread application, promising more efficient heavy metal removal solutions. (Mallakpour & Khadem, 2018). Graphene nanoparticles, particularly graphene oxide (GO) (Ahamed & Loganathan, 2023) and reduced graphene oxide (rGO) (Xu et al., 2022), offer a comparable level of effectiveness in heavy metal removal. Their high surface area and oxygen-containing functional groups make them adept at adsorbing heavy metal ions. The successful use of GO for the removal of copper (Cu) (White et al., 2018) and Nickel (Ni)(Tran et al., 2019) highlights their versatility. It has been also shown that graphene oxide nanoribbons (GONRs) remove As (V) ions and are more adsorbed onto the GONRs than Hg (II) ions and with increasing initial As (V) and Hg (II) ions concentration to 300 ppm, the equilibrium adsorption uptake of the synthesized GONRs increases to 155.61 and 33.02 mg/g for As (V) and Hg (II) ions, respectively through a rapid separation process in just 12 min (Sadeghi et al., 2020). Furthermore, the chemical reduction of GO to form rGO enhances its electrical conductivity, further optimizing its utility in heavy metal remediation. The development of graphene-based nanocomposites by combining these nanoparticles with other materials, such as polymers or other nanoparticles, demonstrates a forward-thinking approach to improving heavy metal removal systems. This approach leverages the strengths of both graphene and the additional materials, resulting in systems that excel in heavy metal ion removal(Ma et al., 2012).

13.3.3 Silver Nanoparticles (AgNPs)

Silver nanoparticles (AgNPs) are increasingly recognized for their role in addressing heavy metal contamination in water. They excel in heavy metal removal through mechanisms like adsorption, catalysis, and aggregation due to their high surface area. AgNPs can also be modified for selectivity and regeneration. However, concerns about their environmental impact and toxicity require careful research and adherence to regulations when using them for wastewater treatment. Consideration of specific heavy metal types, concentrations, and local guidelines is crucial. Consulting experts is advisable for responsible AgNPs application in water treatment (Palani et al., 2023; Sharma & Sharma, 2022).

Silver nanoparticles synthesized using freshwater algal extract can effectively remove heavy metal ions, with a higher adsorption capacity observed for lead (Pb II) compared to cadmium (Cd II). The maximum adsorption capacity for Pb (II) using silver nanoparticles was found to be 23.14 mg/g (Negi et al., 2021). The study investigated the potential of silver nanoparticles (Ag NPs) in removing As (V) from aqueous solutions through adsorption. Various experiments were conducted, including assessments of the optimal dosage, kinetics, pH effects, ultrasonication, and exposure to light, as well as an evaluation of performance under continuous flow conditions. It was observed that Ag NPs exhibited effective As(V) removal capabilities when the Ag:As (V) ratio was less than 1:10. Notably, this removal process was particularly efficient in natural sunlight and under neutral pH conditions. Moreover, in a continuous flow setup, Ag NPs achieved an overall removal rate of up to 72 percent. Additionally, the study compared the performance of Ag NPs to that of other nanoparticles, such as cerium oxide, titanium oxide, magnetic ferric oxide, and manganese oxide, previously reported in the literature for heavy metal removal. Ag NPs demonstrated superior As (V) removal capacity in shorter timeframes when compared to the previously studied nanoparticles(Shaukat et al., 2022).

A significant development in the field of nanotechnology, researchers have successfully created a versatile nanocomposite consisting of chitosan, silver nanoparticles (Ag NPs), copper nanoparticles (Cu NPs), and carbon nanotubes (CNTs). One of the most striking findings of the study was the remarkable efficiency of the multifunctional composite in removing metal ions from water. While chitosan achieved a 60 percent adsorption efficiency for copper (Cu), cadmium (Cd), and lead (Pb) ions at a 5 g L^{-1} adsorbent dose, the multifunctional composite achieved an astounding 100 percent adsorption efficiency with a significantly reduced adsorbent amount of just 1 g L^{-1}. The multifunctional composite's remarkable speed in removing metal ions in just 10 minutes, its adaptability to different materials, its capacity for regeneration using EDTA solution, and its ability to undergo up to five cycles of efficient adsorption and desorption processes, offer promising prospects for more efficient and sustainable metal ion removal, addressing global water quality concerns (Alsabagh et al., 2015).

Researchers have developed an economical adsorbent, (BLP)/(AgNPs), by combining silver nanoparticles with banana leaves powder. This composite efficiently removes Zn (II), Pb (II), and Fe (III) ions from water, achieving removal rates of 79 percent, 88 percent, and 91percent in just 40 minutes. The Langmuir model best represents the adsorption data, highlighting maximum capacities of 190 mg/g for Zn (II), 244 mg/g for Pb (II), and 228 mg/g for Fe (III) ions. Thermodynamic analysis confirms the process is spontaneous and endothermic, while kinetics favor a pseudo-second-order model. This innovation promises eco-friendly metal ion removal for improved water quality worldwide (Darweesh et al., 2022). This study presents a straightforward method for synthesizing nano-fibrillated cellulose from fruit agricultural residues, yielding nanocellulose particles with diameters between 44–50 nm. Alkaline treatment effectively removed lignin and hemicellulose, as confirmed by FTIR analysis. Transmission electron microscopy (TEM) confirmed the presence of spherical nano-fibrillated cellulose particles measuring 44 nm. The study also successfully synthesized and characterized silver nanoparticles using orange peel extract as a reducing agent, with particle sizes of 15–16 nm at 25°C. Furthermore, the combination of silver and nano-fibrillated cellulose demonstrated efficient removal of heavy metal ions cadmium and chromium, achieving removal efficiencies of 83.49 percent and 32.20 percent, respectively (Tavker et al., 2021). Kenaf core was the raw biomass used for the fabrication of nanobiocomposite. The kenaf-based activated carbon (KAC), silver-based nanocomposite denoted as KAC-CS-AgNPs, was developed using a rapid and green synthesis method. This nanobiocomposite effectively addresses the remediation of toxic pollutants, such as Cu^{2+}, Pb^{2+}, and carcinogenic CR dye (Mandal et al., 2023).

Researchers developed cellulose infused with silver and zinc oxide nanoparticles (NPs) to remove heavy metal ions from water. Cotton infused with NPs was subjected to separate exposures in water containing Hg^{2+}, Cr^{3+}, Co^{2+}, Pb^{2+}, and Ni^{2+} at 25°C (pH

5.5) for 400 minutes. The NPs-infused cellulose effectively adsorbed various metal ions, with Hg^{2+} exhibiting the highest adsorption capacity (A. Ali et al., 2018). These examples underscore AgNPs versatility in efficiently removing various heavy metals from wastewater, providing valuable insights into their potential applications in water treatment and pollution mitigation efforts.

13.3.4 Magnetic Nanoparticles (MNPs)

Magnetic nanoparticles (MNPs) have emerged as a highly effective and versatile technology for the removal of heavy metals from water due to their unique magnetic properties and high surface area (Gao, 2019). Here are a few key examples of how magnetic nanoparticles are used for heavy metal removal.

Adsorption: Magnetic nanoparticles can be functionalized with specific ligands (Hong et al., 2020) or coatings (Chhetri et al., 2022) that have a strong affinity for heavy metal ions. When introduced into contaminated water, the MNPs adsorb heavy metal ions onto their surfaces. Common functional groups include thiol, amine, or carboxyl groups (T. Luo et al., 2021). Once the adsorption process is complete, an external magnetic field is applied to separate the loaded MNPs quickly and efficiently from the treated water, leaving behind purified water (Divate et al., 2022).

Plant-mediated magnetic nanoparticles (PMNPs) are a promising solution for the removal of heavy metals from environmental sources. These nanoparticles are synthesized through a green and sustainable approach, utilizing plant extracts rich in bioactive compounds as reducing and stabilizing agents (Das et al., 2018). The resulting PMNPs, often functionalized with magnetic materials, exhibit both strong heavy metal-binding capabilities and magnetic responsiveness. When introduced into contaminated water or soil, PMNPs adsorb heavy metal ions efficiently. They can be easily separated and recovered due to their magnetic properties, making them reusable after regeneration processes. PMNPs offer an eco-friendly and effective method for addressing heavy metal pollution in environmental remediation and wastewater treatment (I. Ali et al., 2017).

Nano-adsorbent Composites: Researchers have developed composite materials by combining MNPs with other nanomaterials like graphene or carbon nanotubes (Shukla et al., 2021). These composites exhibit enhanced adsorption capacities and selectivity for heavy metal ions. The synergistic effects of MNPs and other nanomaterials make them powerful tools for heavy metal removal (R. Wang et al., 2020).

Magnetic Nanoparticle Aggregates: Another approach involves inducing the aggregation of MNPs through the addition of specific agents. These aggregates can capture heavy metal ions through physical adsorption and chemical precipitation. Magnetic separation is then employed to recover the MNP aggregates along with the adsorbed heavy metals (Gutierrez et al., 2017).

The magnetic nanoparticles have become valuable tools in the field of heavy metal removal from water due to their magnetic recoverability and high adsorption capacities. By functionalizing MNPs or incorporating them into composites, researchers are continually improving their performance and selectivity for specific heavy metals.

These innovative applications hold significant promise for efficient and environmentally friendly water treatment processes, reducing heavy metal contamination and improving water quality (Xiaowang Liu et al., 2009).

13.2.5 Titanium Dioxide (TiO_2) Nanoparticles

The photodegradation capabilities of titanium dioxide (TiO_2) nanoparticles have been greatly recognized in the field of organic pollution, functioning as highly effective photocatalysts. Furthermore, they have been shown to be effective in removing heavy metals (Youssef & Malhat, 2014). A study developed a nanocomposite adsorbent, RS-HTO, by loading hydrated titanium oxide (HTO) nanoparticles onto rice straw (RS) to remove Cu^{2+} from water. RS-HTO had stable performance across different pH levels and ionic strengths. Cu^{2+} removal efficiency exceeded 99 percent at pH 7.5 with an initial concentration of 10 mg/L. Coexisting cations had minimal impact. Langmuir adsorption and pseudo-second-order kinetics described the process. In a fixed-bed setup, Cu^{2+} concentration in simulated wastewater was reduced to < 0.2 mg/L, and RS-HTO could be easily regenerated with HCl–NaCl solution (Y. Chen et al., 2020).

The study explores the use of titanium dioxide nanowires (TiO_2) as an innovative adsorbent for removing heavy metal ions (Pb^{2+}, Cu^{2+}, Fe^{3+}, Cd^{2+}, and Zn^{2+}) from contaminated water. TiO_2 nanowires exhibit high absorption efficiency, with Pb2+ showing the highest removal rate at 97.06 percent, while Zn^{2+} has the lowest removal rate at 35.18 percent. The study suggests that incorporating TiO_2 nanowires into water filters could be a promising and economically feasible technology for water purification, addressing the ecotoxicological hazard of heavy metal contamination in drinking water (Youssef & Malhat, 2014).

The study introduces a novel method for analyzing cadmium, cobalt, copper, nickel, and lead in water using starch-coated titanium dioxide nanoparticles as a nano sorbent. The method has been optimized for accuracy and speed. It successfully detected these elements in tap water samples with a 95 percent confidence level and achieved low detection limits of 0.05 μg/L for cadmium, 0.28 μg/L for cobalt, 1.90 μg/L for copper, 3.10 μg/L for nickel, and 0.11 μg/L for lead (3σ, $N = 10$). Additionally, the technique is simple, easy to use, and environmentally friendly (Baysal et al., 2018).

Recently, researchers explored the microwave-enhanced sorption (MES) approach to remove heavy metals from water, employing microwave-synthesized TiO_{2}-chitosan nanoparticles. The MES technique, proven to be both environmentally friendly and rapid, exhibited a maximum adsorption capacity of 1800 $\mu mol/g^{-1}$ for Cd (II). Furthermore, 86.80–88.01 percent and 72.56–70.67 percent of Cu (II) and Cd (II) could be efficiently removed from water within 60–70 seconds using the MES technique (Sinha & Chakma, 2019).

However, it is worth noting that one of the primary challenges associated with titanium dioxide nanoparticles lies in their intricate production processes and difficulties related to separation. This complexity often poses challenges when attempting to separate TiO_2 nanoparticles from water following their application in a slurry suspension (Magrì et al., 2018).

13.2.6 Zinc Oxide (ZnO) Nanoparticles

Zinc oxide nanoparticles have gained their popularity as adsorbents for heavy metals due to their high surface area, low cost, and extraordinary removal capacity. Nanosized zinc oxides have been reported to treat Cr (VI), Cu (II), Ni (II), and so on. Sheela et al. studied the removal performances of ZnO nanoparticles towards Zn (II), Cd (II), and Hg (II) by using the batch method. The studies have shown that the maximum adsorption capacities of ZnO nanoparticles were 357, 387, and 714 mg/g respectively by calculating from the Langmuir model, indicating a highly competitive adsorbent for these heavy metals. The effect of pH was investigated over the range of 4.0–8.0 (Stietiya & Wang, 2014).

The result indicated that pH significantly influenced the adsorption behaviours, which could be explained by the surface charge of the ZnO and the degree of speciation of sorbents. Besides, a study carried out by Ghiloufi et al. compared the adsorption differences between ZnO nanoparticles and calcium-doped ZnO nanoparticles towards Cr (VI), Cd (II), and Ni (II). The results indicated that the incorporation of Ca in ZnO nanoparticles could improve the uptake effect of these heavy metals. Somu et al. prepared zinc oxide nanoparticles by exploiting casein as the reducing and capping agent. The casein-capped ZnO nanoparticles with an average size of 10 nm were employed to treat wastewater containing three metals in batch mode. Adsorption data fitted the Langmuir model well and the adsorption capacities towards Cd (II), Pd (II), and Co (II) were 156.74, 194.93, 67.93 mg/g^{-1} respectively (Tabesh et al., 2018) (Siddiquah et al., 2018).

13.2.7 Other Nanomaterials for the Removal of Contaminants

Additional nanomaterials employed in heavy metal contaminant removal include aluminium oxide-based, magnesium oxide-based, cerium oxide-based, and zirconium oxide-based nanomaterials.

Aluminium Oxides-based Nanomaterials

Aluminium oxide-based nanomaterials represent a prominent class of metal adsorbents widely employed for the removal of heavy metals due to their cost-effectiveness and high decontamination efficiency. Aluminium oxide exists in various crystalline structures, such as α, γ, θ, η, among which γ-alumina stands as the most frequently utilized form. γ-Al_2O_3 nanoparticles possess considerable potential as adsorbents due to their expansive specific surface area, outstanding adsorption capacity, mechanical robustness, and suitability for low-temperature modification (Stietiya & Wang, 2014).

Notably, Tabesh et al. recently synthesized γ-Al_2O_3 nanoparticles using a modified sol-gel method and assessed their adsorption performance toward Pb (II) and Cd (II). These nanoparticles exhibited an average size ranging from 6 to 13 nm and achieved a maximum removal efficiency of 97 percent for Pb (II) and 87 percent for Cd (II) (Tabesh et al., 2018).

The adsorption of Pb (II) and Cd (II) on γ-Al_2O_3 nanoparticles conformed to the Freundlich isotherm, with respective adsorption capacities of 47.08 mg/g^{-1} and 17.22 mg/g^{-1}. Furthermore, a study investigating the impact of phosphate (PO4), citrate, and humic acid (HA) on the adsorption of Al_2O_3 nanoparticles toward Zn (II) and Cd (II) revealed that PO_4 and HA enhanced the adsorption of both Zn (II) and Cd (II), whereas citrate reduced the adsorption of Zn (II) in a single-metal system(Stietiya & Wang, 2014).

Apart from the heavy metals, Al_2O_3 nanomaterials have also demonstrated effective removal of Hg (II), As (III), Cu (II), Ni (II), Cr (VI), among others, showcasing their promising removal capabilities across various elements (X. Wang et al., 2015).

Magnesium Oxides-based Nanomaterials

Magnesium oxide nanoparticles stand out as a highly promising solution for the removal of heavy metals from various environmental contexts, thanks to their exceptional attributes. These nanoparticles offer an extraordinary adsorption capacity, making them particularly effective at sequestering heavy metal contaminants. Additionally, their cost-effectiveness, non-toxic nature, widespread availability, and eco-friendly characteristics contribute to their appeal as a go-to adsorbent (Rafiq et al., 2014).

Zirconium Oxides-based Nanomaterials

Nanoscale zirconium oxides represent a promising class of metallic oxide adsorbents for the removal of heavy metals from wastewater. Their appeal lies in the abundance of -OH groups on their surfaces and their substantial surface areas. Additionally, nanosized zirconium oxides exhibit remarkable chemical stability and exceptional adsorption affinities for heavy metals such as Pb (II), Zn (II), and Cd (II) (Pica, 2021).

13.3 RECENT ADVANCEMENT

In the realm of contemporary environmental remediation, advanced nanomaterials play a pivotal role in the quest for efficient heavy metal removal from wastewater and contaminated soils. These sophisticated materials represent a technological frontier, offering a repertoire of capabilities that transcend traditional remediation methods. Let us explore in greater detail some of the pivotal advanced nanomaterials molding this landscape.

Modified Nanoscale Zero-valent Iron (nZVI):

Engineered for enhanced reactivity and performance, modified nanoscale zero-valent iron (nZVI) stands out as a powerhouse in heavy metal removal. This engineered version of zero-valent iron particles demonstrates a remarkable ability to adsorb (Novikov et al. 2016)(Novikov et al. 2016)heavy metal ions, facilitating their efficient removal from environmental matrices.

Graphene-based Composites

Graphene-based composites harnessing the extraordinary properties of graphene, take center stage in the pursuit of effective heavy metal remediation. With its extensive surface area and unique electronic characteristics, graphene serves as a versatile platform for designing composites that excel in adsorbing and immobilizing heavy metal ions(Novikov et al. 2016).

Nano-engineered Carbon Nanotubes (CNTs)

The tubular wonders of nanoengineered carbon nanotubes (CNTs) offer a structurally advantageous framework for heavy metal removal. Their unique geometry and surface properties make them highly effective in adsorbing heavy metals, contributing to the overall efficiency of nanotechnology-based remediation strategies(Wayu et al. 2013).

Metal-organic Frameworks (MOFs)

Metal-organic frameworks (MOFs) represent another innovative class of nanomaterials contributing to the arsenal against heavy metal contamination. These porous structures, composed of metal ions or clusters coordinated with organic ligands, provide tailored sites for efficient heavy metal adsorption.

These advanced nanomaterials are not confined to a singular form; rather, they exhibit versatility in structural design. Nanofiberss, nanocomposites (Liu et al. 2009) hydrogels (Thakur et al. 2018) [73], and nanofiltration membranes (Covaliu-Mierlă, Păunescu, and Iovu 2023) [83] are just a few examples of the diverse structural adaptations these materials can undergo, expanding their applicability across various environmental contexts.

In practical applications, these nanomaterials leverage their inherent characteristics to adsorb and immobilize heavy metal ions selectively. Ligand functionalization further enhances their specificity, contributing to the precision required in targeted heavy metal removal efforts. The choice of a particular nanomaterial hinges on several considerations, including cost-effectiveness, the feasibility of synthesis methods, and the specific heavy metals earmarked for removal.

As research in this field continues to evolve, ongoing efforts are dedicated to pushing the boundaries of innovation. Researchers explore novel synthesis techniques, delve into the synergistic effects of combining different nanomaterials in composites, and fine-tune the application methods for optimal efficiency. The quest for safer, more sustainable, and highly effective large-scale implementation of these advanced nanomaterials in heavy metal remediation remains at the forefront of environmental science and engineering.

Advanced nanomaterials like modified nanoscale zero-valent iron (nZVI)(B. Wang et al., 2021), graphene-based composites (White et al., 2018), nanoengineered carbon nanotubes (CNTs) (Hasbullah et al., 2019), metal-organic frameworks (MOFs), and others are utilized for the efficient removal of heavy metals from sources such as wastewater and contaminated soils. These nanomaterials offer high surface area and selectivity, and they can be functionalized or integrated into various forms like nanofibers, nanocomposites (Magrì et al., 2018), hydrogels (Sinha & Chakma, 2019), and nanofiltration membranes (Zekić et al., 2018). Their applications are based on their ability to adsorbed and immobilize heavy metal ions, often with enhanced selectivity through ligand functionalization. The choice of nanomaterial depends on factors such as cost, synthesis feasibility, and the specific heavy metals to be removed, with ongoing research driving advancements in this field (Jain, 2022; J. Luo et al., 2021; Yu et al., 2021).

13.4 DRAWBACKS

Nanoparticles have gained considerable attention in wastewater treatment due to their unique properties and potential to address various contaminants effectively (Manager & Minerals, 2016). However, their use is not without drawbacks. One significant drawback is the potential environmental and health risks associated with nanoparticle release into aquatic ecosystems (Shoaib et al., 2022). When nanoparticles are introduced into wastewater, there is a risk that they may not be completely removed during treatment processes, leading to their discharge into receiving waters. Once in the environment, nanoparticles can persist for extended periods and may have toxic effects on aquatic organisms, disrupting ecosystems and potentially entering the food chain (Berekaa, 2016; Gehrke et al., 2015; Roy, 2014).

Additionally, the long-term impact of nanoparticles on human health is still a subject of ongoing research and debate. There are concerns that the inhalation or ingestion of nanoparticles, either directly from wastewater treatment facilities or indirectly through contaminated water sources, could pose health risks, including inflammation, cellular damage, and potential toxicity (Lofrano et al., 2017).

Another drawback is the cost associated with producing and using nanoparticles in wastewater treatment. The production of nanoparticles often involves complex and energy-intensive processes, making them more expensive than traditional treatment methods. Furthermore, their efficient and sustainable recovery and reuse in treatment processes remain challenging, contributing to higher operational costs (Manager & Minerals, 2016).

Lastly, the regulatory framework surrounding the use of nanoparticles in wastewater treatment is still evolving, with limited standardized guidelines in place. This regulatory uncertainty can lead to challenges in ensuring the safe and responsible deployment of nanoparticles in wastewater treatment systems (Qu et al., 2013).

13.5 CONCLUSION

The chapter highlights the critical need to combat heavy metal contamination in wastewater and explores nanotechnology-based solutions. Nanomaterials like nZVI, CNTs, GO, AgNPs, MNPs, and TiO2 nanoparticles offer efficient removal capabilities, presenting a promising alternative to traditional methods. While their success is evident, challenges such as environmental impact and scalability require careful consideration. This underscores the importance of ongoing research to fine-tune these strategies, ensuring their effectiveness and safety for large-scale implementation. The chapter contributes valuable insights into the potential of nanotechnology in addressing the complex issue of heavy metal contamination, urging continued efforts to advance these innovative solutions.

REFERENCES

Abu-Dalo, M. A., Al-Rosan, S. A., & Albiss, B. A. (2021). Photocatalytic degradation of methylene blue using polymeric membranes based on cellulose acetate impregnated with ZnO nanostructures. *Polymers*, *13*(19) , 3451. https://doi.org/10.3390/polym13193451

Ahamed, A. J., & Loganathan, K. (2023). Reduced graphene oxide as effective adsorbent for removal of heavy metals in groundwater of Amaravathi River basin, Tamil Nadu. *Geology, Ecology, and Landscapes*, *7*(1), 59–68. https://doi.org/10.1080/24749508.2021.1923273

Al-Enezi, G., Hamoda, M. F., & Fawzi, N. (2004). Ion exchange extraction of heavy metals from wastewater sludges. *Journal of Environmental Science and Health – Part A Toxic/Hazardous Substances and Environmental Engineering*, *39*(2), 455–464. https://doi.org/10.1081/ESE-120027536

Al-Rifai, A., Aqel, A., Wahibi, L. Al, ALOthman, Z. A., & Badjah-Hadj-Ahmed, A. Y. (2018). Carbon nanotube-based benzyl polymethacrylate composite monolith as a solid phase extraction adsorbent and a stationary phase material for simultaneous extraction and analysis of polycyclic aromatic hydrocarbon in water. *Journal of Chromatography A*, *1535*, 17–26. https://doi.org/10.1016/j.chroma.2018.01.011

Ali, A., Mannan, A., Hussain, I., Hussain, I., & Zia, M. (2018). Effective removal of metal ions from aqueous solution by silver and zinc nanoparticles functionalized cellulose: Isotherm, kinetics and statistical supposition of process. *Environmental Nanotechnology, Monitoring and Management*, *9*(May 2017), 1–11. https://doi.org/10.1016/j.enmm.2017.11.003

Ali, I., & Gupta, V. K. (2007). Advances in water treatment by adsorption technology. *Nature Protocols*, *1*(6), 2661–2667. https://doi.org/10.1038/nprot.2006.370

Ali, I., Peng, C., Naz, I., Khan, Z. M., Sultan, M., Islam, T., & Abbasi, I. A. (2017). Phytogenic magnetic nanoparticles for wastewater treatment: A review. *RSC Advances*, *7*(64), 40158–40178. https://doi.org/10.1039/c7ra04738j

Alsabagh, A. M., Fathy, M., & Morsi, R. E. (2015). Preparation and characterization of chitosan/silver nanoparticle/copper nanoparticle/carbon nanotube multifunctional nanocomposite for water treatment: Heavy metals removal; kinetics, isotherms and competitive studies. *RSC Advances*, *5*(69), 55774–55783. https://doi.org/10.1039/c5ra07477k

Ayalew, Z. M., Guo, X., & Zhang, X. (2022). Synthesis and application of polyethyleneimine (PEI)-based composite/nanocomposite material for heavy metals removal from wastewater: A critical review. *Journal of Hazardous Materials Advances*, *8*(June), 100158. https://doi.org/10.1016/j.hazadv.2022.100158

Baby, R., Hussein, M. Z., Abdullah, A. H., & Zainal, Z. (2022). Nanomaterials for the treatment of heavy metal contaminated water. *Polymers*, *14*(3), 1–17. https://doi.org/10.3390/polym14030583

Baysal, A., Kuznek, C., & Ozcan, M. (2018). Starch coated titanium dioxide nanoparticles as a challenging sorbent to separate and preconcentrate some heavy metals using graphite furnace atomic absorption spectrometry. *International Journal of Environmental Analytical Chemistry*, *98*(1), 45–55. https://doi.org/10.1080/03067319.2018.1427741

Berekaa, M. M. (2016). Nanotechnology in wastewater treatment; influence of nanomaterials on microbial systems. *International Journal of Current Microbiology and Applied Sciences*, *5*(1), 713–726. https://doi.org/10.20546/ijcmas.2016.501.072

Borji, H., Ayoub, G. M., Bilbeisi, R., Nassar, N., & Malaeb, L. (2020). How effective are nanomaterials for the removal of heavy metals from water and wastewater? *Water, Air, and Soil Pollution*, *231*(7), 330. https://doi.org/10.1007/s11270-020-04681-0

Briffa, J., Sinagra, E., & Blundell, R. (2020). Heavy metal pollution in the environment and their toxicological effects on humans. *Heliyon, 6*(9), e04691. https://doi.org/10.1016/j.heliyon.2020.e04691

Chadha, U., Selvaraj, S. K., Vishak Thanu, S., Cholapadath, V., Abraham, A. M., Zaiyan M., M., Manoharan, M., & Paramsivam, V. (2022). A review of the function of using carbon nanomaterials in membrane filtration for contaminant removal from wastewater. *Materials Research Express, 9*(1), 012003. https://doi.org/10.1088/2053-1591/ac48b8

Charerntanyarak, L. (1999). Heavy metals removal by chemical coagulation and precipitation. *Water Science and Technology, 39*(10–11), 135–138. https://doi.org/10.1016/S0273-1223(99)00304-2

Chen, J., Qiu, F., Xu, W., Cao, S., & Zhu, H. (2015). Recent progress in enhancing photocatalytic efficiency of TiO_2-based materials. *Applied Catalysis A: General, 495*, 131–140. https://doi.org/10.1016/j.apcata.2015.02.013

Chen, Y., Shi, H., Guo, H., Ling, C., Yuan, X., & Li, P. (2020). Hydrated titanium oxide nanoparticles supported on natural rice straw for Cu(II) removal from water. *Environmental Technology and Innovation, 20*, 101143. https://doi.org/10.1016/j.eti.2020.101143

Cheriyamundath, S., & Vavilala, S. L. (2021). Nanotechnology-based wastewater treatment. *Water and Environment Journal, 35*(1), 123–132. https://doi.org/10.1111/wej.12610

Chhetri, T., Cunningham, G., Suresh, D., Shanks, B., Kannan, R., Upendran, A., & Afrasiabi, Z. (2022). Wastewater treatment using novel magnetic nanosponges. *Water (Switzerland), 14*(3), 1–12. https://doi.org/10.3390/w14030505

Covaliu-Mierlă, C. I., Păunescu, O., & Iovu, H. (2023). Recent advances in membranes used for nanofiltration to remove heavy metals from wastewater: A review. *Membranes, 13*(7), 643. https://doi.org/10.3390/membranes13070643

Darweesh, M. A., Elgendy, M. Y., Ayad, M. I., Ahmed, A. M. M., Kamel Elsayed, N. M., & Hammad, W. A. (2022). A unique, inexpensive, and abundantly available adsorbent: Composite of synthesized silver nanoparticles (AgNPs) and banana leaves powder (BLP). *Heliyon, 8*(4), e09279. https://doi.org/10.1016/j.heliyon.2022.e09279

Das, P., Ghosh, S., Ghosh, R., Dam, S., & Baskey, M. (2018). *Madhuca longifolia* plant mediated green synthesis of cupric oxide nanoparticles: A promising environmentally sustainable material for waste water treatment and efficient antibacterial agent. *Journal of Photochemistry and Photobiology B: Biology, 189*, 66–73. https://doi.org/10.1016/j.jphotobiol.2018.09.023

Divate, S. B., Chavan, N. B., & Kadu, P. S. (2022). A review: Nanomaterials for removal of heavy metal from wastewater. *Journal of Pharmaceutical Negative Results, 13*(7), 7878–7881. https://doi.org/10.47750/pnr.2022.13.S07.955

Doyle, S., Meade, E., Fowley, C., & Garvey, M. (2020). A comprehensive review of current environmental pollutants of pharmaceutical, agricultural and industrial origin. *European Journal of Experimental Biology, 10*(4), 1–13. https://doi.org/10.36648/2248-9215.10.2.102

Gao, F. (2019). An overview of surface-functionalized magnetic nanoparticles: Preparation and application for wastewater treatment. *ChemistrySelect, 4*(22), 6805–6811. https://doi.org/10.1002/slct.201900701

Gehrke, I., Geiser, A., & Somborn-Schulz, A. (2015). Innovations in nanotechnology for water treatment. *Nanotechnology, Science and Applications, 8*, 1–17. https://doi.org/10.2147/NSA.S43773

Ghasemzadeh, G., Momenpour, M., Omidi, F., Hosseini, M. R., Ahani, M., & Barzegari, A. (2014). Applications of nanomaterials in water treatment and environmental remediation. *Frontiers of Environmental Science and Engineering, 8*(4), 471–482. https://doi.org/10.1007/s11783-014-0654-0

Gontrani, L., Pulci, O., Carbone, M., Pizzoferrato, R., & Prosposito, P. (2021). Detection of heavy metals in water using graphene oxide quantum dots: An experimental and theoretical study. *Molecules*, *26*(18), 1–14. https://doi.org/10.3390/molecules26185519

Gutierrez, A. M., Dziubla, T. D., & Hilt, J. Z. (2017). Recent advances on iron oxide magnetic nanoparticles as sorbents of organic pollutants in water and wastewater treatment. *Reviews on Environmental Health*, *32*(1–2), 111–117. https://doi.org/10.1515/reveh-2016-0063

Hasbullah, H., Sabri, N. S. M., Said, N., Rosid, S. M., Roslan, M. I., Ismail, A. F., Jye, L. W., & Yusof, N. (2019). Nanoengineered materials for water and wastewater treatments. In *Nanotechnology in Water and Wastewater Treatment: Theory and Applications,* Editors: Amimul Ahsan and Ahmad Fauzi Ismail (pp. 303–335). Elsevier. https://doi.org/10.1016/B978-0-12-813902-8.00016-2

Hasdi, N. D. (2020). Reviewing methods to prepare activated carbon from various sources. *Nanoscale Research Letters*, *14*(341), 1–17.

Hashem, E. A. (2014). Nanotechnology in water treatment, case study: Egypt. *Journal of Economics and Development Studies*, *2*(3), 244–259. https://doi.org/10.15640/jeds.v2n3a18

Hong, J., Xie, J., Mirshahghassemi, S., & Lead, J. (2020). Metal (Cd, Cr, Ni, Pb) removal from environmentally relevant waters using polyvinylpyrrolidone-coated magnetite nanoparticles. *RSC Advances*, *10*(6), 3266–3276. https://doi.org/10.1039/c9ra10104g

Ibrahim, R. K., Hayyan, M., AlSaadi, M. A., Hayyan, A., & Ibrahim, S. (2016). Environmental application of nanotechnology: Air, soil, and water. *Environmental Science and Pollution Research*, *23*(14), 13754–13788. https://doi.org/10.1007/s11356-016-6457-z

Islam, M. A., Jacob, M. V., & Antunes, E. (2021). A critical review on silver nanoparticles: From synthesis and applications to its mitigation through low-cost adsorption by biochar. *Journal of Environmental Management*, *281*(July 2020), 111918. https://doi.org/10.1016/j.jenvman.2020.111918

Islam, M. S., Sazawa, K., Sugawara, K., & Kuramitz, H. (2023). Electrochemical biosensor for evaluation of environmental pollutants toxicity. *Environments – MDPI*, *10*(4), 63. https://doi.org/10.3390/environments10040063

Jain, R. (2022). Recent advances of magnetite nanomaterials to remove arsenic from water. *RSC Advances*, *12*(50), 32197–32209. https://doi.org/10.1039/d2ra05832d

Jjagwe, J., Olupot, P. W., Menya, E., & Kalibbala, H. M. (2021). Synthesis and application of granular activated carbon from biomass waste materials for water treatment: A review. *Journal of Bioresources and Bioproducts*, *6*(4), 292–322. https://doi.org/10.1016/j.jobab.2021.03.003

Ken, D. S., & Sinha, A. (2020). Recent developments in surface modification of nano zero-valent iron (nZVI): Remediation, toxicity and environmental impacts. *Environmental Nanotechnology, Monitoring and Management*, *14*(July 2019), 100344. https://doi.org/10.1016/j.enmm.2020.100344

Kiruba Daniel, S. C. G., Malathi, S., Balasubramanian, S., Sivakumar, M., & Anitha Sironmani, T. (2014). Multifunctional silver, copper and zero valent iron metallic nanoparticles for wastewater treatment. In *Application of Nanotechnology in Water Research,* Editor: Ajay Kumar Mishra (pp. 435–457). Wiley. https://doi.org/10.1002/9781118939314.ch15

Koul, B., & Taak, P. (2018). Nanobioremediation. In *Biotechnological Strategies for Effective Remediation of Polluted Soils*, Editors: Bhupendra Koul and Pooja Taak (pp. 197–220). Springer Singapore. https://doi.org/10.1007/978-981-13-2420-8_8

Krishna, R. H., Chandraprabha, M. N., Samrat, K., Krishna Murthy, T. P., Manjunatha, C., & Kumar, S. G. (2023). Carbon nanotubes and graphene-based materials for adsorptive removal of metal ions – A review on surface functionalization and related adsorption mechanism. *Applied Surface Science Advances*, *16*(June), 100431. https://doi.org/10.1016/j.apsadv.2023.100431

Kumar, R., Rauwel, P., & Rauwel, E. (2021). Nanoadsorbants for the removal of heavy metals from contaminated water: Current scenario and future directions. *Processes*, *9*(8), 1379. https://doi.org/10.3390/pr9081379

Liosis, C., Papadopoulou, A., Karvelas, E., Karakasidis, T. E., & Sarris, I. E. (2021). Heavy metal adsorption using magnetic nanoparticles for water purification: A critical review. *Materials*, *14*(24), 7500. https://doi.org/10.3390/ma14247500

Liu, Xiaowang, Hu, Q., Fang, Z., Zhang, X., & Zhang, B. (2009). Magnetic chitosan nanocomposites: A useful recyclable tool for heavy metal ion removal. *Langmuir*, *25*(1), 3–8. https://doi.org/10.1021/la802754t

Liu, Xuyang, Ray, J. R., Neil, C. W., Li, Q., & Jun, Y. S. (2015). Enhanced colloidal stability of CeO_2 nanoparticles by ferrous ions: Adsorption, redox reaction, and surface precipitation. *Environmental Science and Technology*, *49*(9), 5476–5483. https://doi.org/10.1021/es506363x

Lofrano, G., Libralato, G., & Brown, J. (2017). *Nanotechnologies for Environmental Remediation: Applications and Implications*. Springer Cham. https://doi.org/10.1007/978-3-319-53162-5

Luo, J., Fu, K., Yu, D., Hristovski, K. D., Westerhoff, P., & Crittenden, J. C. (2021). Review of advances in engineering nanomaterial adsorbents for metal removal and recovery from water: Synthesis and microstructure impacts. *ACS ES&T Engineering*, *1*(4), 623–661. https://doi.org/10.1021/acsestengg.0c00174

Luo, T., Yang, C., Tian, X., Luo, W., Nie, Y., & Wang, Y. (2021). Application of iron oxide nanomaterials for the remsoval of heavy metals. In *Handbook of Nanomaterials and Nanocomposites for Energy and Environmental Applications: Volumes 1–4*, Editors: Boris Ildusovich Kharisov, Leticia Myriam Torres-Martinez, Oxana Vasilievna Kharissova, (pp. 2067–2092). Springer International Publishing. https://doi.org/10.1007/978-3-030-36268-3_76

Ma, X., Tao, H., Yang, K., Feng, L., Cheng, L., Shi, X., Li, Y., Guo, L., & Liu, Z. (2012). A functionalized graphene oxide–iron oxide nanocomposite for magnetically targeted drug delivery, photothermal therapy, and magnetic resonance imaging. *Nano Research*, *5*(3), 199–212. https://doi.org/10.1007/s12274-012-0200-y

Magrì, D., Caputo, G., Perotto, G., Scarpellini, A., Colusso, E., Drago, F., Martucci, A., Athanassiou, A., & Fragouli, D. (2018). Titanate fibroin nanocomposites: A novel approach for the removal of heavy-metal ions from water. *ACS Applied Materials and Interfaces*, *10*(1), 651–659. https://doi.org/10.1021/acsami.7b15440

Mallakpour, S., & Khadem, E. (2018). Carbon nanotubes for heavy metals removal. In *Composite Nanoadsorbents,* Editor: George Z. Kyzas and Athanasios C. Mitropoulos (pp. 181–210). Elsevier Inc. https://doi.org/10.1016/B978-0-12-814132-8.00009-5

Manager, A. G., & Minerals, K. (2016). Applications of nanotechnology in waste water treatment: A review. *Imperial Journal of Interdisciplinary Research (IJIR)*, *2*(11), 613-26.

Mandal, S., Hwang, S., Marpu, S. B., Omary, M. A., Prybutok, V., & Shi, S. Q. (2023). Bioinspired synthesis of silver nanoparticles for the remediation of toxic pollutants and enhanced antibacterial activity. *Biomolecules*, *13*(7), 1054. https://doi.org/10.3390/biom13071054

Masarwa, A., Meyerstein, D., Daltrophe, N., & Kedem, O. (1997). Compact accelerated precipitation softening (CAPS) as pretreatment for membrane desalination II. Lime softening with concomitant removal of silica and heavy metals. *Desalination*, *113*(1), 73–84. https://doi.org/10.1016/S0011-9164(97)00116-1

Negi, S., Singh, V., & Rawat, J. (2021). Green synthesis of silver nanoparticles using microalgal extract and its application in metal ion removal from aqueous solution. *Journal of Experimental Biology and Agricultural Sciences*, *9*(2), 214–230. https://doi.org/10.18006/2021.9(2).214.230

Palani, G., Trilaksana, H., Sujatha, R. M., Kannan, K., Rajendran, S., Korniejenko, K., Nykiel, M., & Uthayakumar, M. (2023). Silver nanoparticles for waste water management. *Molecules*, *28*(8), 3520. https://doi.org/10.3390/molecules28083520

Pandey, G. (2018). Prospects of nanobioremediation in environmental cleanup. *Oriental Journal of Chemistry*, *34*(6), 2838–2850. https://doi.org/10.13005/ojc/340622

Park, C. M., Wang, D., & Su, C. (2018). Recent developments in engineered nanomaterials for water treatment and environmental remediation. In *Handbook of Nanomaterials for Industrial Applications,* Editor: Chaudhery Mustansar Hussain, (pp. 849–882). Elsevier. https://doi.org/10.1016/B978-0-12-813351-4.00048-1

Pica, M. (2021). Treatment of wastewaters with zirconium phosphate based materials: A review on efficient systems for the removal of heavy metal and dye water pollutants. *Molecules*, *26*(8), 2392. https://doi.org/10.3390/molecules26082392

Pohl, A. (2020). Removal of heavy metal ions from water and wastewaters by sulfur-containing precipitation agents. *Water, Air, and Soil Pollution*, *231*(10), 1–17. https://doi.org/10.1007/s11270-020-04863-w

Puthukkara P., A. R., Jose T., S., & lal S., D. (2021). Plant mediated synthesis of zero valent iron nanoparticles and its application in water treatment. *Journal of Environmental Chemical Engineering*, 9(1), 104569. https://doi.org/10.1016/j.jece.2020.104569

Qu, X., Alvarez, P. J. J., & Li, Q. (2013). Applications of nanotechnology in water and wastewater treatment. *Water Research*, *47*(12), 3931–3946. https://doi.org/10.1016/j.watres.2012.09.058

Rafiq, Z., Nazir, R., Durr-E-Shahwar, Shah, M. R., & Ali, S. (2014). Utilization of magnesium and zinc oxide nano-adsorbents as potential materials for treatment of copper electroplating industry wastewater. *Journal of Environmental Chemical Engineering*, *2*(1), 642–651. https://doi.org/10.1016/j.jece.2013.11.004

Rizwan, M., Singh, M., Mitra, C. K., & Morve, R. K. (2014). Ecofriendly application of nanomaterials: Nanobioremediation. *Journal of Nanoparticles*, *2014*, 1–7. https://doi.org/10.1155/2014/431787

Roy, A. (2014). *Nanotechnology in Industrial Wastewater Treatment. Water Intelligence Online* (Vol. 13). IWA Publishing. https://doi.org/10.2166/9781780406886

Sadeghi, M. H., Tofighy, M. A., & Mohammadi, T. (2020). One-dimensional graphene for efficient aqueous heavy metal adsorption: Rapid removal of arsenic and mercury ions by graphene oxide nanoribbons (GONRs). *Chemosphere*, *253*, 126647. https://doi.org/10.1016/j.chemosphere.2020.126647

Sakhi, D., Rakhila, Y., Elmchaouri, A., Abouri, M., Souabi, S., & Jada, A. (2019). Optimization of coagulation flocculation process for the removal of heavy metals from real textile wastewater. *Advances in Intelligent Systems and Computing*, *913*, 257–266. https://doi.org/10.1007/978-3-030-11881-5_22

Salehi, M., Sharafoddinzadeh, D., Mokhtari, F., Esfandarani, M. S., & Karami, S. (2021). Electrospun nanofibers for efficient adsorption of heavy metals from water and wastewater. *Clean Technologies and Recycling*, *1*(1), 1–33. https://doi.org/10.3934/ctr.2021001

Samarghandi, M. R., Dargahi, A., Shabanloo, A., Nasab, H. Z., Vaziri, Y., & Ansari, A. (2020). Electrochemical degradation of methylene blue dye using a graphite doped PbO_2 anode: Optimization of operational parameters, degradation pathway and improving the biodegradability of textile wastewater. *Arabian Journal of Chemistry*, *13*(8), 6847–6864. https://doi.org/10.1016/j.arabjc.2020.06.038

Sharma, U., & Sharma, J. G. (2022). Nanotechnology for the bioremediation of heavy metals and metalloids. *Journal of Applied Biology and Biotechnology*, *10*(5), 34–43. https://doi.org/10.7324/JABB.2022.100504

Shaukat, S., Hassani, M. A., Yadgari, M. Y., Ullah, S., Iqbal, M. S., Khan, F., Bibi, S. D., Ullah, Z., Hassani, M. A., Rukh, G., Sabir, M., Hussain, S. A., Arshad, N., Ali, S., Izhar, S. K., & Afsar, S. (2022). Green synthesis of silver nanoparticles and its application towards As(V) removal from aqueous systems. *Digest Journal of Nanomaterials and Biostructures*, *17*(4), 1385–1398. https://doi.org/10.15251/DJNB.2022.174.1385

Shi, W., Li, W., Nguyen, W., Chen, W., Wang, J., & Chen, M. (2022). Advances of metal organic frameworks in analytical applications. *Materials Today Advances*, *15*, 100273. https://doi.org/10.1016/j.mtadv.2022.100273

Shoaib, R. M., Abdel-Razik, A. B., Ibrahim, M. M., Al-Kordy, M. A., & Taha, E. H. (2022). Impact of engineered nano silver on plant parasitic nematode and measurement of DNA damage. *Egyptian Journal of Chemistry*, *65*(4), 43–51. https://doi.org/10.21608/EJCHEM.2021.99035.4605

Shukla, S., Khan, R., & Daverey, A. (2021). Synthesis and characterization of magnetic nanoparticles, and their applications in wastewater treatment: A review. *Environmental Technology and Innovation*, *24*, 101924. https://doi.org/10.1016/j.eti.2021.101924

Siddiquah, A., Hashmi, S. S., Mushtaq, S., Renouard, S., Blondeau, J. P., Abbasi, R., Hano, C., & Abbasi, B. H. (2018). Exploiting in vitro potential and characterization of surface modified zinc oxide nanoparticles of *Isodon rugosus* extract: Their clinical potential towards HepG2 cell line and human pathogenic bacteria. *EXCLI Journal*, *17*, 671–687. https://doi.org/10.17179/excli2018-1327

Singh, J., Dutta, T., Kim, K. H., Rawat, M., Samddar, P., & Kumar, P. (2018). "Green" synthesis of metals and their oxide nanoparticles: Applications for environmental remediation. *Journal of Nanobiotechnology*, *16*(1), 84. https://doi.org/10.1186/s12951-018-0408-4

Sinha, V., & Chakma, S. (2019). Advances in the preparation of hydrogel for wastewater treatment: A concise review. *Journal of Environmental Chemical Engineering*, *7*(5), 103295. https://doi.org/10.1016/j.jece.2019.103295

Stietiya, M. H., & Wang, J. J. (2014). Zinc and cadmium adsorption to aluminum oxide nanoparticles affected by naturally occurring ligands. *Journal of Environmental Quality*, *43*(2), 498–506. https://doi.org/10.2134/jeq2013.07.0263

Tabesh, S., Davar, F., & Loghman-Estarki, M. R. (2018). Preparation of γ-Al_2O_3 nanoparticles using modified sol-gel method and its use for the adsorption of lead and cadmium ions. *Journal of Alloys and Compounds*, 730, 441–449. https://doi.org/10.1016/j.jallcom.2017.09.246

Tarekegn, M. M., Hiruy, A. M., & Dekebo, A. H. (2021). Nano zero valent iron (nZVI) particles for the removal of heavy metals (Cd^{2+}, Cu^{2+} and Pb^{2+}) from aqueous solutions. *RSC Advances*, *11*(30), 18539–18551. https://doi.org/10.1039/d1ra01427g

Tavker, N., Yadav, V. K., Yadav, K. K., Cabral-Pinto, M. M. S., Alam, J., Shukla, A. K., Ali, F. A. A., & Alhoshan, M. (2021). Removal of cadmium and chromium by mixture of silver nanoparticles and nano-fibrillated cellulose isolated from waste peels of *Citrus sinensis*. *Polymers*, *13*(2), 1–14. https://doi.org/10.3390/polym13020234

Tran, L. T., Tran, H. V., Le, T. D., Bach, G. L., & Tran, L. D. (2019). Studying Ni(II) adsorption of magnetite/graphene oxide/chitosan nanocomposite. *Advances in Polymer Technology*, *2019*(3), 1–9. https://doi.org/10.1155/2019/8124351

Wang, B., Deng, C., Ma, W., & Sun, Y. (2021). Modified nanoscale zero-valent iron in persulfate activation for organic pollution remediation: a review. *Environmental Science and Pollution Research*, *28*(26), 34229–34247. https://doi.org/10.1007/s11356-021-13972-w

Wang, R., Chen, K., Peng, S., Wang, Q. Y., Huang, S. H., Zhou, Q. H., Wang, J., & Lin, J. (2020). Environmentally friendly magnetic nanoparticles for efficient removal of methylene blue and Cr(VI) from water. *Nano*, *15*(10), 1–14. https://doi.org/10.1142/S17932 9202050126X

Wang, X., Zhan, C., Kong, B., Zhu, X., Liu, J., Xu, W., Cai, W., & Wang, H. (2015). Self-curled coral-like γ-Al_2O_3 nanoplates for use as an adsorbent. *Journal of Colloid and Interface Science*, *453*, 244–251. https://doi.org/10.1016/j.jcis.2015.03.065

White, R. L., White, C. M., Turgut, H., Massoud, A., & Tian, Z. R. (2018). Comparative studies on copper adsorption by graphene oxide and functionalized graphene oxide nanoparticles. *Journal of the Taiwan Institute of Chemical Engineers*, *85*, 18–28. https://doi.org/10.1016/j.jtice.2018.01.036

Xiang, H., Min, X., Tang, C. J., Sillanpää, M., & Zhao, F. (2022). Recent advances in membrane filtration for heavy metal removal from wastewater: A mini review. *Journal of Water Process Engineering*, *49*, 103023. https://doi.org/10.1016/j.jwpe.2022.103023

Xu, X., Zeng, J., Wu, Y., Wang, Q., Wu, S., & Gu, H. (2022). Preparation and application of graphene-based materials for heavy metal removal in tobacco industry: A review. *Separations*, *9*(12), 401. https://doi.org/10.3390/separations9120401

Yadav, K. K., Singh, J. K., Gupta, N., & Kumar, V. (2017). A review of nanobioremediation technologies for environmental cleanup: A novel biological approach. *Journal of Materials and Environmental Science*, *8*(2), 740–757.

Yadav, Krishna Kumar, Singh, J. K., Gupta, N., & Kumar, V. (2017). A review of nanobioremediation technologies for environmental cleanup: A novel biological approach. *Journal of Materials and Environmental Science*, *8*(2), 740–757. www.jmaterenvironsci.com/

Yang, J., Hou, B., Wang, J., Tian, B., Bi, J., Wang, N., Li, X., & Huang, X. (2019). Nanomaterials for the removal of heavy metals from wastewater. *Nanomaterials*, *9*(3). https://doi.org/10.3390/nano9030424

Youssef, A. M., & Malhat, F. M. (2014). Selective removal of heavy metals from drinking water using titanium dioxide nanowire. *Macromolecular Symposia*, *337*(1), 96–101. https://doi.org/10.1002/masy.201450311

Yu, G., Wang, X., Liu, J., Jiang, P., You, S., Ding, N., Guo, Q., & Lin, F. (2021). Applications of nanomaterials for heavy metal removal from water and soil: A review. *Sustainability (Switzerland)*, *13*(2), 1–14. https://doi.org/10.3390/su13020713

Zekić, E., Vuković, Ž., & Halkijević, I. (2018). Application of nanotechnology in wastewater treatment. *Gradjevinar*, *70*(4), 315–323. https://doi.org/10.14256/JCE.2165.2017

Zhao, C., Liu, G., Tan, Q., Gao, M., Chen, G., Huang, X., Xu, X., Li, L., Wang, J., Zhang, Y., & Xu, D. (2023). Polysaccharide-based biopolymer hydrogels for heavy metal detection and adsorption. *Journal of Advanced Research*, *44*, 53–70. https://doi.org/10.1016/j.jare.2022.04.005

Zhao, X., Liu, W., Cai, Z., Han, B., Qian, T., & Zhao, D. (2016). An overview of preparation and applications of stabilized zero-valent iron nanoparticles for soil and groundwater remediation. *Water Research*, *100*, 245–266. https://doi.org/10.1016/J.WATRES.2016.05.019

14 Emerging Nanofiber Technology for the Purification of Wastewater Treatment Facilities for Elimination of Ions

H. Mittal and O.S. Kushwaha

14.1 INTRODUCTION

The primary constituents of nanofiber materials are two-dimensional nanosheets, one-dimensional nanowires, nanorods, nanofibers, nanotubes, and zero-dimensional nanoparticles, also known as quantum dots [1]. The research and academic studies on one-dimensional nanofibers are developing rapidly, which may be correlated to their high demand in preparing composites and membranes for advanced applications, including water treatment. Presently, there is no indication that the rate at which synthesis methods are developed and the application of nanofibers and their nanocomposites are being published will slow down anytime soon [2], [3], [4], [5]. Several challenges must be addressed and solved to move beyond the current stage of nanofiber syntheses and applications towards reality in commercial and industrial settings. First and foremost, it is crucial to create novel nanofiber synthesis techniques that combine the most significant features of all current and emerging technologies to produce high-quality nanofibers at high commercial throughput. Secondly, the next major challenge after producing nanofibers is identifying their practical uses that will change [6]. Two of their most prominent features are their remarkably high porosity and surface area-to-volume ratio, which make them a powerful and attractive candidate for many cutting-edge applications. Investigating the possible functional uses of nanofibers in areas including energy production and storage, water and environmental treatment, healthcare, and biomedical engineering has received much attention in recent years. This has happened while the synthesis and characterisation of nanofibers have advanced quickly [7].

When wastewater treatment entails different ways of wastewater treatment, especially treatment of several wastewaters like dairy wastewater, pharmaceutical

DOI: 10.1201/9781003454199-14

wastewater, and so forth, treatment could occur through mixed and isolated remedies. This treatment method comprises an activated sludge process, trickling filters, microbial fuel cells, and so forth [8]. For sustainable and affordable processing of such wastewater, it is necessary first to do the impact assessment, life cycle assessment, and techno-economic analysis to clarify which treatment method will be most efficient, affordable, and sustainable for its pre- and post-processing.

Both pure and adsorbent- or catalyst-functionalized nanofibrous materials have been actively prepared to remove or separate sub-micrometre pollutants and contaminants from liquid and gas environments. These preparations are based on various physical and chemical techniques – notably adsorption and ultrafiltration techniques. Using nanofibrous structures as filtering materials or adsorbents has dramatically boosted the adsorption capacity and separation efficiency. This results in lower operating energy consumption, higher separation flux, and better particle removal capability [9], [10]. With their high surface-to-volume ratio, high specific surface area, and continuous matrix structure made up of tiny pores, nanofibrous structures enable highly efficient separation processes when surface interaction is the primary factor driving pollutant and contaminant separation, as in the case of air filtration. Nanofiber applications for water treatment and environmental remediation have made great strides, particularly in chemical sensing, photocatalysis, and ultrafiltration. High-porosity nanofibers and their network architectures often improve particulate separation in air and water purification.

Because of their lengthy structure and huge specific surface area, nanofibers also have a high capacity for particle adsorption, which enhances the filtration efficiency and sensing capacities of devices based on nanofibers. The necessity of keeping an eye on environmental quality and ensuring that people have access to clean water and air has increased in tandem with the rise in dangerous pollution in recent years. Nanofibers are a viable alternative for water treatment and environmental remediation, but their performance needs to increase dramatically to satisfy these rising demands [11]. Creating and using nanofibers with adjustable physicochemical characteristics and qualities is vital to achieving this. Nanofibers are a viable alternative for water treatment and environmental remediation, but their performance needs to increase to satisfy these rising demands dramatically. The creation and use of nanofibers with adjustable physicochemical characteristics and attributes are vital.

After a thorough analysis of conventional water treatment technologies, there is a dire need to establish several sustainable and efficient wastewater technologies following the UNSDG goals, which involve trickling filters, nanofiber emerging technology, microbial fuel cells, and so forth. The main scope is first to observe which types of nanofibers are produced and developed for wastewater technologies, like electrospinning nanofiber technology or other purification techniques through emerging nanotechnology methods for wastewater treatments. For such techniques, it is necessary first to discuss the different materials used in nanofiber technology, which involves natural and synthetic composites, ceramics, and carbon-based materials. The shape and structure for quantitative performances are essential, and they involve porous, core-shell, and hollow-based nanofibers. Such nanofibers have other multidisciplinary applications involving adsorption, photocatalysis, filtration, and sensors. Nanofiber wastewater treatment applications usually consist of membrane distillation, forward osmosis, capacitive deionisation, reverse osmosis, solar steam generation, containment removal, and diffusion removal technologies.

14.1.1 Electrospinning Nanofiber Technology for Wastewater Treatment

Electrospinning is among the most well-known and widely applied techniques for producing one-dimensional nanofibers. Recently, several variations to the conventional electrospinning method were also created. These consist of co-electrospinning, sometimes referred to as co-axial electrospinning, multi-needle, and needleless techniques [12], [13]. The needleless and multi-needle methods increase the productivity of traditional electrospinning. However, core-shell and multilayer composite nanofibrous structures with additional functions and better quality are made via co-axial electrospinning. Nanofibrous materials are appealing for their ability to solve water- and element-related environmental issues. Using specifically designed electrospun nanofibers for wastewater and water treatment has recently gained popularity [14], [15]. Nanofibers can be included in nonwoven structures, stand-alone membranes, support layers, or surface modification layers to increase the usefulness of composite materials. The complete process diagram of electrospinning nanofiber technology is showcased in Figure 14.1 [16], [17].

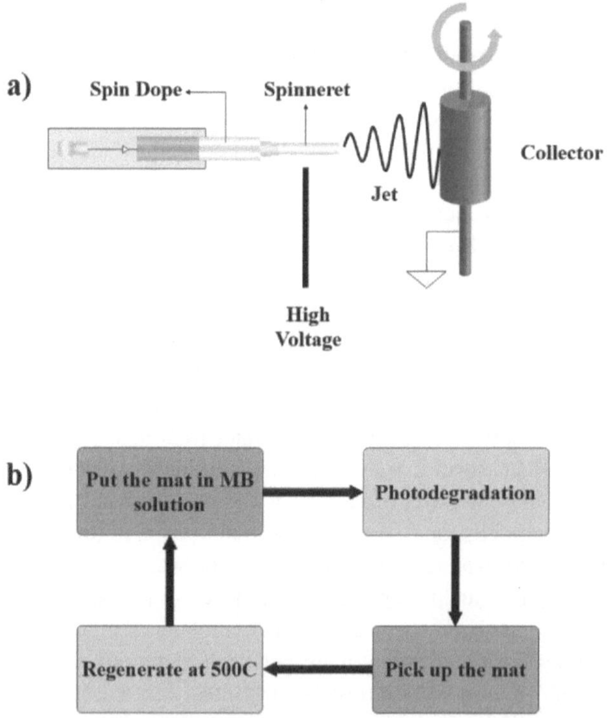

FIGURE 14.1 (a) Schematic Process Diagram of Electrospinning Method; and (b) Schematic Illustrative Flowchart of Electrospinning of Nanocomposite Materials (mats) (compiled by authors).

Research on improving the structure and composition of nanofiber membranes through electrospinning process optimisation of material, procedure, and surrounding parameters has been ongoing. Around the world, membrane technology is crucial to maintaining water security. This results from its excellent performance, cost-effectiveness, and efficiency in treating various water sources to the required level of water quality [18]. The most widely used membrane separation techniques are ultrafiltration, nanofiltration, microfiltration, and reverse osmosis. Despite being the most readily accessible membranes on the market, polymeric membranes still have many issues with fouling, degradation, and overall stability. Especially when using electrospun nanofibers [18]. Over the past 20 years, research has grown notably on nanofibers as materials for water and wastewater treatment (membranes, electrodes, adsorbents, etc.). Much research has been done on the synthesis of nanofibers, surface modification, incorporation of nanomaterials, and enhancement of mechanical and thermal stability, among other topics.

14.1.2 Introduction: Exigent Need for Advancements in the Purification of Wastewater Technologies

Recent oil spills and industrial wastewater releases have seriously compromised water resources. Apart from causing ecological disequilibrium, water contamination directly harms human well-being. High specific surface area, high porosity, and high size homogeneity nanofibers are expected to benefit materials used in wastewater treatment. It has been established that nanofibers can be produced using various techniques, including drawing, electrospinning, template synthesis, phase separation, interfacial polymerisation, and self-assembly [19]. Specifically, electrospinning is perhaps the most appealing technology for producing nanofibers because of its low cost and simplicity. Many published articles attest to the significant reputation of electrospun nanofibers in the wastewater treatment industry. Electrospun nanofiber membranes consisting of a single polymer usually had poor separation capacity. Raw material and electrospun membrane surface changes are commonly used to promote membrane separation behaviour. It has been shown that altering the membrane's surface by oxidation, thermal crosslinking, plasma treatment, and low temperatures are effective methods of enhancing membrane performance [20].

Water scarcity has worsened in recent years due to accidents involving oil spills and the direct release of industrial wastewater, harming the environment. This problem requires a water/oil separation method that is effective. Conventional separation procedures, such as biological treatment, condensation, and air flotation, have certain advantages. However, most also have drawbacks, like low separation efficiency, high energy consumption, a complicated process, and easy generation of secondary pollution. Filtration and adsorption have recently been demonstrated to be effective techniques for treating oil because of their high adsorption efficiency and simplicity of usage [17].

Life on Earth is in immediate danger due to the growing untreated release of hazardous organic contaminants into natural water bodies. Using straightforward, environmentally safe procedures is essential to get rid of pesticides, surfactants, halogens, polycyclic aromatic hydrocarbons, bacteria, dyes, and other contaminants. In recent years, the electrospinning process has touched on the study and development of water treatment applications, which has become essential. A variety of nanofibers can be produced depending on the application, such as carbon, ceramic, and polymeric (synthetic and natural) nanofibers – porous, core-shell/hollow fibres with random or aligned assembly via an electrospinning process [21], [22].

14.2 TYPES OF NANOFIBERS PRODUCED FOR WASTEWATER TECHNOLOGIES

14.2.1 Natural and Synthetic Polymer Nanofibers

It is possible to electrospin any polymer with a sufficiently high molecular weight into nanofibers. Electrospinning has successfully formed over a hundred synthetic and natural polymers into nanofibers. Nanofibers were created from natural polymers such as cellulose, chitosan, collagen, gelatin, silk, DNA (deoxyribonucleic acid), fibrinogen, hyaluronic acid, and alginate. Furthermore, they have unique functional groups that allow them to adsorb certain heavy metals in wastewater treatment through various processes, including electrostatic attraction, chelation, and ion exchange [23]. However, the solubility and biodegradation of a few natural polymers in an aqueous solution are negative. They were limited to being the most used in the purification of water. Thus, this opened the door for using artificial polymers in the water treatment industry. However, their tendency to inflate or collapse in the presence of moisture limited their use in water treatment applications [24]. These water-insoluble synthetic polymers include polybutylene terephthalate (PBT), polyvinylidene fluoride (PVDF), polymer vinyl acetate (PVAc), polyvinyl butyral (PVB), polyethylene-co-vinyl acetate (PEV), polystyrene (PS), polysulfone (PSF), polyAN, polybutylene succinate-co-butylene terephthalate (PBST), polyhydroxyalkanoates (PHA), polybutylene succinate (PBS), polyimide (PI), polymethyl methacrylate (PMMA), Polyvinyl Chloride (PVC), polytrimethylene terephthalate (PTT), polycarbonate (PC), polybenzimidazole (PBI), polymer vinyl acetate (PVAc), polyvinyl butyral (PVB), and polyethylene-co-vinyl acetate (PEV).

14.2.2 Composites: Ceramics and Carbon Nanofibers

Likewise, there is widespread recognition and application of the advantages of electrospun ceramic nanofibers in water remediation methods. Ceramics include metal oxides, non-oxides (such as carbides, nitrides, silicides, or borides), and their composites. Three basic steps are involved in producing ceramic nanofibers: (i) making a stable colloidal suspension (the sol) from a sol–gel precursor, a polymer, and a solvent; (ii) creating composite nanofibers via electrospinning; and (iii) producing ceramic nanofibers by calcination or solvent extraction to remove the organic component. The free-standing ceramic nanofibrous mats' intrinsic brittleness is one persistent technological challenge [25], [26], [27], [28], [29]. Large-area CNF sheets with excellent mechanical strength can be readily manufactured. Additionally, they are ideal for filtration due to their superior chemical resistance compared to polymeric filters [30], [31], [32].

14.2.3 Porous, Core-Shell, and Hollow Nanofibers

Among the several electrospinning classifications, coaxial electrospinning has received much attention since it has made better nanofibers, like hollow, core-shell, and axially aligned nanofibers [33]. The shell solution (a polymeric solution comprising a polymer, a catalyst, a solvent, and a sol–gel precursor) is transferred to the outer spinneret. On the other hand, the core solution (mineral oil) is given to the inner spinneret. Under high voltage conditions, coaxial fibres congregate on the metal substrate. Depending on whether the cores are removed selectively, these coaxial nanofibers, which have walls of inorganic/polymer composites or ceramics, can be categorised as hollow or core-shell nanofibers. Porous nanofibers have been made using two different techniques: (i) eliminating a particular component from the fibres and (ii) quickly cooling the fibres before they fully solidify to cause the polymer and solvent to phase separation [34].

14.3 APPLICATIONS OF NANOFIBERS IN WATER AND WASTEWATER TREATMENT

14.3.1 Adsorption

Adsorption is a simple, economical, and effective way to treat contaminated water compared to other treatment technologies. Highly porous nanofibers are known for their large surface area, flexible surface functions, bespoke pore structure, and self-standing characteristics, which make them effective adsorbents for eliminating impurities from drinking water and wastewater [35], [36], [37], [38], [39].

14.3.2 Photocatalysis

Photocatalysis is one of the most widely used, affordable, and ecologically acceptable methods for removing microorganisms, dangerous organic contaminants, and heavy metals from wastewater treatment. Semiconducting metal oxides (NFs), such as TiO2, ZnO, Fe2O3, CeO2, and WO3, have typically been used as catalysts in photocatalysis to degrade heavy metals and hazardous organic compounds [40], [41], [42], [43]. A multifunctional nanofibrous mat combining functional TiO_2 with Ag nanoparticles showed exceptional photocatalytic activity of methylene blue degradation by 80.0 percent and 99.9 percent bacteria inactivation during 30 minutes of sun irradiation [44].

14.3.3 Filtration

One of the more established techniques for removing microorganisms and other suspended particulates from water is filtration, which uses permeable material and pressure as a driving force. Even though the membrane has significantly improved to increase its efficiency, chemical stability and fouling are the most critical issues that must be resolved. Because several types of nanofiber materials have been industrialised, it is, therefore, essential to alter the characteristics of membranes to improve the efficiency of water filtering [45], [46]. The filtering system is classified as reverse osmosis, forward osmosis, ultrafiltration, nanofiltration, or microfiltration

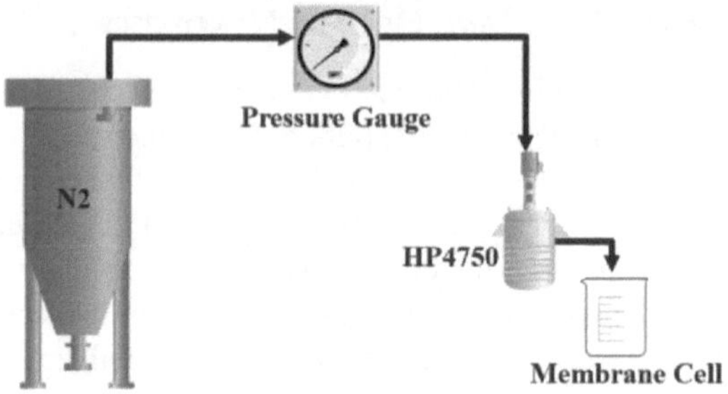

FIGURE 14.2 Filtration Process by Nanofibrous Membrane (compiled by authors).

based on the size of the pores. Among these, the ability of nanofibrous membranes to remove microorganisms, suspended particles, and other materials while allowing water to flow through them was explored for MF and UF applications [47]. Electrospun nanofibers have gained increased attention from researchers because of their prefiltering capacity to remove oversized particles (submicron to micron size) before the raw water is fed into the membrane system. A comprehensive experimental setup of filtration of the nanofibrous membrane is shown in Figure 14.2 [45], [46], [47].

14.3.4 Sensors

Chemical sensors are used in molecule and ion recognition and selective diagnosis. Much research is being done to increase the sensitivity, robustness, and endurance of chemical sensor materials. The sensing material inserted into the nanofiber considerably increases the sensitivity of pollutant detection and chemical interaction by improving detection sensitivity and response time and lowering the concentration detection value (lowest detection value (LOD)) [48], [49]. One of the many benefits of 1D nanofibers over other nanomaterials is their vast surface area, porosity, lack of contamination, and ease of posttreatment. Electrospun nanofibers have inherent properties that render them highly sensitive in detecting particular pollutants.

14.4 NANOFIBERS FOR PURIFICATION IN DESALINATION AND WASTE TREATMENT

14.4.1 Membrane Distillation (MD)

Since phase change is the method of operation for MD, a temperature differential between the feed and cold sides is required. MD is a hybrid thermal/membrane separation process. Employing electrospinning technology can enable nanofiber membrane structure to design all these features.

Because of these intriguing and desirable features, research on nanofiber membranes for MD has increased dramatically over the past ten years. The structure and characteristics of the membrane are crucial to the process [13], [50], [51], [52]. Compared to current microfiltration membranes, nanofiber membranes for MD have shown competitive overall performance and encouraging results thus far. Most reported nanofiber membranes for MD have not yet undergone long-term performance testing, nor have they addressed scaling and fouling issues under challenging conditions [53], [54], [55].

14.4.2 Forward Osmosis (Membrane Support Layers)

Two fluids with different osmotic pressures are separated by FO using a semipermeable membrane. The osmotic pressure differential causes the water from the feed solution to be driven through the membrane, diluting the draw solution nanofiber membranes for MD and resolving scaling and fouling problems when exposed to demanding conditions [56], [57], [58]. A selective membrane with low internal concentration polarisation (ICP) is ideal. There have been encouraging developments in FO membranes and techniques over the last ten years. In general, FO membranes comprise a middle layer, a delicate selective layer, and a backing/support layer for mechanical stability [59]. In order to decrease the ICP effects in FO, research has recently focused on using nanofiber layers as a backing or middle layer. Even though nanofiber membranes in FO show promising results in high fluxes and lower ICP influence, manufacturing the thick polyamide top layer remains challenging since the FO membrane requires a thin selective layer. This is because the structure of the nanofiber layer is not smooth; consequently, there is a high probability that the chosen top layer may delaminate since not all of the surface is securely bound to the valleys or fractures of the overlapping nanofibers. Moreover, the PA layer strongly tends to flow through the larger-than-average nanofiber membrane gaps during the polyamide interfacial polymerisation. An experimental setup of membrane support layer application of nanofibers via forward osmosis is shown in Figure 14.3 [56], [57], [58].

14.4.3 Capacitive Deionization (CDI)

Capacitive deionisation is based on electrosorption, which involves applying an electric potential, usually less than 2 V, to two oppositely charged electrodes. This

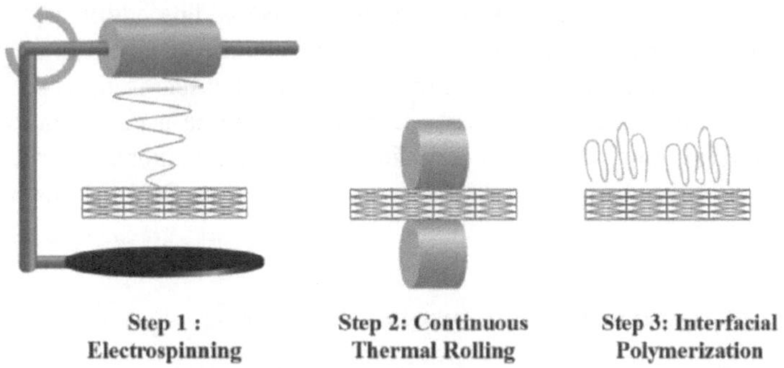

FIGURE 14.3 Schematics of the Three-Step Process of Membrane Support Layers for Forward Osmosis (compiled by authors).

causes cations and anions from a saline solution to be adsorbed on the electrodes. The material of the electrode significantly influences the electrosorption behaviour. It needs to possess the following qualities in order to be the ideal material for CDI: (a) a sizable surface area, (b) a sizable capacitance, (c) excellent conductivity, (d) strong electrochemical stability, and e) simplicity in production. Considering these electrode requirements, highly specific surface area and carbonised electrospun nanofibers are advantageous electrode materials. Furthermore, electrospun nanofibers are more appealing as electrode materials due to the presence of functional moieties to them. Nanomaterials' numerous intriguing features and capabilities, particularly nanofibers, make them potentially valuable materials for CDI electrodes and may enhance desalination efficiency. However, the lack of economic analysis in much of the research may raise the cost of these materials. Also, more research is required to determine the stability of such nanofibers and nanomaterials.

14.4.4 Reverse Osmosis (RO)

As RO is regarded as the state of the art in desalination, new desalination facilities continue to use it worldwide. The desalination process uses much energy despite its high efficiency. The membrane is critical to the RO process because it offers the barrier needed for selective permeation. A suitable membrane should support the high pressure applied to it while having a high degree of mechanical stability and selectivity [59]. Thin film composites, which feature a thick, thin selective layer and a porous polymeric support layer, are the most often used type of membranes in reverse osmosis (RO).

14.4.5 Solar Steam Generation

Since desalination is still often an energy-intensive process, solar desalination enters the picture to reduce energy costs while maintaining good process efficiency. In this sense, "solar desalination" refers to the process of separating water molecules from salts and other impurities through evaporation and condensation while using solar light as a heat source [60], [61], [62], [63], [64], [65]. Pore size, porosity, surface charges, and hydrophilicity are the three most crucial factors to consider while selecting a membrane for filtration. By altering the fibre size of the nanofiber membrane, pores can be made larger, smaller or more porous. By adding additives, modifying electrospinning settings, or variation of polymer concentration, electrospun nanofiber can be easily manipulated to change fibre size from nanoscale to microscale. Thus, electrospun nanofiber membranes have been reported to be used in microfiltration (MF) and ultrafiltration (UF) procedures in various investigations. Charged impurities can be eliminated from nanofibers by varying their surface charges by adding charged polymers.

14.4.6 Contaminant Removal through Nanofibers

Very few techniques are available for in situ groundwater treatment because of challenging building requirements, unstable sampling and detection, and complex geological conditions. Permeable reactive barrier (PRB) technology and direct injection of reducing agents, oxidisers, or adsorbents are examples of traditional therapy. Nanofibers can also be utilised as PRB materials, which can be injected into an aquifer to treat groundwater and stop plumes – because they can be shaped into membranes. There are two approaches to applying nanofiber for groundwater remediation, depending on how important a role it plays in the treatment process. Using nanofiber is one method of eliminating contaminants from groundwater. An approach would be to immobilise the technology and materials utilised in groundwater remediation using nanofiber as a carrier. Furthermore, membranes made of nanofiber can be used in groundwater remediation procedures. Pore size, porosity, surface charges, and hydrophilicity are the three most crucial factors to consider while selecting a membrane for filtration. By altering the fibre size of the nanofiber membrane, pores can be made larger, smaller or more porous. By adding additives, modifying electrospinning settings, or varying polymer concentration, electrospun nanofiber can be easily manipulated to change fibre size from nanoscale to microscale [59]. Thus, electrospun nanofiber membranes have been reported to be used in microfiltration (MF) and ultrafiltration (UF) procedures in various investigations. Charged impurities can be eliminated from nanofibers by varying their surface charges by adding charged polymers.

Interestingly, for the best attraction and immobilisation of pollutants during adsorption, nanofibers having electrical charges opposite those of the contaminants should be added. However, nanofibers are expected to have the same electrical charge as contaminants throughout the membrane filtering process. This would enhance the rejection of contaminants and lessen membrane fouling brought on by the foulants' affinity through the electrostatic repulsion effect. Due to their intriguing and desirable properties – which include a large specific surface area, controllable fibre sizes, ease of fabrication, high flexibility to surface modification, and strong compatibility with other technologies – nanofibers have great potential for water treatment and groundwater remediation. Nonetheless, several problems with nanofiber design and usage still need to be fixed. These comprise, among other things, their often-poorer mechanical qualities, less established stability and long-term performance, and issues with mass manufacturing [66].

14.5 Nanofibers for Water Treatment for Different Ion Removal Techniques

Emerging nanofiber technologies have showcased great potential and performance for wastewater treatment and, in recent times, show an advanced roadmap towards sustainable nanofiber applications towards eliminating ions; nanofiber applications in thorough detail have also been showcased in Figure 14.4 [59]. Table 14.1 discusses the applications and advancements of membrane bioreactors, oil-water separation, dye wastewater treatment and removal of metal ions; whereas Table 14.2 specifies the adsorbent, adsorbate, and adsorbate of adsorbent value for different types of nanoparticles used in wastewater treatment.

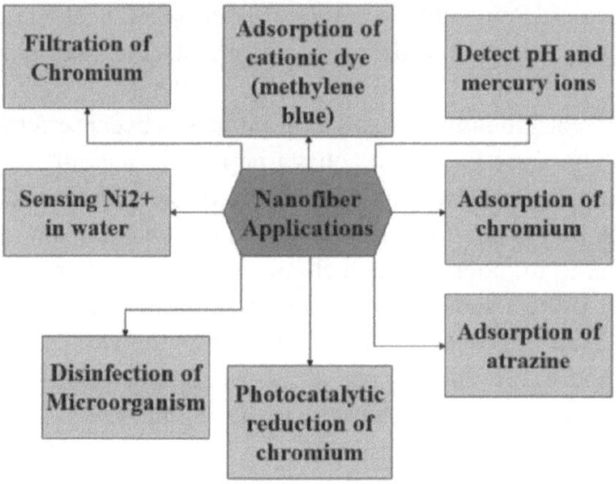

FIGURE 14.4 Various applications of nanofibers in wastewater treatment (compiled by authors).

TABLE 14.1
Applications and Advancements of Nanofibers for Water Treatment

Applications	Advancements	Reference
Nanofibers as Membrane for Membrane Bioreactor (MBR)	To investigate the performance of nanofibrous membrane	[67]
	Improved the nanofibrous PA membrane performance by heat-treating membranes	[68]
	Development of electrospun membrane using PAN	[69]
Nanofibers as Membrane or Adsorbent for Oil-Water Separation	Development of Gravity driven membrane separation, an energy-efficient versatile technology	[70]
	Coating the silver nanocluster on the nanofibrous PAN membranes, followed by super hydrophobisation	[70]
	Developing Elastic polyurethane and chromatic polydiacetylenes for electrospinning	[71]
	The hydrophobicity of the PVDF nanofiber membrane is improved by incorporating ZnO nanoparticles and fluorination post-treatment.	[72]
	Electrospun Membrane formation from carbonaceous materials	[73]
	Coating of cellulose-ionic liquid solution on electrospun PVDF membrane to modify its surface properties.	[74]
	Modification of polysulfone electrospun membrane surface	[75]

TABLE 14.1 (Continued)
Applications and Advancements of Nanofibers for Water Treatment

Applications	Advancements	Reference
Nanofibers as Membrane or Adsorbent for Dye Wastewater Treatment	Fabrication of b-cyclodextrin-based electrospun membranes	[76]
	Development of membrane having 99 % selectivity towards cationic dyes	[77]
	Fabrication of branch-like carboxylated MWCNTs/ Chitosan nanofiber membranes	[78]
	PDMS/PVDF nanofiber membrane for high flux and rejection rate	[79]
Nanofibers as Adsorbent for Removal of Heavy Metal Ions from Water/Wastewater	Fabrication of chitosan-rectorite nanospheres embedded polystyrene (PS) membranes to adsorb Ca(II) ions	[34]
	Development of hierarchical composite membranes by applying polymerisation of polyaniline on PS nanofiber membranes	[80]
	Development of amino-rich hydrothermal carbon-coated electrospun PAN membranes	[81], [82]

TABLE 14.2
Adsorbent, Adsorbate and Adsorbate of Adsorbent Value for Different Types of Nanoparticles Used in Wastewater Treatment

Adsorbent	Adsorbate	Adsorbate (mg/g) of adsorbent	Reference
CNTs	Cd^{2+}	1.1 mg/g	[83]
Oxidised CNTs	Cu^{2+}	0.0404 mmol/g	[84]
	Cd^{2+}	0.0260 mmol/g	[84]
	Zn^{2+}	0.0152 mmol/g	[84]
	Ni^{2+}	0.0311 mmol/g	[84]
Oxidised MWCNTs	Pb^{2+}	91 mg/g	[85]
HNO_3 oxidised MWCNTs	Th^{4+}	57.14 µmol/g (293k)	[86]
Graphene Oxide (GO) aerogels	Cu^{2+}	17.73 mg/g (10°C)	[87]
Reduced Graphene Oxide, RGO-MnO_2, RGO-Ag	Hg^{2+}	>10L/g	[88]
Graphene Oxide Nanosheets	Cd^{2+}	106.3 mg/g	[89]
	Co^{2+}	**68.2 mg/g**	**[89]**
Bamboo Waste Granulated Activated Carbon)	RB 5	39.9 mg of dye/g	[90]
Microwave Induced MWCNTs	Cu^{2+}	99 mg/g	[91]
Nanoporous Granular Activated Carbon	Basic Dye	77 mg of dye/g	[92]
GNS^{PFS}	Pb^{2+}	406 mg/g	[93]

(continued)

TABLE 14.2 (Continued)
Adsorbent, Adsorbate and Adsorbate of Adsorbent Value for Different Types of Nanoparticles Used in Wastewater Treatment

Adsorbent	Adsorbate	Adsorbate (mg/g) of adsorbent	Reference
	Cd^{2+}	73 mg/g	[93]
MWCNTs	Cr (VI)	2.35 mg/g	[94]
Iron Oxide Nanoparticles	As (V)	136.98 mg/g	[95]
	Cr (VI)	136.98 mg/g	[95]
Nanocomposites	Cu^{2+}	1.26 mg/g	[96]
	Cr (VI)	1.26 mg/g	[96]
Graphene Nanosheet	Ni (II)	167.2 mg/g	[97]
Carbon-based Nano sorbents	Ni^{2+}	260.7 mg/g	[98]
Electrospun Nanofibrous Membranes	Cr (VI)	70 mg/g	[99]

14.6 CONCLUSIONS

In the time of high emissions and with global warming and unsustainable conventional approaches towards multidisciplinary applications, it is essential to pivot towards sustainable approaches towards as many applications as possible. Such sustainable approaches also involve emerging nanofiber technology for the purification of wastewater treatment. As there are many limitations for natural and synthetic polymer nanofibers involving solubility and biodegradation, composites, ceramics and carbon nanofibers are used. The significant emerging nanofiber technology mainly involves electrospun nanofibers. These fibres are concluded to have cost-effective, efficient and sustainable applications involving adsorption, photocatalysis, filtration, and sensors. Electrospun nanofibers showcase excellent longevity, robustness, and sensitivity, making them an ideal alternative for sensing applications. Similarly, the ability to remove microorganisms, suspended particles, and other materials makes nanofibers suitable for filtration. Nanofibers for purification in desalination and waste treatment is one of the essential characteristics showcased by nanofibers, which involves membrane distillation, forward osmosis, capacitive deionisation, solar steam generation, contaminant removal, and reverse osmosis. The main reason behind such desalination and waste treatment is that nanofibers have great potential for water treatment and groundwater remediation due to their intriguing and desirable properties, which include large specific surface areas, controllable fibre sizes, ease of fabrication, high flexibility to surface modification, and strong compatibility with other technologies.

ACKNOWLEDGEMENTS

The authors equally acknowledge the Indian Institute of Technology, Madras and University School of Chemical Technology, and Guru Gobind Singh Indraprastha University for providing excellent research facilities and a supportive atmosphere.

The authors would also like to acknowledge MOKSH Research and Development for the support facilitating this research.

REFERENCES

[1] S. Pan, J. Feng, B. Safaei, Z. Qin, F. Chu, and D. Hui, "A comparative experimental study on damping properties of epoxy nanocomposite beams reinforced with carbon nanotubes and graphene nanoplatelets," *Nanotechnol Rev*, vol. 11, no. 1, pp. 1658–1669, April 2022, doi: 10.1515/ntrev-2022-0107

[2] R. P. Singh *et al.*, "Nucleic acid nanotechnology: Trends, opportunities and challenges," *Curr Pharm Biotechnol*, vol. 24, no. 1, pp. 50–60, January 2023, doi: 10.2174/1389201023666220520103325

[3] K. Mylvaganam and L. C. Zhang, "Ballistic resistance capacity of carbon nanotubes," *Nanotechnology*, vol. 18, no. 47, p. 475701, November 2007, doi: 10.1088/0957-4484/18/47/475701

[4] R. P. Singh and O. S. Kushwaha, "International Council of Materials Education," *Journal of Materials Education*, vol. 35, pp. 79–119, 2013.

[5] O. S. Kushwaha, C. V. Avadhani, and R. P. Singh, "Preparation and characterization of self-photostabilizing UV-durable bionanocomposite membranes for outdoor applications," *Carbohydr Polym*, vol. 123, pp. 164–173, June 2015, doi: 10.1016/j.carbpol.2014.12.062

[6] H. Mittal, S. Verma, A. Bansal, and O. Singh Kushwaha, "Low-carbon hydrogen economy perspective and net zero-energy transition through proton exchange membrane electrolysis cells (PEMECs), anion exchange membranes (AEMs) and wind for green hydrogen generation," *Qeios*, 2024, doi: 10.32388/9V7LLC

[7] A. D. Sekar and M. Manickam, "Current trends of electrospun nanofibers in water and wastewater treatment," In *Energy, Environment, and Sustainability*, Springer Nature, pp. 469–485, 2019. doi: 10.1007/978-981-13-3259-3_21

[8] H. Mittal and O. S. Kushwaha, "Policy implementation roadmap, diverse perspectives, challenges, solutions towards low-carbon hydrogen economy," *Green and Low-Carbon Economy*, March 2024, doi: 10.47852/bonviewGLCE42021846

[9] G. Zhao, L. Shi, G. Yang, X. Zhuang, and B. Cheng, "3D fibrous aerogels from 1D polymer nanofibers for energy and environmental applications," *J Mater Chem A Mater*, vol. 11, no. 2, pp. 512–547, 2023, doi: 10.1039/D2TA05984C

[10] H. Mittal and O. S. Kushwaha, "Machine learning in commercialized coatings," In *Functional Coatings*, R.K Arya et al., Wiley, pp. 450–474, 2024. doi: 10.1002/9781394207305.ch17`

[11] P. Wang, H. Lv, X. Cao, Y. Liu, and D.-G. Yu, "Recent progress of the preparation and application of electrospun porous nanofibers," *Polymers (Basel)*, vol. 15, no. 4, p. 921, February 2023, doi: 10.3390/polym15040921

[12] H. Liu *et al.*, "Recent progress of electrospun herbal medicine nanofibers," *Biomolecules*, vol. 13, no. 1, p. 184, January 2023, doi: 10.3390/biom13010184

[13] J. Cui, F. Li, Y. Wang, Q. Zhang, W. Ma, and C. Huang, "Electrospun nanofiber membranes for wastewater treatment applications," *Sep Purif Technol*, vol. 250, 111716, November 2020. doi: 10.1016/j.seppur.2020.117116

[14] L. Svobodová, T. Lederer, and L. Kriklavova, "A review study of nanofiber technology for wastewater treatment." [Online]. Available: www.researchgate.net/publication/264885983

[15] L. D. Tijing, M. Yao, J. Ren, C. H. Park, C. S. Kim, and H. K. Shon, "Nanofibers for water and wastewater treatment: Recent advances and developments," In *Energy, Environment, and Sustainability*, Xuan-Thanh Bui, Chart Chiemchaisri, Takahiro Fujioka and Sunita Varjani, Springer Nature, pp. 431–468, 2019. doi: 10.1007/978-981-13-3259-3_20

[16] N. Olaru and L. Olaru, "Electrospinning of cellulose acetate phthalate from different solvent systems," *Ind Eng Chem Res*, vol. 49, no. 4, pp. 1953–1957, February 2010, doi: 10.1021/ie901427f

[17] A. Nadaf *et al.*, "Recent update on electrospinning and electrospun nanofibers: Current trends and their applications," *RSC Adv*, vol. 12, no. 37, pp. 23808–23828, 2022, doi: 10.1039/D2RA02864F

[18] Y. L. Cheng, J. G. Kim, H. Bin Kim, J. H. Choi, Y. Fai Tsang, and K. Baek, "Occurrence and removal of microplastics in wastewater treatment plants and drinking water purification facilities: A review," *Chem Eng J*, vol. 410, p. 128321, April 2021, doi: 10.1016/j.cej.2020.128381

[19] Y. Yang, Y. Du, J. Zhang, H. Zhang, and B. Guo, "Structural and functional design of electrospun nanofibers for hemostasis and wound healing," *Adv Fiber Mater*, vol. 4, no. 5, pp. 1027–1057, October 2022, doi: 10.1007/s42765-022-00178-z

[20] O. S. Kushwaha, C. V. Avadhani, and R. P. Singh, "Photo-oxidative degradation of polybenzimidazole derivative membrane," *Adv Mater Lett*, vol. 4, no. 10, pp. 762–768, October 2013, doi: 10.5185/amlett.2013.3432

[21] V. Vinayagam *et al.*, "Recent developments on advanced oxidation processes for degradation of pollutants from wastewater with focus on antibiotics and organic dyes," *Environ Res*, vol. 240, p. 117500, January 2024, doi: 10.1016/j.envres.2023.117500

[22] V. Vinayagam, N. Kumar Kishor Kumar, K. N. Palani, S. Ganesh, O. S. Kushwaha, and A. Pugazhendhi, "Recent breakthroughs on the development of electrodeionization systems for toxic pollutants removal from water environment," *Environ Res*, vol. 241, p. 117549, January 2024, doi: 10.1016/j.envres.2023.117549

[23] V. Vinayagam *et al.*, "Sustainable adsorbents for the removal of pharmaceuticals from wastewater: A review," *Chemosphere*, vol. 300, p. 134597, August 2022, doi: 10.1016/j.chemosphere.2022.134597

[24] S. Zahmatkesh, M. Hajiaghaei-Keshteli, A. Bokhari, S. Sundaramurthy, B. Panneerselvam, and Y. Rezakhani, "Wastewater treatment with nanomaterials for the future: A state-of-the-art review," *Environ Res*, vol. 216, p. 114652, January 2023, doi: 10.1016/j.envres.2022.114652

[25] R. Riffat and T. Husnain, *Fundamentals of Wastewater Treatment and Engineering*, CRC Press, 2022. doi: 10.1201/9781003134374

[26] R. B. Ashok, C. V. Srinivasa, and B. Basavaraju, "Dynamic mechanical properties of natural fiber composites – A review," *Adv Compos Hybrid Mater*, vol. 2, no. 4, pp. 586–607, December 2019, doi: 10.1007/s42114-019-00121-8

[27] P. H. F. Pereira *et al.*, "Vegetal fibers in polymeric composites: A review," *Polímeros*, vol. 25, no. 1, pp. 9–22, February 2015, doi: 10.1590/0104-1428.1722

[28] A. Saravanan, P. Thamarai, P. S. Kumar, and G. Rangasamy, "Recent advances in polymer composite, extraction, and their application for wastewater treatment: A review," *Chemosphere*, vol. 308, p. 136368, December 2022, doi: 10.1016/j.chemosphere.2022.136368

[29] J.-K. Kim and Y.-W. Mai, *Engineered Interfaces in Fiber Reinforced Composites*, Elsevier, 1998.

[30] W. Meng, Y. Huang, Y. Fu, Z. Wang, and C. Zhi, "Polymer composites of boron nitride nanotubes and nanosheets," *J Mater Chem C*, vol. 2, no. 47, pp. 10049–10061, 2014, doi: 10.1039/C4TC01998A

[31] K. O. Odesanya, R. Ahmad, M. Jawaid, S. Bingol, G. O. Adebayo, and Y. H. Wong, "Natural fibre-reinforced composite for ballistic applications: A review," *J Polym Environ*, vol. 29, no. 12, pp. 3795–3812, December 2021, doi: 10.1007/s10924-021-02169-4

[32] M. T. Byrne and Y. K. Gun'ko, "Recent advances in research on carbon nanotube-polymer composites," *Adv Mat*, vol. 22, no. 15, pp. 1672–1688, April 2010, doi: 10.1002/adma.200901545

[33] S. F. Ahmed *et al.*, "Strategies to improve membrane performance in wastewater treatment," *Chemosphere*, vol. 306, p. 135527, November 2022, doi: 10.1016/j.chemosphere.2022.135527

[34] M. Vaselbehagh, H. Karkhanechi, R. Takagi, and H. Matsuyama, "Biofouling phenomena on anion exchange membranes under the reverse electrodialysis process," *J Memb Sci*, vol. 530, pp. 232–239, May 2017, doi: 10.1016/j.memsci.2017.02.036

[35] A. Heydari-Gorji, Y. Belmabkhout, and A. Sayari, "Polyethylenimine-impregnated mesoporous silica: Effect of amine loading and surface alkyl chains on CO_2 adsorption," *Langmuir*, vol. 27, pp. 12411–12416, 2011, doi: 10.1021/la202972t

[36] S. Sircar and T. C. Golden, "Purification of hydrogen by pressure swing adsorption," *Sep Sci Technol*, vol. 35, pp. 667–687, 2000, doi: 10.1081/ss-100100183

[37] C.-H. Yu, C.-H. Huang, and C.-S. Tan, "A review of CO_2 capture by absorption and adsorption," Volume 12, issue 5, 745-769, *Aerosol Air Qual Res*, 2012, doi: 10.4209/aaqr.2012.05.0132

[38] C. Gan *et al.*, "Effect of porous zinc–biochar nanocomposites on Cr(VI) adsorption from aqueous solution," *RSC Adv*, 5(44), 35107-35115, 2015, doi: 10.1039/c5ra04416b

[39] E. Ostuni, R. G. Chapman, R. E. Holmlin, S. Takayama, and G. M. Whitesides, "A survey of structure–property relationships of surfaces that resist the adsorption of protein," *Langmuir*, vol. 17, no. 18, pp. 5605–5620, September 2001, doi: 10.1021/la010384m

[40] M. Matsuoka *et al.*, "Photocatalysis for new energy production: Recent advances in photocatalytic water splitting reactions for hydrogen production," *Catal Today*, vol. 122, pp. 51–61, 2007, doi: 10.1016/j.cattod.2007.01.042

[41] P. Wongwanwattana, P. Krongkitsiri, P. Limsuwan, and U. Tipparach, "Fabrication and photocatalysis of nanostructured TiO_2 for solar hydrogen production," *Ceram Int*, vol. 38, pp. S517–S519, 2012, doi: 10.1016/j.ceramint.2011.05.066

[42] M. Navlani-García, D. Salinas-Torres, K. Mori, Y. Kuwahara, and H. Yamashita, "Photocatalytic approaches for hydrogen production via formic acid decomposition," Heterogeneous Photocatalysis, Springer, Mario J. Muñoz-Batista, Alexander Navarrete Muñoz, Rafael Luque. pp. 193–223, 2020, doi: 10.1007/978-3-030-49492-6_6

[43] Z. Sun *et al.*, "Enhanced CO_2 photocatalytic reduction on alkali-decorated graphitic carbon nitride," *Appl Catal B Environ*, vol. 216, pp. 146–155, 2017, doi: 10.1016/j.apcatb.2017.05.064

[44] H. Uthayakumar, P. Radhakrishnan, K. Shanmugam, and O. S. Kushwaha, "Growth of MWCNTs from *Azadirachta indica* oil for optimization of chromium(VI) removal efficiency using machine learning approach," *Environ Sci Pollut Res*, vol. 29, no. 23, pp. 34841–34860, May 2022, doi: 10.1007/s11356-021-17873-w

[45] X. Ji *et al.*, "Advances in particulate matter filtration: Materials, performance, and application," *Green Energy Environ*, vol. 8, no. 3, pp. 673–697, June 2023, doi: 10.1016/j.gee.2022.03.012
[46] H. Xiang, X. Min, C.-J. Tang, M. Sillanpää, and F. Zhao, "Recent advances in membrane filtration for heavy metal removal from wastewater: A mini review," *J Water Process Eng*, vol. 49, p. 103023, October 2022, doi: 10.1016/j.jwpe.2022.103023
[47] Y. Zhang *et al.*, "Continuous air purification by aqueous interface filtration and absorption," *Nature*, vol. 610, no. 7930, pp. 74–80, October 2022, doi: 10.1038/s41586-022-05124-y
[48] D. Strawser, J. Thangavelautham, and S. Dubowsky, "A passive lithium hydride based hydrogen generator for low power fuel cells for long-duration sensor networks," *Int J Hydrogen Energy*, vol. 39, no. 19, pp. 10216–10229, June 2014, doi: 10.1016/j.ijhydene.2014.04.110
[49] H. Ding, J. Guan, P. Lu, S. J. Mihailov, C. T. Kingston, and B. Simard, "Boron nitride nanotubes for optical fiber sensor applications," *Optical Fiber Sensors Conference 2020 Special Edition*, 2020.
[50] X. Tongwen and Y. Weihua, "Fundamental studies of a new series of anion exchange membranes: Membrane preparation and characterization," *J Memb Sci*, vol. 190, no. 2, pp. 159–166, September 2001, doi: 10.1016/S0376-7388(01)00434-3
[51] A. Rahimpour and S. S. Madaeni, "Polyethersulfone (PES)/cellulose acetate phthalate (CAP) blend ultrafiltration membranes: Preparation, morphology, performance and antifouling properties," *J Memb Sci*, vol. 305, no. 1–2, pp. 299–312, November 2007, doi: 10.1016/j.memsci.2007.08.030
[52] S. Sawada *et al.*, "Solid polymer electrolyte water electrolysis systems for hydrogen production based on our newly developed membranes, Part I: Analysis of voltage–current characteristics," *Prog Nucl Energy*, vol. 50, pp. 443–448, 2008, doi: 10.1016/j.pnucene.2007.11.029
[53] D. Aili *et al.*, "Porous poly(perfluorosulfonic acid) membranes for alkaline water electrolysis," *J Memb Sci*, vol. 493, pp. 589–598, 2015, doi: 10.1016/j.memsci.2015.06.057
[54] G. Merle, M. Wessling, and K. Nijmeijer, "Anion exchange membranes for alkaline fuel cells: A review," *J Memb Sci*, vol. 377, no. 1–2, pp. 1–35, July 2011, doi: 10.1016/j.memsci.2011.04.043
[55] T. Huang *et al.*, "Hydrogen crossover through microporous anion exchange membranes for fuel cells," *J Power Sources*, vol. 527, p. 231143, April 2022, doi: 10.1016/j.jpowsour.2022.231143
[56] O. S. Kushwaha, C. V. Avadhani, N. S. Tomer, and R. P. Singh, "Accelerated degradation study of highly resistant polymer membranes for energy and environment applications," *Adv Chem Sci*, pp. 19–30, 2014.
[57] O. S. Kushwaha, C. V. Avadhani, and R. P. Singh, "Effect of UV rays on degradation and stability of high performance polymer membranes," *Adv Mater Lett*, vol. 5, no. 5, pp. 272–279, May 2014, doi: 10.5185/amlett.2014.10533
[58] J. Parrondo, C. G. Arges, M. Niedzwiecki, E. B. Anderson, K. E. Ayers, and V. Ramani, "Degradation of anion exchange membranes used for hydrogen production by ultrapure water electrolysis," *RSC Adv*, vol. 4, no. 19, p. 9875, 2014, doi: 10.1039/c3ra46630b
[59] L. Zeng and T. S. Zhao, "Integrated inorganic membrane electrode assembly with layered double hydroxides as ionic conductors for anion exchange membrane water electrolysis," *Nano Energy*, vol. 11, pp. 110–118, January 2015, doi: 10.1016/j.nanoen.2014.10.019

[60] M. Mokhtar *et al.*, "Solar-assisted post-combustion carbon capture feasibility study," *Appl Energy*, vol. 92, pp. 668–676, 2012, doi: 10.1016/j.apenergy.2011.07.032
[61] C. A. Rodriguez, M. A. Modestino, D. Psaltis, and C. Moser, "Design and cost considerations for practical solar-hydrogen generators," *Energy Environ Sci*, vol. 7, no. 12, pp. 3828–3835, 2014, doi: 10.1039/C4EE01453G
[62] M. A. Modestino and S. Haussener, "An integrated device view on photo-electrochemical solar-hydrogen generation," *Annu Rev Chem Biomol Eng*, vol. 6, no. 1, pp. 13–34, July 2015, doi: 10.1146/annurev-chembioeng-061114-123357
[63] R. P. Singh and O. S. Kushwaha, "Polymer solar cells: An overview," *Macromol Symp*, vol. 327, no. 1, pp. 128–149, May 2013, doi: 10.1002/masy.201350516
[64] M. G. Walter *et al.*, "Solar water splitting cells," *Chem Rev*, vol. 110, no. 11, pp. 6446–6473, 2010, doi: 10.1021/cr1002326
[65] M. S. Prévot and K. Sivula, "Photoelectrochemical tandem cells for solar water splitting," *J Phys Chem C*, vol. 117, no. 35, pp. 17879–17893, 2013, doi:10.1021/jp405291g
[66] R. Thiruvenkatachari, S. Su, H. An, and X. X. Yu, "Post combustion CO_2 capture by carbon fibre monolithic adsorbents," *Prog Energy Combust Sci*, vol. 35, no. 5, pp. 438–455, 2009, doi: 10.1016/j.pecs.2009.05.003
[67] D. Bjorge *et al.*, "Performance assessment of electrospun nanofibers for filter applications," *Desalination*, vol. 249, no. 3, pp. 942–948, December 2009, doi: 10.1016/j.desal.2009.06.064
[68] M. R. Bilad, P. Westbroek, and I. F. J. Vankelecom, "Assessment and optimization of electrospun nanofiber-membranes in a membrane bioreactor (MBR)," *J Memb Sci*, vol. 380, no. 1–2, pp. 181–191, September 2011, doi: 10.1016/j.memsci.2011.07.003
[69] G. Moradi, S. Zinadini, L. Rajabi, and S. Dadari, "Fabrication of high flux and anti-fouling mixed matrix fumarate-alumoxane/PAN membranes via electrospinning for application in membrane bioreactors," *Appl Surf Sci*, vol. 427, pp. 830–842, January 2018, doi: 10.1016/j.apsusc.2017.09.039
[70] O. Arslan, Z. Aytac, and T. Uyar, "Superhydrophobic, hybrid, electrospun cellulose acetate nanofibrous mats for oil/water separation by tailored surface modification," *ACS Appl Mater Interfaces*, vol. 8, no. 30, pp. 19747–19754, August 2016, doi: 10.1021/acsami.6b05429
[71] X. Li, M. Wang, C. Wang, C. Cheng, and X. Wang, "Facile immobilization of Ag nanocluster on nanofibrous membrane for oil/water separation," *ACS Appl Mater Interfaces*, vol. 6, no. 17, pp. 15272–15282, September 2014, doi: 10.1021/am503721k
[72] L. Chen *et al.*, "Robust and elastic superhydrophobic breathable fibrous membrane with in situ grown hierarchical structures," *J Memb Sci*, vol. 547, pp. 93–98, February 2018, doi: 10.1016/j.memsci.2017.10.023
[73] C. Liu, X. Li, B. Ma, A. Qin, and C. He, "Removal of water contaminants by nano-scale zero-valent iron immobilized in PAN-based oxidized membrane," *Appl Surf Sci*, vol. 321, pp. 158–165, December 2014, doi: 10.1016/j.apsusc.2014.09.202
[74] M. H. Tai, P. Gao, B. Y. L. Tan, D. D. Sun, and J. O. Leckie, "Highly efficient and flexible electrospun carbon–silica nanofibrous membrane for ultrafast gravity-driven oil–water separation," *ACS Appl Mater Interfaces*, vol. 6, no. 12, pp. 9393–9401, June 2014, doi: 10.1021/am501758c
[75] F. E. Ahmed, B. S. Lalia, and R. Hashaikeh, "A review on electrospinning for membrane fabrication: Challenges and applications," *Desalination*, vol. 356, pp. 15–30, January 2015, doi: 10.1016/j.desal.2014.09.033
[76] M. Obaid, N. A. M. Barakat, O. A. Fadali, M. Motlak, A. A. Almajid, and K. A. Khalil, "Effective and reusable oil/water separation membranes based on modified

polysulfone electrospun nanofiber mats," *Chem Eng J*, vol. 259, pp. 449–456, January 2015, doi: 10.1016/j.cej.2014.07.095

[77] R. Zhao, Y. Wang, X. Li, B. Sun, and C. Wang, "Synthesis of β-cyclodextrin-based electrospun nanofiber membranes for highly efficient adsorption and separation of methylene blue," *ACS Appl Mater Interfaces*, vol. 7, no. 48, pp. 26649–26657, December 2015, doi: 10.1021/acsami.5b08403

[78] A. Ghaffar, L. Zhang, X. Zhu, and B. Chen, "Porous PVdF/GO nanofibrous membranes for selective separation and recycling of charged organic dyes from water," *Environ Sci Technol*, vol. 52, no. 7, pp. 4265–4274, April 2018, doi: 10.1021/acs.est.7b06081

[79] J. Shi *et al.*, "Simultaneous electrospinning and spraying toward branch-like nanofibrous membranes functionalised with carboxylated MWCNTs for dye removal," *Mater Lett*, vol. 166, pp. 26–29, March 2016, doi: 10.1016/j.matlet.2015.12.024

[80] H. Tu *et al.*, "Chitosan-rectorite nanospheres immobilized on polystyrene fibrous mats via alternate electrospinning/electrospraying techniques for copper ions adsorption," *Appl Surf Sci*, vol. 426, pp. 545–553, December 2017, doi: 10.1016/j.apsusc.2017.07.159

[81] J. J. Alcaraz-Espinoza, A. E. Chávez-Guajardo, J. C. Medina-Llamas, C. A. S. Andrade, and C. P. de Melo, "Hierarchical composite polyaniline–(electrospun polystyrene) fibers applied to heavy metal remediation," *ACS Appl Mater Interfaces*, vol. 7, no. 13, pp. 7231–7240, April 2015, doi: 10.1021/acsami.5b00326

[82] R. Zhao, X. Li, B. Sun, H. Ji, and C. Wang, "Diethylenetriamine-assisted synthesis of amino-rich hydrothermal carbon-coated electrospun polyacrylonitrile fiber adsorbents for the removal of Cr(VI) and 2,4-dichlorophenoxyacetic acid," *J Colloid Interface Sci*, vol. 487, pp. 297–309, February 2017, doi: 10.1016/j.jcis.2016.10.057

[83] Y.-H. Li, S. Wang, Z. Luan, J. Ding, C. Xu, and D. Wu, "Adsorption of cadmium(II) from aqueous solution by surface oxidized carbon nanotubes," *Carbon N Y*, vol. 41, no. 5, pp. 1057–1062, 2003, doi: 10.1016/S0008-6223(02)00440-2

[84] Z. Gao, T. J. Bandosz, Z. Zhao, M. Han, and J. Qiu, "Investigation of factors affecting adsorption of transition metals on oxidized carbon nanotubes," *J Hazard Mater*, vol. 167, no. 1–3, pp. 357–365, August 2009, doi: 10.1016/j.jhazmat.2009.01.050

[85] H. Wang, A. Zhou, F. Peng, H. Yu, and J. Yang, "Mechanism study on adsorption of acidified multiwalled carbon nanotubes to Pb(II)," *J Colloid Interface Sci*, vol. 316, no. 2, pp. 277–283, December 2007, doi: 10.1016/j.jcis.2007.07.075

[86] C. Chen, X. Li, D. Zhao, X. Tan, and X. Wang, "Adsorption kinetic, thermodynamic and desorption studies of Th(IV) on oxidized multi-wall carbon nanotubes," *Colloids Surf A Physicochem Eng Asp*, vol. 302, no. 1–3, pp. 449–454, July 2007, doi: 10.1016/j.colsurfa.2007.03.007

[87] X. Mi, G. Huang, W. Xie, W. Wang, Y. Liu, and J. Gao, "Preparation of graphene oxide aerogel and its adsorption for Cu^{2+} ions," *Carbon N Y*, vol. 50, no. 13, pp. 4856–4864, November 2012, doi: 10.1016/j.carbon.2012.06.013

[88] T. S. Sreeprasad, S. M. Maliyekkal, K. P. Lisha, and T. Pradeep, "Reduced graphene oxide–metal/metal oxide composites: Facile synthesis and application in water purification," *J Hazard Mater*, vol. 186, no. 1, pp. 921–931, February 2011, doi: 10.1016/j.jhazmat.2010.11.100

[89] G. Zhao, J. Li, X. Ren, C. Chen, and X. Wang, "Few-layered graphene oxide nanosheets as superior sorbents for heavy metal ion pollution management," *Environ Sci Technol*, vol. 45, no. 24, pp. 10454–10462, December 2011, doi: 10.1021/es203439v

[90] A. A. Ahmad and B. H. Hameed, "Fixed-bed adsorption of reactive azo dye onto granular activated carbon prepared from waste," *J Hazard Mater*, vol. 175, no. 1–3, pp. 298–303, March 2010, doi: 10.1016/j.jhazmat.2009.10.003

[91] N. M. Mubarak, J. N. Sahu, E. C. Abdullah, N. S. Jayakumar, and P. Ganesan, "Novel microwave-assisted multiwall carbon nanotubes enhancing Cu(II) adsorption capacity in water," *J Taiwan Inst Chem Eng*, vol. 53, pp. 140–152, August 2015, doi: 10.1016/j.jtice.2015.02.016

[92] V. Meshko, L. Markovska, M. Mincheva, and A. E. Rodrigues, "Adsorption of basic dyes on granular activated carbon and natural zeolite," *Water Res*, vol. 35, no. 14, pp. 3357–3366, October 2001, doi: 10.1016/S0043-1354(01)00056-2

[93] D. Dang, P. Wu, C. He, Z. Xie, and C. Duan, "Homochiral metal–organic frameworks for heterogeneous asymmetric catalysis," *J Am Chem Soc*, vol. 132, no. 41, pp. 14321–14323, October 2010, doi: 10.1021/ja101208s

[94] D. T. M. Huong *et al.*, "Removal of cationic dye waste by nanofiber membrane immobilized with waste proteins," *Int J Biol Macromol*, vol. 164, pp. 3873–3884, December 2020, doi: 10.1016/j.ijbiomac.2020.09.020

[95] H. Liang, B. Xu, and Z. Wang, "Self-assembled 3D flower-like α-Fe_2O_3 microstructures and their superior capability for heavy metal ion removal," *Mater Chem Phys*, vol. 141, no. 2–3, pp. 727–734, September 2013, doi: 10.1016/j.matchemphys.2013.05.070

[96] M. H. Dehghani *et al.*, "Removal of noxious Cr(VI) ions using single-walled carbon nanotubes and multi-walled carbon nanotubes," *Chem Eng J*, vol. 279, pp. 344–352, November 2015, doi: 10.1016/j.cej.2015.04.151

[97] G. Xie *et al.*, "A facile chemical method to produce superparamagnetic graphene oxide–Fe_3O_4 hybrid composite and its application in the removal of dyes from aqueous solution," *J Mater Chem*, vol. 22, no. 3, pp. 1033–1039, 2012, doi: 10.1039/C1JM13433G

[98] F. M. Machado *et al.*, "Adsorption of Reactive Red M-2BE dye from water solutions by multi-walled carbon nanotubes and activated carbon," *J Hazard Mater*, vol. 192, no. 3, pp. 1122–1131, September 2011, doi: 10.1016/j.jhazmat.2011.06.020

[99] I. Tlili and T. A. Alkanhal, "Nanotechnology for water purification: Electrospun nanofibrous membrane in water and wastewater treatment," *J Water Reuse Desalin*, vol. 9, no. 3, pp. 232–248, September 2019, doi: 10.2166/wrd.2019.057

15 Wastewater Treatment, Recycling, and Reuse – Past, Present, and Future

A Circular Economy Perspective

Innocent Tayari Mwizerwa, Mansuur Husein, Khaoula Houssini, Susan Praise, and Joseline Kateita

15.1 INTRODUCTION

Wastewater treatment has become an enormous global concern due to increased exponential population growth and natural resources consumption. While wastewater was originally disposed in different regions of the world, in the recent decades it has emerged as a valuable source to conserve water, protect the environment, and generate revenue from the materials recovered from wastewater. Thus, in addition to the water itself, which is becoming an increasingly valued resource, wastewater is abundant in energy, nutrients, and other intriguing materials (Salgot & Folch, 2018; Song et al., 2018). Water resource recovery facilities (WRRFs), which use biochemical, physical, physicochemical, and biological conversion processes to extract value from once underutilized streams, are replacing wastewater treatment plants as industry leaders recognize the potential of wastewater (Chonde & Raut, 2017; Hall & Sutton, 2016; Kato et al., 1997; Salgot & Folch, 2018; A. L. Smith et al., 2012).

Municipal wastewater sludge was initially merely disposed off in the 1970s but by the early 1990s, biosolids were being utilized in a positive way, such as for the production of fertilizer. Applications for bioenergy started by the early 2000s, and the recovery of other resources have further developed in the last decades. Christopher Gasson, publisher of *Global Water Intelligence*, stated in 2018 that a "golden decade" for wastewater treatment was about to begin (DIANA et al., 2020).

This chapter is aimed at revealing the existing literature and identifying the current status, opportunities, successes and policies associated with the implementation of a circular economy (CE) in the wastewater sector.

The successful integration of the circular economy into the wastewater treatment industry has the potential to reduce environmental impacts, reduce the public

 DOI: 10.1201/9781003454199-15

health risk, mitigate air pollution and climate change consequences, control water pollution, recover valuable resources, ensure food security and boost economic growth.

15.2 WASTEWATER TREATMENT AND CIRCULAR ECONOMY

Circular economy is a model in which materials are never wasted and nature is renewed. It is a system where the use of products and materials are kept in circularity via processes of maintenance, reuse, refurbishment, refabrication, recycling, and composting.

Hence, the concept of a circular economy (CE) has gained widespread popularity, aiming not only to be a theoretical idea but also to function as an economic model replacing the current linear take-make-use-dispose system. The primary objective of a CE model is to prolong the life of materials and products within the system by converting waste into valuable resources, optimizing processes for efficiency, and extending the usability of products and materials (Zając & Avdiushchenko, 2020). The implementation of a CE model serves to minimize the extraction of resources from natural systems while simultaneously reducing waste and pollution.

In the realm of wastewater treatment, the role of the CE becomes particularly significant. Water stands as a crucial resource essential not only for human and organisms survival but also for various economic sectors. The availability of fresh water, both in quantity and quality, plays a pivotal role in fostering and sustaining economic development. However, this development has led to the overexploitation and pollution of water bodies, exacerbated by the worsening quantity and quality of fresh water due to climate change-induced droughts (Hernández-Chover et al., 2023).

Specifically, in the urban water cycle, CE holds relevance in the wastewater treatment sector. Reclaimed wastewater emerges as an alternative water resource, ensuring water availability while alleviating pressure on water bodies, thereby ensuring sustainability. Additionally, valuable byproducts, such as nutrient-rich wastewater sludge, can be recovered and utilized as fertilizers and biogas through effective wastewater treatment, resulting in profitable benefits (Guerra-Rodríguez et al., 2020; Sfez et al., 2019).

There is a growing emphasis on achieving sustainability in industrial production processes, encompassing sectors like wastewater treatment. This involves encouraging the reduction of material and energy exploitation in various production processes. By removing or reducing different waste streams through valorization and establishing closed-loop material cycles, sustainability goals can be met. Currently, attention is focused on the energy usage of wastewater treatment facilities, which account for a substantial portion of total energy consumption, nearing 44 percent (Zarei, 2020).

Efforts to enhance sustainability involve addressing technological gaps and sizing issues, where factors such as carrying capacity, influent Chemical Oxygen Demand, and adherence to standards significantly impact energy use and efficiency (Chu et al., 2017). Wastewater treatment plants present opportunities for considerable energy reduction, achieved through sustainable wastewater treatment choices. Lower energy consumption not only results in reduced greenhouse gas emissions but also positions wastewater treatment plants as potential sources of alternative resources, including reclaimed water, energy, and nutrients (Henriques & Catarino, 2017; Hernández-Chover et al., 2023).

15.3 WASTEWATER ENERGY RECOVERY FOR BIOFUELS

An inventive and environmentally responsible solution to the problems of wastewater management and the creation of renewable energy is the use of wastewater for the production of biofuels (Martin et al., 2011). The organic matter and nutrients found in wastewater can be used to produce biofuels, which are renewable energy sources made from biological resources (Lopes et al., 2018; K. Smith et al., 2019).

In order to create liquid or gaseous fuels that can replace fossil fuels, organic matter must be converted into biofuels. Wastewater serves as a valuable feedstock for the generation of biofuels, especially the portion that is rich in organic matter. By reducing reliance on fossil fuels and promoting a more sustainable and integrated approach to wastewater treatment and energy production, this strategy supports circular economy (Quaschning, 2019). In the following sub-sections, we describe comprehensive methods for using wastewater for the production of biofuels.

15.3.1 Biogas Generation

Biogas is a renewable fuel produced through bacterial decomposition of organic wastes in the absence of oxygen. The gas produced by this waste-to-energy process rises to the digester's top, where it is collected and used to create steam that can be used in various processes or to create power. This lowers energy expenditures for a business and might possibly generate revenue. Some facilities are capable of becoming net energy producers, or net zero (Leite et al., 2016; Owusu & Asumadu-Sarkodie, 2016).

Biogas primarily consists of methane (CH_4) and carbon dioxide (CO_2), accompanied by trace amounts of other gases such as nitrogen (N_2) and hydrogen sulfide (H_2S) (Yao et al., 2020).

Methane, a potent greenhouse gas, constitutes the major component of biogas. However, the release of biogas into the atmosphere, where it contributes to global warming, can be mitigated by capturing and utilizing it as a fuel (Owusu & Asumadu-Sarkodie, 2016). The conversion of biogas into compressed natural gas (CNG) or bio methane allows it to be used as a renewable energy source for heat and electricity production, as well as for fueling vehicles or natural gas pipelines (Opare, 2018).

Anaerobic digestion, a process employed in wastewater treatment, involves the breakdown of organic molecules by microbes without the use of oxygen. This naturally occurring biological process is particularly effective in the absence of oxygen. Wastewater treatment facilities frequently utilize anaerobic digestion to break down organic particles like sewage sludge or biosolids, converting them into usable energy in the form of biogas (Leyva-Díaz et al., 2016).

By tapping into the energy potential of wastewater, anaerobic digestion reduces greenhouse gas emissions while providing a renewable energy source for heat and power generation. Implementing anaerobic digestion technology in wastewater treatment facilities contributes to achieving energy sustainability and promoting a more circular approach to waste management (DIANA et al., 2020).

The main sources of organic matter in wastewater, often found in high concentrations, include human waste, food waste, and other organic materials flushed or released into the sewage system. During anaerobic digestion, methanogens, bacteria capable of breaking down organic material without oxygen, generate biogas as a byproduct (Peng et al., 2019). The composition of biogas primarily comprises methane (CH_4) and carbon dioxide (CO_2), with additional traces of nitrogen (N_2) and hydrogen sulfide (H_2S) (Vassalle et al., 2020). While methane is a potent greenhouse gas, its capture and utilization as a fuel present an effective strategy to prevent its release into the atmosphere and mitigate its contribution to global warming.

15.3.2 Algae-based Biofuels

Algae-based biofuels are a form of renewable energy which use algae for source of energy-rich oils. These can be grown in wastewater, and offer more oil compared to conventional crops. Algae-based biofuels aid to reduce overdependence on oil for transportation.

Wastewater can offer nutrients like nitrogen and phosphorus that are needed for the production of algae. Algae are capable of producing biomass, which can then be processed to provide lipids, carbohydrates, or proteins, by efficiently converting sunlight and nutrients (Colzi Lopes et al., 2018). Biodiesel may be made from algae-derived lipids, and bioethanol can be made from carbohydrates. High growth rates, low acreage needs, and the capacity to grow in non-potable water sources like wastewater are all benefits of algae-based biofuels (S. Malik et al., 2023).

15.3.3 Bio-oil Production

This is a chemical reaction referred to as transesterification that changes natural fats and oils to form Fatty Acid Methyl Esters (FAME) or biodiesel. The microalgal biomass is a consecutively process of cultivation, harvest, oil extraction, and conversion of algal lipids to form advanced biofuels.

Using high temperatures and pressure, the thermochemical conversion process known as "hydrothermal liquefaction" turns organic materials, such as wastewater sludge, into bio-oil. Further processing can turn this bio-oil into renewable diesel or jet fuel. Hydrothermal liquefaction is a potential technology for treating sewage sludge and producing biofuels since it is especially suitable for wet or high-moisture feedstocks (Amin, 2009).

15.3.4 Microbial Fuel Cells

Microbial fuel cells (MFCs) directly transform organic waste from wastewater into power by using the metabolic activity of microorganisms (Wu et al., 2022). By decomposing organic substances, the microorganisms release electrons that can be collected and used to generate an electrical current. MFCs have the ability to produce power and treat wastewater at the same time, while being in the early phases of research.

There are various advantages to using wastewater in the manufacture of biofuels. First, it offers a substitute for traditional fossil fuels, cutting down on greenhouse gas emissions and reliance on limited resources. By using the organic content of wastewater and lowering its environmental impact, it also aids in addressing the problem of wastewater management. The concept of circular economy is also supported by the creation of biofuel from wastewater.

However, it is crucial to take into account a few challenges and factors related to the use of wastewater for the production of biofuels. Wastewater composition can vary, necessitating constant monitoring to maintain process stability and optimization of the conversion processes (Hamawand et al., 2016). Contaminants like heavy metals or persistent organic pollutants may have an impact on the viability and quality of the production of biofuels. Therefore, appropriate pretreatment and quality control procedures are required to guarantee the efficiency and safety of wastewater-based biofuel production (Shao et al., 2018). These biofuels, including biogas, biodiesel, and bioethanol, can be utilized as renewable energy sources for transportation, heating, or electricity generation, lowering greenhouse gas emissions and reducing reliance on fossil fuels (Amin, 2009; Brennan & Owende, 2010; Raheem et al., 2018).

15.4 WATER RECOVERY

There are growing pressures to recover water from wastewater in many water-scarce locations (Obeta & Nwankwo, 2015; Wang et al., 2018). Guidelines for the safe use of wastewater were developed by the World Health Organization (WHO), the Food and Agriculture Organization of the United Nations (FAO) (Caruso & Slembrouck, 2019; Mekouar, 2013), and the United Nations Environment Program (UNEP) in 2006 (Warning, 2011). Additionally, a wide number of governments developed their own guidelines for using recovered water (Krantzberg & Hartley, 2018; O.A. Malik et al., 2015).

Advanced treatment techniques can be used to remove impurities from wastewater, which contains a variety of toxins and pollutants, and turn it into pure drinking water. To ensure that the water satisfies strict quality standards for human consumption, this advanced treatment goes above and beyond traditional wastewater treatment techniques (Kehrein et al., 2020).

Usually, there are several treatment phases involved in turning wastewater into potable water. Depending on the particular technology and desired level of water quality, these steps may change (Saravanan et al., 2021). Wastewater can be treated and reused to create high-quality water that can be used for a variety of non-potable purposes, including irrigation, business operations, and toilet flushing. Before being used again, effluent may also be released into surface water bodies like rivers or groundwater aquifers after being thoroughly purified for indirect potable purposes (Salgot & Folch, 2018).

15.5 ORGANIC FERTILIZER

Organic fertilizers are naturally produced materials added to soil or to improve nutrients and foster plant growth. They are composed of carbon and other natural elements, such as hydrogen and oxygen, useful for plant nourishment. Nitrogen, phosphorus, potassium, and other micronutrients that are required for agricultural output can be found in human excreta. These nutrients are vital for many biological processes as well as the growth and development of living things. They can, however, result in environmental issues such eutrophication when used in excess (Monfet et al., 2017).

Organic fertilizers can be recovered via chemical methods like chemical precipitation or chemical stripping, or they can just be used with the wastewater or sewage sludge. Reusing sewage sludge, however, comes with dangers because it contains significant levels of harmful substances such heavy metals, environmental persistent pharmaceutical pollutants, and other chemicals (Deblonde et al., 2011). It can be advantageous to separate the excreta fractions of wastewater (such as toilet waste) from the rest of the wastewater flows because the majority of fertilizing nutrients are contained in excreta (Gaur et al., 2018). Because of this, the volume that needs to be treated before using recovered nutrients in agricultural output is reduced and the risk of unwanted chemicals is decreased. While phosphorus is present in the form of orthophosphates and organic phosphates, domestic wastewater primarily comprises nitrogen in the form of organic compounds, ammonia, and nitrates (Zhou et al., 2019).

15.6 BIOPRODUCTS

Bioproducts are materials, chemicals and energy made from renewable biological material. Through various biotechnological processes, wastewater can be used as a source of raw materials for the manufacturing of bioplastics and other value-added bioproducts. We may develop sustainable substitutes for current materials and fuels by utilizing the organic chemicals and nutrients found in wastewater. This would lessen our dependency on fossil fuels and encourage a more environmentally friendly and cyclical system of production and consumption (Abuabdou et al., 2020).

Bioplastics are one of the bioproducts that can be produced from wastewater. Materials that are either biodegradable or made from renewable resources are referred to as bioplastics. Organic molecules from wastewater can be used to create bio-based polymers through fermentation or other biochemical procedures. These bioplastics provide a more environmentally friendly option to traditional plastics made from fossil fuels and can be utilized in a variety of applications, such as packaging, textiles, and consumer goods (Yadav et al., 2020).

Additionally, wastewater can be a source of additional high-value bioproducts. For instance, the microbial biomass created during the processes of treating wastewater can be used as a nutrient-rich feed for livestock or as a source of enzymes and microbial compounds with numerous industrial uses. In order to manage nutrients sustainably, nutrients from wastewater, such as nitrogen and phosphorus, can also be collected and used as fertilizers or soil amendments in agriculture (Arcos-Hernández et al., 2015; Chong et al., 2021).

15.7 METALS AND MINERALS

Wastewater may include trace levels of metals and minerals that can be recovered for recycling or reuse (Chen et al., 2011). We can recover these critical resources in order to reduce our reliance on primary extraction, and support sustainable resource management practices by using the proper technologies and techniques (Kromah & Zhang, 2021). This benefits not only the environment but it creates the basis for the development of circular economy practices.

Metal use in manufacturing processes, corrosion of metal infrastructure, and even natural occurrences are all caused due to the presence of metals and minerals in wastewater. Despite the fact that these metals and minerals are normally found in low quantities, their combined presence in significant amounts of wastewater can make their recovery practical and profitable (Máthé-Gáspár & Anton, 2005; Petelka et al., 2019).

Depending on their unique qualities and the intended end-use, different techniques are used to recover metals and minerals from wastewater. Precipitation, adsorption, ion exchange, membrane technology, and electrochemical approaches are few of the recovery processes uses nowadays. The necessary metals and minerals are carefully targeted and separated from the wastewater using these procedures, enabling their extraction and subsequent reuse or recycling (Shan et al., 2017; Shao et al., 2018).

15.8 BIOSOLIDS

The solid waste left over after wastewater treatment, sometimes referred to as biosolids, can be treated and applied as a soil amendment for farming and land reclamation, enhancing soil fertility and supplying nutrients (Lu et al., 2012).

The solid waste product produced during wastewater treatment, known as biosolids, can be processed and applied as a soil amendment in farming and land restoration. They offer priceless organic matter, nutrients, and trace elements that increase soil fertility, structure, and plant growth. Utilizing biosolids in a controlled and responsible way enhance sustainable waste management strategies and it reduces the need for synthetic fertilizers improving soil health and environmental sustainability (Chang & Kim, 2005).

Biosolids are a solid byproduct of wastewater treatment procedures. The solid material removed during wastewater treatment, known as sewage sludge, is treated and stabilized to produce these biosolids. Organic material, nutrients, and trace elements found in biosolids can be beneficially reused in a variety of uses, mostly in agriculture (Chang & Kim, 2005).

Biosolids go through a number of treatment stages to make sure they adhere to legal requirements and are secure for use. These treatments include techniques that stabilize organic matter and lessen the presence of pathogens and hazardous elements, such as anaerobic digestion, composting, or heat drying. The resulting biosolids are nutrient-rich, being high in organic matter, nitrogen, and phosphorus (Melia et al., 2017).

There are various advantages for using biosolids as a soil supplement. First of all, biosolids are a source of organic matter, which helps to build soil structure and increase water and nutrient retention. Natural fertilizer, organic matter works by gradually releasing nutrients to plants over time. Additionally, by restoring nutrients like nitrogen and phosphorus, which are crucial for plant growth, biosolids can improve soil fertility (Nguyen, 2018). By lowering the demand for synthetic fertilizers and the environmental impact of their manufacturing and usage, biosolids are applied to agricultural lands. As organic matter from the biosolids is deposited in the soil, biosolids also help to sequester carbon from the atmosphere by lowering greenhouse gas emissions (Semblante et al., 2015). Additionally, the use of biosolids in land reclamation projects can aid in the restoration of deteriorated soils and encourage the development of vegetation in areas impacted by mining operations, building sites, or landfills. In these difficult conditions, biosolids supply vital nutrients and organic matter that help restore soil fertility and support plant growth (Yadav et al., 2020).

To ensure the protection of both human health and the environment, it is crucial to emphasize that the usage of biosolids necessitates careful management and adherence to regulatory criteria. To verify the absence of pathogens and pollutants within acceptable limits, extensive testing and monitoring are carried out. To avoid any potential negative effects on soil, water quality, and adjacent ecosystems, proper application procedures, such as timing, dosage, and site-specific considerations, are crucial (Song et al., 2018; Water, 2017). Figure 15.1 below illustrates a summary of major technologies for resource recovery from wastewater treatment facilities. End products of this treatment technique become valuable materials for other domestic, agricultural and industrial application such as biogas, nutrients and clean water.

15.9 BIOCHAR AND HEAT RECOVERY

In wastewater treatment facilities, biochar and heat recovery are two resource recovery techniques that can be used to reduce waste volume, enhance environmental sustainability, and produce useful products (biochar) or energy (heat). In terms of wastewater management, these strategies adhere to the concepts of the circular economy and resource efficiency (Song et al., 2018).

A technique called pyrolysis, which involves heating organic materials (wastewater treatment sludge) without oxygen, produces biochar, a carbon-rich, porous substance. As a stable form of carbon, the resulting biochar can be applied to a variety of environmental and agricultural activities. Wastewater treatment sludge is collected and preprocessed to eliminate pollutants that may interfere with pyrolysis and moisture using sludge dewatering processes such as centrifugation and mechanical pressing (Kehrein et al., 2020). The prepared sludge is then put through pyrolysis in a regulated setting with less oxygen. This procedure entails heating (incinerating) the sludge to high temperatures for a predetermined period (usually between 300°C and 700°C) (Sipra et al., 2018; Vassalle et al., 2020; Yang et al., 2016).

The organic components of the sludge are thermally broken-down during pyrolysis, producing biochar, a solid byproduct that is mostly made of carbon. The soil can be supplemented with biochar, a stable form of carbon, to absorb carbon, reduce greenhouse gas emissions, and increase soil carbon content (Amann et al., 2018; Xiao et al., 2018).

Additionally, the biochar enhances the structure of the soil and increases nutrient and moisture retention, which increases agricultural productivity and lowers the need for chemical fertilizers. Heavy metals and organic pollutants can be adsorbed and immobilized by biochar, which lessens their mobility and negative effects on the

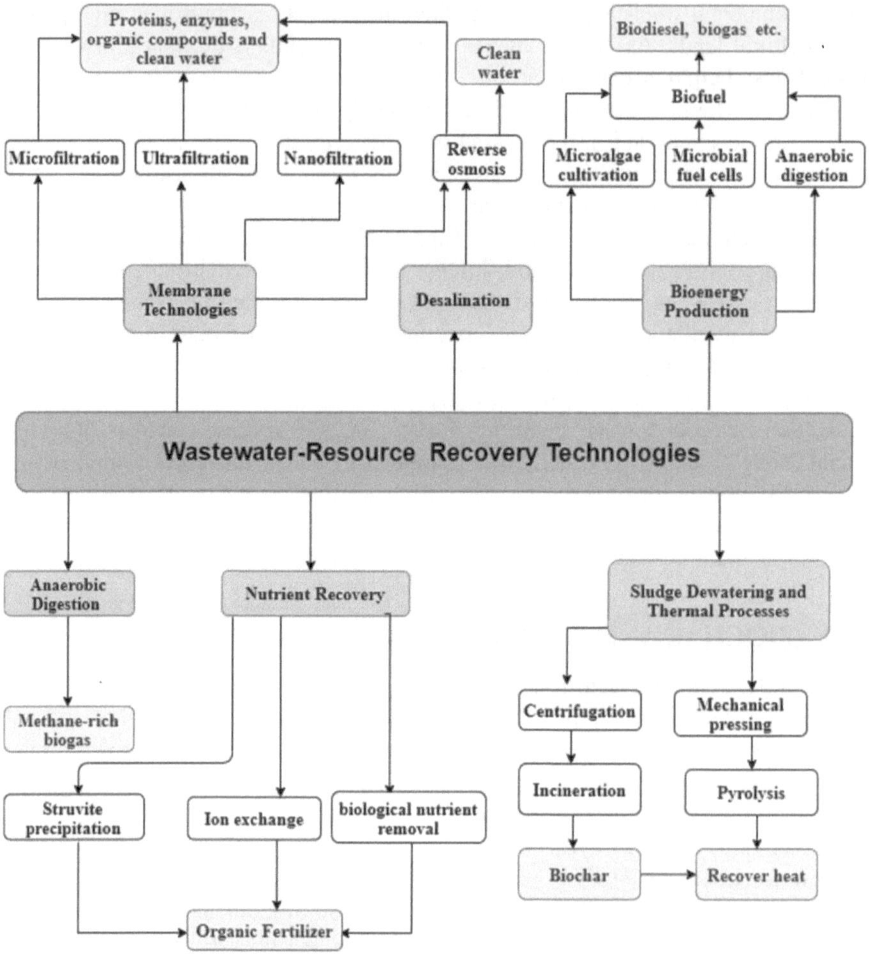

FIGURE 15.1 Flow chart diagram showing various wastewater resources recovery technologies with their by-products (compiled by authors).

environment. Additionally, it decreases the amount of wastewater treatment sludge, improving the effectiveness of waste management (Sharma & Melkania, 2017).

The process of heat recovery entails capturing and using the thermal energy produced during various operations at wastewater treatment facilities, such as the handling of sludge. Anaerobic digestion, which uses microbes to break down organic materials (Peng et al., 2019; Shi et al., 2018), is one of the main methods for treating wastewater sludge. As a consequence of this action, heat is produced. Heat produced by anaerobic digestion is collected via heat exchangers. The heat from the sludge can be used to heat other areas of the treatment facility or be transferred to the incoming influent wastewater.

By utilizing the heat produced during the treatment process, heat recovery lowers the energy consumption of wastewater treatment facilities. This minimizes the facilities' dependency on external energy sources, which lowers costs. reducing greenhouse gas emissions and supporting environmental sustainability in the process (Qiu et al., 2019; Sharma & Melkania, 2017).

15.10 WASTEWATER RECOVERY AND CIRCULAR ECONOMY PRACTICES

Water shortages have become a major global threat which has been further exacerbated by global warming and climate change. Water scarcity in arid- and semi-arid regions along with the global efforts to secure future water security resulted in increasing efforts to consider water reuse as an alternative source (Hirwa et al., 2022). This requires a global shift from a linear approach to water resources to a holistic circular approach based on circular economy (CE) practices. Wastewater reuse is one potential alternative to enhance water security, sustainability and resilience. In addition, wastewater reuse is beneficial to irrigation and agricultural system, industrial processes, groundwater replenishment, and environmental restoration (Mannina et al., 2022). Although there is a growing attention on the benefit of wastewater reuse and recycling practices, to date only few countries have the technical and financial capacities to integrate CE practices in their water resource system and to efficiently reclaim waste water.

15.11 CIRCULAR ECONOMY INTEGRATION WITH WASTEWATER

Circular economy for wastewater is a concept that aims to transform wastewater from a waste to a resource, by recovering and reusing water, nutrients, energy and organic matter from wastewater. Implementing circular economy for wastewater and water recovery can reduce the environmental impacts and costs of wastewater treatment, while creating economic and social benefits.

Existing circular economy practices for wastewater include the water reuse (Capodaglio, 2017; Ghafourian et al., 2022; Rugaimukamu et al., 2022), which involves reusing treated wastewater for various purposes, such as irrigation, industry, or even drinking. Also, nutrient recovery is one of the circular economy practices for wastewater which involves extracting valuable nutrients, such as nitrogen and phosphorus, from wastewater and sludge as well as using them as fertilizers or bio-products. In addition, energy recovery from waste has become a dominant practice. It involves capturing and utilizing biogas from anaerobic digestion of sludge, or producing biofuels from microalgae grown in wastewater. Moreover, organic matter recovery technique is used to producing organic fertilizers or soil amendments from sludge or biosolids, which can enhance soil fertility and carbon sequestration. Figure 15.2 below illustrates the system diagram of CE practices for wastewater treatment.

These practices demonstrate how circular economy for wastewater can optimize water resources and extract valuable resources from wastewater, while mitigating emissions and enhancing resilience to climate change.

Water disposal can be further optimized and recycled in wastewater treatment plants and through water recovery practices discussed in the above section. Clean

FIGURE 15.2 Circular economy practices for wastewater recovery (compiled by authors).

energy can be adopted for wastewater treatment plants and for recovery technologies and finally the treated and recovered water can be further reintroduced for residential and industrial use.

The implementation of CE practices for wastewater treatment and water recovery provides a wide number of economic and environmental benefits that can be summarized as follows:

(i) Increasing water resources by reclaiming and reusing wastewater for various purposes, such as irrigation, industry, or even drinking.
(ii) Reducing greenhouse gas emissions by capturing and utilizing biogas from anaerobic digestion of sludge, or by producing biofuels from microalgae grown in wastewater.
(iii) Enhancing soil fertility by producing organic fertilizers from sludge or biosolids, which can also reduce the dependence on chemical fertilizers.
(iv) Creating economic value by generating revenue from selling the recovered products or services, or by reducing the operational costs of wastewater treatment.

Therefore, circular economy can contribute to wastewater treatment by transforming it from a costly and polluting service to a self-sustaining and value-adding one.

Smol et al. (2020) presented a comprehensive framework delineating the necessary measures for implementing a circular economy (CE) in the water sector, covering six key technological, organizational, and societal dimensions. This has been illustrated in Figure 15.3 below depicting rethink, reduction, reclamation, reuse, recycle, recovery as the 6 dimensions for implementing CE in the water sector.

The adoption of decentralization emerges as a strategic approach to address sustainability challenges in wastewater management systems, focusing on on-site wastewater treatment and local recycling of resources found in domestic wastewater (Capodaglio, 2017). Decentralized wastewater management specifically targets the

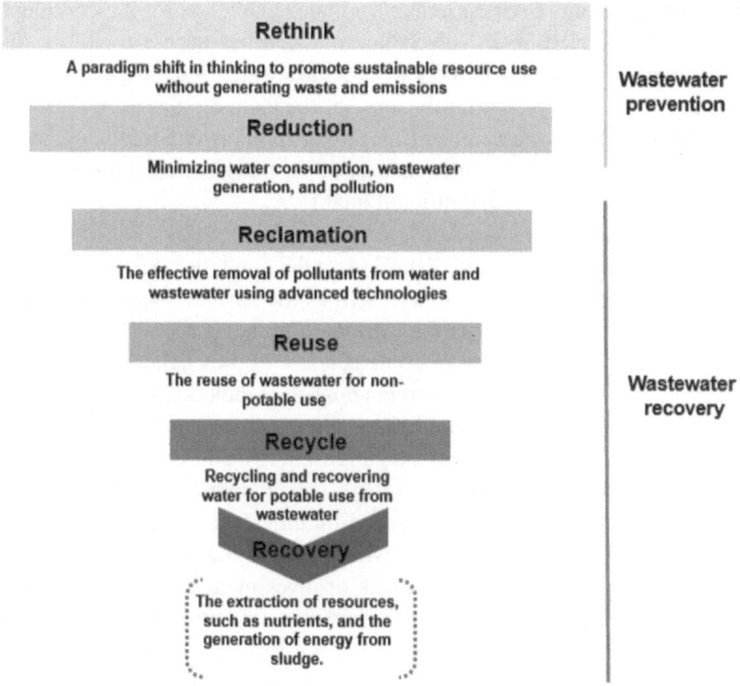

FIGURE 15.3 Model for circular economy implementation in the water sector (compiled by authors).

treatment and disposal of relatively small wastewater volumes at or near their source, typically originating from individual households or clusters of dwellings within close proximity (approximately less than 3-5 km). These systems are not connected to a central sewer system leading to a regional wastewater treatment plant (WWTP).

In contrast to centralized facilities, which may be financially and technically challenging for developing countries, the decentralized approach to wastewater treatment, incorporating on-site and/or cluster systems, is gaining prominence. This approach offers management flexibility, with a range of both simple and sophisticated technologies available (Massoud et al., 2009). Notably, decentralized systems for greywater reuse (GWR), Wastewater recycling (WWR), and rainwater harvesting (RWH) enable the treatment of rainwater and greywater at the source. This reduces the costs associated with water transportation and collection, which can constitute a significant portion, exceeding 60 percent, of the total budget for wastewater management, particularly in low-density population communities (Ghafourian et al., 2022).

15.12 WORLD POLICIES ON WASTEWATER TREATMENT

Wastewater treatment and recovery technologies differ worldwide based on technological development, industrialization and the policy framework adopted at national level for water resource management and disposal.

Wastewater treatment in Africa remains a major challenge for the continent. Only 17 percent of the population in sub-Saharan Africa has access to safely managed sanitation services, and only 4 percent of the wastewater is treated before discharge (Rugaimukamu et al., 2022). Currently some of the technologies used for wastewater treatment in Africa are waste stabilization ponds and conventional activated sludge, which are low-cost and low-maintenance options, but have limitations in terms of land use, energy consumption and effluent quality.

The Sanitation and Wastewater Atlas of Africa (UNEP, 2011) issued by the United Nations Environment Programme, GRID-Arendal, and the African Development Bank is a comprehensive resource that describes the current situation of wastewater management and sanitation across the African continent, as well as the opportunities and challenges for improving it. To date, some of the challenges for wastewater treatment and water recovery in the African continent include, (i.) high organic loads and uncontrolled waste input, which affect the efficiency and performance of the treatment processes; (ii.) power cuts and high energy costs, which limit the operation and maintenance of the treatment plants; (iii.) increasing wastewater flow rates and limited land availability, which pose difficulties for scaling up the treatment capacity; (iv.) lack of skilled workers, which hinder the proper management and monitoring of the treatment plants; (v.) lack of investments and funding, which prevent the upgrading and expansion of the treatment infrastructure, and, finally, (vi.) low public awareness and acceptance, which reduce the demand and potential for wastewater reuse. These challenges require more attention and action from policy makers, researchers, practitioners and communities to improve the sanitation and wastewater situation in Africa.

Over the past four decades, China has undergone a rapid transformation in wastewater treatment, evolving from a state characterized by a scarcity of technology and capacity to attaining global super capacity and technological prowess (Zhang et al., 2023). In 2015, the government introduced the Action Plan for Prevention and Control of Water Pollution, which, for the first time, emphasized the oversight of industrial water treatment and reclamation. The plan set forth requirements for assessing industrial wastewater treatment comprehensively, including the formulation of specialized wastewater treatment plans for ten key industries (such as coking, printing and dyeing, and pharmaceuticals) and industrial agglomerations.

In recent years, sewage treatment has been a substantial contributor to energy consumption in China, with the overall societal energy consumption proportion increasing annually. China's sewage treatment technology predominantly encompasses conventional pretreatment and conventional advanced treatment, representing two distinct treatment units. The choice of treatment methods varies based on sewage quality and other influencing factors. However, irrespective of whether physical or biochemical treatment is employed, the process consumes a significant amount of resources and energy (Li et al., 2023).

Resource recovery is considered a valuable method for circular economy integration in wastewater treatment. For instance, Europe has an advanced policy framework for wastewater treatment and water recovery. There are different policies and regulations for wastewater management and reuse in Europe. These include: the EU Water Framework Directive (European Parliament, 2023), which aims to protect and improve the quality of surface and groundwater sources, the Urban Wastewater Treatment Directive (ECC, 1991), which sets standards for the collection and treatment of urban wastewater to prevent pollution; and the Water Reuse Regulation (EC, 2020), which establishes minimum requirements for the reuse of treated wastewater for agricultural irrigation.

Similarly, the United States developed a comprehensive policy framework for wastewater treatment in the last decades with the aim to control water discharge and to manage water resources (UN, 2017). For instance, the Clean Water Act (EPA, 1972), regulates the discharge of pollutants into U.S. waters and sets wastewater standards for different sectors; the National Pollutant Discharge Elimination System (NPDES) (EPA, 2023), which issues permits to wastewater dischargers and treatment facilities and establishes discharge limits and conditions; and the Safe Drinking Water Act (EPA, 2023), which protects public health by regulating the nation's public drinking water supply and its sources.

In 2017, India developed a National Policy on Faecal Sludge and Septage Management (Ministry of Urban Development, 2017), which aims to improve the management of wastewater from onsite sanitation systems, such as septic tanks and pit latrines, and promotes safe reuse of treated sludge.

Brazil enacted a National Policy on Sanitation in 2007, which establishes the principles, objectives and instruments for the provision of sanitation services, including wastewater treatment, and defines the roles and responsibilities of various actors and institutions (Borges et al., 2022).

These policies show how diverse countries and regions are addressing the challenges and opportunities of wastewater treatment in a sustainable way.

15.13 CONCLUSION AND FUTURE OUTLOOK

15.13.1 Conclusion

The ideals of sustainability, resource conservation, and the circular economy are all supported by recovering and reusing these resources from wastewater in addition to easing the burden on freshwater resources. It is vital to remember that resource recovery from wastewater depends on a number of variables, including wastewater quality, treatment technology, local restrictions, and infrastructure. The implementation of circular economy practices for wastewater treatment and water recovery provides a wide number of economic and environmental benefits, including the increase of water resources by reclaiming and reusing wastewater, the reduction of greenhouse gas emissions and other hazardous materials in the ecosystem, enhancing soil fertility by producing organic fertilizers from sludge or biosolids and creating economic value from the valorization of wastewater.

Sustainability, resource conservation, and the circular economy are all supported by recovering and reusing these resources from wastewater in addition to easing the burden on freshwater resources. It is vital to remember that resource recovery from wastewater depends on a number of variables, including wastewater quality, treatment technology, local restrictions, and infrastructure.

Overall, wastewater treatment and recovery technologies differ globally based on technological development, industrialization and the policy framework adopted at national level for water resource management and disposal. Sustainable management of water resources and efficient wastewater recovery need pragmatic efforts that have to be addressed by policymakers and stakeholders worldwide.

15.13.2 Future Outlook

In summary, we explored the under-exploited resources in wastewater, including bioplastics, enzymes, metals, and minerals that hold significant promise for a more sustainable and circular approach to resource management and sustainability. However, substantial efforts are needed to overcome technical, economic, and environmental challenges to make their reclamation economically viable. By prioritizing research, collaboration, and knowledge sharing, we can pave the way for the responsible and efficient recovery of these valuable resources, contributing to a more sustainable future.

Based on circular economy practices for wastewater treatment we suggest the following policy recommendations. First, it is crucial to reduce pollution and minimize the release of hazardous chemicals and materials into water sources. Second, major efforts have to be undertaken to increase the treatment capacity and efficiency of wastewater globally. Third, there is a major need for implementing regulations for recycling and reusing wastewater for various purposes, such as irrigation, energy production, fertilizer and nutrient recovery, and environmental restoration. Last, policymakers and stakeholders should engage actively to invest in wastewater infrastructure and innovation, and strengthening the legal, institutional and financial frameworks for wastewater management. Such efforts will ultimately provide the basis for a global sustainable wastewater treatment and consequentially enhance circularity of water resources.

ACKNOWLEDGEMENTS

This study has been supported by the National Natural Science Foundation of China (Project No. 21906016), and the State Key Laboratory of Pollution Control and Resource Reuse Foundation (No. PCRRF19007).

REFERENCES

Abuabdou, S. M. A., Ahmad, W., Aun, N. C., & Bashir, M. J. K. (2020). A review of anaerobic membrane bioreactors (AnMBR) for the treatment of highly contaminated landfill leachate and biogas production: Effectiveness, limitations and future perspectives. *Journal of Cleaner Production*, *255*, 120215.

Amann, A., Zoboli, O., Krampe, J., Rechberger, H., Zessner, M., & Egle, L. (2018). Environmental impacts of phosphorus recovery from municipal wastewater. *Resources, Conservation and Recycling*, *130*, 127-139. https://doi.org/https://doi.org/10.1016/j.resconrec.2017.11.002

Amin, S. (2009). Review on biofuel oil and gas production processes from microalgae. *Energy Conversion and Management*, *50*(7), 1834-1840. https://doi.org/10.1016/j.enconman.2009.03.001

Arcos-Hernández, M., Montaño-Herrera, L., Murugan Janarthanan, O., Quadri, L., Anterrieu, S., Hjort, M., Alexandersson, T., Karlsson, A., Karabegovic, L., & Magnusson, P. (2015). Value-added bioplastics from services of wastewater treatment. *Water Practice and Technology*, *10*(3), 546-555.

Borges, M. C. P., Abreu, S. B., Lima, C. H. R., Cardoso, T., Yonamine, S. M., Araujo, W. D. V., Silva, P. R. S., Machado, V. B., Moraes, V., Silva, T. J. B., Reis, V. A., Santos, J. V. R., Reis, M. L., Canamary, É. A., Vieira, G. C., & Meireles, S. (2022). The Brazilian National System for Water and Sanitation Data (SNIS): Providing information on a municipal level on water and sanitation services. *Journal of Urban Management*, *11*(4), 530-542. https://doi.org/10.1016/J.JUM.2022.08.002

Brennan, L., & Owende, P. (2010). Biofuels from microalgae – A review of technologies for production, processing, and extractions of biofuels and co-products. *Renewable and Sustainable Energy Reviews*, *14*, 557-577. https://doi.org/10.1016/j.rser.2009.10.009

Capodaglio, A. G. (2017). Integrated, decentralized wastewater management for resource recovery in rural and peri-urban areas. *Resources*, *6*(2), 22. https://doi.org/10.3390/RESOURCES6020022

Caruso, D., & Slembrouck, J. (2019). *Food and Agriculture Organization of the United Nations for a world without hunger Fisheries and Aquaculture Department Osphronemus goramy (Lacépède, 1801).*

Chang, I.-S., & Kim, S.-N. (2005). Wastewater treatment using membrane filtration –Effect of biosolids concentration on cake resistance. *Process Biochemistry*, *40*(3-4), 1307-314.

Chen, J., Yang, Z., Su, Y., Han, F., & Monts, D. L. (2011). Phytoremediation of heavy metal/metalloid-contaminated soils. *Contaminated Soils: Environmental Impact, Disposal and Treatment*, 181-206.

Chonde, S., & Raut, P. (2017). Treatment of dairy wastewater by *Moringa oleifera* seeds. *World Journal of Pharmaceutical Research*, *6*(8), 1484-1493.

Chong, J. W. R., Khoo, K. S., Yew, G. Y., Leong, W. H., Lim, J. W., Lam, M. K., Ho, Y.-C., Ng, H. S., Munawaroh, H. S. H., & Show, P. L. (2021). Advances in production of bioplastics by microalgae using food waste hydrolysate and wastewater: A review. *Bioresource Technology*, *342*, 125947.

Chu, S., Cui, Y., & Liu, N. (2017). The path towards sustainable energy. *Nature Materials*, *16*(1), 16-22.

Colzi Lopes, A., Valente, A., Iribarren, D., & González-Fernández, C. (2018). Energy balance and life cycle assessment of a microalgae-based wastewater treatment plant: A focus on alternative biogas uses. *Bioresource Technology*, *270*, 138-146. https://doi.org/10.1016/j.biortech.2018.09.005

Deblonde, T., Cossu-Leguille, C., & Hartemann, P. (2011). Emerging pollutants in wastewater: A review of the literature. *International Journal of Hygiene and Environmental Health*, *214*, 442- 448. https://doi.org/10.1016/j.ijheh.2011.08.002

Diana, G. E., Gupta, R., Goyal, R. K., & Ana, F. A. (2020). Implemented models of circular management of water. *Agricultural Management/Lucrari Stiintifice Seria I, Management Agricol*, *22*(3).

EC. (2020). The Water Reuse Regulation – EU 2020/741. https://eur-lex.europa.eu/legal-content/EN/TXT/HTML/?uri=CELEX:32020R0741

ECC. (1991). The Urban Wastewater Treatment Directive. *Official Journal of the European Communities*.

EPA. (1972). Federal Water Pollution Control Act. https://www.rivernetwork.org/resource/read-the-full-clean-water-act/

EPA. (2023). National Pollutant Discharge Elimination System (NPDES). www.epa.gov/npdes

European Parliament. (2023). Water Protection and Management. www.europarl.europa.eu/factsheets/en/sheet/74/water-protection-and-management

Gaur, N., Narasimhulu, K., & Y., PydiSetty. (2018). Recent advances in the bio-remediation of persistent organic pollutants and its effect on environment. *Journal of Cleaner Production*, *198*, 1602-1631. https://doi.org/10.1016/j.jclepro.2018.07.076

Ghafourian, M., Nika, C. E., Mousavi, A., Mino, E., Al-Salehi, M., & Katsou, E. (2022). Economic impact assessment indicators of circular economy in a decentralised circular water system – Case of eco-touristic facility. *Science of the Total Environment*, *822*, 153602. https://doi.org/10.1016/J.SCITOTENV.2022.153602

Guerra-Rodríguez, S., Oulego, P., Rodríguez, E., Singh, D. N., & Rodríguez-Chueca, J. (2020). Towards the implementation of circular economy in the wastewater sector: Challenges and opportunities. *Water*, *12*(5), 1431.

Hall, E. R., & Sutton, P. M. (2016). State-of-the-Art review. Parameters governing permeate flux in an anaerobic membrane bioreactor treating low-strength municipal wastewaters: A literature review. *Water Environment Research*,*78*(8), 887-896. https://doi.org/10.2175/106143005X72858

Hamawand, I., Sandell, G., Pittaway, P., Chakrabarty, S., Yusaf, T., Chen, G., Seneweera, S., Al-Lwayzy, S., Bennett, J., & Hopf, J. (2016). Bioenergy from cotton industry wastes: A review and potential. *Renewable and Sustainable Energy Reviews*, *66*, 435-448. https://doi.org/https://doi.org/10.1016/j.rser.2016.08.033

Henriques, J., & Catarino, J. (2017). Sustainable value – An energy efficiency indicator in wastewater treatment plants. *Journal of Cleaner Production*, *142*, 323-330.

Hernández-Chover, V., Castellet-Viciano, L., Fuentes, R., & Hernández-Sancho, F. (2023). Circular economy and efficiency to ensure the sustainability in the wastewater treatment plants. *Journal of Cleaner Production*, *384*, 135563.

Hirwa, H., Peng, Y., Zhang, Q., Qiao, Y., Leng, P., Tian, C., Yang, G., Muhirwa, F., Diop, S., Kayiranga, A., Li, F., & Chen, G. (2022). Virtual water transfers in Africa: Assessing topical condition of water scarcity, water savings, and policy implications. *Science of the Total Environment*, *835*, 155343. https://doi.org/10.1016/J.SCITOTENV.2022.155343

Kato, M. T., Field, J. A., & Lettinga, G. (1997). The anaerobic treatment of low strength wastewaters in UASB and EGSB reactors. *Water Science and Technology*, *36*(6-7), 375-82.

Kehrein, P., Van Loosdrecht, M., Osseweijer, P., Garfí, M., Dewulf, J., & Posada, J. (2020). A critical review of resource recovery from municipal wastewater treatment plants – Market supply potentials, technologies and bottlenecks. *Environmental Science: Water Research & Technology*, *6*(4), 877-910.

Krantzberg, G., & Hartley, P. (2018). Feasible policy development and implementation for the destruction of endocrine disruptors in wastewater. *The Science of the Total Environment*, *631-632*, 246-251. https://doi.org/10.1016/j.scitotenv.2018.03.031

Kromah, V., & Zhang, G. (2021). Aqueous adsorption of heavy metals on metal sulfide nanomaterials: Synthesis and application. *Water*, *13*(13), 1843.

Leite, W. R. M., Gottardo, M., Pavan, P., Belli Filho, P., & Bolzonella, D. (2016). Performance and energy aspects of single and two phase thermophilic anaerobic digestion of waste activated sludge. *Renewable Energy*, *86*, 1324-1331. https://doi.org/10.1016/j.renene.2015.09.069

Leyva-Díaz, J. C., Muñío, M. M., González-López, J., & Poyatos, J. M. (2016). Anaerobic/anoxic/oxic configuration in hybrid moving bed biofilm reactor-membrane bioreactor for nutrient removal from municipal wastewater. *Ecological Engineering*, *91*, 449-458. https://doi.org/https://doi.org/10.1016/j.ecoleng.2016.03.006

Li, D., Wang, Z., Yang, Y., Liu, H., Fang, S., & Liu, S. (2023). Research status and development trend of wastewater treatment technology and its low carbonization. *Applied Sciences*, *13*(3), 1400. https://doi.org/10.3390/APP13031400

Lopes, A. C., Valente, A., Iribarren, D., & González-Fernández, C. (2018). Energy balance and life cycle assessment of a microalgae-based wastewater treatment plant: A focus on alternative biogas uses. *Bioresource Technology*, *270*, 138-146.

Lu, Q., He, Z. L., & Stoffella, P. J. (2012). Land application of biosolids in the USA: A review. *Applied and Environmental Soil Science*, *2012*(2), 201462.

Malik, O. A., Hsu, A., Johnson, L. A., & de Sherbinin, A. (2015). A global indicator of wastewater treatment to inform the Sustainable Development Goals (SDGs). *Environmental Science & Policy*, *48*, 172-185. https://doi.org/https://doi.org/10.1016/j.envsci.2015.01.005

Malik, S., Kishore, S., Bora, J., Chaudhary, V., Kumari, A., Kumari, P., Kumar, L., & Bhardwaj, A. (2023). A comprehensive review on microalgae-based biorefinery as two-way source of wastewater treatment and bioresource recovery. *CLEAN - Soil, Air, Water*, *51*(3), 2200044.

Mannina, G., Gulhan, H., & Ni, B. J. (2022). Water reuse from wastewater treatment: The transition towards circular economy in the water sector. *Bioresource Technology*, *363*, 127951. https://doi.org/10.1016/J.BIORTECH.2022.127951

Martin, I., Pidou, M., Soares, A., Judd, S., & Jefferson, B. (2011). Modelling the energy demands of aerobic and anaerobic membrane bioreactors for wastewater treatment. *Environmental Technology*, *32*(9), 921-932.

Massoud, M. A., Tarhini, A., & Nasr, J. A. (2009). Decentralized approaches to wastewater treatment and management: Applicability in developing countries. *Journal of Environmental Management*, *90*(1), 652-659. https://doi.org/10.1016/J.JENVMAN.2008.07.001

Máthé-Gáspár, G., & Anton, A. (2005). Phytoremediation study: Factors influencing heavy metal uptake of plants. *Acta Biologica Szegediensis*, *49*, 69-70.

Mekouar, M. A. (2013). 15. Food and Agriculture Organization (FAO). *Yearbook of International Environmental Law*, *24*(1), 587-602.

Melia, P. M., Cundy, A. B., Sohi, S. P., Hooda, P. S., & Busquets, R. (2017). Trends in the recovery of phosphorus in bioavailable forms from wastewater. *Chemosphere*, *186*, 381-395. https://doi.org/https://doi.org/10.1016/j.chemosphere.2017.07.089

Ministry of Urban Development. (2017). The National Policy on Faecal Sludge and Septage Management (FSSM).

Monfet, E., Aubry, G., & Ramirez, A. A. (2017). Nutrient removal and recovery from digestate: a review of the technology. *Biofuels*, *9*, 247-262. https://doi.org/10.1080/17597269.2017.1336348

Nguyen, T. (2018). Composting of sewage sludge with a simple aeration method and its utilization as a soil fertilizer. *Environmental Management*, *63*, 455-465. https://doi.org/10.1007/s00267-017-0963-8

Obeta, M. C., & Nwankwo, C. F. (2015). Factors responsible for rural residential water supply shortage in southeastern Nigeria. *Journal of Environmental Geography*, *8*(3-4), 21-32.

Opare, S. (2018). Adaptation to climate change impacts: Coping strategies of an indigenous community in Ghana to declining water supply. *Climate and Development*, *10*(1), 73-83.

Owusu, P. A., & Asumadu-Sarkodie, S. (2016). A review of renewable energy sources, sustainability issues and climate change mitigation. *Cogent Engineering*, *3*(1), 1167990.

Peng, W., Lü, F., Hao, L., Zhang, H., Shao, L., & He, P. (2019). Digestate management for high-solid anaerobic digestion of organic wastes: A review. *Bioresource Technology*, *297*, 122485. https://doi.org/10.1016/j.biortech.2019.122485

Petelka, J., Abraham, J., Bockreis, A., Deikumah, J., & Zerbe, S. (2019). Soil heavy metal(loid) pollution and phytoremediation potential of native plants on a former gold mine in Ghana. *Water, Air, & Soil Pollution*, *230*, 267. https://doi.org/10.1007/s11270-019-4317-4

Qiu, L., Deng, Y. F., Wang, F., Davaritouchaee, M., & Yao, Y. Q. (2019). A review on biochar-mediated anaerobic digestion with enhanced methane recovery. *Renewable and Sustainable Energy Reviews*, *115*(April), 109373. https://doi.org/10.1016/j.rser.2019.109373

Quaschning, V. V. (2019). *Renewable Energy and Climate Change*. John Wiley & Sons.

Raheem, A., Prinsen, P., Vuppaladadiyam, A. K., Zhao, M., & Luque, R. (2018). A review on sustainable microalgae based biofuel and bioenergy production: Recent developments. *Journal of Cleaner Production*, *181*, 42-59. https://doi.org/10.1016/j.jclepro.2018.01.125

Rugaimukamu, Q., Wang, H., Huang, R., & Xie, L. (2022). Wastewater treatment needs more attention. *Nature Africa 2022*. https://doi.org/10.1038/d44148-022-00071-2

Salgot, M., & Folch, M. (2018). Wastewater treatment and water reuse. *Current Opinion in Environmental Science & Health*, *2*, 64-74. https://doi.org/https://doi.org/10.1016/j.coesh.2018.03.005

Saravanan, A., Kumar, P. S., Jeevanantham, S., Karishma, S., Tajsabreen, B., Yaashikaa, P. R., & Reshma, B. (2021). Effective water/wastewater treatment methodologies for toxic pollutants removal: Processes and applications towards sustainable development. *Chemosphere*, *280*, 130595.

Semblante, G. U., Hai, F. I., Huang, X., Ball, A. S., Price, W. E., & Nghiem, L. D. (2015). Trace organic contaminants in biosolids: Impact of conventional wastewater and sludge processing technologies and emerging alternatives. *Journal of Hazardous Materials*, *300*, 1-17.

Sfez, S., De Meester, S., Vlaeminck, S. E., & Dewulf, J. (2019). Improving the resource footprint evaluation of products recovered from wastewater: A discussion on appropriate allocation in the context of circular economy. *Resources, Conservation and Recycling*, *148*, 132-144.

Shan, T. C., Matar, M. Al, Makky, E. A., & Ali, E. N. (2017). The use of *Moringa oleifera* seed as a natural coagulant for wastewater treatment and heavy metals removal. *Applied Water Science*, *7*(3), 1369-1376. https://doi.org/10.1007/s13201-016-0499-8

Shao, W., Ebaid, R., Abomohra, A. E.-F., & Shahen, M. (2018). Enhancement of *Spirulina* biomass production and cadmium biosorption using combined static magnetic field. *Bioresource Technology*, *265*, 163-169. https://doi.org/https://doi.org/10.1016/j.biortech.2018.06.009

Sharma, P., & Melkania, U. (2017). Biochar-enhanced hydrogen production from organic fraction of municipal solid waste using co-culture of *Enterobacter aerogenes* and *E. coli*. *International Journal of Hydrogen Energy*, *42*(30), 18865-18874. https://doi.org/https://doi.org/10.1016/j.ijhydene.2017.06.171

Shi, L., Simplicio, W. S., Wu, G., Hu, Z., Hu, H., & Zhan, X. (2018). Nutrient recovery from digestate of anaerobic digestion of livestock manure: A review. *Curr Pollution Report,* **4**, 74–83 (2018). https://doi.org/10.1007/s40726-018-0082-z

Sipra, A. T., Gao, N., & Sarwar, H. (2018). Municipal solid waste (MSW) pyrolysis for biofuel production: A review of effects of MSW components and catalysts. *Fuel Processing Technology*, *175*, 131-147. https://doi.org/https://doi.org/10.1016/j.fuproc.2018.02.012

Smith, A. L., Stadler, L. B., Love, N. G., Skerlos, S. J., & Raskin, L. (2012). Perspectives on anaerobic membrane bioreactor treatment of domestic wastewater: A critical review. *Bioresource Technology*, *122*, 149-159.

Smith, K., Guo, S., Zhu, Q., Dong, X., & Liu, S. (2019). An evaluation of the environmental benefit and energy footprint of China's stricter wastewater standards: Can benefit be increased? *Journal of Cleaner Production*, *219*, 723-733. https://doi.org/https://doi.org/10.1016/j.jclepro.2019.01.204

Smol, M., Adam, C., & Preisner, M. (2020). Circular economy model framework in the European water and wastewater sector. *Journal of Material Cycles and Waste Management*, *22*(3), 682-697. https://doi.org/10.1007/S10163-019-00960-Z/FIGURES/4

Song, X., Luo, W., Hai, F. I., Price, W. E., Guo, W., Ngo, H. H., & Nghiem, L. D. (2018). Resource recovery from wastewater by anaerobic membrane bioreactors: Opportunities and challenges. *Bioresource Technology*, *270*, 669-677.

UN. (2017). The United Nations World Water Development Report 2017: Wastewater: The Untapped Resource. Geneva: UN Water.

UNEP. (2011). UNEP Year Book 2011: Emerging Issues in Our Global Environment. UNEP/Earth print.

UNEP. (2020). The Sanitation and Wastewater Atlas of Africa. www.unep.org/resources/publication/sanitation-and-wastewater-atlas-africa

Vassalle, L., Díez-Montero, R., Machado, A. T. R., Moreira, C., Ferrer, I., Mota, C. R., & Passos, F. (2020). Upflow anaerobic sludge blanket in microalgae-based sewage treatment: Co-digestion for improving biogas production. *Bioresource Technology*, *300*, 122677.

Wang, S., Fu, B., Chen, H., & Liu, Y. (2018). Regional development boundary of China's Loess Plateau: Water limit and land shortage. *Land Use Policy*, *74*, 130-136. https://doi.org/https://doi.org/10.1016/j.landusepol.2017.03.003

Wu, S., Zhang, X., Lu, P., & Zhang, D. (2022). Copper removal and elemental sulfur recovery from fracturing flowback water in a microbial fuel cell with an extra electrochemical anode. *Chemosphere*, *303*, 135128.

Xiao, S., Dong, H., Geng, Y., & Brander, M. (2018). An overview of China's recyclable waste recycling and recommendations for integrated solutions. *Resources, Conservation and Recycling*, *134*, 112-120. https://doi.org/https://doi.org/10.1016/j.resconrec.2018.02.032

Yadav, B., Pandey, A., Kumar, L. R., & Tyagi, R. D. (2020). Bioconversion of waste (water)/residues to bioplastics – A circular bioeconomy approach. *Bioresource Technology*, *298*, 122584.

Yang, C., He, Z., Guo, Z., Zhou, A., Wang, A., & Liu, W. (2016). Effects of temperature on hydrolysis performance and short-chain fatty acids production during thermophilic micro-aerobic fermentation of waste activated sludge. *Desalination and Water Treatment*, *57*(28), 13183-13189. https://doi.org/10.1080/19443994.2015.1056838

Yao, Y., Huang, G., An, C., Chen, X., Zhang, P., Xin, X., Jian Shen, & Agnew, J. (2020). Anaerobic digestion of livestock manure in cold regions: Technological advancements and global impacts. *Renewable and Sustainable Energy Reviews*, *119*, 109494. https://doi.org/https://doi.org/10.1016/j.rser.2019.109494

Zając, P., & Avdiushchenko, A. (2020). The impact of converting waste into resources on the regional economy, evidence from Poland. *Ecological Modelling*, *437*, 109299.

Zarei, M. (2020). Wastewater resources management for energy recovery from a circular economy perspective. *Water-Energy Nexus*, *3*, 170-185.

Zhang, C., Quan, B., Tang, J., Cheng, K., Tang, Y., Shen, W., Su, P., & Zhang, C. (2023). China's wastewater treatment: Status quo and sustainability perspectives. *Journal of Water Process Engineering*, *53*, 103708. https://doi.org/10.1016/J.JWPE.2023.103708

Zhou, Z., Ruan, D., Jiang, L.-M., Yang, Y., Ge, H., & Wang, L. (2019). Comparison on treatment strategy for chemical cleaning wastewater: Pollutants removal, process design and techno-economic analysis. *Journal of Environmental Management*, *235*, 161-168. https://doi.org/https://doi.org/10.1016/j.jenvman.2019.01.053

16 Treatment Techniques Used for the Removal of Pollutants in Hospital Wastewater

Mohd. Aamir Mazhar, Pervez Alam, Sirajuddin Ahmed, and Kafeel Ahmad

16.1 INTRODUCTION

Good quality water is everyone's need, and life without it is unimaginable (Assob, et al., 2012). Still, large sections of society are deprived of access to clean and safe drinking water. Wide use of antibiotics (Abs) is leading to contamination of water bodies at various levels. Pharmaceuticals (PhCs) use needs to be restricted to some extent, as seen by the widespread presence of PhCs in the waterbodies. PhCs are used by humans in amounts like those of numerous insecticides. Significant amounts of PhCs residues are clearly seen entering the aquatic system from a variety of sources. Hence, there are pressing needs to monitor and regulate the release of pharmaceuticals into the ecosystem, particularly the water bodies. The choice of a combination treatment or a separate treatment will undoubtedly depend on deterministic factors, such as the type of environment and the amount of dosage that the relevant medication center has recommended (Verlicchi, Galletti & Masotti, 2010). Smart healthcare facilities stock their PhCs wastewater (that may contain PhCs) and separate rainwater collection for later treatment needs. Pre-treatment, biological treatment, particularly membrane biological reactor (MBR), advanced oxidation process (AOP), and other treatment methods for hospital wastewater (HWW) are required. To eliminate the desired amount of the targeted micropollutant, this must be heavily incorporated. The implementation of this plan will also depend on available technologies and operational circumstances.

Environmental restrictions, regional economics, and local laws are among the factors that have influenced HWW treatment in that area. AOP, membrane filters, reverse osmosis (RO), and nanofiltration processes are in the adoption phase with reference to HWW treatment in many developing nations (Rodriguez-Mozaz, et al., 2015). The potential for several antibiotics to be ineffective for HWW has been emphasized in studies. The micropollutant strategy's overall efficacy will fall between 1 and 5 log units (Verlicchi, Al Aukidy & Zambello, 2015). Absorbable

 DOI: 10.1201/9781003454199-16

organic halides (AOXs) are one of the most persistent environmental contaminants. They typically have a propensity to accumulate within the food chain and may be hazardous to both aquatic and terrestrial species. Certain halogens can occur naturally in several prescription medications and their metabolites, which can contribute to AOX emissions. Disinfectants, cleaning products, and tablets containing chlorine, which results in the formation of disinfection byproducts (Mazhar, et al., 2020, 2024). These disinfectants are far less significant than solvent discharge from laboratories. Soon, treatment processes will need to be stepped to work out sustainable and competitively priced strategies to face the challenge posed by disinfectants and antibiotic-resistant microorganisms (Shah, et al., 2024).

16.2 BIOLOGICAL TREATMENT TECHNIQUES FOR HWW

Biological treatment is being widely used for wastewater because of its low capital investment and operation and maintenance costs. The limitation of this treatment is the presence of contaminants in the effluent that can inhibit the growth of bacteria. The biological processes are either aerobic or anaerobic.

16.2.1 Treatment by Activated Sludge Process

For municipal wastewater treatment plants (WWTPs), the activated sludge process (ASP) is typically used. The sludge and wastewater are combined with microorganisms in this treatment process to remove nutrients that oxidize carbonous living stuff and produce gaseous matter like ammonia and other compounds. The trash is combined with the culture of acclimated microorganisms in the aerated tank to begin the treatment process. The sludge is adequately acclimated to handle any shock loading that could happen during the treatment procedure. To create effective effluent in WWTP, ASP is designed to minimize organic pollutants and suspended solids. Several PhCs have been found in different treatment plants. With longer retention time, it has been discovered that the removal efficiency during ASP might range from 30–70 percent. Seasonal fluctuations, temperature, pH, dissolved oxygen (DO), natural load, and microbial community all have an impact on performance (Castiglioni, et al., 2006).

According to some research, numerous PhCs, including ibuprofen, bezafibrate, naproxen, and estrogens may biodegrade in aerobic and anaerobic circumstances (Kim, et al., 2014). Inhibited by handling active medicinal substances at higher concentrations, ASP results in limited therapeutic efficacy. To ensure that the desired removal efficiency of a particular parameter is achieved, the settling time in the aeration tank is adjusted prior to starting a solid/liquid separation process in the clarifier. The collected settling activated sludge is initially recycled by returning it to the aeration tank to handle a difficult and high concentration of depolluting microorganisms. Some researchers believe that co-metabolic biodegradations may be crucial to the efficient removal of micro-pollutants from the activated sludge process of HWW because their concentrations may be too low to cause an immediate rise in substrate (Fischer & Majewsky, 2014).

16.2.2 Treatment by Sequencing Batch Reactor

A sophisticated and contemporary method of treating activated sludge for organic matter is the sequencing batch reactor (SBR). These methods are presently being used in a few systems. However, these must be considered for treatment, particularly with several types of wastewaters. In SBR, cycles are independently filled, react, settle, draw, and remain at rest. An SBR performs all the processes in a single reactor, which distinguishes it from a standard treatment activated sludge system. When compared to the activated sludge method in the SBR process, the concentration of sludge or biomass may not decrease. SBR is beneficial for the nutrition and organic removal process. It has been utilized for WWTP for many years due to this advantage. Separate aeration tanks and natural occurrences do not seem to be required in the case of SBR, an inactivated sludge treatment technology, and there is no sludge return. When there is a lack of land or a smaller region, such a strategy is typically used (Ileri, et al., 2003).

Removal rates of 82 percent for biological oxygen demand (BOD), 88 percent for COD, 96 percent for NH_3, and 98 percent for suspended solids (SS) from residential and pharmaceutical waste products were observed in a study using SBR operated after four hours of aeration and 60 min SRT (Ileri, et al., 2003). In a different study, poor treatment of the SBR technology led to reduced COD efficiencies of between 63 and 69 percent (Rodriguez-Mozaz, et al., 2015). BOD_5 efficiency was 81 percent with 4 hours of aeration and 90 minutes of SRT. As a result, it is possible to reach the best removal efficiency in less than four hours of aeration time and 60 minutes of SRT. The average BOD_5 removal was found to be 82 percent, and the average COD removal was found to be half a mile. The average rate of NH_3 removal effectiveness was found to be 96 percent. On average, SS elimination was found to be 98 percent. SBR has been found to be an effective technique for treating HWW. Anaerobic sequencing batch reactors (AnSBR) are used in a few experiments to treat HWW in the presence of oxygen while the environment remains dark. Fatty acids are used under various dosage circumstances. PhCs can be treated up to 40 mg/L, but after that point, the microbial community becomes hazardous. Due to the usage of AnSBR, the microbial network could see the influence of dosages on Abs. Due to the microbial system, it changed and now exhibits stability under high antibiotic sensitivity and long-term operation using the resources of homoacetogenesis along with hydrogenotrophic methanogenesis.

16.2.3 Treatment by Membrane Biological Reactor

Abs elimination happens mostly through biodegradation, sorption to sludge, photodegradation, and volatilization during wastewater treatment employing a membrane biological reactor (MBR). Low pressure membranes are used in the MBR and ASP treatments to prevent pollution. This method is recognized as the most effective technology for treating HWW because it reduces sludge, produces low suspended particles, and effectively removes pathogens (Sipma, et al., 2010). Long retention times allow for the removal of nitrogen and promote the growth of nitrifying bacteria while also removing micro-pollutants more effectively. In comparison to the

ASP system, the MBR has a lower VSS/TSS ratio. Pollutants physical properties are removed in the MBR system, which then releases high-quality effluents. High chemical oxygen demand (COD) elimination in MBR offers microorganisms a stable habitat for effective micropollutant removal. Because of the MBR process's remarkable efficiency in removing micropollutants during the past ten years, it has become a viable alternative to ASP. In comparison to ASP the removal efficiency is greater by 15–42 percent (Tran, et al., 2016). However, it was also shown that the clearance rates of certain Abs in both methods are often the same.

However, volatilization has no effect on Henry values below 10^{-5} (Stevens-Garmon, et al., 2011), and because of the high turbidity, biodegradation is barely detectable (Trinh, et al., 2016). Versioloursis injections come in a variety of forms in the MBR treatment program. A fungus performs biological functions when it is sterilized. Numerous studies are being conducted in this area, nevertheless. For instance, there are both anoxic and aerobic reactors (Luo, et al., 2014). An MBR can more cheaply treat a variety of micropollutants. However, the procedure can be time-consuming because of membrane fouling (Pavithra, et al., 2017; Mackuľak, et al., 2015; Meijide, et al., 2016). With a removal effectiveness of up to 85 percent, bio-entrapped membrane-salt marsh sediment membrane bio-reactors are also employed for HWW treatment (Khalfbadam, et al., 2017). Anaerobic membrane bio-reactors are modern, creative solutions for varying loads (Lin, et al., 2013). In a study, a removal efficiency of 47.2 percent for COD and poor removal efficiency for organics were achieved (Dvořák, et al., 2016). With the addition of methanol, it is possible to increase the COD removal effectiveness by up to 78 percent by adjusting the organic load (Svojitka, et al., 2017).

More than 65 microcontaminants were analyzed in SP-HPLC-MS/MS at the European-Swiss PILLS pilot project plant, with sludge retention times (SRT) up to 50 days. However, due to the low percentage of iodinated contrast media degradation, a removal effectiveness of just 22 percent was attained (Kovalova, et al., 2012). In another study, wash-off of faeces for HWW was 80 percent effective at removing 40 micropollutants (OECD, 2016). However, research suggests that following MBR treatment, oestrogen and cytostatic should be discontinued (Maletz, et al., 2013; Mahnik, et al., 2007). Even after sustaining SRT for 42 to 300 days, elimination can occasionally occur due to bio-degradation (Lenz, et al., 2007). Due to the adsorption process, a low HRT of 3 hours has been seen in several investigations (Prasertkulsak, et al., 2016). MBR treatment requires polishing methods such as ozonation, activated carbon, photo-degradation, and so forth, to fully eliminate micropollutants fully. The MBR process is more expensive, and it is then either followed by ozonation, packed activated carbon, or filtration (Nielsen, et al., 2013).

16.2.4 Treatment by Up-flow Anaerobic Sludge Blanket Reactor

Due to their higher biomass content, variety of microbial species, low cost, adaptability to pH and temperature, and flexibility, up-flow anaerobic sludge blankets (UASB) are commonly employed (Mazhar, et al., 2021). An increased organic loading rate results in up to 91 percent more COD elimination. The efficiency is severely impacted when the loading rate increases to over 2.09 kg COD/m^3/d at pH 8, which favors sulfate-lowering microorganisms over methanogens in reactors (Li, et al., 2015). To increase COD efficiency and remove some PhCs residues, UASB reactors are also utilized as pre-treatment, incorporating different doses of Abs under high organic loads up to 21.02 kg COD/m^3/d (Yi, et al., 2017). There are several diverse types of UASB reactors in use, like the expanded granular sludge blanket (EGSB), which separates sludge from granules by utilizing a high up-flow velocity. The effectiveness of COD removal from reactors using psychrophilic anaerobic digestion was reported to be up to 70 percent (Yi, et al., 2017).

EGSB and aerobic reactors are being employed in a large-scale oxytetracycline production wastewater treatment facility in China. The up-flow anaerobic stage reactor (UASR), which combines the characteristics of an anaerobic baffled reactor and a UASBR, is another comparable type of reactor. Separating acidogenesis and methanogenesis has advantages since it reduces biomass runoff and allows for quick hydraulic recovery and stress loading. Tylosin and avilamycin antibiotics were added to HWW as a pre-treatment, and the viability of UASR was evaluated. The separation of acidogenesis from methanogenesis from sludge was discovered during observation with varied organic loading rates (OLR) on UASR, which also evaluated the reactor's stability. At low OLRs, COD removal effectiveness can reach up to 75 percent; however, by raising OLRs to 3.73 kg COD/m^3/d and reducing HRT from 2 to 4 days, COD removal efficiency can reach 45 percent (Chelliapan, et al., 2011).

16.3 PHYSIO-CHEMICAL TREATMENT TECHNIQUES FOR HWW

HWW shares some physiochemical properties with residential wastewater in terms of chemical composition. According to municipal regulations, dissolved organic carbon, BOD, COD, suspended particles, and chlorides are on the greater side in HWW, but PhCs are substantially greater in HWW (Verlicchi, et al., 2010 and Kovalova, et al., 2012). Studies on the existence of micropollutants in HWW are currently concentrated mostly due to low biodegradability, genes that are resistant to antibiotics, and allied effects like cancer, mutagens, and EDCs (Fent, Weston & Caminada, 2006). There are currently very few PhCs standards in the HWW. AOP adjustments for natural pollutants or dealing with various surface-based media for treatment have recently inspired a few recent developments in the field of water treatment. Since these mixture treatment systems can produce inbound hydroxyl radicals, AOPs can be classified. Herbal and inorganic mixtures can be eliminated by a mixture with a high oxidation capacity. AOPs may be preferred depending on the outcome of the oxidizing agent or the process for using it in its production.

Since the PhCs are still not being treated in traditional WWT, there is considerable concern. There are other methods available today, such as treating HWW at the source or combining it with domestic wastewater before processing. During the degradation of PhCs, pseudo first-order kinetics were noticed. Therefore, the best way to prevent dilution is to treat at the source. Therefore, distinct treatment for HWW is now being examined (Joss, et al., 2006; Langenhoff, et al., 2013; Nielsen, et al., 2013). Typically, HWW treatments begin with a pre-treatment phase to break up the solid fraction, followed by a natural therapy to remove most of the dissolved organic carbon, nutrients, and a few PhCs. To sum up, it represents a physical and chemical technique for removing and sanitizing effluents. Some physiochemical treatment techniques depend on the HWWs that are being treated. It displays the application space and recommended procedure. Ozonation and UV/H_2O_2 are advised to handle solutions with low flow rates and little organic natural carbon. High flow rate effluents may benefit from biological treatments. For effluents with low flow charges, organic removal is recommended as an alternative. For medium organic content and intermediate flow rates, the combination of an AOP with biological treatment is advised.

16.3.1 Treatment by Activated Carbon

The use of activated carbon in adsorption to remediate pollutants, particularly HWW, is an incredibly old technique. It was discovered to be able to eliminate endocrine disruptors (Liu, Kanjo & Mizutani, 2009). Different types of activated carbon were used in numerous therapy studies to provide effective results (Yu, et al., 2014; Mailler, et al., 2014). In adsorption, chemical compounds that are soluble and gaseous are bound to the surface by molecular-level attraction. The porous structure that arises from carbon's activation improves its capacity for adsorption.

More than 70 endocrine disruptors were examined for absorbability, chemical structure, and adsorption in a study using activated carbon, with successful findings (Nakanishi, et al., 2002). Treatment of endocrine disruptors has been found to benefit from a variety of variable factors, including adsorption capacity, specific surface area and micropore volume. Their lifespan also varies based on the type of carbon and service life. Due to its large pore volume, the coal-based plant is the most effective. A variation in adsorption capability for several types of endocrine disruptors between real and simulated wastewater was also noted during the investigation (Snyder, et al., 2007).

Treating wastewater may be an energy-intensive procedure, thus cost is a crucial factor that needs to be taken into consideration before it is put into practice. Although the process of biodegradation is inexpensive, the activated sludge approach is popular worldwide. According to some studies, normal biological waste treatment procedures can effectively remove copious amounts of endocrine-disrupting chemicals (EDCs), however, these procedures vary widely. Numerous factors, including temperature, hydraulic retention time (HRT), sludge age, and nitrification setting, are essential for the removal of oestrogen during biological wastewater treatment. The optimization of operational parameters during the removal of EDCs is a widespread practice in conventional waste treatment. It is possible to use biological waste treatment technologies to efficiently remove the EDC from surface water bodies using HWW. Physical and advanced oxidation cannot be used for treatment on a large scale. Physical processes in some circumstances imply that complex chemical reactions may potentially be desirable for simple operation and high removal effectiveness.

16.3.2 Treatment by Carbon Nanotubes

By adjusting crucial factors, including temperature and pH, single-, double-, and multi-walled carbon nanotubes are employed to remove PhCs. In the pH range of

3–7, the adsorption capacity rises by up to 70 percent (García-Mateos, et al., 2015). Up to a maximum removal effectiveness of 10 to 95 percent, carbon nanotubes exhibit improved Abs removal (Suriyanon, et al., 2015). For the treatment of HWW, carbon nanotubes are a new and very promising adsorption technology (Liu, et al., 2014; Liu and Wang, 2013). These can also be used with graphene and graphene oxide to increase surface area and efficiency. However, other factors need to be addressed, such as minimizing the steric effect of micropores, regeneration, recycling, and lowering production costs. PhCs may also pose a serious threat to aquatic life because of the coexistence of adsorbents. Nano-absorbents like TiO_2, FeO, aluminium, and so forth, remove a lot of AbS in a step-by-step process that starts with adsorption on an external surface and ends with antiparticle diffusion (Li, et al., 2017; Beiginejad, et al., 2015). Many metals, including arsenic, lead, mercury, cadmium, and chromium, have demonstrated high adsorption capability due to their high surface area. The similar uses for numerous additional nanomaterials are listed in Table 16.1.

The usage of carbon nanotubes in a study by Wang et al. on wastewater and municipal wastewater led to a high clearance rate of PhCs in the pH range of 4.0 to 10.0 (Wang, et al., 2016). The fact that the mixed ions and many other thermodynamic

TABLE 16.1
Properties of Nanoparticles Used

Nanomaterials	Properties	Advantages
Biosensors	Efficient Sensitivity, optimum surface energy, quickly reactive, effective surface conductivity with good surface to volume ratio	Small size, friendly to users
Nanoparticles of Zinc oxide	Transparency in optics and the luminescence effect	Environmentally friendly, easy to synthesis
Cellulose nanoparticles	Lightweight with high strength	Compostable, substitutes toxic materials
Silicon nanoparticles	Superior electrical and optoelectronic properties	Cheap biocompatible
Graphene oxide Graphene	Strong mechanical efficiency, optimal surface area	Effective surface adjustability with minimal water permeability
Carbon nanotubes	Optimal surface area with the capacity for π-π interactions	Low price range, simplicity of use, and functionality
Magnetic nanoparticles	Super paramagnet, excellent surface area, and densimetric separation	Low cost, biocompatible, and easily synthesized
Nanoparticles with mixed and metallic oxides	Adsorption ability, chemical permanence, and effective specific area	Employed as inexpensive, hazardous magneto-optical devices

Source: Khan et al., 2020.

parameters hardly ever affect the adsorption pattern indicates that exothermic and spontaneous adsorption occur. According to research on the mechanism controlling the adsorption rate, mass transfer from outside the body controls the entire process. With recovery possibilities exceeding 93 percent, the adsorbent is effectively desorbed in an ethanol/sodium hydrate solution. Thereafter, carbon nanotubes may be recycled at least eight times with good overall efficiency. In the future, kinetics and thermodynamics can both be useful when planning and thinking through the HWW treatment process.

16.3.3 Treatment by UV/H_2O_2

When radiation is absorbed, the usual mechanism for the photolysis of H_2O_2 is the random conversion of the peroxide molecules into hydroxyl radicals, with a yield of two HO radicals. When an O-O bond is broken, it takes 213 kJ/mol of energy, which suggests low wavelength radiation (200–280 nm). A new way of thinking was made possible by the removal of micropollutants from aqueous solutions by UV/H_2O_2 treatment. The OH radical is crucial in the breakdown of PhCs with unity quantum yield. A number of Abs are susceptible to removal by UV, H_2O_2, or both (Jung, et al., 2012; He, et al., 2012).

In a study, the degrading effect of UV light was demonstrated at low pressure and at polychromatic medium pressure (Gerrity, et al., 2010). The electrophilic addition tendency of organic pollutants is decreased. The formation of free radicals is impacted by any issue with UV radiation or low transmission to waste-content fabric. The lightweights must therefore be routinely cleaned. The instrument is sensitive to the pH scale, and basicity must be preserved. Hydroxide must be used close to carbonate since carbonate reacts with radicals (Dobrin, et al., 2013). Although it shows promise in mineralizing natural contaminants, there are risks involved, such as the awful ground assimilation capabilities of UV light. Numerous additional porous materials, including those made of silica, effectively remove PhCs and Abs (Yi, et al., 2017). These often increase a specific surface area and hence, the capacity for absorption.

16.3.4 Treatment by Fenton Reagent

The Fenton process, which can be improved by photoreduction of Fe^{+2} and Fe^{+3}, includes the interaction of iron with hydrogen peroxide to form its radical. Numerous studies indicate that Photo-Fenton may be an effective treatment for HWW (Wen, et al., 2016; Pavithra, et al, 2017). For certain PhCs, this procedure took 50 minutes to fully mineralize. After 40 minutes, penicillin was eliminated using advanced oxidation, pH=3, COD, and Fe^{2+}/H_2O_2. Dark Fe^{2+}/H_2O_2 at pH=3 has a higher total organic carbon (TOC) removal efficiency than dark Fenton-like Fe^{3+}/H_2O_2 (Arslan-Alaton & Dogruel, 2004).

The Fe^{3+} hydroxyl complexes regenerate adequate Fe^{+2} in the absence of organic ligands by absorbing UV rays. Fe^{3+} readily reacts with other natural ligands to form complexes, frequently with polydentate ligands. When combined with ozonation, the Photo-Fenton process is cost-effective for treating PhCs and is used to treat membrane-rejected effluents with excessive quantities of micropollutants. (Pavithra, et al, 2017). However, there are certain drawbacks, such as pH adjustment, sludge disposal, expensive H_2O_2 consumption, and catalyst use. The Photo-Fenton technique holds promise for AOP in the removal of diverse types of PhCs. Like HWW treatment, the key difficulty is successfully treating micropollutants in accordance with local or national legislation while taking the cost into consideration. The Fenton procedure is a more straightforward approach that is also used to remove numerous micropollutants that are dissolved in water or wastewater. In a few instances, it is unnecessary to remove the soluble iron species from the waste that has been treated. Everything is based on the regulatory effluent limitations imposed by local and national authorities. Most research has demonstrated that hydrogen ions are useful for Photo-Fenton processes. However, Fe^{3+} with greater molar absorption coefficients has seen satisfactory results in many places with aquo-complexes within the broad UV spectrum, at the same time of radiation as much as 580 nm. Iron athletics in the environment requires the introduction of chelation complexes with natural ligands to control iron transport, specification, and accessibility, particularly in moderate water bodies. The method is cost-effective even if the chemical rate for hydrogen ions needs to be modified with considerable care, especially for the entire treatment procedure. Pilot-scale research and techniques for concentrating the parabolic superstar approach have improved because of the excessive Photo-Fenton efficiency in the removal of micropollutants. For the low cost approach, the employment of various processes has been carefully explored, and there are numerous alternatives in this direction (Klamerth, et al., 2013).

The pilot-scale study has taken into account the complete removal of the excess prescription medications typically found in HWW, including antibiotics, non-steroidal anti-inflammatory medicines, analgesics, hormones, x-ray steady-state contrast media and others, which has been quite satisfactory (Karaolia, et al., 2014). The Photo-Fenton method for HWW treatment for prescription pharmaceuticals inside the concentrate has been used in several pilot studies, and LCMS has been applied for a practical examination of the mechanics together with nurse financial assessment (Miralles-Cuevas, et al., 2017). The pre-treatment of effluents has made it possible for Photo-Fenton to operate more affordably and with better starting megacycles in accordance with per second concentrations, which has resulted in a reduction in the length of the photoreactor and the quantity of reagents needed per kiloliter of the treated effluents. These results support the assumption that high-organic-load commercial wastewaters require fundamentally unique operational concepts from the equipment of AOPs to treat exceptionally low quantities of pollutants, such as those present in HWW.

16.3.5 Treatment by Ozone

In this process, ozone either directly or indirectly oxidizes the micro-pollutants, with the hydroxyl radical serving as a potent oxidant. It only reacts with anions of N, P, O, and S as well as double chain bonds, -OH, $-CH_3$, and $-OCH_3$. However, indirect reactions take place in alkaline circumstances. In a study, ozone concentrations between 5 and 15 mg/L were maintained, and it was shown that some medications had clearance efficiencies of 96 percent (Tiwari & Bose, 2007). Through its reaction during its era, ozone can initiate and take part in a wide variety of organic and inorganic substances. Ozone molecules have three atoms each. It is unstable at room temperature and can be detected by people, even at low concentrations. The use of pH, temperature, float charge, bubble length, ozone stress, and ionic strength, controls the ozone change effectiveness from gasoline to liquid (Tiwari & Bose, 2007).

In recent years, numerous treatment facilities in Europe have undergone upgrades that combine physical and biological processes with the efficient removal of micropollutants. The conversion of the micro-pollutants into unknown compounds with unknown toxicity has been discovered (Joss, et al., 2006; Stadler, et al., 2012). However, they are of inferior quality and have negligible antimicrobial and estrogenic effects. Additionally, it was noted that the end result can contain carcinogenic chemicals such as formaldehyde, ketones, phenols, nitromethanes, and others (Carbajo, et al., 2015). Pre-treatment, membrane bioreactors, and post-treatment using ozone technologies were used in a pilot-scale HWW project. In general, the amount of dissolved organic carbon required for the clearance of the medicines was 1.08 mg/mg O_3 (Blackbeard, et al., 2016). In general, higher levels of DOC will result in higher ozone doses to treat HWW. In one study, it was found that its content in the HWW treatment will range from 0.50 to 4.7 mg O_3/mg of DOC (Hansen, et al., 2016). Additionally, it was discovered that a higher pH (5–9) in HWW treatment has a detrimental influence on removal effectiveness, while a lower pH shows a significant function in its longevity by raising it from 1 to 10 minutes. However, there are certain issues with this technique, high production costs, heat generation, and its poor conversion factor are a few of them (1 kg of O_3 production needs 12 kWh) (Hollender, et al., 2009). The retrofitting ozonation degree may desire to increase small to medium scale WWTP for HWW's upgrade value by up to 50 percent, and improve energy consumption by up to 30 percent (Eggen, et al., 2014). It is evocative of how radical species can react on a regular basis after combining with water, both directly as molecular fuel and indirectly through oxidants.

16.4 TREATMENT BY COMBINATION OF TECHNIQUES

It is clear that particular treatment methods are less effective, as was stated in the section prior concerning the various PhCs from HWW and their treating efficacy. Therefore, combining diverse procedures is popular, and numerous research works have been published in this regard (Grandclément, et al., 2017; Oller, Malato &Sánchez-Pérez, 2011). The therapeutic approaches used by various nations and how well they work together will be covered in this section. HWW often comprises a number of micropollutants that are released from various medical and non-medical treatment activities by both 200–1200 liters/bed/day, such as from OPD waste, laboratory waste, operating room waste, radiology rooms, and the kitchen (Emmanuel et al., 2005; Sharma & Sanghi, 2013). It may or may not have the properties of household wastewater, but it definitely has heavy metals, Abs, and a disinfectant (Verlicchi, et al., 2012). However, these are either dumped into the environment in their natural state or mixed with household garbage (Verlicchi, et al., 2010). As a result, numerous health issues have surfaced. These include the formation of antibiotic-resistant bacteria, the failure of Abs, and the appearance of Abs-resistant genes (Ahmad, et al., 2014 and Alam, et al., 2013). The end-of-pipe method or a separate treatment facility is still used today to treat medications in many underdeveloped nations (Verlicchi, Al Aukidy & Zambello, 2015).

Depending on the country's economic situation, different HWW therapies may be combined in different sequences. In India, full-scale plants like combined ASP are common (Prabhasankar, et al., 2016), combined ASP is followed by sand filtration, chlorination (Chitnis et al., 2004), coagulation is followed by filtration, chlorination (Gautam, Kumar & Sabumon, 2007). To accurately assess the methodology and efficacy of utilizing optimization approaches, since most of the research in this field was published on a pilot/lab scale, and only a handful were related to full scale treatment plants, there is still much room for improvement. It is important to thoroughly research variables such as the design of a full scale with characteristics, temperature conditions, and the viability of HWW economically. Trends in this area include keeping an eye on things, accepting innovative ideas, and looking toward the future. It is well known that individual treatment methods do not remove micropollutants very effectively. As a result, many combinations of newer alternative technologies are currently being studied. For example, the use of activated carbon can enhance the removal of less biologically decomposed substances through adsorption (Alvarino et al., 2016a). Additionally, it was discovered that floating biomass could speed up the biodegradation of micropollutants. With an emphasis on sand biofilm technologies, biofilters with fixed biofilms are also being studied (Casas & Bester, 2015). While intriguing effects were discovered by cultivating a variety of microorganisms within the system, many significant parameters, such as SRT, also imposed performance risks. Thus, a larger biomass concentration can be accommodated by the biofilm reactor.

In the hybrid biofilm activated sludge processes, bacteria with modest growth rates boosted removal with ammonia and nitrite-oxidizing bacteria. A promising technology for in-tank biofilm formation is the moving bed biofilm reactor (MBBR). This can be improved with nitrification and denitrification, thanks to ASP (Casas & Bester, 2015). Due to their biodegradation paths, a few research that combined MBBR and MBR, in which physical parameters were examined with varying detention times, indicated encouraging trends. However, as covered in the preceding sections, MBR operates with some physical difficulties. UASB combined with aerobic MBR, and low HRT is another good option. Compared to the usual technique, this technology can effectively remove micropollutants (Alvarino et al., 2016b). The cost of the technology used, the types of wastewaters, the temperature, and, most importantly, the financial situation for carrying out these projects can all play a significant part in determining the best treatment method.

16.5 CONCLUSION

There are no strict regulations for HWW treatment, and it is frequently combined with residential wastewater in many nations. This chapter has covered many technologies used for HWW treatment based on biological and physio-chemical techniques. Different removal procedures would be required since HWW is uneven. Due to sophisticated analyses, targeted Abs detection, effects on waterbodies flow rates, and treatment methods that vary seasonally, HWW treatment has many challenges. It will be possible to effectively eradicate Abs by using sustainable, cost-effective, and energy-efficient technology after researching the risks associated with HWW treatment. When determining the appropriate metabolic pathways, it is necessary to consider the long-term risk they pose from acute and repeated exposure. Antibiotic-resistant systems are raised in WWTP, so close attention must also be paid to monitoring. Numerous trials combining various therapy modalities were taken into consideration. Due to the seriousness of the risk involved, several nations are paying close attention to this issue. The production and disposal of HWW can now be governed by a robust national strategy with appropriate legislation, thanks to further technological improvement. Given how nature is being destroyed and harmful impacts are being imposed, it is imperative to act promptly and strongly in this area.

REFERENCES

Ahmad, M., Khan, A.U., Wahid, A., Farhan, M., Butt, Z.A., Ahmad, F., 2014. Urban wastewater as hotspot for antibiotic and antibiotic resistant bacteria spread into the aquatic environment. *Asian J. Chem.* 26, 579–582. https://doi.org/10.14233/ajchem.2014.15906

Alam, M.Z., Aqil, F., Ahmad, I., Ahmad, S., 2013. Incidence and transferability of antibiotic resistance in the enteric bacteria isolated from hospital wastewater. *Braz. J. Microbiol.* 44, 799–806. https://doi.org/10.1590/S1517-83822013000300021

Alvarino, T., Komesli, O., Suarez, S., Lema, J.M., Omil, F., 2016a. The potential of the innovative SeMPAC process for enhancing the removal of recalcitrant organic micropollutants. *J. Hazard. Mater.* 308, 29–36. https://doi.org/10.1016/j.jhazmat.2016.01.040

Alvarino, T., Suárez, S., Garrido, M., Lema, J.M., Omil, F., 2016b. A UASB reactor coupled to a hybrid aerobic MBR as innovative plant configuration to enhance the removal of organic micropollutants. *Chemosphere* 144, 452–458. https://doi.org/10.1016/j.chemosphere.2015.09.016

Arslan-Alaton, I., Dogruel, S., 2004. Pre-treatment of penicillin formulation effluent by advanced oxidation processes. *J. Hazard. Mater.* 112, 105–113. https://doi.org/10.1016/j.jhazmat.2004.04.009

Assob, J.C.N., Nde, P.F., Nsagha, D.S., Njimoh, D.L., Nfor, O., Njunda, A.L., Kamga, H.L.F., 2012. The incidence of feco-oral parasites in street-food vendors in Buea, south-west region Cameroon. *Afr. Health Sci.* 12, 376–380. https://doi.org/10.4314/ahs.v12i3.20

Beiginejad, H., Amani, A., Nematollahi, D., Khazalpour, S., 2015. Thermodynamic study of the electrochemical oxidation of some aminophenol derivatives: Experimental and theoretical investigation. *Electrochim. Acta* 154, 235–243. https://doi.org/10.1016/j.electacta.2014.12.042

Blackbeard, J., Lloyd, J., Magyar, M., Mieog, J., Linden, K.G., Lester, Y., 2016. Demonstrating organic contaminant removal in an ozone-based water reuse process at full scale. *Environ. Sci. Water Res. Technol.* 2, 213–222. https://doi.org/10.1039/C5EW00186B

Carbajo, J.B., Petre, A.L., Rosal, R., Herrera, S., Letón, P., García-Calvo, E., Fernández-Alba, A.R., Perdigón-Melón, J.A., 2015. Continuous ozonation treatment of ofloxacin: Transformation products, water matrix effect and aquatic toxicity. *J. Hazard. Mater.* 292, 34–43. https://doi.org/10.1016/j.jhazmat.2015.02.075

Casas, M.E., Bester, K., 2015. Can those organic micro-pollutants that are recalcitrant in activated sludge treatment be removed from wastewater by bio film reactors (slow sand filters)? *Sci. Total Environ.* 506–507, 315–322. https://doi.org/10.1016/j.scitotenv.2014.10.113

Castiglioni, S., Bagnati, R., Fanelli, R., Pomati, F., Calamari, D., Zuccato, E., 2006. Removal of pharmaceuticals in sewage treatment plants in Italy. *Environ. Sci. Technol.* 40, 357–363. https://doi.org/10.1021/es050991m

Chelliapan, S., Wilby, T., Yuzir, A., Sallis, P.J., 2011. Influence of organic loading on the performance and microbial community structure of an anaerobic stage reactor treating pharmaceutical wastewater. *Desalination* 271, 257–264. https://doi.org/10.1016/j.desal.2010.12.045

Chitnis, V., Chitnis, S., Vaidya, K., Ravikant, S., Patil, S., Chitnis, D.S., 2004. Bacterial population changes in hospital effluent treatment plant in central India. *Water Res.* 38, 441–447. https://doi.org/10.1016/j.watres.2003.09.038

Dobrin, D., Bradu, C., Magureanu, M., Mandache, N.B., Parvulescu, V.I., 2013. Degradation of diclofenac in water using a pulsed corona discharge. *Chem. Eng. J.* 234, 389–396. https://doi.org/10.1016/j.cej.2013.08.114

Dvořák, L., Gómez, M., Dolina, J., Černín, A., 2016. Anaerobic membrane bioreactors – A mini review with emphasis on industrial wastewater treatment: Applications, limitations and perspectives. *Desalin. Water Treat.* 57, 19062–19076. https://doi.org/10.1080/19443994.2015.1100879

Eggen, R.I.L., Hollender, J., Joss, A., Schärer, M., Stamm, C., 2014. Reducing the discharge of micropollutants in the aquatic environment: The benefits of upgrading wastewater treatment plants. *Environ. Sci. Technol.* 48, 7683–7689. https://doi.org/10.1021/es500907n

Emmanuel, E., Perrodin, Y., Keck, G., Blanchard, J.M., Vermande, P., 2005. Ecotoxicological risk assessment of hospital wastewater: A proposed framework for raw effluents discharging into urban sewer network. *J. Hazard. Mater.* 117, 1–11. https://doi.org/10.1016/j.jhazmat.2004.08.032

Fent, K., Weston, A.A., Caminada, D., 2006. Ecotoxicology of human pharmaceuticals. *Aquat. Toxicol.* 76, 122–159. https://doi.org/10.1016/j.aquatox.2005.09.009

Fischer, K., Majewsky, M., 2014. Cometabolic degradation of organic wastewater micropollutants by activated sludge and sludge-inherent microorganisms. *Appl. Microbiol. Biotechnol.* 98, 6583–6597. https://doi.org/10.1007/s00253-014-5826-0

García-Mateos, F.J., Ruiz-Rosas, R., Marqués, M.D., Cotoruelo, L.M., Rodríguez-Mirasol, J., Cordero, T., 2015. Removal of paracetamol on biomass-derived activated carbon: Modeling the fixed bed breakthrough curves using batch adsorption experiments. *Chem. Eng. J.* 279, 18–30. https://doi.org/10.1016/j.cej.2015.04.144

Pavithra, K. G., Senthil Kumar, P., Sundar Rajan, P., Saravanan, A., Naushad, M. (2017). Sources and impacts of pharmaceutical components in wastewater and its treatment process: a review. *Korean Journal of Chemical Engineering*, *34*, 2787–2805. https://doi.org/10.1007/s11814-017-0255-2

Gautam, A.K., Kumar, S., Sabumon, P.C., 2007. Preliminary study of physico-chemical treatment options for hospital wastewater. *J. Environ. Manage.* 83, 298–306. https://doi.org/10.1016/j.jenvman.2006.03.009

OECD (2016), *Development Co-operation Report 2016: The Sustainable Development Goals as Business Opportunities*, OECD Publishing, Paris, https://doi.org/10.1787/dcr-2016-en.

Gerrity, D., Stanford, B.D., Trenholm, R.A., Snyder, S.A., 2010. An evaluation of a pilot-scale nonthermal plasma advanced oxidation process for trace organic compound degradation. *Water Res.* 44, 493–504. https://doi.org/10.1016/j.watres.2009.09.029

Grandclément, C., Seyssiecq, I., Piram, A., Wong-Wah-Chung, P., Vanot, G., Tiliacos, N., Roche, N., Doumenq, P., 2017. From the conventional biological wastewater treatment to hybrid processes, the evaluation of organic micropollutant removal: A review. *Water Res.* 111, 297–317. https://doi.org/10.1016/j.watres.2017.01.005

Hansen, K.M.S., Spiliotopoulou, A., Chhetri, R.K., Escolà Casas, M., Bester, K., Andersen, H.R., 2016. Ozonation for source treatment of pharmaceuticals in hospital wastewater – Ozone lifetime and required ozone dose. *Chem. Eng. J.* 290, 507–514. https://doi.org/10.1016/j.cej.2016.01.027

He, X., Pelaez, M., Westrick, J.A., O'Shea, K.E., Hiskia, A., Triantis, T., Kaloudis, T., Stefan, M.I., de la Cruz, A.A., Dionysiou, D.D., 2012. Efficient removal of microcystin-LR by UV-C/H_2O_2 in synthetic and natural water samples. *Water Res.* 46, 1501–1510. https://doi.org/10.1016/j.watres.2011.11.009

Hollender, J., Zimmermann, S.G., Koepke, S., Krauss, M., McArdell, C.S., Ort, C., Singer, H., von Gunten, U., Siegrist, H., 2009. Supplementary Information. Elimination of organic micropollutants in a municipal wastewater treatment plant upgraded with a full scale post-ozonation followed by sand filtration. *Environ. Sci. Technol.* 43, 7862–7869. https://doi.org/10.1021/es9014629

Ileri, R., Sengil, I.A., Kulac, S., Damar, Y., 2003. Treatment of mixed pharmaceutical industry and domestic wastewater by sequencing batch reactor. *J. Environ. Sci. Health A Tox. Hazard. Subst. Environ. Eng.* 38, 2101–2111. https://doi.org/Doi 10.1081/Ese-120023336

Joss, A., Zabczynski, S., Göbel, A., Hoffmann, B., Löffler, D., McArdell, C.S., Ternes, T.A., Thomsen, A., Siegrist, H., 2006. Biological degradation of pharmaceuticals in municipal wastewater treatment: Proposing a classification scheme. *Water Res.* 40, 1686–1696. https://doi.org/10.1016/j.watres.2006.02.014

Jung, Y.J., Kim, W.G., Yoon, Y., Kang, J.W., Hong, Y.M., Kim, H.W., 2012. Removal of amoxicillin by UV and UV/H_2O_2processes. *Sci. Total Environ.* 420, 160–167. https://doi.org/10.1016/j.scitotenv.2011.12.011

Karaolia, P., Michael, I., García-Fernández, I., Agüera, A., Malato, S., Fernández-Ibáñez, P., Fatta-Kassinos, D., 2014. Reduction of clarithromycin and sulfamethoxazole-resistant *Enterococcus* by pilot-scale solar-driven Fenton oxidation. *Sci. Total Environ.* 468–469, 19–27. https://doi.org/10.1016/j.scitotenv.2013.08.027

Khalfbadam, H.M., Ginige, M.P., Sarukkalige, R., Kayaalp, A.S., Cheng, K.Y., 2017. Sequential solid entrapment and in situ electrolytic alkaline hydrolysis facilitated reagent-free bioelectrochemical treatment of particulate-rich municipal wastewater. *Water Res.* 117, 18–26. https://doi.org/10.1016/j.watres.2017.03.045

Khan, N.A., Khan, S.U., Ahmed, S., Farooqi, I.H., Yousefi, M., Mohammadi, A.A., Changani, F., 2020. Recent trends in disposal and treatment technologies of emerging-pollutants – A critical review. *TrAC – Trends Anal. Chem.* 122, 115744. https://doi.org/10.1016/j.trac.2019.115744

Kim, M., Guerra, P., Shah, A., Parsa, M., Alaee, M., Smyth, S.A., 2014. Removal of pharmaceuticals and personal care products in a membrane bioreactor wastewater treatment plant. *Water Sci. Technol.* 69, 2221–2229. https://doi.org/10.2166/wst.2014.145

Klamerth, N., Malato, S., Agüera, A., Fernández-Alba, A., 2013. Photo-Fenton and modified photo-Fenton at neutral pH for the treatment of emerging contaminants in wastewater treatment plant effluents: A comparison. *Water Res.* 47, 833–840. https://doi.org/10.1016/j.watres.2012.11.008

Kovalova, L., Siegrist, H., Singer, H., Wittmer, A., McArdell, C.S., 2012. Hospital wastewater treatment by membrane bioreactor: Performance and efficiency for organic micropollutant elimination. *Environ. Sci. Technol.* 46, 1536–1545. https://doi.org/10.1021/es203495d

Langenhoff, A., Inderfurth, N., Veuskens, T., Schraa, G., Blokland, M., Kujawa-Roeleveld, K., Rijnaarts, H., 2013. Microbial removal of the pharmaceutical compounds ibuprofen and diclofenac from wastewater. *Biomed Res. Int.* 2013. https://doi.org/10.1155/2013/325806

Lenz, K., Mahnik, S.N., Weissenbacher, N., Mader, R.M., Krenn, P., Hann, S., Koellensperger, G., Uhl, M., Knasmüller, S., Ferk, F., Bursch, W., Fuerhacker, M., 2007. Monitoring, removal and risk assessment of cytostatic drugs in hospital wastewater. *Water Sci. Technol.* 56, 141–149. https://doi.org/10.2166/wst.2007.828

Li, W., Niu, Q., Zhang, H., Tian, Z., Zhang, Y., Gao, Y., Li, Y.Y., Nishimura, O., Yang, M., 2015. UASB treatment of chemical synthesis-based pharmaceutical wastewater containing rich organic sulfur compounds and sulfate and associated microbial characteristics. *Chem. Eng. J.* 260, 55–63. https://doi.org/10.1016/j.cej.2014.08.085

Li, W., Wang, J., He, G., Yu, L., Noor, N., Sun, Y., Zhou, X., Hu, J., Parkin, I.P., 2017. Enhanced adsorption capacity of ultralong hydrogen titanate nanobelts for antibiotics. *J. Mater. Chem. A* 5, 4352–4358. https://doi.org/10.1039/C6TA09116D

Lin, H., Peng, W., Zhang, M., Chen, J., Hong, H., Zhang, Y., 2013. A review on anaerobic membrane bioreactors: Applications, membrane fouling and future perspectives. *Desalination* 314, 169–188. https://doi.org/10.1016/j.desal.2013.01.019

Liu, F.F., Zhao, J., Wang, S., Du, P., Xing, B., 2014. Effects of solution chemistry on adsorption of selected pharmaceuticals and personal care products (PPCPs) by graphenes and carbon nanotubes. *Environ. Sci. Technol.* 48, 13197–13206. https://doi.org/10.1021/es5034684

Liu, Y., Wang, J., 2013. Degradation of sulfamethazine by gamma irradiation in the presence of hydrogen peroxide. *J. Hazard. Mater.* 250–251, 99–105. https://doi.org/10.1016/j.jhazmat.2013.01.050

Liu, Z.-h., Kanjo, Y., Mizutani, S., 2009. Removal mechanisms for endocrine disrupting compounds (EDCs) in wastewater treatment – Physical means, biodegradation, and chemical advanced oxidation: A review. *Sci. Total Environ.* 407, 731–748. https://doi.org/10.1016/j.scitotenv.2008.08.039

Luo, Y., Guo, W., Ngo, H.H., Nghiem, L.D., Hai, F.I., Zhang, J., Liang, S., Wang, X.C., 2014. A review on the occurrence of micropollutants in the aquatic environment and their fate and removal during wastewater treatment. *Sci. Total Environ.* 473–474, 619–641. https://doi.org/10.1016/j.scitotenv.2013.12.065

Mackuľak, T., Mosný, M., Grabic, R., Golovko, O., Koba, O., Birošová, L., 2015. Fenton-like reaction: A possible way to efficiently remove illicit drugs and pharmaceuticals from wastewater. *Environ. Toxicol. Pharmacol.* 39, 483–488. https://doi.org/10.1016/j.etap.2014.12.016

Mahnik, S.N., Lenz, K., Weissenbacher, N., Mader, R.M., Fuerhacker, M., 2007. Fate of 5-fluorouracil, doxorubicin, epirubicin, and daunorubicin in hospital wastewater and their elimination by activated sludge and treatment in a membrane-bio-reactor system. *Chemosphere* 66, 30–37. https://doi.org/10.1016/j.chemosphere.2006.05.051

Mailler, R., Gasperi, J., Coquet, Y., Deshayes, S., Zedek, S., Cren-Olivé, C., Cartiser, N., Eudes, V., Bressy, A., Caupos, E., Moilleron, R., Chebbo, G., Rocher, V., 2014. Study of a large scale powdered activated carbon pilot: Removals of a wide range of emerging and priority micropollutants from wastewater treatment plant effluents. *Water Res.* 72, 315–330. https://doi.org/10.1016/j.watres.2014.10.047

Maletz, S., Floehr, T., Beier, S., Klümper, C., Brouwer, A., Behnisch, P., Higley, E., Giesy, J.P., Hecker, M., Gebhardt, W., Linnemann, V., Pinnekamp, J., Hollert, H., 2013. In vitro characterization of the effectiveness of enhanced sewage treatment processes to eliminate endocrine activity of hospital effluents. *Water Res.* 47, 1545–1557. https://doi.org/10.1016/j.watres.2012.12.008

Mazhar, M.A., Khan, N.A., Ahmed, S., Khan, A.H., Hussain, A., Rahisuddin, Changani, F., Yousefi, M., Ahmadi, S., Vambol, V., 2020. Chlorination disinfection by-products in municipal drinking water – A review. *J. Clean. Prod.* 273, 123159. https://doi.org/10.1016/j.jclepro.2020.123159

Mazhar, M.A., Khan, N.A., Khan, A.H., Ahmed, S., Siddiqui, A.A., Husain, A., Rahisuddin, Tirth, V., Islam, S., Shukla, N.K., Changani, F., Yousefi, M., Hassaballa, A.E., Radwan, N., 2021. Upgrading combined anaerobic–aerobic UASB-FPU to UASB-DHS system: Cost comparison and performance perspective for developing countries. *J. Clean. Prod.* 284, 124723. https://doi.org/10.1016/j.jclepro.2020.124723

Mazhar, M.A., Madhav, S., Ahmed, S., Kumar, P., Springer, C., 2024. Drinking water chlorination and disinfection by-products: Formation, history, and regulations, in: Madhav, S., Mazhar, M.A., Ahmed, S., Kumar, P., Mishra, P.K. (Eds.), *Drinking Water Disinfection By-Products: Sources, Fate and Remediation.* Springer Nature Switzerland, Cham, pp. 21–34. https://doi.org/10.1007/978-3-031-49047-7_2

Meijide, J., Gómez, J., Pazos, M., Sanromán, M.A., 2016. Degradation of thiamethoxam by the synergetic effect between anodic oxidation and Fenton reactions. *J. Hazard. Mater.* 319, 43–50. https://doi.org/10.1016/j.jhazmat.2016.02.064

Miralles-Cuevas, S., Oller, I., Agüera, A., Sánchez Pérez, J.A., Malato, S., 2017. Strategies for reducing cost by using solar photo-Fenton treatment combined with nanofiltration to remove microcontaminants in real municipal effluents: Toxicity and economic assessment. *Chem. Eng. J.* 318, 161–170. https://doi.org/10.1016/j.cej.2016.06.031

Nakanishi, A., Tamai, M., Kawasaki, N., Nakamura, T., Tanada, S., 2002. Adsorption characteristics of bisphenol A onto carbonaceous materials produced from wood chips as organic waste. *J. Colloid Interface Sci.* 252, 393–396. https://doi.org/10.1006/jcis.2002.8387

Nielsen, U., Hastrup, C., Klausen, M.M., Pedersen, B.M., Kristensen, G.H., Jansen, J.L.C., Bak, S.N., Tuerk, J., 2013. Removal of APIs and bacteria from hospital wastewater by MBR plus O_3, O_3+ H_2O_2, PAC or ClO_2. *Water Sci. Technol.* 67, 854–862. https://doi.org/10.2166/wst.2012.645

Oller, I., Malato, S., Sánchez-Pérez, J.A., 2011. Combination of advanced oxidation processes and biological treatments for wastewater decontamination – A review. *Sci. Total Environ.* 409, 4141–4166. https://doi.org/10.1016/j.scitotenv.2010.08.061

Pavithra, K.G., Senthil Kumar, P., Sundar Rajan, P., Saravanan, A., Naushad, M., 2017. Sources and impacts of pharmaceutical components in wastewater and its treatment process: A review. *Korean J. Chem. Eng.* 34, 2787–2805. https://doi.org/10.1007/s11814-017-0255-2

Prabhasankar, V.P., Joshua, D.I., Balakrishna, K., Siddiqui, I.F., Taniyasu, S., Yamashita, N., Kannan, K., Akiba, M., Praveenkumarreddy, Y., Guruge, K.S., 2016. Removal rates of antibiotics in four sewage treatment plants in South India. *Environ. Sci. Pollut. Res.* 23, 8679–8685. https://doi.org/10.1007/s11356-015-5968-3

Prasertkulsak, S., Chiemchaisri, C., Chiemchaisri, W., Itonaga, T., Yamamoto, K., 2016. Removals of pharmaceutical compounds from hospital wastewater in membrane bioreactor operated under short hydraulic retention time. *Chemosphere* 150, 624–631. https://doi.org/10.1016/j.chemosphere.2016.01.031

Rodriguez-Mozaz, S., Chamorro, S., Marti, E., Huerta, B., Gros, M., Sànchez-Melsió, A., Borrego, C.M., Barceló, D., Balcázar, J.L., 2015. Occurrence of antibiotics and antibiotic resistance genes in hospital and urban wastewaters and their impact on the receiving river. *Water Res.* 69, 234–242. https://doi.org/10.1016/j.watres.2014.11.021

Shah, M., Mazhar, M.A., Ahmed, S., Lew, B., Khalil, N., 2024. Recent trends in controlling the disinfection by-products before their formation in drinking water: A review, in: Madhav, S., Mazhar, M.A., Ahmed, S., Kumar, P., Mishra, P.K. (Eds.), *Drinking Water Disinfection By-Products: Sources, Fate and Remediation*. Springer Nature Switzerland, Cham, pp. 177–192. https://doi.org/10.1007/978-3-031-49047-7_9

Sharma, S.K., Sanghi, R., 2013. *Wastewater Reuse and Management*. Springer, Dordrecht. https://doi.org/10.1007/978-94-007-4942-9

Sipma, J., Osuna, B., Collado, N., Monclús, H., Ferrero, G., Comas, J., Rodriguez-Roda, I., 2010. Comparison of removal of pharmaceuticals in MBR and activated sludge systems. *Desalination* 250, 653–659. https://doi.org/10.1016/j.desal.2009.06.073

Snyder, S.A., Adham, S., Redding, A.M., Cannon, F.S., DeCarolis, J., Oppenheimer, J., Wert, E.C., Yoon, Y., 2007. Role of membranes and activated carbon in the removal of endocrine disruptors and pharmaceuticals. *Desalination* 202, 156–181. https://doi.org/10.1016/j.desal.2005.12.052

Stadler, L.B., Ernsto, A.S., Aga, D.S., Love, N.G., 2012. Micropollutant fate in wastewater treatment: Redefining "Removal." *Environ. Sci. Technol.* 46, 10485–10486. https://doi.org/10.1021/es303478w

Stevens-Garmon, J., Drewes, J.E., Khan, S.J., McDonald, J.A., Dickenson, E.R.V., 2011. Sorption of emerging trace organic compounds onto wastewater sludge solids. *Water Res.* 45, 3417–3426. https://doi.org/10.1016/j.watres.2011.03.056

Suriyanon, N., Permrungruang, J., Kaosaiphun, J., Wongrueng, A., Ngamcharussrivichai, C., Punyapalakul, P., 2015. Selective adsorption mechanisms of antilipidemic and non-steroidal anti-inflammatory drug residues on functionalized silica-based porous materials in a mixed solute. *Chemosphere* 136, 222–231. https://doi.org/10.1016/j.chemosphere.2015.05.005

Svojitka, J., Dvořák, L., Studer, M., Straub, J.O., Frömelt, H., Wintgens, T., 2017. Performance of an anaerobic membrane bioreactor for pharmaceutical wastewater treatment. *Bioresour. Technol.* 229, 180–189. https://doi.org/10.1016/j.biortech.2017.01.022

Tiwari, G., Bose, P., 2007. Determination of ozone mass transfer coefficient in a tall continuous flow counter-current bubble contactor. *Chem. Eng. J.* 132, 215–225. https://doi.org/10.1016/j.cej.2006.12.025

Tran, N.H., Chen, H., Reinhard, M., Mao, F., Gin, K.Y.H., 2016. Occurrence and removal of multiple classes of antibiotics and antimicrobial agents in biological wastewater treatment processes. *Water Res.* 104, 461–472. https://doi.org/10.1016/j.watres.2016.08.040

Trinh, T., van den Akker, B., Coleman, H.M., Stuetz, R.M., Drewes, J.E., Le-Clech, P., Khan, S.J., 2016. Seasonal variations in fate and removal of trace organic chemical contaminants while operating a full-scale membrane bioreactor. *Sci. Total Environ.* 550, 176–183. https://doi.org/10.1016/j.scitotenv.2015.12.083

Verlicchi, P., Al Aukidy, M., Galletti, A., Petrovic, M., Barceló, D., 2012. Hospital effluent: Investigation of the concentrations and distribution of pharmaceuticals and environmental risk assessment. *Sci. Total Environ.* 430, 109–118. https://doi.org/10.1016/j.scitotenv.2012.04.055

Verlicchi, P., Al Aukidy, M., Zambello, E., 2015. What have we learned from worldwide experiences on the management and treatment of hospital effluent? – An overview and a discussion on perspectives. *Sci. Total Environ.* 514, 467–491. https://doi.org/10.1016/j.scitotenv.2015.02.020

Verlicchi, P., Galletti, A., Masotti, L., 2010. Management of hospital wastewaters: The case of the effluent of a large hospital situated in a small town. *Water Sci. Technol.* 61, 2507–2519. https://doi.org/10.2166/wst.2010.138

Verlicchi, P., Galletti, A., Petrovic, M., Barceló, D., 2010. Hospital effluents as a source of emerging pollutants: An overview of micropollutants and sustainable treatment options. *J. Hydrol.* 389, 416–428. https://doi.org/10.1016/j.jhydrol.2010.06.005

Wang, Y., Ma, J., Zhu, J., Ye, N., Zhang, X., Huang, H., 2016. Multi-walled carbon nanotubes with selected properties for dynamic filtration of pharmaceuticals and personal care products. *Water Res.* 92, 104–112. https://doi.org/10.1016/j.watres.2016.01.038

Wen, Y., Yi, J., Zhao, S., Jiang, S., Chi, Y., Liu, K., 2016. Non-thermal plasma treatment of *Radix aconiti* wastewater generated by traditional Chinese medicine processing. *J. Environ. Sci. (China)* 44, 99–108. https://doi.org/10.1016/j.jes.2015.10.028

Yi, Q., Zhang, Y., Gao, Y., Tian, Z., Yang, M., 2017. Anaerobic treatment of antibiotic production wastewater pretreated with enhanced hydrolysis: Simultaneous reduction of COD and ARGs. *Water Res.* 110, 211–217. https://doi.org/10.1016/j.watres.2016.12.020

Yu, W., Xu, L., Qu, J., Graham, N., 2014. Investigation of pre-coagulation and powder activate carbon adsorption on ultrafiltration membrane fouling. *J. Memb. Sci.* 459, 157–168. https://doi.org/10.1016/j.memsci.2014.02.005

17 Machine Learning for Water Quality Soft Sensing in Wastewater Treatment

State of the Art and Future Directions

Adil Masood, Aman Srivastava, Mohammed Majeed Hameed, Mohd. Saquib Tanweer, Kafeel Ahmad, and Siti Fatin Mohd Razali

17.1 INTRODUCTION

Wastewater generated by anthropogenic activities is a pressing concern, ranking as a primary source of pollution for watercourses and aquifers [1, 2]. The escalating volumes, laden with pollutants, necessitate thorough and state-of-the-art wastewater treatment measures to address society's growing environmental challenges. Wastewater treatment plants are one such facility, often used to reduce the pollution load of urban water bodies [3]. In recent years, advancements in analytics, soft sensing, data transmission, computing, and data management have created opportunities for information technology to enhance the management of these treatment plants, promising improved urban infrastructure efficiency. At present, advanced control and optimization systems have become essential in functioning wastewater treatment plants, allowing efficient plant management [4]. Monitoring and controlling effluent variables like biochemical oxygen demand (BOD) and suspended solids (SS) in a treatment plant process is complex [5]. Typically, these parameters are determined offline with uncertain results, or they demand costly sensors [6], imposing significant operational limitations on the process. Soft sensors enable the estimation of offline primary variables by leveraging real-time secondary variables [7]. These sensors amalgamate data processing, mathematical modeling, and software methodologies, and have ascended to prominence as an invaluable substitute for conventional methodologies when it comes to the acquisition of critical process parameters, internal state estimation, process surveillance, and fault detection.

 DOI: 10.1201/9781003454199-17

Moreover, the soft-sensing based on machine learning models prioritizes optimizing control operations when factoring in the dynamic behavior of the wastewater treatment process. This helps minimize energy consumption within different process units, promoting resource efficiency and sustainable operation. In the realm of wastewater treatment plant (WWTP) soft-sensing models for effluent quality, machine learning-based techniques have found extensive utilization within prevailing soft-sensing models. These techniques are employed to effectively model and forecast effluent quality [8]. For example, some researchers reported the application of neural networks, Fuzzy techniques, and genetic algorithm-based approaches for modeling effluent concentration [9]. The results suggest that neural network-based data-driven soft sensors offer better efficiency and cost-effectiveness to plant operators in recognizing the plant operational states compared to other ML techniques. Some researchers utilized an optimized extreme learning machine-based model to predict effluent BOD concentrations for a wastewater treatment plant [10]. The study results revealed that the optimized extreme learning machine (ELM) model presented better accuracy and improved stability than the base ELM model in accurately predicting BOD concentration. Others presented an SVM model based on an improved optimization algorithm for accurately forecasting the water quality indicators of effluent from a wastewater treatment facility [11]. The model was developed on the inputs of 23 water quality parameters. It was observed that the developed model showed both heightened accuracy and reduced computational time in forecasting effluent water quality parameters. Also, some authors designed hybrid neural network models to predict two Anammox process parameters. The results reported that ML models were efficient tools for estimating water quality parameters [12, 13] and would help reduce the need for expensive and untrustworthy equipment used to evaluate treatment plant nutrient levels [14, 15].

In this chapter, an extensive review of the soft-sensing models for WWTP is undertaken, encompassing a comprehensive range from simple ML learning models to more complex modeling approaches. The primary objective of this review work is to inspire researchers to delve deeper into ML-based soft-sensing research. Simultaneously, this study delivers valuable insights to practitioners, aiding them in discerning the optimal models tailored to their specific WWTP and related domain requirements. This comprehensive overview aims to foster advancements in soft-sensing techniques and enhance their practical utility across wastewater treatment and associated domains.

17.2 OVERVIEW OF MACHINE LEARNING MODELS

This section discusses the most frequent predictive models used in wastewater treatment. Table 17.1 furnishes a comprehensive synthesis of research endeavors employing AI models for the purpose of soft sensing in wastewater treatment's water quality assessment. The references of the studies are listed alongside the models used for comparison. Also, each study's output or target variables are specified, representing the water quality parameters of interest. Additionally, various statistical metrics have been employed to evaluate the performance of the models.

Table 17.1 summarizes various research papers and their models (machine learning approaches) for different prediction tasks in the field of environmental engineering

TABLE 17.1
Shows the Reviews of Water Quality Soft Sensing Models in Wastewater Treatment, Highlighting the Best Models

Comparable models	Output/Target	Statistics	References
DeepRecurrent Neural Network	TSS, S_{ALK}, S_{NH}, S_{ND}	RMSE, MAPE	[16]
Monitoring Models based on BLS, and OICA	BOD_5	-	[17]
RG,PCA-R,PLS, and **ANN**	E. coli	R^2, MAE, MAPE	[18]
SBiLSTM,LSTM, BiLSTM	DCI	RMSE, R^2	[19]
ANN, LSTM	S_{Ntot}, S_{NH}	MAPE, R^2, RMSE	[20]
ANN,PCA-R, **PCA-ANN,** MLR	PET	RMSE	[21]
CAE-DFC,SVR, ANN, RF	S_p, S_{Ntot}	R^2	[22]
SVR,RFN-SVR, **FCM-RFN-SVR**	H_2S	MSE, R^2	[23]
PSO-SVR, RBFNN, BPNN	COD	RMSE, R^2, MAE	[24]
GK-LSSVM	BOD_5	RMSE	[25]
SVR	DO	MSE, RMSE, R^2	[26]
GP, SVR, ANN	lipase activity d bacterial production of poly *	R, R^2, RMSE	[27]
OK,ANN,SVR, GPR	PC	NRMSE, RMSE, R	[28]
ANFIS	DO	Graphical comparison	[29]
ANN, **ANFIS**	Industrial dehumanizer column compositions	MAE, MAPE, RMSE, R^2	[30]
GA-ANFIS, ANFIS	COD, effluent S_{Ntot}, effluent TP, Sp	RMSE, MAPE, R^2	[31]
ANFIS,**RF**,M5P tree	Aeration efficiency	MAPE, R, MSE, RMSE, MAE, R^2	[32]
GEP,MLR,ANN, k-nearest neighbors, GPR,regression trees	BOD_5, COD	RMSE, MSE, RAE, R^2	[33]
RF	long-term stream nitrogen concentrations	-	[34]
KPCA-OCSVM	identification of anomalies in the influent conditions (ICs) of wastewater treatment plants: clustering analysis**	TPR, AUC, FPR, Accuracy, F1Score, Precision.	[35]
PCA-GPR, **PPCA-GPR**	BOD	RMSE, R, MAE	[36]
ELM, ANN	Coagulant dosing	MAPE	[37]
KNN	COD, SS, SNtot, Sp	RMSE	[38]
SVR, ANN	S_{Ntot}	NSE, d_r, R^2	[39]

and water quality assessment. The models include Deep Recurrent Neural Network (DRNN), Support Vector Regression (SVR), Artificial Neural Networks (ANN), and others. The target variables range from water quality parameters like Biological Oxygen Demand (BOD_5), Chemical Oxygen Demand (COD), and E. coli to process efficiency and anomaly detection. Evaluation metrics used in the papers include Root Mean Square Error (RMSE), Mean Absolute Error (MAPE), Correlation of Determination (R^2), and others. Overall, the table provides a concise overview of different modeling approaches and their applications in environmental engineering research.

17.2.1 ANN-based Soft Sensors

In the realm of machine learning applied to water quality soft sensing in wastewater treatment plants (WWTPs), an Artificial Neural Networks (ANN)-based soft sensor designates a soft sensing system that employs ANNs as the core machine learning framework for real-time or near-real-time estimation or prediction of critical water quality parameters or variables within the wastewater treatment process. More specifically, an ANN-based soft sensor is a computational model that employs ANNs to perform soft sensing tasks in wastewater treatment [16–21]. The literature describes that soft sensing involves estimating or predicting critical water quality-related variables (e.g., chemical concentrations, pollutant levels, or process performance indicators) using indirect measurements or sensor data [16]. Essential components of an ANN-based soft sensor typically include (1) Input Data: These are the data inputs provided to the ANN, which can consist of measurements from various sensors within the wastewater treatment plant, historical data, or other relevant process information; (2) ANN: are computational constructs inspired by the architecture and functioning of biological neural networks (Ismael et al. 2023). Comprising interconnected nodes, or neurons, arranged in layers that include input, hidden, and output layers (as illustrated in Figure 17.1), ANNs possess the capacity to discern intricate patterns and correlations within input data; (3) Training Data: To develop an ANN-based soft sensor, a dataset with known input-output pairs is used to train the network. This dataset should include a range of input data and corresponding actual values of the water quality parameter to be estimated; (4) Model Training: During training, the ANN learns to map the input data to the desired water quality parameter's output. This process involves adjusting the network's weights and biases through iterative optimization techniques (e.g., backpropagation) to minimize prediction errors; (5) Model Validation: Once trained, the ANN-based soft sensor is validated using independent datasets to assess its accuracy and generalization performance. This ensures that the model can provide reliable estimates in real-world scenarios; (6) Real-Time Estimation: Following successful training and validation, the deployed ANN-based soft sensor stands prepared to furnish real-time or near-real-time estimations of the target water quality parameter. These estimations are derived from contemporary sensor measurements.

ANN-based soft sensors have gained popularity in wastewater treatment because they can handle nonlinear relationships and adapt to changing process conditions. They offer a data-driven approach to soft sensing, allowing for more accurate and

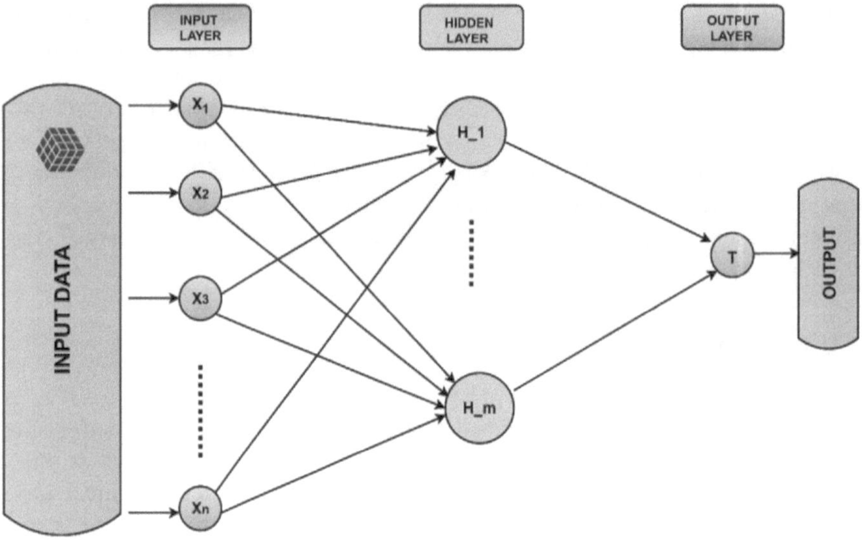

FIGURE 17.1 Typical structure of an Artificial Neural Network (ANN).

flexible monitoring and control of wastewater treatment processes. Several studies in the recent decade have applied ANN-based soft sensors to enhance the diverse functionality of water and wastewater treatment plants. Some researchers identified that the intricate nature of this wastewater treatment process introduces challenges, particularly in monitoring elusive effluent indicators like 5-day Biological Oxygen Demand (BOD_5), impeding the evaluation of effluent quality [26]. In response to these challenges, the researchers introduced a novel approach involving Multi-objective Pigeon-inspired Optimization (MOPIO). They amalgamated MOPIO with Over-complete Independent Analysis (OICA) and the Broad Learning System (BLS), collectively denominated as MOPIO-OBLS, to address these concerns. The approach capitalizes on OICA to bolster the feature extraction capabilities of the BLS, thus enhancing its capacity to capture non-Gaussian data characteristics typical of sewage data. Empirical outcomes underscore the efficacy of MOPIO-OBLS, exhibiting a reduced Root Mean Square Error (RMSE) in contrast to BLS, Support Vector Machine (SVM), and Recurrent Neural Network (RNN). This highlights the performance optimization achieved through MOPIO. In a related study, researchers devised an Effluent Concentration and Alarm Prediction System (ECAPS) tailored for effluent concentration forecasting in wastewater treatment plants. ECAPS, driven by an Artificial Neural Network (ANN)-based soft sensor system, particularly focused on predicting ammonium and total nitrogen concentrations within the Benchmark Simulation Model N.2 (BSM2) framework [20]. Two LSTM-based structures were employed for predictions. Results demonstrated high-accuracy predictions for ammonium, with 87 percent, 89 percent, and 94 percent probability of violation detection at high, medium, and low control levels, respectively. High control yields around 86 percent violation detection for total nitrogen, and ROC curves indicate nearly perfect performance, ensuring 100 percent violation detection.

In summary, ANN-based soft sensor systems offer a powerful tool for accurately and adaptively monitoring water quality in wastewater treatment processes. Their proficiency in managing intricate, non-linear, and non-Gaussian datasets, coupled with their real-time monitoring capabilities, renders them a valuable resource in ensuring the efficacy of wastewater treatment and upholding environmental preservation. Readers can refer to yet other recent research on the applications of ANN-based soft sensor systems [16, 18, 19, 21].

17.2.2 SVR-based Soft Sensors

An SVR-based soft sensor, denoting Support Vector Regression, represents a soft sensing framework that harnesses SVR as the fundamental machine learning methodology [40, 41]. Its purpose is the real-time or near-real-time estimation or prediction of pivotal water quality parameters or variables within a wastewater treatment process. More specifically, an SVR-based soft sensor is a computational model that utilizes SVR, a machine learning algorithm, to perform soft sensing tasks in the context of wastewater treatment [22–28]. The critical components of an SVR-based soft sensor typically include (1) Input Data, the same as defined for an ANN-based soft sensor; (2) SVR: SVR is a regression technique that uses Support Vector Machines (SVMs) to model and predict continuous output variables. SVR aims to find a hyperplane (or multiple hyperplanes) that best fits the data while minimizing prediction errors(Masood et al. 2023b) (Figure 17.2); (3) Kernel Function: SVR often employs a kernel function to map the input data into a higher-dimensional feature space, making it possible to find a linear regression model in this transformed space. Common kernel functions include linear, polynomial, Radial Basis Function (RBF), and sigmoid [42]; (4) Training Data: To develop an SVR-based soft sensor, the same approach is employed as was defined for an ANN-based soft sensor; (5) Model Training: During training, the SVR model learns to find the hyperplane(s) that best fit the input-output relationship. This involves optimizing the model's parameters and selecting appropriate kernel functions (6) Model Validation: After training, the SVR-based soft sensor is validated in the same

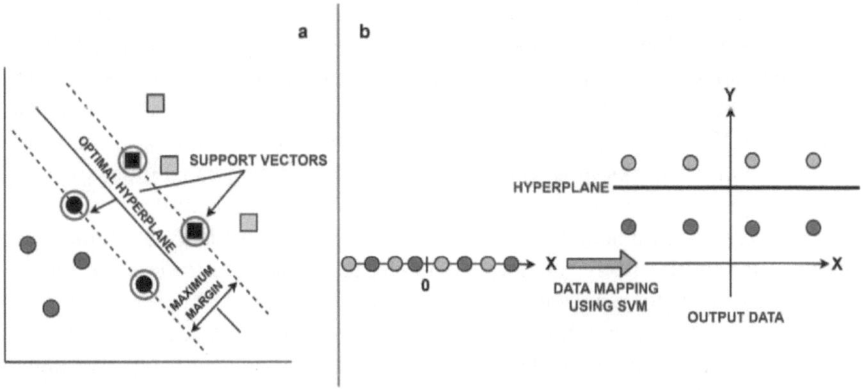

FIGURE 17.2 Support vectors with an optimal hyperplane.

way as was defined for an ANN-based soft sensor; (7) Real-Time Estimation: Once trained and validated, the SVR-based soft sensor can be deployed in the same way as was defined for an ANN-based soft sensor.

SVR-based soft sensors are valued in wastewater treatment applications because they can handle nonlinear relationships between input variables and water quality parameters. They can provide accurate predictions even when underlying processes are complex and nonlinear. Several studies in the recent decade have applied SVR-based soft sensors to enhance the diverse functionality of water and wastewater treatment plants. Some researchers hybridized the Gustafson-Kessel (GK) algorithm and the Least Squares Support Vector Machines (LS-SVM) to develop a GK-LSSVM to predict BOD_5 concentration in wastewater [25]. Evaluated against SVM, Relevance Vector Machine (RVM), and Fast-RVM models using the RMSE metric, GK-LSSVM achieved an RMSE of 10.8, outperforming the other three models. This algorithm provided better BOD5 prediction accuracy and required less computation time. Experimental results affirm the proposed method's effectiveness and reliability in soft sensing applications. In other investigations, machine learning techniques, specifically Ordinary Kriging (OK), ANN, and SVR, were harnessed to develop a practical soft-sensing approach for batch systems with minimal tuning requirements [28]. Despite the inherent dynamic nature of batch processes being tracked through indirect online measurements, these models were evaluated. OK exhibited the best performance with an RMSE of 1.19, a Normalized RMSE (NRMSE) of 2.37 percent, and a high Correlation Coefficient (CC) of 0.9956. The SVR model also yielded reasonable results with an RMSE of 2.84, NRMSE of 5.66 percent, and a CC of 0.9730, showcasing the effectiveness of both approaches. Along similar lines, several recent researchers have studied the application of SVR-based soft sensors in wastewater treatment applications and related industrial processes [22–24, 26, 27].

17.2.3 ANFIS-based Soft Sensors

An ANFIS-based (Adaptive Neuro-Fuzzy Inference System) soft sensor is a soft sensing system that utilizes the ANFIS framework as the underlying machine learning technique to estimate or predict important water quality parameters or variables in real-time or near-real-time within a wastewater treatment process. An ANFIS-based soft sensor is a computational model that incorporates the principles of ANFIS for performing soft sensing tasks in the context of wastewater treatment [29–32]. The key components of an ANFIS-based soft sensor typically include (1) Input Data (2) ANFIS: ANFIS is a hybrid computational framework that combines the adaptability of neural networks (neuro) and the linguistic modeling capabilities of Fuzzy Logic (Fuzzy). ANFIS models use fuzzy inference systems where the membership functions and rules are adapted and optimized based on training data; (3) Membership Functions: In ANFIS, fuzzy membership functions are defined to describe the linguistic terms used to represent input and output variables. These membership functions are adapted during training to fit the data; (4) Fuzzy Rules: ANFIS models employ Fuzzy Rules that define the relationships between input and output variables. The structure and parameters of these rules are adjusted during the training process (5) Training Data; (6) Model Training: During training, the ANFIS model adapts its membership functions and Fuzzy Rules to minimize prediction errors and accurately represent the underlying input-output relationships (7) Model Validation; (8) Real-Time Estimation: Following the phases of training and validation, the ANFIS-based soft sensor is poised for deployment, facilitating the provision of real-time or near-real-time estimations of the target water quality parameter, drawing upon contemporaneous sensor measurements.

ANFIS-based soft sensors are attractive in wastewater treatment applications because they combine the capabilities of neural networks for capturing complex patterns with Fuzzy Logic for handling linguistic variables. This hybrid approach can effectively model and predict water quality parameters in dynamic and uncertain wastewater treatment processes. Over the past decade, numerous investigations have employed ANFIS-based soft sensors to augment the multifaceted operations of water and wastewater treatment facilities and their affiliated processes. In a published scholarly work, a novel approach was introduced, combining Genetic Algorithm with Adaptive Neuro-Fuzzy Inference System (GA-ANFIS), to estimate nutrient concentrations in the effluent of a full-scale biological wastewater treatment plant [31]. Two distinct sets of measured process variables were employed: influent parameters (e.g., flow rate, TSS, BOD, COD, TN, TP) and effluent parameters (effluent COD, effluent TN, effluent TP). In contrast to standalone ANFIS modeling, the incorporation of Genetic Algorithm with Adaptive Neuro-Fuzzy Inference System (GA-ANFIS) demonstrates markedly superior performance and robust generalization capabilities. This is notably evident through substantial enhancements in RMSE, Mean Absolute Percentage Error (MAPE), and R-squared values for the prediction of effluent COD, TN, and TP. These findings underscore the efficacy of GA-ANFIS-based soft sensors in elucidating nutrient removal mechanisms within wastewater treatment systems. Other researchers presented a soft Fault-Tolerant Control (sFTC) approach tailored for decentralized management of Dissolved Oxygen (DO) levels within aerated compartments of an activated sludge process. This strategy serves to rectify sensor anomalies, encompassing biases and gradual drifts [29]. The approach consists of three fundamental modules: an ANFIS-based soft sensor serving as a contingency for the sensor-controller block, a module dedicated to identifying and isolating faults, and an online reconfiguration law. Leveraging correlations in system variables, particularly the oxygen transfer coefficient, this approach ensured process efficiency during sensor failures. Several soft sensors, individually allocated to distinct Dissolved Oxygen (DO) control loops, preserved a decentralized control methodology, and the fault identification system promptly transitioned to substitute surrogates. Implemented in the BSM1 model, this approach demonstrated faster response times than prior methods. Furthermore, similar applications of ANFIS-based soft sensors in wastewater treatment and associated industrial processes have been observed in recent times [29–32].

17.2.4 Other Machine-learning Techniques

In the domain of machine learning applied to soft sensing for water quality in wastewater treatment, various alternative machine learning methodologies are employed for soft sensing, distinct from those elaborated upon in the preceding sections. To enlist a few, it includes Decision Trees (DT): They are a supervised machine learning technique used for classification and regression tasks. They create a tree-like structure of decisions and their possible consequences. Decision trees can be used to model relationships between sensor data and water quality parameters in soft sensing for wastewater treatment, providing interpretable models [33]. Random Forests (RF): These constitute an ensemble learning method that combines multiple decision trees to improve predictive accuracy and reduce overfitting. RF can be applied to soft sensing by aggregating the predictions of numerous decision trees, enhancing the robustness of the model in handling complex wastewater treatment data [34]. The k-Nearest Neighbors (k-NN) method represents a machine learning approach, suitable for both supervised and unsupervised tasks involving classification and regression. Its predictive mechanism relies on either the majority class or the average derived from the k-nearest data points residing within the feature space. k-NN can be employed for soft sensing by using similarity metrics to find the nearest data points with known water quality parameters and make predictions based on their values [38]. Principal Component Analysis (PCA): This is an unsupervised dimensionality reduction technique that transforms high-dimensional data into a lower-dimensional representation while preserving the most critical information. PCA can be used for feature extraction and dimensionality reduction in soft sensing, simplifying the modeling process and improving computational efficiency[35]. Gaussian Process Regression (GPR): A Bayesian non-parametric regression technique models the relationship between input variables and outputs as a distribution over functions. GPR can be applied to soft sensing tasks to capture uncertainty in predictions and provide probabilistic estimates of water quality parameters [36]. Extreme Learning Machines (ELM) denote feedforward neural networks characterized by a sole hidden layer composed of neurons that possess random initialization [43]. They are trained using analytical methods and are known for their fast training speed [41, 44–46]. ELM can be used for soft sensing, mainly when computational efficiency is a concern, as they typically require less training time than traditional neural networks [37]. Gradient Boosting Machines (GBM): These constitute an ensemble learning method that builds a predictive model stage-wise by optimizing for the residual errors of the previous model. GBM can be employed in soft sensing to create strong predictive models by iteratively improving model accuracy, making them suitable for complex wastewater treatment processes [47]. Clustering Algorithms (e.g., K-Means): They are unsupervised techniques used to group similar data points together based on their feature similarity. Clustering can be useful in soft sensing for identifying patterns or segments within the wastewater treatment data, potentially revealing distinct behaviors or conditions [48]. These represent a subset of machine learning methodologies applicable within the soft sensing framework for wastewater treatment. The selection of an appropriate technique hinges upon the unique attributes of the data and the objectives of the soft sensing endeavor, each method presenting its own set of strengths and constraints.

17.3 ASSESSMENT OF THE REVIEWED PAPERS

Researchers have targeted wastewater parameters such as BOD5, COD, nitrogen concentrations, and others in the reviewed studies using different time prediction models, including ANN, ANFIS, LSTM, SVR, and PCA-based models. The choice of an appropriate prediction model depends on data availability, whereas ANN-based models are preferred when data size is sufficient. At the same time, SVR is a better choice when data is limited. Understanding this distinction helps interpret why ANN may perform better in some cases and why SVR may outperform ANN in others. Furthermore, it is imperative to emphasize that models founded on deep learning necessitate a dataset of substantial size to ensure effective training. Overall, selecting the suitable prediction model is crucial in accurately estimating wastewater parameters and relies on the data availability and characteristics.

Furthermore, researchers have employed a variety of AI-based models in their studies, including deep learning techniques that are particularly effective for handling and predicting time series variables. Examples of these techniques include LSTM, Deep Recurrent Neural Network, and BiLSTM. These models excel in capturing spatial and temporal patterns in data, thereby enhancing the efficiency of wastewater parameter forecasting. In addition to deep learning models, classical AI models such as ANN, SVR, and ANFIS have also been extensively investigated. Unlike the aforementioned deep learning models, these models can be utilized for simulating both time series and non-time series data. The studies have reported promising results using these classical AI models. Overall, the research indicates that a combination of deep learning techniques and classical AI models can contribute to improved wastewater parameter forecasting. The selection of an appropriate model is contingent upon the inherent characteristics of the data as well as the precise objectives delineated in the study.

Besides, in the present review, it was observed that the researchers have employed several statistical performance metrics to assess the models' accuracy. Upon analyzing the data presented in Figure 17.3, it was found that a significant proportion of the reviewed studies, specifically more than 26.6 percent, utilized RMSE as their chosen metric for evaluation. Following closely behind was R^2, which accounted for approximately 22.2 percent of the studies, while MSPE constituted 17.7 percent of the total. This indicates that researchers in the field of soft sensing for wastewater

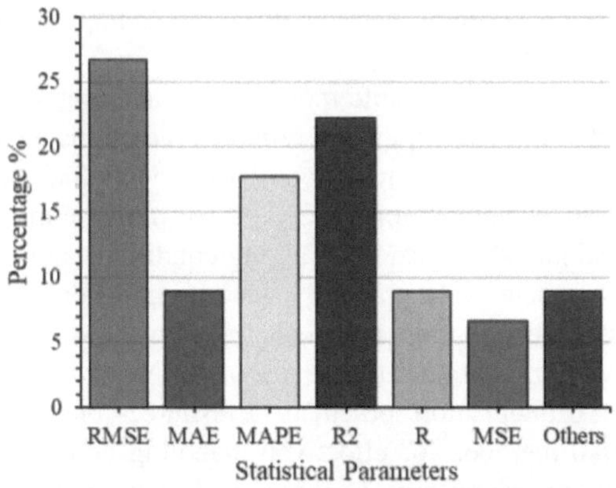

FIGURE 17.3 Analysis (in percentage) of the distribution of statistical indices employed in reviewed papers on soft sensing for water quality in wastewater treatment.

treatment employ a variety of statistical performance metrics to assess model accuracy, reflecting the complexity and multifaceted nature of their evaluation criteria. Overall, the choice of performance metrics in soft sensing research for wastewater treatment is not uniform, reflecting the nuanced nature of the field and the importance of selecting metrics that align with specific research objectives and requirements.

17.4 CHALLENGES AND FUTURE SCOPE IN UTILIZING MACHINE-LEARNING-BASED SOFT SENSORS

Using Soft Sensors based on ANN, SVR, ANFIS in specific, and other machine learning in general in various applications comes with certain challenges and scope for improvement. Considering the limitations, soft sensors rely heavily on data for training and validation. Limited or poor-quality data can lead to inaccurate predictions. ANNs, especially deep networks, can be computationally intensive and require substantial resources for training and inference. ANN models may overfit the training data, leading to poor generalization [49] when applied to new or unseen data.

Furthermore, proper tuning of hyperparameters in ANNs and SVR models is crucial but can be time-consuming and require expertise[50]. Besides, the black-box nature of ANNs can make interpreting how the model arrives at its predictions challenging, which may be a concern in some applications[51]. ANNs, SVR, and ANFIS may struggle with data that exhibits strong non-Gaussian characteristics or outliers. Therefore, ensuring high-quality data for training and validation is an ongoing challenge. Data preprocessing and cleaning are essential but can be resource-intensive.

Moreover, choosing the most appropriate model (ANN, SVR, ANFIS, etc.) for a specific application can be challenging and may require experimentation. Achieving robust model generalization to diverse operating conditions is challenging, especially in dynamic environments. Hence, adapting soft sensors to real-time changes and sensor failures requires efficient appropriate learning strategies. Also, deep ANNs demand substantial computational power, which may not be available in all settings. Considering the limitations and challenges of employing soft sensors in wastewater treatment processes, it is imperative to discuss the possible scope of improvement. Expanding and enhancing training datasets through techniques like data augmentation can improve model robustness.

Combining the strengths of multiple models, such as ensembles of ANNs, SVR, and ANFIS, can lead to more accurate and reliable predictions. Developing methods to interpret and explain the decisions of complex models like ANNs can increase their usability in critical applications [52]. Furthermore, implementing adaptive soft sensors that continuously adjust to changing conditions and sensor faults could be useful for real-time sensing. Apparently, incorporating domain knowledge into model development can lead to more physically meaningful soft sensors. At the same time, optimizing the computational efficiency of ANNs and exploring lightweight architectures can make them more practical in resource-constrained environments. Besides, research into methods for effectively handling non-Gaussian data in soft sensing models is an emerging area of research. In summary, Soft Sensors based on ANN, SVR, ANFIS, and other machine learning techniques offer valuable tools for various applications. However, they come with data-related challenges, model complexity, and interpretability. Ongoing research and innovation can address these limitations and expand the scope of soft sensor applications in diverse domains.

While some of the techniques discussed in this discourse have not yet found widespread use as soft sensors in wastewater treatment, they have established themselves as fundamental machine learning methodologies in various closely related disciplines. For example, these methods have been extensively applied in modeling a spectrum of parameters intricately tied to water flow dynamics. This includes intricate aspects such as evaporation rates, evapotranspiration processes, vapor pressure deficits, and water productivity in agriculture [53]. Additionally, climate change's influence on wastewater treatment necessitates adaptive solutions, and soft sensors based on advanced ML techniques offer a promising avenue. They enable treatment plants to respond dynamically to changing conditions, optimize performance, and contribute to environmental sustainability in the face of evolving climate challenges. These multidisciplinary applications underscore their versatility and robustness in addressing complex hydrological phenomena. This insight underscores the potential for knowledge transfer across domains. The expertise amassed in modeling water-related processes can be instrumental in pioneering advanced soft-sensing solutions for wastewater treatment. This prospect holds the promise of further enriching the field and advancing its scientific frontiers, ultimately contributing to more efficient and sustainable wastewater management practices.

17.5 CONCLUSION

This chapter contributes to the comprehensive understanding of Machine Learning (ML)-based water quality soft sensing within the context of wastewater treatment processes (WWTPs). Through an exhaustive review of existing literature, the study explored the application of various ML techniques, including Artificial Neural Networks (ANNs), Adaptive Neuro-Fuzzy Inference Systems (ANFIS), and Support Vector Regression (SVR), in the domain of soft sensing modeling. These ML techniques offer a pragmatic approach, obviating the need for exhaustive knowledge of wastewater treatment systems. Their user-friendliness, adaptability, and ability to mitigate extended time delays make them invaluable tools in the realm of WWTP management. The findings underscore that, among these ML techniques, ANN stands out as the most commonly applied method for soft sensing in water quality monitoring within WWTPs. Furthermore, the Root Mean Square Error (RMSE) emerges as the prevailing statistical metric for assessing the performance of ML-based soft sensing models. However, it is essential to acknowledge that, like any scientific endeavor, the application of ML-based soft sensing in WWTPs is not without its limitations and challenges. These challenges include data quality, resource-intensive model complexity, limited interpretability, generalization across conditions and WWTPs, real-time adaptation, and demanding computational resources for deep ANNs. Despite these challenges, this study has provided a foundational reference point for addressing the existing obstacles in ML-based soft sensing modeling, thereby fostering their resolution and advancement in this evolving field. The study thus anticipates that future strides in multi-parameter sensors and the Internet of Things (IoT) will play a pivotal role in furthering the capabilities of real-time monitoring for large-scale

WWTPs. These developments will empower plant operators with the tools needed to swiftly identify equipment malfunctions, optimize energy utilization, and mitigate greenhouse gas emissions. In conclusion, the synergy of machine learning and water quality soft sensing holds great promise in enhancing the efficiency and sustainability of wastewater treatment processes, contributing to a cleaner and healthier environment for all.

Abbreviations

BOD_5	Biological Oxygen Demand
DRNN	Deep Recurrent Neural Network
ODRNN	Over-Complete Deep Recurrent Neural Network
TSS	Total Suspended Solids
S_{NH}	Ammonia Nitrogen
S_{ND}	Soluble Biodegradable Organic Nitrogen
S_{ALK}	Alkalinity
OICA	Overcomplete Independent Component Analysis
BLS	Broad Learning System
RG	Ridge Regression
PCA-R	Principal Component Regression and Partial Least Squares
PLS	Partial Least Squares
R^2	Correlation of Determination
RMSE	Root Mean Square Error
MAPE	Mean Absolute Percentage Error
SBiLSTM	Supervised Bidirectional Long Short-Term Memory
BiLSTM	Bidirectional Long Short-Term Memory
DCL	Debutanizer Column Process
S_{Ntot}	Total Nitrogen
PET	Polyethylene Terephthalate
CAE-DFC	Convolutional Autoencoder (CAE) Integrated with Deep Fully Connected Layers
Sp	Total Phosphorus
ANN	Neural Network
RF	Random Forest
SVR	Support Vector Regression
RFN-SVR	Recursive Finite Newton Algorithm for Training the Support Vector Regression
FCM-RFN-SVR	Fuzzy C-Means Clustering with the Recursive Finite Newton Algorithm for Training the Support Vector Regression
BPNN	Back Propagation Neural Network
RBFNN	Radial Basis Function Neural Network
COD	Chemical Oxygen Demand
PSO-SVR	Particle Swarm Optimization-Based Support Vector Regression
GK-LSSVM	Gustafson-Kessel Algorithm and Least Squares Support Vector Machines
GP	Genetic Programming

OK	Ordinary Kriging
NRMSE	Normalized Root Mean Square Error
GPR	Gaussian Process Regression
PC	Penicillin Concentration
GA-ANFIS	Genetic Algorithm with Adaptive Neuro-Fuzzy Inference System
GEP	Gene Expression Programming
KPCA-OCSVM	Kernel Principal Components Analysis Based on One-class Support Vector Machine
PPCA	Probabilistic Principle Component Analysis
$\mathbf{d_r}$	Efficiency Criteria
NSE	Nash–Sutcliffe

REFERENCES

1. Ahmed M, Rizvi DR, Masood S, et al (2020) Machine learning approach for predicting the quality of water. *Int J Adv Sci Technol* 29:275–282.
2. Tanweer MS, Iqbal Z, Alam M (2022) Experimental insights into mesoporous polyaniline-based nanocomposites for anionic and cationic dye removal. *Langmuir* 38:8837–8853. https://doi.org/10.1021/acs.langmuir.2c00889
3. Shanmugam K, Gadhamshetty V, Tysklind M, et al (2022) A sustainable performance assessment framework for circular management of municipal wastewater treatment plants. *J Clean Prod* 339:130657. https://doi.org/https://doi.org/10.1016/j.jclepro.2022.130657
4. Wang D, Thunéll S, Lindberg U, et al (2022) Towards better process management in wastewater treatment plants: Process analytics based on SHAP values for tree-based machine learning methods. *J Environ Manage* 301:113941. https://doi.org/https://doi.org/10.1016/j.jenvman.2021.113941
5. Bir R, Tanweer MS, Singh M, Alam M (2022) Multifunctional ternary NLP/ZnO@l-cysteine-grafted-PANI bionanocomposites for the selective removal of anionic and cationic dyes from synthetic and real water samples. *ACS Omega* 7:44836–44850. https://doi.org/10.1021/acsomega.2c04936
6. Hameed M, Sharqi SS, Yaseen ZM, et al (2017) Application of artificial intelligence (AI) techniques in water quality index prediction: A case study in tropical region, Malaysia. *Neural Comput Appl* 28:893–905. https://doi.org/10.1007/s00521-016-2404-7
7. Alvi M, French T, Cardell-Oliver R, et al (2022) Cost effective soft sensing for wastewater treatment facilities. *IEEE Access* 10:55694–55708. https://doi.org/10.1109/ACCESS.2022.3177201
8. Sundui B, Ramirez Calderon OA, Abdeldayem OM, et al (2021) Applications of machine learning algorithms for biological wastewater treatment: Updates and perspectives. *Clean Technol Environ Policy* 23:127–143. https://doi.org/10.1007/s10098-020-01993-x
9. Fernandez de Canete J, Del Saz-Orozco P, Baratti R, et al (2016) Soft-sensing estimation of plant effluent concentrations in a biological wastewater treatment plant using an optimal neural network. *Expert Syst Appl* 63:8–19. https://doi.org/https://doi.org/10.1016/j.eswa.2016.06.028
10. Yu P, Cao J, Jegatheesan V, Du X (2019) A real-time BOD estimation method in wastewater treatment process based on an optimized extreme learning machine. *Appl Sci* 9:523.

11. Zhou M, Zhang Y, Wang J, et al (2022) Water quality indicator interval prediction in wastewater treatment process based on the improved BES-LSSVM algorithm. *Sensors* 22:422.
12. Mamat N, Razali SFM (2023) Comparisons of various imputation methods for incomplete water quality data: A case study of the Langat River, Malaysia. *Jurnal Kejuruteraan* 35:191–201.
13. Abba SI, Linh NTT, Abdullahi J, et al (2020) Hybrid machine learning ensemble techniques for modeling dissolved oxygen concentration. *IEEE Access* 8:157218–157237. https://doi.org/10.1109/ACCESS.2020.3017743
14. Nawaz A, Arora AS, Yun CM, et al (2019) Data authorization and forecasting by a proactive soft sensing tool – Anammox based process. *Ind Eng Chem Res* 58:9552–9563. https://doi.org/10.1021/acs.iecr.9b00722
15. Mamat N, Mohd Razali SF, Hamzah FB (2023) Enhancement of water quality index prediction using support vector machine with sensitivity analysis. *Front Environ Sci* 10:1–21. https://doi.org/10.3389/fenvs.2022.1061835
16. Chang P, Li Z (2021) Over-complete deep recurrent neutral network based on wastewater treatment process soft sensor application. *Appl Soft Comput* 105:107227. https://doi.org/https://doi.org/10.1016/j.asoc.2021.107227
17. Chang P, Bao X, Meng F, Lu R (2023) Multi-objective pigeon-inspired optimized feature enhancement soft-sensing model of wastewater treatment process. *Expert Syst Appl* 215:119193. https://doi.org/https://doi.org/10.1016/j.eswa.2022.119193
18. Foschi J, Turolla A, Antonelli M (2021) Soft sensor predictor of *E. coli* concentration based on conventional monitoring parameters for wastewater disinfection control. *Water Res* 191:116806. https://doi.org/https://doi.org/10.1016/j.watres.2021.116806
19. Lui CF, Liu Y, Xie M (2022) A supervised bidirectional long short-term memory network for data-driven dynamic soft sensor modeling. *IEEE Trans Instrum Meas* 71:1–13. https://doi.org/10.1109/TIM.2022.3152856
20. Pisa I, Santín I, Vicario JL, et al (2019) ANN-based soft sensor to predict effluent violations in wastewater treatment plants. *Sensors* 19:1280.
21. Thiruneelakandan A, Kaur G, Vadnala G, et al (2022) Measurement of oxygen content in water with purity through soft sensor model. *Measurement: Sensors* 24:100589. https://doi.org/https://doi.org/10.1016/j.measen.2022.100589
22. Ba-Alawi AH, Heo S, Aamer H, et al (2023) Development of transparent high-frequency soft sensor of total nitrogen and total phosphorus concentrations in rivers using stacked convolutional auto-encoder and explainable AI. *J Water Process Eng* 53:103661. https://doi.org/https://doi.org/10.1016/j.jwpe.2023.103661
23. Gholami AR, Shahbazian M (2015) Soft sensor design based on fuzzy C-Means and RFN_SVR for a stripper column. *J Nat Gas Sci Eng* 25:23–29. https://doi.org/https://doi.org/10.1016/j.jngse.2015.04.014
24. Fu X, Zheng Q, Jiang G, et al (2023) Water quality prediction of copper–molybdenum mining-beneficiation wastewater based on the PSO-SVR model. *Front Environ Sci Eng* 17:98. https://doi.org/10.1007/s11783-023-1698-9
25. Luo F, Qiao X, Liao W (2017) Soft-sensing modeling based on GK-LSSVM method for online predictions of BOD in activated sludge process. In: 2017 International Conference on Robotics and Automation Sciences (ICRAS), pp. 134–138.
26. Nong X, Lai C, Chen L, et al (2023) Prediction modelling framework comparative analysis of dissolved oxygen concentration variations using support vector regression coupled with multiple feature engineering and optimization methods: A case study in China. *Ecol Indic* 146:109845. https://doi.org/https://doi.org/10.1016/j.ecolind.2022.109845

27. Sharma S, Tambe SS (2014) Soft-sensor development for biochemical systems using genetic programming. *Biochem Eng J* 85:89–100. https://doi.org/https://doi.org/10.1016/j.bej.2014.02.007
28. Shokry A, Vicente P, Escudero G, et al (2018) Data-driven soft-sensors for online monitoring of batch processes with different initial conditions. *Comput Chem Eng* 118:159–179. https://doi.org/https://doi.org/10.1016/j.compchemeng.2018.07.014
29. Belchior CAC, Araújo RAM, Souza FAA, Landeck JAC (2018) Sensor-fault tolerance in a wastewater treatment plant by means of ANFIS-based soft sensor and control reconfiguration. *Neural Comput Appl* 30:3265–3276. https://doi.org/10.1007/s00521-017-2901-3
30. Fatima SA, Ramli N, Taqvi SAA, Zabiri H (2021) Prediction of industrial debutanizer column compositions using data-driven ANFIS- and ANN-based approaches. *Neural Comput Appl* 33:8375–8387. https://doi.org/10.1007/s00521-020-05593-0
31. Liu H, Huang M, Yoo C (2013) A fuzzy neural network-based soft sensor for modeling nutrient removal mechanism in a full-scale wastewater treatment system. *Desalin Water Treat* 51:6184–6193.
32. Puri D, Sihag P, Sadeghifar T, et al (2023) Soft computing-based model development for estimating the aeration efficiency through Parshall flume and Venturi flumes. *Multiscale Multidiscip Model Exp Des* 6:401–413. https://doi.org/10.1007/s41939-023-00153-0
33. Aghdam E, Mohandes SR, Manu P, et al (2023) Predicting quality parameters of wastewater treatment plants using artificial intelligence techniques. *J Clean Prod* 405:137019. https://doi.org/https://doi.org/10.1016/j.jclepro.2023.137019
34. Sheikholeslami R, Hall JW (2023) Global patterns and key drivers of stream nitrogen concentration: A machine learning approach. *Sci Total Environ* 868:161623. https://doi.org/https://doi.org/10.1016/j.scitotenv.2023.161623
35. Cheng T, Dairi A, Harrou F, et al (2019) Monitoring influent conditions of wastewater treatment plants by nonlinear data-based techniques. *IEEE Access* 7:108827–108837. https://doi.org/10.1109/ACCESS.2019.2933616
36. Xiong W, Shi X (2018) Soft sensor modeling with a selective updating strategy for Gaussian process regression based on probabilistic principle component analysis. *J Franklin Inst* 355:5336–5349. https://doi.org/https://doi.org/10.1016/j.jfranklin.2018.05.017
37. Deng X, Lin C (2016) Application of ELM to predict the coagulant dosing in water treatment plants. *Water Supply* 17:1053–1061. https://doi.org/10.2166/ws.2016.203
38. Kim M, Kim Y, Kim H, et al (2016) Evaluation of the k-nearest neighbor method for forecasting the influent characteristics of wastewater treatment plant. *Front Environ Sci Eng* 10:299–310. https://doi.org/10.1007/s11783-015-0825-7
39. Guo H, Jeong K, Lim J, et al (2015) Prediction of effluent concentration in a wastewater treatment plant using machine learning models. *J Environ Sci* 32:90–101. https://doi.org/https://doi.org/10.1016/j.jes.2015.01.007
40. Masood A, Niazkar M, Zakwan M, Piraei R (2023) A machine learning-based framework for water quality index estimation in the Southern Bug River. *Water (Basel)* 15:3543.
41. Adnan RM, Mostafa RR, Dai H-L, et al (2023) Enhancing accuracy of extreme learning machine in predicting river flow using improved reptile search algorithm. *Stoch Environ Res Risk Assess* 37:3063–3083. https://doi.org/10.1007/s00477-023-02435-y
42. Khalaf JA, Majeed AA, Aldlemy MS, et al (2021) Hybridized deep learning model for Perfobond Rib Shear Strength Connector prediction. *Complexity* 2021:6611885. https://doi.org/10.1155/2021/6611885

43. Hameed MM, Razali SFM, Mohtar WHMW, et al (2023) Machine learning models development for accurate multi-months ahead drought forecasting: Case study of the Great Lakes, North America. *PLoS One* 18:e0290891. https://doi.org/10.1371/JOURNAL.PONE.0290891
44. Hameed MM, Mohd Razali SF, Wan Mohtar WHM, Yaseen ZM (2023) Improving multi-month hydrological drought forecasting in a tropical region using hybridized extreme learning machine model with Beluga Whale Optimization algorithm. *Stoch Environ Res Risk Assess* 37:4963–4989. https://doi.org/10.1007/s00477-023-02548-4
45. Hameed MM, AlOmar MK, Al-Saadi AAA, AlSaadi MA (2022) Inflow forecasting using regularized extreme learning machine: Haditha reservoir chosen as case study. *Stoch Environ Res Risk Assess* 36:4201–4221. https://doi.org/10.1007/s00477-022-02254-7
46. Alomar MK, Khaleel F, Aljumaily MM, et al (2022) Data-driven models for atmospheric air temperature forecasting at a continental climate region. *PLoS One* 17:e0277079.
47. Xing Y, Zheng Z, Sun Y, Agha Alikhani M (2021) A review on machine learning application in biodiesel production studies. *Int J Chem Eng* 2021:2154258. https://doi.org/10.1155/2021/2154258
48. Deng X, Chen Y, Wang P, Cao Y (2020) Soft sensor modeling for unobserved multimode nonlinear processes based on modified kernel partial least squares with latent factor clustering. *IEEE Access* 8:35864–35872. https://doi.org/10.1109/ACCESS.2020.2974783
49. Masood A, Hameed MM, Srivastava A, et al (2023) Improving $PM_{2.5}$ prediction in New Delhi using a hybrid extreme learning machine coupled with snake optimization algorithm. *Sci Rep* 13:21057. https://doi.org/10.1038/s41598-023-47492-z
50. Masood A, Ahmad K (2022) Data-driven predictive modeling of $PM_{2.5}$ concentrations using machine learning and deep learning techniques: A case study of Delhi, India. *Environ Monit Assess* 195:60. https://doi.org/10.1007/s10661-022-10603-w
51. Masood A, Ahmad K (2023) Prediction of $PM_{2.5}$ concentrations using soft computing techniques for the megacity Delhi, India. *Stoch Environ Res Risk Assess* 37:625–638. https://doi.org/10.1007/s00477-022-02291-2
52. Ismael BH, Khaleel F, Ibrahim SS, et al (2023) Permeation flux prediction of vacuum membrane distillation using hybrid machine learning techniques. *Membranes (Basel)* 13:900.
53. Elbeltagi A, Raza A, Hu Y, et al (2022) Data intelligence and hybrid metaheuristic algorithms-based estimation of reference evapotranspiration. *Appl Water Sci* 12:152. https://doi.org/10.1007/s13201-022-01667-7

18 Advancing Sustainable Wastewater Treatment

Unleashing the Potential of Nanosponges for Effective Remediation

Bidisha Ghosh, Suranjana Sarkar, and Semanti Ghosh

Abbreviations	Meaning
3D	3-Dimensional
Ag	Silver
Al	Aluminium
Au	Gold
BOD	Biological Oxygen Demand
Cd	Cadmium
Co	Cobalt
COD	Chemical Oxygen Demand
EDTA	Ethylenediaminetetraacetic acid
Hg	Mercury
MD	Molecular Dynamics
MF	Microfiltration
NF	Nanofiltration
Nm	Nanometers
NP	Nanoparticle
Pd	Palladium
RO	Reverse Osmosis
TOC	Total Organic Carbon
TSS	Total Suspended Solid
UF	Ultrafiltration

DOI: 10.1201/9781003454199-18

Abbreviations	Meaning
UNEP	United Nations Environment Program
UNEP	United Nations Environment Program
WWAP	World Water Assessment Program
WWAP	World Water Assessment Program
Zn	Zinc

18.1 INTRODUCTION

Water, a fundamental resource for life on Earth, is facing an increasingly severe threat – water pollution. The significance of water in sustaining ecosystems, human health, and various societal activities cannot be overstated. However, the quality of our water sources is steadily deteriorating due to a global environmental crisis that transcends geographical boundaries and socio-economic disparities [1]. The roots of this crisis can be traced back to the relentless expansion of human activities over the past century. Rapid industrialization, urban development, and intensified agriculture have significantly altered the chemical composition of our water bodies. An array of pollutants, ranging from hazardous chemicals and heavy metals to microplastics and excessive nutrients, have infiltrated rivers, lakes, oceans, and groundwater. The consequences of this pollution are dire, affecting not only aquatic ecosystems but also the well-being of millions who rely on these water sources for drinking, agriculture, and recreation [2]. The primary objective of this chapter is to offer an exhaustive analysis of the multifaceted issue of water pollution. We aim to delve into the various sources and categories of pollutants that taint our water bodies. By examining their origins, transport mechanisms, and ecological impacts, we will gain a comprehensive understanding of the complexities involved. Additionally, we will scrutinize the evolving regulatory frameworks and technological innovations aimed at mitigating and preventing water pollution. To address this global challenge effectively, comprehending the intricate interplay of contributing factors is indispensable [3]. As we embark on this exploration of water pollution, we must recognize the urgency of addressing this crisis. Preserving and restoring clean water is not merely a moral obligation but a vital necessity for the well-being of present and future generations.

18.1.1 Water Pollution

Aquatic ecosystems represent environments where biotic components interact with the abiotic elements of aquatic surroundings. These ecosystems are commonly categorized into two primary types: marine ecosystems and freshwater ecosystems. The issue of water pollution is a global concern that has escalated both in developed and developing nations [4]. Globally, a concerning 80 percent of municipal wastewater is discharged into water bodies without treatment, and industries are accountable for releasing millions of tons of heavy metals, solvents, toxic sludge, and other waste materials into aquatic systems annually [4–7]. Pollutants can cause direct harm, leading to immediate and recognizable damage to the environment, or indirect harm by disturbing the delicate balance of the biological food web, which may only become apparent over extended periods [7].

18.1.1.1 Point Source Water Pollution

Point-source pollution stems from emissions originating from a single, identifiable source, often exemplified by industrial facilities. Point source pollution pertains to the release of pollutants from specific sources characterized by fixed discharge points, primarily including industrial wastewater pipelines and urban sewage systems [8].

18.1.1.2 Nonpoint Source Water Pollution

Chemicals can find their way into waterways through either point source or nonpoint source pollution. Nonpoint-source pollution encompasses numerous small sources that collectively contribute to substantial pollution. For instance, when rain or irrigation water traverses the land, it can pick up pollutants like fertilizers, herbicides, and insecticides, subsequently carrying them into rivers, lakes, reservoirs, coastal waters, or groundwater [8].

18.1.1.3 Types of Water Pollution

The major contaminants in the aquatic bodies include chemical dyes, heavy metals [9] untreated industrial and sewage discharge [10], agricultural runoff including fertilizer, pesticides and animal waste [11–12] oil spills [13] from pipeline and storage facility, mining waste [14], construction waste [15], microplastics [16], pharmaceutical disposals [17], and microbial and faecal matter contamination. All of these pollutants have severe detrimental effects on the environmental niche, leading to harmful consequences.

18.1.1.4 Effect of Pollutants in Water

18.1.1.4.1 Heavy Metals

Heavy metals have the capacity to interact with nuclear proteins and DNA, resulting in specific site damage. This damage manifests in two main categories: "direct" and "indirect."

In the case of "direct" damage, heavy metals induce conformational changes in biomolecules due to their presence. Conversely, "indirect" damage occurs when heavy metals trigger the generation of reactive oxygen and nitrogen species, encompassing hydroxyl radicals, superoxide radicals, hydrogen peroxide, nitric oxide, and other endogenous oxidants [18].

Heavy metals also have the potential to disrupt metal-mediated calcium homeostasis, leading to the activation of various calcium-dependent systems, including endonucleases. The formation of free radicals has been extensively studied in relation to metals such as iron, copper, nickel, chromium, and cadmium. Metal-mediated free radicals can induce mutagenic DNA base alterations, underlining the connection between oxidative damage and carcinogenicity [19]. Furthermore, metals like cadmium, nickel, and arsenic are known to inhibit DNA repair mechanisms. Oxidative effects on DNA encompass base modification (observed with chromium and nickel), crosslinking (linked to nickel, copper, and iron in the presence of oxidants), strand scission (associated with nickel, cadmium, chromium, and oxidants), and depurination, primarily caused by copper, chromium, and nickel [20] explained in Figure 18.1 below.

18.1.1.4.2 Dye

The contamination of water with dyes can result in various significant issues, including alterations in water quality, such as changes in colour and odour, which can make the water unsafe. Dye contaminants have been associated with a range of health concerns in humans, including allergies, dermatitis, skin irritations, as well as the potential for cancer and mutations. Notably, a significant proportion, roughly 60 to 70 percent, of azo dyes are known to possess toxic and carcinogenic properties. These dyes pose challenges to conventional treatment processes because they are resistant to typical physicochemical degradation methods and have limited biodegradability [21]. The presence of dyes and textile pigments in wastewater results in intense coloration, pH fluctuations, and elevated levels of key pollution indicators such as biochemical oxygen demand (BOD), chemical oxygen demand (COD), total organic carbon (TOC), and suspended solids (TSS) [22].

18.1.1.4.3 Pesticide

Pesticides can exert deleterious effects on the human organism upon exposure, whether via ingestion, inhalation, or dermal contact. These chemical agents, formulated for the eradication or regulation of pests in agricultural and various other contexts, can inflict harm upon human health through diverse mechanisms. Acute pesticide poisoning may manifest with symptoms including, but not limited to, nausea, emesis, diarrhea, vertigo, and cephalalgia [23]. Prolonged or recurring encounters with pesticides have been correlated with graver health implications, encompassing the emergence of diverse malignancies, neuro developmental disorders, reproductive anomalies, and perturbations in the endocrine system. Specific pesticide classifications, notably organophosphates, possess the capacity to hinder acetylcholinesterase, an enzyme of paramount importance in neurological functionality, thereby instigating a cascade of neurological symptoms and, in severe instances, paralysis [24–26].

18.1.1.4.4 Fertilizer

The detrimental effects of chemical fertilizers on both human health and water pollution are of paramount concern in modern agriculture. These fertilizers, while essential for boosting crop yields, pose a significant risk to human health through food contamination, with residues containing potentially hazardous substances and nitrogen-based fertilizers leading to elevated nitrate levels in food and drinking water [27]. Simultaneously, chemical fertilizer runoff into water bodies contributes to water pollution, as excess nutrients like nitrogen and phosphorus contaminate aquatic ecosystems [28].

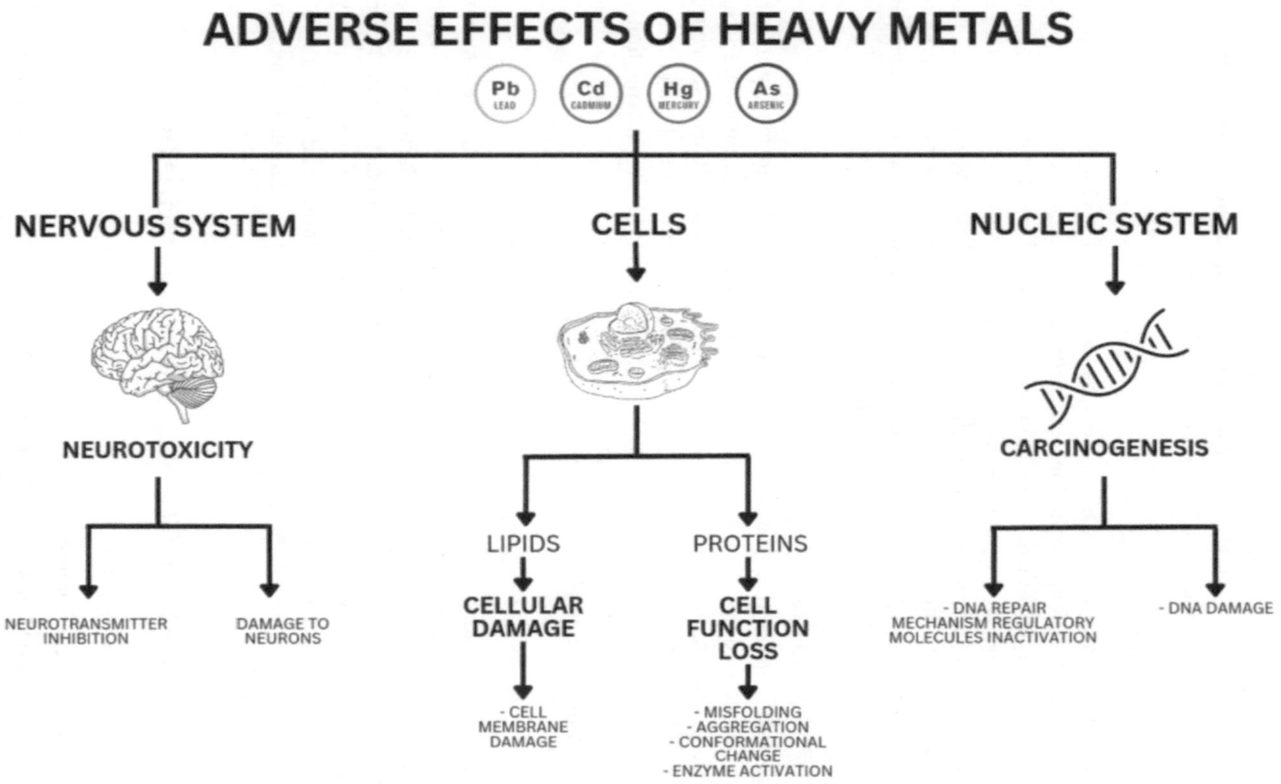

FIGURE 18.1 Effect of heavy metals on biological system (compiled by authors).

18.1.1.4.5 Microplastic

The escalating problem of microplastic pollution in water is a growing concern, as these tiny plastic particles have infiltrated aquatic ecosystems, potentially posing significant risks to both the environment and human health [29]. Human exposure to microplastics occurs through the consumption of water contaminated with these particles, where they can accumulate and persist [30–31].

18.1.1.4.6 Oil Spill

The immediate impacts of oil spills encompass a range of health issues, including respiratory problems, ocular irritation, and dermal symptoms such as skin rashes and chemical burns. Individuals exposed to oil spills may also experience immediate symptoms such as headaches, nausea, dizziness, and a persistent feeling of fatigue [32]. Respiratory disorders, such as chronic bronchitis and aggravated asthma, may persist or worsen over time due to exposure to oil spill contaminants. Furthermore, exposure to certain components of oil spills can have genotoxic effects, potentially leading to DNA damage and an elevated risk of cancer [33].

18.1.1.4.7 Pharmaceutical

The issue of pharmaceutical waste in water pollution is steadily increasing, raising significant concerns due to its potential adverse impacts on both the environment and human health [34]. Among the contaminants commonly found in rivers, substances like paracetamol, caffeine, metformin, fexofenadine, sulfamethoxazole, and metronidazole have been detected in the highest concentrations. The introduction of pharmaceuticals into the water supply primarily occurs through human excretion and the improper disposal of drugs by flushing them down toilets [35].

18.1.1.4.8 Microbial Contamination

Microbial contamination in water poses significant risks to both the environment and human health. It can result in gastrointestinal illnesses, characterized by symptoms such as diarrhea, vomiting, cramps, nausea, headaches, fever, and even fatalities, with higher vulnerability among infants, children, the elderly, and immunocompromised individuals. Additionally, microbial contamination can give rise to serious waterborne diseases [36] like cholera, typhoid, and hepatitis [37] potentially leading to severe health complications and fatalities, further jeopardizing public health.

18.2 CONVENTIONAL TECHNIQUE WATER REMEDIATION

18.2.1 Physical Treatment

18.2.1.1 Membrane Technology

A membrane functions as a selective barrier, separating two phases and regulating the passage of components between them. The history of membrane technology dates back to the eighteenth century, with numerous advancements over time to enhance their suitability for various applications. Membrane can be broadly categorized as isotropic or anisotropic. Isotropic membranes possess a uniform composition and physical structure [38]. When they are microporous, they allow for relatively high permeation fluxes, while nonporous (dense) isotropic membranes have limited applications due to their low permeation fluxes. Microporous isotropic membranes are widely used in microfiltration applications. In contrast, anisotropic membranes exhibit non-uniform characteristics across their surface, consisting of distinct layers with varying structures and compositions [39]. These membranes typically feature a thin selective layer atop a thicker, highly permeable layer, making them particularly valuable in processes such as reverse osmosis (RO) [40].

18.2.1.2 Nanofiltration (NF), Ultrafiltration (UF), and Microfiltration (MF)

Nanofiltration (NF), ultrafiltration (UF), and microfiltration (MF) represent advanced membrane-based wastewater treatment techniques, each characterized by specific pore sizes and separation abilities. Microfiltration employs membranes with larger pores (0.1 to 10 micrometers) to primarily remove suspended solids and bacteria, serving as an effective pre-treatment step. Ultrafiltration, with smaller pores (0.001 to 0.1 micrometers), excels in eliminating suspended solids, bacteria, viruses, proteins, and macromolecules, making it ideal for water purification and concentration. Meanwhile, nanofiltration, featuring intermediate pore sizes (0.001 to 0.01 micrometers), is proficient in removing divalent ions, organic matter, and colour, commonly used for water softening, colour removal, and selective ion rejection in desalination processes [41].These membrane processes play vital roles in treating wastewater, ensuring the removal of contaminants, and finding applications across industries and specialized water treatment needs [42].

18.2.2 Chemical Treatment

Coagulation and flocculation involve the introduction of chemical coagulants, such as aluminium sulfate or ferric chloride, to induce the destabilization of colloidal particles and suspended solids, thereby encouraging their aggregation into larger, settleable flocs [43]. Precipitation mechanisms are employed for the elimination of dissolved metals and ions through the utilization of chemical additives, such as lime or sodium hydroxide, which catalyse the formation of insoluble precipitates [44]. Chemical oxidation processes, encompassing chlorine or ozone treatment, are instrumental in the disintegration of organic compounds and the disinfection of wastewater. Advanced oxidation processes (AOPs) harness the potency of oxidants like hydrogen peroxide and UV light to decompose persistent organic pollutants . Additionally, pH adjustment is implemented through the introduction of acids or bases, serving to [45]. regulate the pH of wastewater and optimize subsequent treatment phases.

18.2.3 Physio-chemical

Ion exchange is a chemical method extensively employed in wastewater treatment to eliminate undesirable dissolved ions from water and wastewater by swapping them with ions of similar charge. This process relies on the utilization of ion exchange resins [46], which attract and replace unwanted ions in the wastewater with more favorable ions, effectively purifying the water before it is discharged [47] called anions. Ion exchange finds application in various water treatment processes, including water softening, water demineralization, dealkalization, deionization, denitrification, and disinfection [48].

18.2.4 Biological Treatment

Biological methods of wastewater treatment through bioremediation capitalize on the unique abilities of microorganisms [49] to degrade and detoxify contaminants present in wastewater. This approach [50] encompasses several techniques, including bioaugmentation, where specific microbial cultures are introduced to enhance the breakdown of targeted pollutants. Bio-stimulation involves altering environmental conditions, such as nutrient levels or oxygen availability, to stimulate the growth and activity of native microorganisms capable of degrading contaminants effectively [51]. Phytoremediation [52] utilizes plants like water hyacinths or reed beds, which not only absorb and accumulate pollutants but also provide a conducive habitat for beneficial microorganisms to flourish and aid in the degradation of contaminants. These biological methods offer eco-friendly and sustainable solutions for wastewater treatment, facilitating the removal of diverse pollutants while minimizing the reliance on chemicals and reducing overall energy consumption in the treatment process.

18.2.5 Disadvantages of the Conventional Techniques

One of the primary drawbacks associated with the filtration method in wastewater treatment is its limited capacity to effectively remove dissolved or extremely fine particulate matter, including colloidal particles and specific dissolved ions [53]. While filtration proves efficient in capturing larger suspended solids, it often struggles to eliminate microscopic contaminants or dissolved pollutants [54], allowing them to pass through the filter media and persist in the treated water.

A notable disadvantage associated with membrane technology in wastewater treatment is its vulnerability to fouling and scaling [55]. Membranes, crucial for separating contaminants from water, are prone to becoming obstructed or coated over time by substances like organic matter, minerals, and biological growth. This fouling and scaling not only diminish the effectiveness of the treatment process but also necessitate frequent membrane cleaning or replacement, leading to heightened maintenance costs [56] and system downtime. Furthermore, the energy demands of high-pressure membrane filtration [57] can be substantial, rendering it an energy-intensive method compared to certain other wastewater treatment techniques.

A significant risk associated with the adsorption method [58] in wastewater treatment is its limited applicability in effectively removing a wide range of contaminants. Although adsorption is proficient at capturing specific organic compounds and pollutants from water, its effectiveness can diminish when dealing with high concentrations of contaminants or substances that do not readily adhere to the chosen adsorbent material. Furthermore, as the adsorbent surfaces become saturated over time, it necessitates the regeneration or replacement of the adsorbent

[59], leading to increased operational costs [55]. Additionally, adsorption is often selective in its pollutant removal, meaning it may not comprehensively target all types of pollutants in complex wastewater streams, requiring the integration of multiple treatment methods for thorough purification.

Another disadvantage associated with chemical methods in wastewater treatment is the potential for generating harmful byproducts and the necessity for meticulous chemical management [59–60]. While chemicals are commonly employed to precipitate, coagulate, or disinfect contaminants in wastewater, improper handling or dosing can result in the formation of toxic substances or residuals that pose environmental and health hazards. Additionally, the storage and disposal of chemicals can be financially burdensome and require strict adherence to regulations [61]. Moreover, chemical treatments may not prove as effective in treating certain persistent pollutants or micro-pollutants, thereby requiring the integration of multiple treatment processes, which can escalate operational complexity and expenses in wastewater treatment facilities.

A critical setback of the Physicochemical method like the ion exchange methods [62–63] in wastewater treatment is the generation of concentrated brine waste during the regeneration process. When ion exchange resins become saturated with unwanted ions and need regeneration, a concentrated brine solution is typically produced, containing high levels of the ions [46] that were removed from the wastewater. Managing and disposing of this brine waste can be environmentally challenging and costly. Additionally, the process may be less effective in treating very high concentrations of ions [64] or in situations where selectivity for specific ions is required, as it involves exchanging ions with others of similar charge, which may not completely remove all target contaminants [65]. This limitation may necessitate the use of additional treatment steps to achieve the desired water quality, potentially increasing the complexity and expense of the overall treatment process.

A significant problem associated with biological methods [66] in wastewater treatment is their susceptibility to variations in environmental conditions, such as temperature, pH levels, and the composition of influent wastewater [67]. These fluctuations can disrupt the activity of microorganisms responsible for breaking down organic matter and pollutants, leading to decreased treatment efficiency and potential system upsets. Moreover, biological treatment processes often demand longer retention times and occupy more physical space [68], necessitating the construction of larger treatment facilities and resulting in higher infrastructure costs. Additionally, maintaining optimal conditions for microbial activity requires continuous monitoring and the expertise of skilled operators, contributing to increased operational expenses [69]

18.3 NANO-REMOVAL OF POLLUTANTS FROM THE AQUEOUS ENVIRONMENT

18.3.1 Causative Disruptors of Aqueous Habitat, a Highlight

Nanomaterials can be considered as a present-day alternative way or reclamation of wastewater. As evident from the literature surveys, particles of the nano regime, due to their extensive properties of having ultra smaller dimensions ranging between 1–100nm [70], have attracted a novel way for treating waste water. These nanomaterials are capable of absorbing organic pollutants owing to their large surface-to-volume ratio [71]; their reactivity towards a particular organic pollutant coupled with their affinity towards the same. This has paved an excellent breakthrough in the arena of wastewater remediation. Of the alarming components that over the years have disrupted the aqueous habitat are the persistent organic pollutants (POPs), mostly sources from unscrupulous anthropogenic activities. These include unrestrained dumping of toxic chemicals sourcing from industrial effluents. These POPs enlist a number of phenol derivatives like heptachlor, para-nitrophenol and also a series of endocrine disruptors which include bisphenols, phthalate esters [72]. Consumption of these toxic phenolic compounds poses detrimental effects encompassing deteriorating metabolic function, loss in cardiac activity [73], impaired reproductive function [74] with ancillary alarm of having carcinogenic potential as well [75]. Phenolic derivatives are used in vast areas of urban modernization right from their use in textile industries [76], pesticides and pharmaceuticals. These chemicals percolate into the aqueous ecosystem while the insoluble ones leach into the sediments [77]. Reference can also be cited of a vast number of textile dyes [78] as well as of the heavy metals [79] that are used for a number of applications. Conventional techniques as discussed in the earlier section mostly have met with a number of disadvantages, the crucial ones being uneconomical and lacking biocompatibility. So, an effective strategy can be designed by the use of nanostructured materials that have the properties of absorbing the organic pollutants remediate and converting them into products that are safe to be used for further applications.

18.3.2 Use of Metal Nanoparticles in the Reclamation

18.3.2.1 Noble use of the Noble Metals – A Quick Ride on their Decontamination Potential

Metal nanoparticles, typically noble metal nanoparticles like gold and silver, have been in vogue since ancient times and there are considerable observations that show not only they are used in jewelry but their usage is also pronounced in medical applications [80]. With the advent of modern civilization, and also with the development of technologies, the catalytic potential of metal nanoparticles is mostly harnessed. Photocatalysis and advanced oxidation processes [81] are by far the most effective technologies used in recent days to decontaminate the water habitat from obnoxious persistent organic pollutants (POPs) [82]. Heterogeneous catalysis [83] by are by far the best aid to utilize the potential of noble metal nanoparticles for removing a number of organic pollutants with additional features of recyclability [84]. Noble nanoparticles, either when used alone or in combination as in bimetallics [85–86] and in trimetallics [87] carry out different pivotal features of removing heavy metal ions [88], dyes [89] and organic pollutants [90]. Au nanostructures have excellent optical properties, which is generally seen as surface plasmon resonance arising out of the collective oscillations on the surface of the metal [91] with ancillary properties of being stable in the environment, high reduction potential [92] and less prone to oxidation as compared with other noble metals like gold and silver [93]. Au nps

TABLE 18.1
Variously Modified Au Nanoparticles and Their Role in Remediating Contaminants from Wastewater

Capping agent/ Modifying agent	Type of contaminant removed/ remediated	References
4 amino-thio phenol	Removal of Hg, Cd, Co, Pd ions	[103]
Lemon peel extract	Removal of Nickel ions	[104]
PDA functionalized PVA@Au	Removal of Rhodamine B dyes	[105]
Alpinia nigra leaf extract	Methyl orange and Rhodamine B	[106]
Polydimethylsiloxane@Au	Thioanisole	[107]
Sausevieria roxburghiana extract	4 nitrophenol	[108]

carry out different catalyzing reactions by virtue of their catalytic properties that are generally seen in the forms of hydrogenation [94] and dehydrogenation [95], hydrocarbon oxidation [96] reactions. Gold nanoparticles are surface functionalized with a number of polymers [97–98], biological ligands [52, 99] that overall increase the repertoire of the metal towards catalysis. The surface functionalization of the gold nanoparticles with the different organic polymers and biogenic extracts renders them not only stabilization by the stimulation of steric repulsion between the individual nanoparticles promoting longer shelf life [100-101] but also to minimize cytotoxicity [102]. Table 18.1 shows the different properties of Au nanoparticles synthesized using various surface coating agents for decontaminating the aqueous environment.

Silver nanoparticles (Ag nps) on the other hand are quite economic compared to that of gold nanoparticles and have been routinely used for their marked antibacterial and antibiofilm action. Highly carcinogenic dyes of textile importance are seen to be removed by Ag nps, either alone or in combination with other metallic nanoparticles like TiO_2 or other nanocomposites of metal oxides and sulphides [109–112] alongside phenol [113–114], derivatives of phenol [115–117], also antibiotics [118] and heavy metals [119–120] can be adsorbed and simultaneously removed from the aquatic environment when silver synthesized through various chemical and biological methods are employed for removal.

Silver nanoparticles (Ag nps) are excellent enough in decontaminating water from the yoke of a number of pathogenic microorganisms – a pertinent cause for water pollution. Ag nanoparticles have been fabricated in a number of ways that have shown cidal activity against coliforms and antibiotic-resistant bacteria. Silver nanoparticles impregnated in paper have been shown to have significant loss of viability of *E. coli* and *E. faecalis* when a solution of bacteria is passed through silver decorated blotting papers [121]. Being a comparatively cheaper material, silver nanoparticles fabricated on inert matrices like silica are effective enough not only in the adsorption of various organic compounds and dyes through the π-π interaction but the nanocomposite was seen to have marked anti- fouling and anti-biofilm activities rendering it to be a potent candidate to be used in the water purification set ups [122]. Recent reports emphasized the combined action of hybrid bimetallic Ag–Au nanoparticles where significant decontamination has been seen to be achieved. Cryogels loaded with bimetallic Ag-Au nps [86] has been seen to decontaminate water not only from coliforms but also from organic contaminants like *p*- nitrophenol and Rhodamine B. Organic compounds like HIV drugs disposed into the aqueous environment are seen to be properly degraded through catalytic degradation by the use of bimetallic Ag–Au nanoparticles.

The action of silver nanoparticles on waterborne bacteria is much more pronounced than gold nanoparticles. It has been revealed that Ag nanoparticles bring about cytoderm damage [123] alongside production of reactive oxygen species (ROS) [124]. The cytotoxicity of silver nanoparticles is attributed to the fact that it not only causes an alteration of the membrane permeability of the cell but also results in the deactivation and complete functioning of the respiratory enzymes present in the cell as silver ions have greater affinity towards sulfur containing membrane proteins [125].

18.3.3 Nanosponges, a Gateway Mitigating Toxicity and Reclamation of Water

Selective removal of contaminants has been the foundation of membrane technology in the past few years. Use of membranes have led to the perfect removal of contaminants being desalinating water as well as other contaminants to render water potable for drinking purposes. Membrane technology can be categorized into either polymeric membrane whereby a suitable organic or biological polymer aid in the filtration technique in contrast to inorganic membranes whereby an inorganic material like metal is impregnated into porous material like silica. In spite of the huge demand of the latter polymers have almost won over due to their high mechanical strength, low cost, high thermal properties which are prerequisites for the separation of the contaminants out of water [126]. For this a large number of polymers like polyvinyl pyrrolidone (PVA) [127], Polyethylene glycol (PEG) [128], Polyvinylidene fluoride (PVDF) [129], Poly Ethyl Sulphone (PES) [130], Polyacrylonitrile (PAN) [131], are the organic polymers in use. Apart from this there are a large number of biopolymers like starch [132] chitosan [133], xylan [134], cellulose [135] available that have been fabricated with a number of metallic and non-metallic nano materials. In recent times there has been an advent of cyclodextrin that has been employed as excellent absorbs in trapping organic wastes out of pollutant water. Cyclodextrins form crosslinks to attain a 3D structure which thereby gains porosity [136]. Cyclodextrin cross linking networks thus are characterized by pore size of nano-dimension ranging from 0.7–1.2nm that creates a hydrophobic environment enabling its tendency to absorb organic compounds from water [137]. The cross linking is brought about by a number of cheaper materials like diphenyl carbonate which are safe to the environment along with beta cyclodextrin at a temperature not exceeding 100°C by solvothermal or microwave irradiation processes [138–139]. The different absorptive capability of the nanosponges is summarised below.

18.3.3.1 Nanosponges in the Removal of Dyes

In recent times, nanosponges have been combined with a number of other inorganic materials that effectively removes a number of potentially carcinogenic dyes from the aqueous habitat. It has been observed by Massaro et al. [140] that nanosponges are effective towards both cationic and anionic dyes, however a differential analysis has revealed comparatively better results of the nanosponges towards cationic dyes than anionic dyes [140]. Through electrostatic interactions between EDTA based β cyclodextrin nanosponges were found to remove almost 90 percent of crystal violet from the environment [141]. Cyclodextrins based hydrogels are modifications that consist of a water-based scaffold that removes Methylene Blue dye from the environment [142]. Through host-guest interactions cyclodextrins modified with a number of cross linkers are seen to remove a number of cationic dyes like Safranin and Malachite Green [143]. Not only cyclodextrin, cellulose based nanosponges effectively remove starzarine dyes from water [144].

18.3.3.2 Nanosponges in the Smart Removal of Heavy Metals and Organic Compounds

β cyclodextrin based nanosponges are also credited in the removal of heavy metals from the aqueous realm. Nanosponges prepared through β cyclodextrin and linecaps, a linear polymer of starch obtained from pea removers marked ability to removal Al, Zn, Cd ions from contaminated [145]. Recent reports suggest nano-sponges formed through conjugation of carbon nanotubes along with cyclodextrin leads to the removal of Pb^{2+} and Co^{2+} from real water samples [146].

Nanosponges are diverse in their application of removing various organic compounds and drugs that are discharged into the environment. A number of organic compounds sourcing out of pesticides are removed through cyclodextrin based nanosponges [147]. Hydro-soluble pesticides like paraquat that disrupts the abiotic and biotic components of the water systems are seen to be effectively removed through the adsorptive capability of the nanosponges prepared through PVA-Cyclodextrin cross linkage [148]. Often nanoparticles are decorated with nanosponges that have been seen to bring about significant adsorption of a large number of organic pollutants. Salazar et al. decorated Fe_3O_4 nanoparticles that showed significant activity against chlorophenols [148], neonicotinoid dinotefuran [149]. The adsorption phenomenon is best described following Langmuir isotherms that clearly hints out that the available number of sorption sites provides faster reaction rates than when all the sites are occupied [139].

18.4 MOLECULAR DYNAMICS SIMULATION STUDY OF NANOSPONGES

In recent studies the Molecular Dynamics (MD) simulation is used to study the interacting surfaces in relative motion. The tribological behaviour investigated using micro or nanoscale atomistic or molecular level MD simulation [150]. During MD simulation, according to Newton's second law the equations of motion for all atoms obtain a trajectory of all particles in the system in isothermal and isobaric ensemble. To enhance and determine of thermal conductivity of nano fluids and effects of various factors can be studied by using the Brownian motion of nanoparticles through MD simulation [151]. The supramolecular self-assembly molecular structures are built by using non-covalent interactions. This process is studied by applying a classical force field of MD simulation, contain the bonded intra-molecular interactions, and inter-molecular non-bonded interactions where solvent calculated through explicit or implicit models [152]. The nanosponges are made in ranges of 80 nm to micrometers in diameters. The structure of the novel nanosponges were investigated through explicit solvent and then coarse-grained molecular dynamics (MD) simulations [153].

18.5 CONCLUSION

In conclusion, the examination of nanosponges as an avenue for advancing sustainable wastewater treatment presents a compelling prospect. Their exceptional adsorption capabilities and versatility in the removal of diverse pollutants underscore their potential contributions to environmental remediation. However, the seamless integration of nanosponges into practical applications requires meticulous attention to scalability and environmental considerations, necessitating continuous research and development efforts to foster cleaner and more sustainable water management solutions. The reclamation of wastewater presents formidable challenges, despite the utilization of conventional treatment methods like membrane filtration, chemical treatments such as coagulation and flocculation, ion exchange, and biological interventions like bio-stimulation, bioaugmentation, and phytoremediation. While these methods have demonstrated effectiveness, they also come with inherent limitations. This is where the integration of nanotechnology interventions becomes indispensable. Nano-based removal of pollutants from aqueous environments, particularly employing silver nanoparticles (Ag NPs), has emerged as a promising solution. Ag NPs have proven to be effective in eliminating heavy metals like mercury (Hg), cadmium (Cd), palladium (Pd), and nickel (Ni), as well as various dyes, including Methyl Orange, Rhodamine B, Thioanisole, and 4-nitrophenol. Furthermore, nanosponges represent a groundbreaking approach to mitigating toxicity and reclaiming water resources. These nanostructures play a crucial role in treating water contaminated with heavy metals, dyes, and organic compounds. By harnessing the power of nanotechnology, these interventions offer promising avenues for more efficient and comprehensive wastewater treatment, overcoming the limitations of traditional methods and contributing to the sustainable management of water resources. Additionally, researchers have turned to molecular dynamics (MD) simulations to enhance the thermal conductivity of nanofluids and explore the influence of various factors. By leveraging the Brownian motion of nanoparticles, MD simulations enable a detailed examination of nanoparticle behaviour within nanofluids and their impact on thermal conductivity. Through explicit solvent simulations followed by coarse-grained MD simulations, researchers gain valuable insights into nanosponge structure and behaviour, facilitating the optimization of their properties for specific applications.

In conclusion, while traditional wastewater treatment methods have shown efficacy, they also have limitations in addressing complex water pollution challenges. Nanotechnology interventions offer promising solutions to overcome these limitations. Nano-based removal of pollutants, especially with silver nanoparticles, is highly effective in eliminating heavy metals and dyes from water. Nanosponges offer an innovative approach to mitigate toxicity and reclaim contaminated water resources.

Molecular dynamics simulations provide insights into enhancing nanofluids' thermal conductivity and understanding nanoparticle behaviour. Leveraging these simulations, researchers can optimize nanomaterial design for wastewater treatment.

Declaration of Competing Interest

The authors declare that they have no personal relationships that could have appeared to influence this study reported in this chapter.

ACKNOWLEDGEMENT

The authors are highly thankful to Swami Vivekananda University for constant support and motivation to complete this study.

REFERENCES

1. Niemeyer JC, Medici LO, Correa B, et al. Treated Produced Water in Irrigation: Effects on Soil Fauna and Aquatic Organisms. *Chemosphere.* 2020; 240:124791.
2. Xia C, Li X, Wu Y, Suharti S, Unpaprom Y, Pugazhendhi A. A Review on Pollutants Remediation Competence of Nanocomposites on Contaminated Water. *Environ Res.* 2023; 222:115318.
3. Warren-Vega WM, Campos-Rodríguez A, Zárate-Guzmán AI, Romero-Cano LA. A Current Review of Water Pollutants in American Continent: Trends and Perspectives in Detection, Health Risks, and Treatment Technologies. *Int J Environ Res Public Health.* 2023;20(5):4499.
4. Mateo-Sagasta J, Zadeh SM, Turral H, Burke J. Water pollution from agriculture: a global review. *Food and Agriculture Organization of the United Nations, Rome and the International Water Management Institute on behalf of the Water Land and Ecosystems Research Program,* Colombo. 2017.
5. Wastewater, the Untapped Recourse, WWAP (World Water Assessment Programme). *United Nations World Water Development Report.* 2017.
6. UNEP. *A Snapshot of the World's Water Quality: Towards a global Assessment.* Nairobi: United Nations Environment Programme; 2016.
7. Hampel M, Blasco J, Segner H. Molecular and Cellular Effects of Contamination in Aquatic Ecosystems. *Environ Sci Pollut Res Int.* 2015;22(22):17261–17266.
8. Zhang Z, He W, Shen J, et al. The Driving Forces of Point Source Wastewater Emission: Case Study of COD and NH_4-N Discharges in Mainland China. *Int J Environ Res Public Health.* 2019;16(14):2556.
9. Afrad MSI, Monir MB, Haque ME, Barau AA, Haque MM. Impact of Industrial Effluent on Water, Soil and Rice Production in Bangladesh: A Case of Turag River Bank. *J Environ Health Sci Eng.* 2020;18(2):825–834.
10. Vinti G, Bauza V, Clasen T, et al. Municipal Solid Waste Management and Adverse Health Outcomes: A Systematic Review. *Int J Environ Res Public Health.* 2021;18(8):4331.
11. Massarelli C, Losacco D, Tumolo M, Campanale C, Uricchio VF. Protection of Water Resources from Agriculture Pollution: An Integrated Methodological Approach for the Nitrates Directive 91-676-EEC Implementation. *Int J Environ Res Public Health.* 2021;18(24):13323.
12. Rey-Romero DC, Domínguez I, Oviedo-Ocaña ER. Effect of Agricultural Activities on Surface Water Quality from Páramo Ecosystems. *Environ Sci Pollut Res Int.* 2022;29(55):83169–83190.

13. Silva IA, Almeida FCG, Souza TC, et al. Oil Spills: Impacts and Perspectives of Treatment Technologies With Focus on the use of Green Surfactants. *Environ Monit Assess*. 2022;194(3):143.
14. Suárez-Macías J, Terrones-Saeta JM, Bernardo-Sánchez A, Ortiz-Marqués A, Castañón AM, Corpas-Iglesias FA. Reutilisation of Water Contaminated by Mining Waste for the Encapsulation of Potentially Toxic Elements. *Materials (Basel)*. 2022;15(3):1076.
15. Scanlon MM, Gordon JL, McCoy WF, Cain MF. Water Management for Construction: Evidence for Risk Characterization in Community and Healthcare Settings: A Systematic Review. *Int J Environ Res Public Health*. 2020;17(6):2168.
16. Koelmans AA, Mohamed Nor NH, Hermsen E, Kooi M, Mintenig SM, De France J. Microplastics in Freshwaters and Drinking Water: Critical Review and Assessment of Data Quality. *Water Res*. 2019; 155:410–422.
17. Ortúzar M, Esterhuizen M, Olicón-Hernández DR, González-López J, Aranda E. Pharmaceutical Pollution in Aquatic Environments: A Concise Review of Environmental Impacts and Bioremediation Systems. *Front Microbiol*. 2022; 13:869332.
18. Briffa J, Sinagra E, Blundell R. Heavy Metal Pollution in the Environment and their Toxicological Effects on Humans. *Heliyon*. 2020;6(9):e04691.
19. Rehman K, Fatima F, Waheed I, Akash MSH. Prevalence of Exposure of Heavy Metals and their Impact on Health Consequences. *J Cell Biochem*. 2018;119(1):157–184.
20. Balali-Mood M, Naseri K, Tahergorabi Z, Khazdair MR, Sadeghi M. Toxic Mechanisms of Five Heavy Metals: Mercury, Lead, Chromium, Cadmium, and Arsenic. *Front Pharmacol*. 2021; 12:643972.
21. Sarvajith M, Reddy GKK, Nancharaiah YV. Textile Dye Biodecolourization and Ammonium Removal Over Nitrite in Aerobic Granular Sludge Sequencing Batch Reactors. *J Hazard Mater*. 2018; 342:536–543.
22. Zhao X, Baharinikoo L, Farahani MD, Mahdizadeh B, Farizhandi AAK. Experimental Modelling Studies on the Removal of Dyes and Heavy Metal Ions using $ZnFe_2O_4$ Nanoparticles. *Sci Rep*. 2022;12(1):5987.
23. Syafrudin M, Kristanti RA, Yuniarto A, et al. Pesticides in Drinking Water-A Review. *Int J Environ Res Public Health*. 2021;18(2):468
24. Rajmohan KS, Chandrasekaran R, Varjani S. A Review on Occurrence of Pesticides in Environment and Current Technologies for Their Remediation and Management. *Indian J Microbiol*. 2020;60(2):125–138.
25. Pathak VM, Verma VK, Rawat BS, et al. Current Status of Pesticide Effects on Environment, Human Health and it's Eco-friendly Management as Bioremediation: A Comprehensive Review. *Front Microbiol*. 2022; 13:962619.
26. Oltramare C, Weiss FT, Staudacher P, Kibirango O, Atuhaire A, Stamm C. Pesticides Monitoring in Surface Water of a Subsistence Agricultural Catchment in Uganda using Passive Samplers. *Environ Sci Pollut Res Int*. 2023;30(4):10312–10328.
27. Zapata JG, Vangipuram B, Dalin C, Erfani T. Water Quality and Pollution Trading: A Sustainable Solution for Future Food Production. *ACS ES T Eng*. 2023;3(8):1112–1124.
28. Liu, L., Zheng, X., Wei, X., et al. Excessive Application of Chemical Fertilizer and Organophosphorus Pesticides Induced Total Phosphorus Loss From Planting Causing Surface Water Eutrophication. *Scientific Reports*. 2021; 11(1): 23015.
29. Ziani K, Ioniță-Mîndrican CB, Mititelu M, et al. Microplastics: A Real Global Threat for Environment and Food Safety: A State of the Art Review. *Nutrients*. 2023;15(3):617.
30. Campanale C, Massarelli C, Savino I, Locaputo V, Uricchio VF. A Detailed Review Study on Potential Effects of Microplastics and Additives of Concern on Human Health. *Int J Environ Res Public Health*. 2020;17(4):1212.

31. Lee Y, Cho J, Sohn J, Kim C. Health Effects of Microplastic Exposures: Current Issues and Perspectives in South Korea. *Yonsei Med J.* 2023;64(5):301–308.
32. O'Callaghan-Gordo, C., Orta-Martínez, M., Kogevinas, M. Health Effects of Non-Occupational Exposure to Oil Extraction. *Environmental Health: A Global Access Science Source.* 2016;15:56.
33. Barron MG, Vivian DN, Heintz RA, Yim UH. Long-Term Ecological Impacts from Oil Spills: Comparison of *Exxon Valdez*, *Hebei Spirit*, and Deepwater Horizon. *Environ Sci Technol.* 2020;54(11):6456–6467.
34. Hejna M, Kapuścińska D, Aksmann A. Pharmaceuticals in the Aquatic Environment: A Review on Eco-Toxicology and the Remediation Potential of Algae. *Int J Environ Res Public Health.* 2022;19(13):7717.
35. Pinto I, Simões M, Gomes IB. An Overview of the Impact of Pharmaceuticals on Aquatic Microbial Communities. *Antibiotics (Basel).* 2022;11(12):1700.
36. Alegbeleye OO, Singleton I, Sant'Ana AS. Sources and Contamination Routes of Microbial Pathogens to Fresh Produce During Field Cultivation: A Review. *Food Microbiol.* 2018;73:177–208.
37. Kongprajug A, Denpetkul T, Chyerochana N, Mongkolsuk S, Sirikanchana K. Human Fecal Pollution Monitoring and Microbial Risk Assessment for Water Reuse Potential in a Coastal Industrial-Residential Mixed-Use Watershed. *Front Microbiol.* 2021;12:647602.
38. Obotey Ezugbe E, Rathilal S. Membrane Technologies in Wastewater Treatment: A Review. *Membranes (Basel).* 2020;10(5):89.
39. Ahmed, S. F., Mofijur, M., Nuzhat, S., et al. Recent Developments in Physical, Biological, Chemical, and Hybrid Treatment Techniques for Removing Emerging Contaminants from Wastewater. *Journal of Hazardous Materials.* 2021; *416*: 125912.
40. Aghalari Z, Dahms HU, Sillanpää M, Sosa-Hernandez JE, Parra-Saldívar R. Effectiveness of Wastewater Treatment Systems in Removing Microbial Agents: A Systematic Review. *Global Health.* 2020;16(1):13.
41. Cevallos-Mendoza J, Amorim CG, Rodríguez-Díaz JM, Montenegro MDCBSM. Removal of Contaminants from Water by Membrane Filtration: *A Review. Membranes (Basel).* 2022;12(6):570.
42. Rathi BS, Kumar PS. Application of Adsorption Process for Effective Removal of Emerging Contaminants from Water and Wastewater. *Environ Pollut.* 2021; 280:116995.
43. Zahmatkesh S, Bokhari A, Karimian M, et al. A Comprehensive Review of Various Approaches for Treatment of Tertiary Wastewater with Emerging Contaminants: What do we Know? *Environ Monit Assess.* 2022;194(12):884.
44. Ahmed MB, Zhou JL, Ngo HH, Guo W, Thomaidis NS, Xu J. Progress in the Biological and Chemical Treatment Technologies for Emerging Contaminant Removal from Wastewater: A Critical Review. *J Hazard Mater.* 2017;323(Pt A):274–298.
45. Gasmi A, Ibrahimi S, Elboughdiri N, et al. Comparative Study of Chemical Coagulation and Electrocoagulation for the Treatment of Real Textile Wastewater: Optimization and Operating Cost Estimation. *ACS Omega.* 2022;7(26):22456–22476.
46. Chen YG, Sofińska-Chmiel W, Lv GY, Kołodyńska D, Chen SH. Application of Modern Research Methods for the Physicochemical Characterization of Ion Exchangers. *Materials (Basel).* 2021;14(22):7067.
47. Czupryński, P., Płotka, M., Glamowski, P., et al. An Assessment of an Ion Exchange Resin System for the Removal And Recovery of Ni, Hg, and Cr from Wet Flue Gas Desulphurization Wastewater-a Pilot Study. *RSC Advances.* 2022; 12(9): 5145–5156.

48. Zakaria KK, Farag HA, El-Gayar DA. Removal of Cu^{2+}, Fe^{2+} and SO_4^{2-} Ions from Industrial Wastewater by Ion Exchange Resins Contained in a Rotating Perforated Cylindrical Basket of Different Heights. *Sci Rep*. 2023;13(1):3248.
49. Bouchaala L, Charchar N, Sahraoui H, Gherib A. Assessment of Wastewater Biological Treatment Efficiency and Mapping of WWTPs and LTPs in Algeria. *J Environ Health Sci Eng*. 2021;19(1):1153–1169.
50. Ceretta MB, Nercessian D, Wolski EA. Current Trends on Role of Biological Treatment in Integrated Treatment Technologies of Textile Wastewater. *Front Microbiol*. 2021; 12:651025.
51. Khalidi-Idrissi A, Madinzi A, Anouzla A, et al. Recent Advances in the Biological Treatment of Wastewater Rich in Emerging Pollutants Produced by Pharmaceutical Industrial Discharges. *Int J Environ Sci Technol (Tehran)*. 2023; 1–12.
52. Khan AU, Khan AN, Waris A, Ilyas M, Zamel D. Phytoremediation of pollutants from wastewater: A concise review. *Open Life Sci*. 2022;17(1):488–496.
53. Pachaiappan R, Cornejo-Ponce L, Rajendran R, Manavalan K, Femilaa Rajan V, Awad F. A Review on Biofiltration Techniques: Recent Advancements in the Removal of Volatile Organic Compounds and Heavy Metals in the Treatment of Polluted Water. *Bioengineered*. 2022;13(4):8432–8477.
54. Cybulski J, Witczak A, Pokorska-Niewiada K. Influence of Water Treatment and Wastewater Treatment on the Changes in Residues of Important Elements in Drinking Water. *Molecules*. 2022;27(3):972.
55. Yalcinkaya F, Boyraz E, Maryska J, Kucerova K. A Review on Membrane Technology and Chemical Surface Modification for the Oily Wastewater Treatment. *Materials (Basel)*. 2020;13(2):493.
56. Zulkefli NF, Alias NH, Jamaluddin NS, et al. Recent Mitigation Strategies on Membrane Fouling for Oily Wastewater Treatment. *Membranes (Basel)*. 2021;12(1):26.
57. Othman NH, Alias NH, Fuzil NS, et al. A Review on the Use of Membrane Technology Systems in Developing Countries. *Membranes (Basel)*. 2021;12(1):30.
58. Naim R, Pei Sean G, Nasir Z, Mokhtar NM, Safiah Muhammad NA. Recent Progress and Challenges in Hollow Fiber Membranes for Wastewater Treatment and Resource Recovery. *Membranes (Basel)*. 2021;11(11):839.
59. Al-Maqdi KA, Elmerhi N, Athamneh K, et al. Challenges and Recent Advances in Enzyme-Mediated Wastewater Remediation-A Review. *Nanomaterials (Basel)*. 2021;11(11):3124.
60. Sathya K, Nagarajan K, Carlin Geor Malar G, Rajalakshmi S, Raja Lakshmi P. A Comprehensive Review on Comparison among Effluent Treatment Methods and Modern Methods of Treatment of Industrial Wastewater Effluent from Different Sources. *Appl Water Sci*. 2022;12(4):70. doi:10.1007/s13201-022-01594-7
61. Masudi A, Harimisa GE, Ghafar NA, Jusoh NWC. Magnetite-based Catalysts for Wastewater Treatment. *Environ Sci Pollut Res Int*. 2020;27(5):4664–4682.
62. Czupryński P, Płotka M, Glamowski P, Żukowski W, Bajda T. An Assessment of an Ion Exchange Resin System For The Removal And Recovery of Ni, Hg, and Cr from Wet Flue Gas Desulphurization Wastewater-a Pilot Study. *RSC Adv*. 2022;12(9):5145–5156.
63. Vidu R, Matei E, Predescu AM, et al. Removal of Heavy Metals from Wastewaters: A Challenge from Current Treatment Methods to Nanotechnology Applications. *Toxics*. 2020;8(4):101.
64. Arslanoglu H, Tumen F. A Study on Cations and Color Removal from Thin Sugar Juice by Modified Sugar Beet Pulp. *J Food Sci Technol*. 2012;49(3):319–327.
65. Liu Y, Wang H, Cui Y, Chen N. Removal of Copper Ions from Wastewater: A Review. *Int J Environ Res Public Health*. 2023;20(5):3885.

66. Machineni L. Review on Biological Wastewater Treatment and Resources Recovery: Attached and Suspended Growth Systems. *Water Sci Technol.* 2019;80(11): 2013–2026.
67. Chahal C, van den Akker B, Young F, Franco C, Blackbeard J, Monis P. Pathogen and Particle Associations in Wastewater: Significance and Implications for Treatment and Disinfection Processes. *Adv Appl Microbiol.* 2016; 97:63–119.
68. Abdel-Raouf N, Al-Homaidan AA, Ibraheem IB. Microalgae and Wastewater Treatment. *Saudi J Biol Sci.* 2012;19(3):257–275.
69. Wu L, Yang Y, Chen S, et al. Long-term Successional Dynamics of Microbial Association Networks in Anaerobic Digestion Processes. *Water Res.* 2016; 104:1–10.
70. Calzolai L, Gilliland D, Rossi F. Measuring Nanoparticles Size Distribution in Food and Consumer Products: A Review. *Food Addit Contam Part A Chem Anal Control Expo Risk Assess.* 2012;29(8):1183–1193.
71. Campelo JM, Luna D, Luque R, Marinas JM, Romero AA. Sustainable Preparation of Supported Metal Nanoparticles and their Applications in Catalysis. *ChemSusChem.* 2009;2(1):18–45.
72. Pawełczyk A. Assessment of Health Risk Associated with Persistent Organic Pollutants in Water. *Environ Monit Assess.* 2013;185(1):497–508.
73. Vassilopoulou L, Psycharakis C, Petrakis D, Tsiaoussis J, Tsatsakis AM. Obesity, Persistent Organic Pollutants and Related Health Problems. *Adv Exp Med Biol.* 2017;960:81–110.
74. Damstra T. Potential Effects of Certain Persistent Organic Pollutants and Endocrine Disrupting Chemicals on the Health of Children. *J Toxicol Clin Toxicol.* 2002;40(4):457–465.
75. Stich HF. The Beneficial and Hazardous Effects of Simple Phenolic Compounds. *Mutat Res.* 1991;259(3–4):307–324.
76. Albuquerque BR, Heleno SA, Oliveira MBPP, Barros L, Ferreira ICFR. Phenolic Compounds: Current Industrial Applications, Limitations and Future Challenges. *Food Funct.* 2021;12(1):14–29.
77. Levy W, Pandelova M, Henkelmann B, et al. Persistent Organic Pollutants in Shallow Percolated Water of the Alps Karst System (Zugspitze summit, Germany). *Sci Total Environ.* 2017;579:1269–1281.
78. Al-Tohamy R, Ali SS, Li F, et al. A Critical Review on the Treatment of Dye-Containing Wastewater: Ecotoxicological and Health Concerns of Textile Dyes and Possible Remediation Approaches for Environmental Safety. *Ecotoxicol Environ Saf.* 2022;231:113160.
79. Joseph L, Jun BM, Flora JRV, Park CM, Yoon Y. Removal of Heavy Metals from Water Sources in the Developing World Using Low-Cost Materials: A Review. *Chemosphere.* 2019;229:142–159.
80. Pradeep, T. Noble Metal Nanoparticles for Water Purification: A Critical Review. *Thin solid films.* 2009;517(24):6441–6478.
81. Wang P, Huang B, Dai Y, Whangbo MH. Plasmonic Photocatalysts: Harvesting Visible Light with Noble Metal Nanoparticles. *Phys Chem Chem Phys.* 2012;14(28):9813–9825.
82. Vasseghian Y, Hosseinzadeh S, Khataee A, Dragoi EN. The Concentration of Persistent Organic Pollutants in Water Resources: A Global Systematic Review, Meta-Analysis and Probabilistic Risk Assessment. *Sci Total Environ.* 2021;796:149000.
83. Li X, Yang X, Huang Y, Zhang T, Liu B. Supported Noble-Metal Single Atoms for Heterogeneous Catalysis. *Adv Mater.* 2019;31(50):e1902031.
84. Zhang L, Zheng S, Kang DE, Shin JY, Suh H, Kim I. Synthesis of Multi-Amine Functionalized Hydrogel for Preparation of Noble Metal Nanoparticles: Utilization

as Highly Active and Recyclable Catalysts in Reduction of Nitroaromatics. *RSC Advances*. 2013;3(14):4692–4703.
85. Zaleska-Medynska A, Marchelek M, Diak M, Grabowska E. Noble Metal-Based Bimetallic Nanoparticles: The Effect of the Structure on the Optical, Catalytic and Photocatalytic Properties. *Adv Colloid Interface Sci*. 2016; 229:80–107.
86. Haleem A, Chen SQ, Ullah M, Siddiq M, He WD. Highly Porous Cryogels Loaded with Bimetallic Nanoparticles as an Efficient Antimicrobial Agent and Catalyst for Rapid Reduction of Water-Soluble Organic Contaminants. *Journal of Environmental Chemical Engineering*. 2021;9(6):106510–106510.
87. Hoseini Chopani SM, Asadi S, Heravi MM. Application of Bimetallic and Trimetallic Nanoparticles Supported on Graphene as novel Heterogeneous Catalysts in the Reduction of Nitroarenes, Homo-coupling, Suzuki-Miyaura and Sonogashira Reactions. *Current Organic Chemistry*. 2020;24(19):2216–2234.
88. Aazam ES, Zaheer Z. Gold@Silver Bimetallic Nanoparticles: Fabrication and Removal of Toxic Chromium (VI). *Journal of Materials Science: Materials in Electronics*. 2021;32(8):11043–11058.
89. Kang CW, Kolya H. Green Synthesis of Ag-Au Bimetallic Nanocomposites Using Waste Tea Leaves Extract for Degradation Congo Red and 4-Nitrophenol. *Sustainability*. 2021;13(6):3318.
90. Akilandaeaswari B, Muthu K. One-pot Green Synthesis of Au-Ag Bimetallic Nanoparticles from Lawsonia Inermis Seed Extract and its Catalytic Reduction of Environmental Polluted Methyl Orange and 4-Nitrophenol. *Journal of the Taiwan Institute of Chemical Engineers*. 2021;127:292–301.
91. Kostantinos Pantapasis, Alexandru Mihai Grumezescu. Gold Nanoparticles: Advances in Water Purification Approaches. *Elsevier eBooks*. Published online January 1, 2017:447–477
92. Ke X, Zhang X, Zhao J, Sarina S, Barry J, Zhu H. Selective Reductions Using Visible Light Photocatalysts of Supported Gold Nanoparticles. *Green Chem*. 2013;15(1):236–244.
93. Hui Hun Cho, Do Hyeon Jung, Jun Hyuk Heo, Chae Yeon Lee, Sang Yun Jeong, Jung Heon Lee. Gold Nanoparticles as Exquisite Colorimetric Transducers for Water Pollutant Detection. *ACS Applied Materials & Interfaces*. 2023;15(16):19785–19806.
94. Corma A. Chemoselective Hydrogenation of Nitro Compounds with Supported Gold Catalysts. *Science*. 2006;313(5785):332–334.
95. Fang W, Chen J, Zhang Q, Deng W, Wang Y. Hydrotalcite-Supported Gold Catalyst for the Oxidant-Free Dehydrogenation of Benzyl Alcohol: Studies on Support and Gold Size Effects. *Chemistry – A European Journal*. 2010;17(4):1247–1256.
96. Scirè S, Liotta LF. Supported Gold Catalysts for the Total Oxidation of Volatile Organic Compounds. *Applied Catalysis B: Environmental*. 2012;125:222–246.
97. Scott A, Gupta R, Kulkarni GU. A Simple Water-Based Synthesis of Au Nanoparticle/PDMS Composites for Water Purification and Targeted Drug Release. *Macromolecular Chemistry and Physics*. 2010;211(15):1640–1647.
98. Tsunoyama H, Sakurai H, Ichikuni N, Negishi Y, Tsukuda T. Colloidal gold Nanoparticles as Catalyst for Carbon-Carbon Bond Formation: Application to Aerobic Homocoupling of Phenylboronic Acid in Water. *Langmuir*.
99. Pradhan SP, Swain S, Sa N, et al. Photocatalysis of Environmental Organic Pollutants and Antioxidant Activity of Flavonoid Conjugated Gold Nanoparticles. *Spectrochim Acta A Mol Biomol Spectrosc*. 2022;282:121699.

100. Liu J, Legros S, Ma G, Veinot JG, von der Kammer F, Hofmann T. Influence of Surface Functionalization and Particle Size on the Aggregation Kinetics of Engineered Nanoparticles. *Chemosphere*. 2012;87(8):918–924.
101. Dumur F, Guerlin A, Dumas E, Bertin D, Gigmes D, Mayer CR. Controlled Spontaneous Generation of Gold Nanoparticles Assisted by Dual Reducing and Capping Agents. *Gold Bulletin*. 2011;44(2):119–137.
102. Deepa K, Sridhar A, Panda T. Biogenic Gold Nanoparticles: Current Applications and Future Prospects. *Journal of Cluster Science*. Published online July 7, 2022.
103. Chauhan N, Gupta S, Singh N, et al. Aligned Nanogold Assisted one Step Sensing and Removal of Heavy Metal Ions. *J Colloid Interface Sci*. 2011;363(1):42–50.
104. Barnawi, N., Allehyani, S., & Seoudi, R. Biosynthesis and Characterization of Gold Nanoparticles and its Application in Eliminating Nickel from Water. *Journal of Materials Research and Technology*. 2022;17:537–545.
105. Li X, Yang X, Huang Y, Zhang T, Liu B. Supported Noble-Metal Single Atoms for Heterogeneous Catalysis. *Adv Mater*. 2019;31(50):e1902031.
106. Baruah D, Goswami M, Yadav RNS, Yadav A, Das AM. Biogenic Synthesis of Gold Nanoparticles and their Application in Photocatalytic Degradation of Toxic Dyes. *J Photochem Photobiol B*. 2018;186:51–58.
107. Gupta R, Kulkarni GU. Removal of Organic Compounds from Water by using a Gold Nanoparticle-Poly(Dimethylsiloxane) Nanocomposite Foam. *Chem Sus Chem*. 2011;4(6):737–743.
108. Kumar I, Mondal M, Meyappan V et al. Green One-Pot Synthesis of Gold Nanoparticles using Sansevieria Roxburghiana Leaf Extract for the Catalytic Degradation of Toxic Organic Pollutants. *Materials Research Bulletin*. 2019;117:18–27.
109. Modi S, Pathak B, Fulekar MH. Microbial Synthesized Silver Nanoparticles for Decolorization and Biodegradation of Azo Dye Compound. *J Environ Nanotechnol*. 2015;4(2):37–46.
110. Ramalingam B, Khan MMR, Mondal B, et al. Facile Synthesis of Silver Nanoparticles Decorated Magnetic-Chitosan Microsphere for Efficient Removal of Dyes and Microbial Contaminants. *ACS Sustainable Chemistry & Engineering*. 2015;3(9):2291–2302.
111. David L, Moldovan B. Green Synthesis of Biogenic Silver Nanoparticles for Efficient Catalytic Removal of Harmful Organic Dyes. *Nanomaterials (Basel)*. 2020;10(2):202.
112. Naeem H, Tofil HM, Soliman M, et al. Reduced Graphene Oxide-Zinc Sulfide Nanocomposite Decorated with Silver Nanoparticles for Wastewater Treatment by Adsorption, *Photocatalysis and Antimicrobial Action. Molecules*. 2023;28(3):926.
113. Jilani A, Ansari MO, Rehman G, et al. Phenol Removal and Hydrogen Production from Water: Silver Nanoparticles Decorated on Polyaniline Wrapped Zinc Oxide Nanorods. *Journal of Industrial and Engineering Chemistry*. 2022;109:347–358.
114. Muthu Kumara Pandian A, Rajasimman M, Rajamohan N, Varjani S, Karthikeyan C. Anaerobic Mixed Consortium (AMC) Mediated Enhanced Biosynthesis of Silver Nano Particles (Agnps) and its Application for the Removal of Phenol. *J Hazard Mater*. 2021;416:125717.
115. Hunge YM, Yadav AA, Kang SW, et al. Facile Synthesis of Multitasking Composite of Silver Nanoparticle with Zinc Oxide for 4-Nitrophenol Reduction, Photocatalytic Hydrogen Production, and 4-Chlorophenol Degradation. *Journal of Alloys and Compounds*. 2022;928:167133.
116. Huang Z, Chen G, Zeng G, et al. Toxicity Mechanisms and Synergies of Silver Nanoparticles in 2,4-Dichlorophenol Degradation by Phanerochaete Chrysosporium. *J Hazard Mater*. 2017;321:37–46.

117. Ayad MM, Amer WA, Zaghlol S, et al. Polypyrrole-coated Cotton Fabric Decorated with Silver Nanoparticles for the Catalytic Removal of p-Nitrophenol from Water. *Cellulose*. 2018;25:7393–7407.
118. Lotfollahzadeh R, Yari M, Sedaghat S, et al. Biosynthesis and Characterization of Silver Nanoparticles for the Removal of Amoxicillin from Aqueous Solutions using Oenothera Biennis Water Extract. *Journal of Nanostructure in Chemistry*. 2021;1–14.
119. Samrot AV, Angalene JLA, Roshini SM, et al. Bioactivity and Heavy Metal Removal using Plant Gum Mediated Green Synthesized Silver Nanoparticles. *Journal of Cluster Science*. 2019;30:1599–1610.
120. Bhargava S, Uma V. Rapid Extraction of Cu (II) Heavy Metal from Industrial Waste Water by using Silver Nanoparticles Anchored with Novel Schiff Base. *Separation Science and Technology*. 2019;54(7):1182–1193.
121. Dankovich TA, Gray DG. Bactericidal Paper Impregnated with Silver Nanoparticles for Point-Of-Use Water Treatment. *Environ Sci Technol*. 2011;45(5):1992–1998.
122. Das SK, Khan MM, Parandhaman T, et al. Nano-silica Fabricated with Silver Nanoparticles: Antifouling Adsorbent for Efficient Dye Removal, Effective Water Disinfection and Biofouling Control. *Nanoscale*. 2013;5(12):5549–5560.
123. Gu X, Xu Z, Gu L, Huayu Xu, et al. Preparation and Antibacterial Properties of Gold Nanoparticles: A Review." *Environmental Chemistry Letters*. 2021;19:167–187.
124. Das B, Dash SK, Mandal D, et al. Green Synthesized Silver Nanoparticles Destroy Multidrug Resistant Bacteria Via Reactive Oxygen Species Mediated Membrane Damage. *Arabian Journal of Chemistry*. 2017;10(6):862–876.
125. Krishnaraj C, Jagan EG, Rajasekar S, et al. Synthesis of Silver Nanoparticles using Acalypha Indica Leaf Extracts and its Antibacterial Activity against Water Borne Pathogens. *Colloids Surf B Biointerfaces*. 2010;76(1):50–56.
126. Zahid M, Anum R, Saba A, et al. A Comprehensive Review on Polymeric Nano-Composite Membranes for Water Treatment. *J. Membr. Sci. Technol.* 2018;8(1):1–20.
127. Kusworo TD, Nita A, Enny N, Dani PU, editors. *AIP Conference Proceedings; 2020. PVA Coated Nano Hybrid PES-ZnO Membrane for Natural Rubber Wastewater Treatment*. 2197(1). AIP Publishing; 2020.
128. Masoumeh S, Marjani A, Mahdavian L, et al. Density Functional Theory Study of Dyes Removal from Colored Wastewater by a Nano-Composite of Polysulfone/Polyethylene Glycol. *Journal of Nanostructure in Chemistry*. 2023;13(5):519–532.
129. Lu Y, Li YS, Xiang CB, et al. Effect of Nano-sized Al_2O_3-particle Addition on PVDF Ultrafiltration Membrane Performance. *Journal of Membrane Science*. 2006;276(1–2):162–167.
130. Djoko KT, Utomo DP. Performance Evaluation of Double Stage Process Using Nano Hybrid PES/SiO2-PES Membrane and PES/ZnO-PES Membranes for Oily Waste Water Treatment to Clean Water. *Journal of Environmental Chemical Engineering*. 2017;5(6):6077–6086.
131. Mohy Eldin MS, Aggour YA, El-Aassar MR, et al. Development of Nano-Crosslinked Polyacrylonitrile Ions Exchanger Particles for Dyes Removal. *Desalination and Water Treatment*. 2016;57(9):4255–4266.
132. Fang K, Deng L, Yin J, et al. Recent Advances in Starch-Based Magnetic Adsorbents for the Removal of Contaminants from Wastewater: A Review. *Int J Biol Macromol*. 2022;218:909–929.
133. Rizzi V, Gubitosa J, Fini P, et al. Chitosan Biopolymer from Crab Shell as Recyclable Film to Remove/Recover in Batch Ketoprofen from Water: Understanding the Factors Affecting the Adsorption Process. *Materials (Basel)*. 2019;12(23):3810.

134. Hebatullah HF, Nebsen M, El-Sayed MMH. Multifunctional Chitosan/Xylan-Coated Magnetite Nanoparticles for the Simultaneous Adsorption of the Emerging Contaminants Pb (II), Salicylic Acid, and Congo Red Dye. *Water.* 2023;15(4):829.
135. Guleria A, Kumari G, Lima EC. Cellulose-g-poly-(acrylamide-co-acrylic acid) Polymeric Bioadsorbent for the Removal of Toxic Inorganic Pollutants from Wastewaters. *Carbohydr Polym.* 2020;228:115396.
136. Sherje AP, Dravyakar BR, Kadam D, Jadhav M. Cyclodextrin-based Nanosponges: A Critical Review. *Carbohydr Polym.* 2017;173:37–49.
137. Li D, Ma M. Nanosponges for Water Purification. *Clean Products and Processes.* 2000;2(2):112–116.
138. Li L, Liu H, Li W. et al. One-step Synthesis of an Environment-Friendly Cyclodextrin-Based Nanosponge and its Applications for the Removal of Dyestuff from Aqueous Solutions. *Res Chem Intermed.* 2020;46:1715–1734.
139. Massaro M, Colletti CG, Lazzara G, et al. Synthesis and Characterization of Halloysite–Cyclodextrin Nanosponges for Enhanced Dyes Adsorption. *ACS Sustainable Chemistry & Engineering.* 2017;5(4):3346–3352.
140. Feiping Z, Repo E, Yin D, et al. EDTA-cross-linked β-Cyclodextrin: An Environmentally Friendly Bifunctional Adsorbent for Simultaneous Adsorption of Metals and Cationic Dyes." *Environmental Science & Technology.* 2015;49(17):10570–10580.
141. Zhenyu Y, Wang J, Wang Y, et al. Preparation of a Poly (Acrylic Acid) Based Hydrogel with Fast Adsorption Rate and High Adsorption Capacity for the Removal of Cationic Dyes. *RSC Advances.* 2019;9(37):21075–21085.
142. Shiralipour R, Larki A. Pre-concentration and Determination of Tartrazine Dye from Aqueous Solutions using Modified Cellulose Nanosponges. *Ecotoxicology and Environmental Safety.* 2017;135:123–129.
143. Martwong E, Chuetor S, Junthip J. Adsorption of Cationic Contaminants by Cyclodextrin Nanosponges Cross-Linked with 1,2,3,4-Butanetetracarboxylic Acid and Poly(vinyl alcohol). *Polymers (Basel).* 2022;14(2):342.
144. Rubin Pedrazzo A, Smarra A, Caldera F, et al. Eco-Friendly β-cyclodextrin and Linecaps Polymers for the Removal of Heavy Metals. *Polymers (Basel).* 2019;11(10):1658.
145. Taka AL, Fosso-Kankeu E, Pillay K, Mbianda XY. Removal of Cobalt and Lead Ions from Wastewater Samples using an Insoluble Nanosponge Biopolymer Composite: Adsorption Isotherm, Kinetic, Thermodynamic, and Regeneration Studies. *Environ Sci Pollut Res Int.* 2018;25(22):21752–21767.
146. Martwong E, Chuetor S, Junthip J. Adsorption of Paraquat by Poly(Vinyl Alcohol)-Cyclodextrin Nanosponges. *Polymers (Basel).* 2021;13(23):4110.
147. Romita R, Rizzi V, Semeraro P, et al. Operational Parameters Affecting the Atrazine Removal from Water by using Cyclodextrin based Polymers as Efficient Adsorbents for Cleaner Technologies. *Environmental Technology & Innovation.* 2019; 16: 100454.
148. Salazar S, Guerra D, Yutronic N, Jara P. Removal of Aromatic Chlorinated Pesticides from Aqueous Solution Using β-Cyclodextrin Polymers Decorated with Fe_3O_4 Nanoparticles. *Polymers (Basel).* 2018;10(9):1038.
149. Salazar S, Yutronic N, Jara P. Magnetic β-Cyclodextrin Nanosponges for Potential Application in the Removal of the Neonicotinoid Dinotefuran from Wastewater. *Int J Mol Sci.* 2020;21(11):4079.
150. Srivastava I, Kotia A, Ghosh SK, et al. Recent Advances of Molecular Dynamics Simulations in Nanotribology, *Journal of Molecular Liquids.* 2021;335:116154.

151. Babaei H, Keblinski P, Khodadadi JM. A Proof for Insignificant Effect of Brownian Motion-Induced Micro-Convection on Thermal Conductivity of Nanofluids by Utilizing Molecular Dynamics Simulations, *J. Appl. Phys.* 2013;113:855–919.
152. Frederix PWJM, Patmanidis I, Marrink SJ. Molecular Simulations of Self-Assembling Bio-Inspired Supramolecular Systems and their Connection to Experiments. *Chem Soc Rev*. 2018;47(10):3470–3489.
153. Wang H, Yapa AS, Kariyawasam NL, et al. Rationally Designed Peptide Nanosponges for Cell-Based Cancer Therapy. *Nanomedicine*. 2017;13(8):2555–2564.

Index

E

F

G

H

O

P

Q

R

S

X

Z